THE MAGNETIC FIELD OF THE EARTH'S LITHOSPHERE

Many geologic features of the earth's lithosphere create variations in the earth's magnetic field that can be detected by satellites. The resulting magnetic anomaly maps can provide new insights into the tectonic features and broad structures of the lithosphere.

This book documents the acquisition, reduction, and analysis of satellite magnetic field data in the study of the earth's lithosphere. The isolation of the lithospheric field from fields originating in the earth's core, the ionosphere, and the magnetosphere is discussed in detail, and a summary of the characteristics of each field is included. This work also provides a complete summary of the published maps and the methods used to create them. Rock magnetism concepts and sources of variation in magnetization are discussed to help the reader understand the issues in interpreting the data, and the various interpretation methods, such as forward modeling, are summarized. Analysis results using data from North America, Africa, Australia, Europe, and the oceans illustrate the methodologies of interpretation and the anomalies associated with particular geologic and tectonic settings.

Mapping and interpreting lithospheric fields from satellite magnetic data have required the solution of a new set of problems and the development of a new complement of analytic tools, resulting in a new subdiscipline of geomagnetism. Advanced students and researchers will find that *The Magnetic Field of the Earth's Lithosphere* provides a much needed review of this important topic.

THE MAGNETIC FIELD OF THE EARTH'S LITHOSPHERE

THE SATELLITE PERSPECTIVE

R. A. Langel   W. J. Hinze

CAMBRIDGE
UNIVERSITY PRESS

CAMBRIDGE UNIVERSITY PRESS
Cambridge, New York, Melbourne, Madrid, Cape Town,
Singapore, São Paulo, Delhi, Tokyo, Mexico City

Cambridge University Press
The Edinburgh Building, Cambridge CB2 8RU, UK

Published in the United States of America by Cambridge University Press, New York

www.cambridge.org
Information on this title: www.cambridge.org/9780521189644

First published 1998
First paperback edition 2011

A catalogue record for this publication is available from the British Library

Library of Congress Cataloguing in Publication data
Langel, R. A.
    The magnetic field of the Earth's lithosphere : the satellite
perspective / R. A. Langel, W. J. Hinze.
        p.   cm.
    Includes bibliographical references and index.
    ISBN 0-521-47333-0
    1. Earth – Crust.   2. Geomagnetism – Remote sensing.   3. Artificial
satellites in earth sciences.   1. Hinze, William J.   11. Title.
QE511.L24   1998
538'.78 – dc21   97-51319
                                                                            CIP

ISBN 978-0-521-47333-0 Hardback
ISBN 978-0-521-18964-4 Paperback

Additional resources for this publication at www.cambridge.org/9780521189644

CONTENTS

PREFACE

In a very real sense the mapping and interpretation of lithospheric fields from satellite data have required solutions for a new set of problems and the development of a new complement of analysis tools – in effect, the development of a new subdiscipline of satellite geomagnetism. It is the purpose of this book to document the particular and specialized characteristics and methods of this new subdiscipline.

It is our opinion that this subdiscipline has reached a plateau in its progress, with limited new developments. There are several reasons for this situation. First, satellite magnetic field studies are starving for lack of new data. In spite of extensive planning for new missions to follow *Magsat*, the only satellite mission to be dedicated to observations of lithospheric magnetic anomalies, no new magnetic satellite mission by the National Aeronautics and Space Administration (NASA) or other national agencies of the United States or other nations, suitable for mapping the lithospheric field, has been flown since 1980. Currently planned missions include Denmark's *Ørsted*, South Africa's *SUNSAT*, Argentina's *SAC-C*, and Germany's *CHAMP*. *SUNSAT* is to be launched with *Ørsted* sometime in 1998, and *SAC-C* and *CHAMP* launches are scheduled for March 1999 and mid-1999, respectively. Of these, only *CHAMP* is being designed for lithospheric studies. Second, while many lithospheric studies using POGO and *Magsat* data have been and continue to be published, analysis of these data is difficult and, as a result, somewhat controversial. Third, funding for satellite geomagnetic research in general has become increasingly scarce.

On a more positive note, having reached a plateau we believe that it is an appropriate time to consolidate past gains, put them in perspective, and launch out toward a higher level. We intend this book to provide that consolidation and perspective and so preserve in a single source the principal tools and experience accumulated over the past 15–20 years. The audience to which we write is diverse: nonspecialists trying to determine the role of satellite magnetic studies of the lithosphere, geologists who want to know how such data can aid their interpretations, mathematical geophysicists concerned with magnetization distributions and sources, and geomagnetists, who may view the anomalous fields as "noise" in their analyses of the main or ionospheric fields. Each will read the book differently. The scope and degree of complexity of this book are intended to make it a primary reference for those concerned with acquisition, reduction, inversion, and interpretation of satellite magnetic field data for lithospheric studies. Here we shall describe the contents of the book and provide guidance for the different segments of our audience.

The presentation does not assume any familiarity with satellite data in general or satellite geomagnetic data in particular. It does assume a working familiarity with magnetic fields, such as might be acquired in a graduate or advanced undergraduate class in electricity and magnetism, and a mathematical background that includes some familiarity with Legendre functions, matrix manipulation, least-squares analysis, Fourier analysis, and elementary potential theory. An understanding of some sections will benefit from familiarity with vector spaces and Hilbert space, but the necessary basic concepts are included in appendixes. Some familiarity with geology or geophysics is needed for the last chapter.

We intend to be thorough in our treatment of the methods used for representation, reduction, and inversion of the data, as well as in documenting the published results. Space permits only a general discussion of magnetic anomaly sources and a representative treatment of interpretations, rather than the extended treatment we would like, although an extensive list of references is provided for the reader who seeks additional details.

The content and organization of the material have been dictated by the nature of the data and by the need for systematic presentation. Where consistent with these dictates, we have attempted to provide documentation regarding the history of the development of the subdiscipline. Chapter 1 introduces the value of geomagnetic data for study of the earth, and its lithosphere in particular, points out some of the special characteristics of satellite data, and introduces some basic notation and concepts. Because the lithospheric magnetic field is only one of several magnetic fields present near the earth and comprises only a small proportion of the field, to deal properly with these data one must have an understanding of the other fields and their sources. That is the task of Chapter 2. Chapter 3 returns to the characteristics of satellite geomagnetic data in more detail and describes the specifics of those satellites that have contributed data useful for lithospheric studies. In a sense, then, the first three chapters are all introductory to the main thrust of the book.

A dominant theme in the study of lithospheric fields measured by satellites is the problem of the signal-to-noise ratio. That is, how can the field from the lithosphere be isolated from the main field and the fields of external origin? This is the theme of Chapter 4. Once the anomaly field has been isolated, we are left with a set of data points at various positions and altitudes. Subsequent steps include such operations as gridding, representation by suitable functions, inversion, and continuation. All of these have common mathematical features and can be treated in a unified manner. That is the purpose of Chapter 5. Among the products of most analyses of such data are maps. Almost every researcher has either adopted a map from an earlier study or derived a new map. Chapter 6 tabulates and summarizes all of the published maps known to us. It further reproduces a select subset of these maps, both to illustrate the methods of Chapters 4 and 5 and to provide readers sufficient information from which to choose the map or maps suitable for their purposes and to evaluate the likely accuracy of the maps.

Specific attention to interpretation in terms of an understanding of the lithosphere is the concern of Chapters 7–9. Although a thorough discussion of rock magnetism is beyond the scope of this book, Chapter 7 presents a summary of the likely sources of satellite altitude magnetic anomalies and the factors that might influence the magnetic properties of those sources. The inverse methods of Chapter 5 are not the only tools for interpretation; others include, for example, correlation analysis and forward modeling, and these and others are discussed in Chapter 8.

Many studies containing interpretations of satellite magnetic data have been published. Representatives of those known to us are reviewed in Chapter 9 for selected regions: North America, Africa, Australia, Europe, and the oceans. To place these studies in their regional contexts, the chapter gives a very brief summary of the geology and tectonics of each region and qualitatively describes how the variations in magnetization from one particular model correspond to that geology and tectonics. In that setting, published studies are reviewed.

Some guidance is in order for the nonspecialist reader. Those unconcerned with the details of data acquisition, reduction, and inversion, but who wish to turn directly to examples of data interpretation, may pass directly from the end of Chapter 1 to Chapter 7. Information regarding the accuracy and resolution of the data is found in Chapters 4 and 6, with Chapter 2 furnishing needed background. The various sections of Chapter 4 can generally be read independently. For the reader seeking only the bare essentials, we would recommend Sections 4.0, 4.1, 4.2, the introduction to 4.3, 4.4.1–4.4.3, and 4.10. Similarly, the general flavor of Chapter 6 can be found in Sections 6.0–6.2 and 6.9. Chapter 5 provides a detailed mathematical description of inversion, gridding, and continuation methods. For those not concerned with such methods, only the results, this chapter can be omitted. At the geologic end of the spectrum, Chapter 7 deals with rock magnetic properties. Most crucial for interpretation of satellite magnetic anomaly data are Sections 7.2 and 7.3. Although all of the methodologies described in Chapter 8 are useful, only Section 8.5 is essential as background to Chapter 9.

It is a pleasure to introduce the magneto-gremlin, compliments of Charlie Barton of the Australian Geological Survey. Something of the gremlin resides in all of us who are engaged in geomagnetic studies. He is sometimes bewildered, sometimes elated, but always with a magnet.

A substantial number of figures are reproduced from other sources. Many of these originated in journals or books published by the American Geophysical Union, which allows republication in books such as ours without special permission. For this we are grateful. Although other permissions are acknowledged in the figure legends, here we would also like to acknowledge the republication permissions kindly provided by the American Physical Society, Blackwell Science Ltd., and Terra Scientific Publishing Company.

We are grateful for the aid of others in the preparation of this manuscript. Particular thanks are due to Patrick Taylor for a complete and thoughtful review and to Mike Purucker for detailed and helpful reviews and discussions and for preparation of many of the figures. In the latter task he was ably aided by Joy Conrad and Jairo Santana. Joy Conrad also reviewed the entire manuscript for consistency and readability. Useful reviews of selected chapters were received from Cathy Constable, Nils Olsen, Carol Raymond, Dhananjay Ravat, Kathy Whaler, and Ralph von Frese. We must also acknowledge the patience and encouragement of Cambridge University Press, of NASA's Office of Mission to Planet Earth, which supported R. A. L. well and faithfully, particularly Miriam Baltuck and John LaBrecque, and our home institutions for making the needed resources available and for much encouragement.

R. A. L
W. J. H

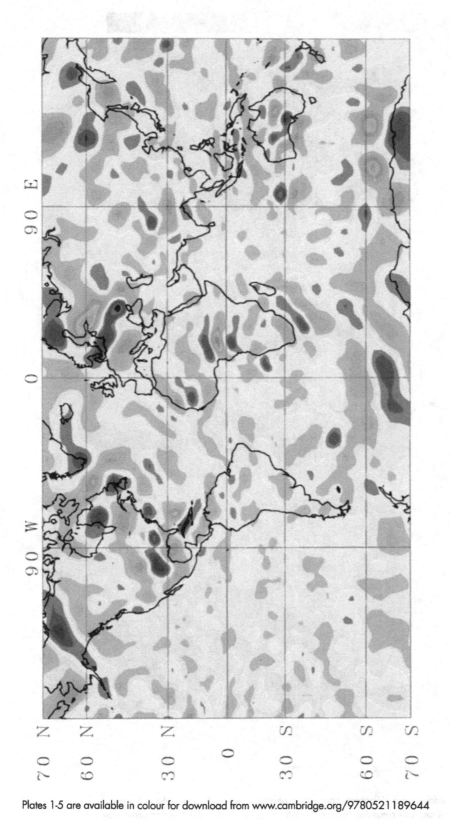
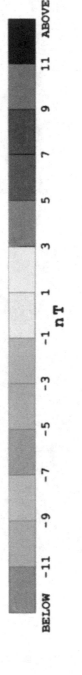

PLATE 1

Combined POGO/*Magsat* scalar anomaly map (31) at 400-km altitude. Van der Grinten projection; units are nanoteslas. (From Arkani-Hamed et al., 1994, with permission.)

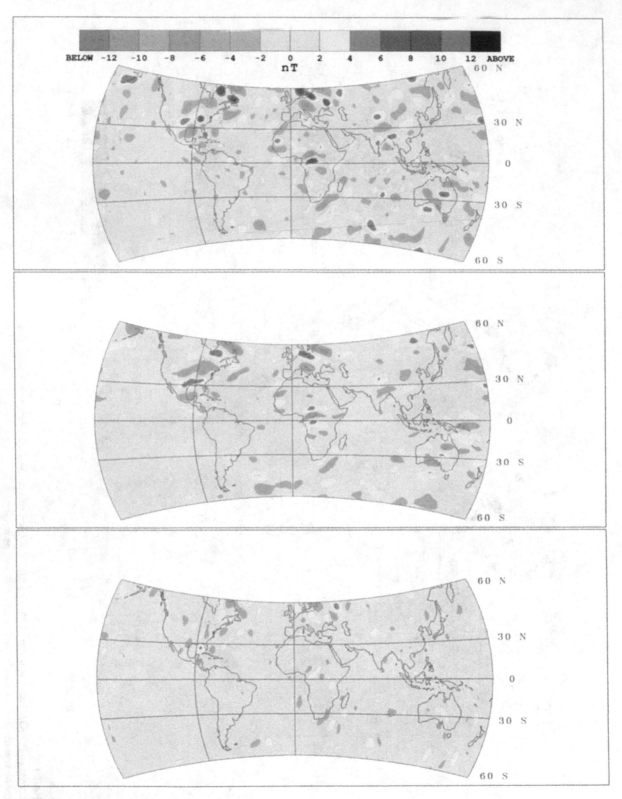

PLATE 2

Combined residual Z (top), X (middle), and Y (bottom) component anomaly maps (29) from *Magsat* dawn and dusk data; based on a SHA analysis retaining degree-15--65 coefficients. Van der Grinten projection. (From Ravat et al., 1995, with permission.)

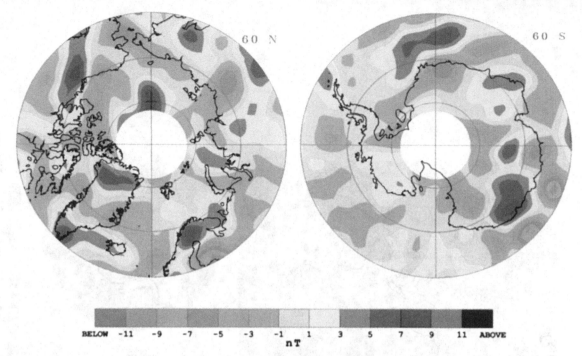

60 N   60 S

BELOW -11 -9 -7 -5 -3 -1 1 3 5 7 9 11 ABOVE
nT

PLATE 3
Combined POGO and *Magsat* polar scalar anomaly maps (31). Equatorward latitude is 50°
or −50°. (From Arkani-Hamed et al., 1994, with permission.)

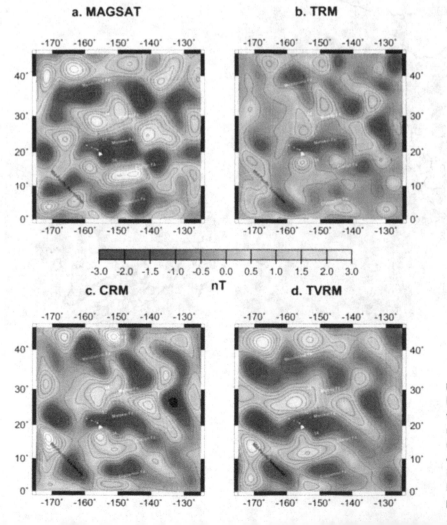

a. MAGSAT   b. TRM

-3.0 -2.0 -1.5 -1.0 -0.5 0.0 0.5 1.0 1.5 2.0 3.0
nT

c. CRM   d. TVRM

PLATE 4
Northeastern Pacific
three-dimensional magnetic
modeling: (a) *Magsat* data,
(b) TRM, (c) CRM, (d) TVRM;
contour lines at 0.5-nT
intervals. (From Yañez and
LaBrecque, 1997, with
permission.)

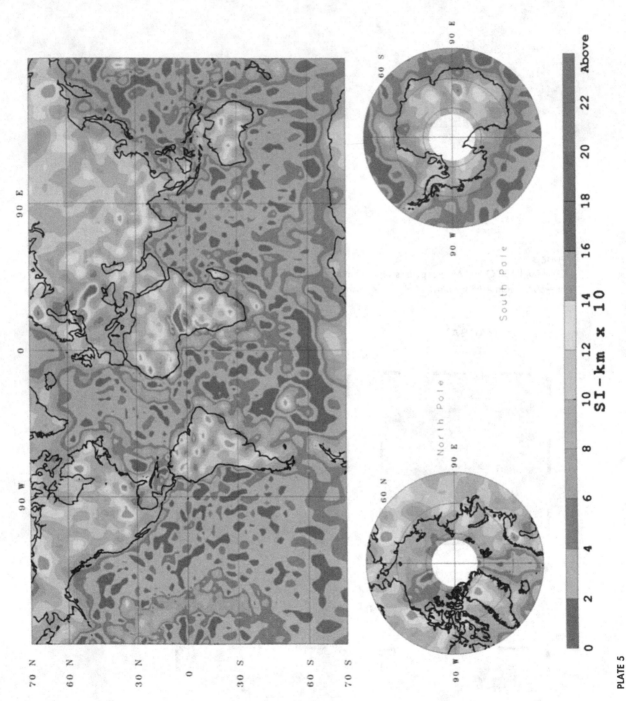

PLATE 5

Susceptibility times thickness times 10, ζ, of the SEMM-1 model. Units are $(10 \cdot SI \cdot km)$. Mercator and polar stereographic projections. (From Purucker et al., 1997a, with permission.)

North Pole

South Pole

SI-km × 10

0 2 4 6 8 10 12 14 16 18 20 22 Above

CHAPTER 1

LITHOSPHERIC MAGNETIC FIELDS AT SATELLITE ALTITUDE

1.0 INTRODUCTION

1.0.1 LOOKING INTO THE EARTH'S INTERIOR

This book describes the study of magnetic fields originating in the crust and, possibly, the upper portion of the mantle of the earth, as measured at satellite altitudes, roughly 150–800 kilometers (km) above the earth's surface. Questions addressed include the significance of measuring magnetic fields for lithospheric investigations, the contributions made by satellite altitude measurements, and the procedures for reducing the magnetic field measurements and isolating and interpreting the lithospheric component. Background is provided for the measurement and analysis of lithospheric fields at satellite altitude, and the results of those measurements over the past few decades are summarized.

Except for information derived from mines and drill holes, knowledge of what is beneath the earth's surface is dependent upon indirect measurements made at or above the earth's surface. Our deepest penetrations into the earth have been little more than 10 km, or roughly 0.2% of the earth's radius, and that depth has been reached in only few locations. Several methods have been developed in the effort to probe the interior of the earth, including laboratory replications of how the earth's increasing temperature and pressure with increasing depth are likely to affect the materials believed to exist at great depths, imaging of the interior with seismic waves from natural or man-made mechanical forces, and measurement and interpretation of gravity, heat flow, and magnetic and electrical fields. Geomagnetic measurements have been used in this regard for three centuries or more, but profound advances in observational and computational technology over the past half century have fostered rapid growth in the breadth of applications and in the utility of the method.

Sources of the geomagnetic field are located both within and external to the earth. There is a natural dichotomy in the magnetic fields deriving from internal

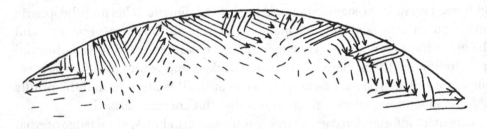

sources between those generated within the earth's core and those originating in the crust and upper mantle. The latter are referred to as magnetic anomaly fields and are the subject of this book. Studies of the magnetic field of the earth's crust were originally used to locate hidden iron ore deposits and other ores associated with the highly magnetic mineral magnetite. Subsequently, the magnetic method has been used to map crustal geology, which has proven to be useful for petroleum exploration and environmental studies, for studies of the structure, thermal state, and tectonic history of the lithosphere, and for numerous other societal and scientific applications.

The surface shell of the earth, the crust, is petrologically and geophysically distinct from the underlying mantle. Characteristically, the rocks of the crust contain variable amounts of the mineral magnetite, which commonly produces a magnetic field mappable at the earth's surface. The earth's crust and the underlying uppermost part of the mantle make up the mechanically strong lithosphere, which is roughly 100 km thick. The contribution of the uppermost mantle to the

geomagnetic field is a matter of some controversy, but is generally considered
to be minor because of the paucity of magnetite in the rocks of the mantle. In
any event, that contribution is limited because the temperature within the earth
increases with depth, and at depths only slightly greater than the average thickness (about 35 km) of the crust in continental regions, the average temperature
reaches values above which most minerals of the lithosphere lose their strong
magnetic properties. However, because of the possibility of a magnetic field
produced in the upper mantle, the discussion in this book includes the entire
lithosphere.

The magnetic field from rocks is determined by their magnetic polarization,
also called magnetization. Sedimentary rocks generally have low magnetizations because of a paucity of magnetite, whereas the underlying crystalline
(metamorphic, igneous) rocks, or basement rocks, can be highly magnetic. Magnetic anomaly maps have provided a primary tool for mapping the structure
and other characteristics of the crystalline rocks of the continental crust. This
mapping ability is particularly crucial in areas where these rocks have limited
exposure or no exposure at the surface because they are covered by sediments
or water. Anomaly maps derived from magnetic data taken with airborne and
shipborne instruments are routinely used in the preparation of geologic and geophysical models of the crust. The dominant magnetic features in oceanic areas
are the familiar striped anomalies that have played a key role in the formulation
of the seafloor spreading and plate tectonic paradigms.

Variations in the magnetic field over distances of 1 km or less can be very
large. Investigations using magnetic data from airborne and shipborne instruments have concentrated mainly on anomalies associated with geologic features
less than about 50 km in minimum surface dimension. However, in the past few
decades there has been increasing interest in studies of the broad-scale anomalies, hundreds of kilometers in extent, that appear in regional compilations
of data from airborne and shipborne instruments. Pioneering studies of such
long-wavelength anomalies may be found in, for example, Pakiser and Zietz
(1965), Zietz et al. (1966), Shuey et al. (1973), Hall (1974), and Krutikhovskaya
and Pashkevich (1977, 1979). Spatial variations in these anomalies reflect variations in crustal and upper-mantle magnetization that in turn are related to
geophysical and/or tectonic history and processes. This is true because magnetization variations not only are caused by differential distributions of magnetic
minerals but also reflect variations in the crustal thickness, the depth to the
Curie isotherm, the chemistry of the rock source and the processes by which
it was formed, the thermal and mechanical history of the rock formation, and
subsequent alterations due to chemical and physical processes.

As with any young subdiscipline, the study of long-wavelength magnetic
anomalies has not been without surprises and controversy. As will be discussed, interpretations of these anomalies have pointed to the existence of

unsuspected highly magnetized sources, presumably in the lower crust. The existence, origin, and location of these sources have been hotly debated. That debate continues, although there is some indication that a consensus is emerging. One extremely valuable result of this debate is that it has spurred renewed laboratory research into rock magnetization, especially for rocks that are likely to occur in the lower crust and upper mantle.

1.0.2 HISTORICAL DEVELOPMENT OF SATELLITE MAGNETIC ANOMALY STUDIES

Magnetic field measurements were among the earliest experiments conducted with rocket- and satellite-borne instruments. Those early data were intended to measure the magnetic field originating in the earth's outer core, its extension into space, and its perturbation by fields due to currents in the plasma environment surrounding the earth. Little thought was given to the use of satellite data for measuring the fields due to rocks. The first attempt to discern lithospheric magnetic fields in satellite data yielded the map of the United States that was derived from *Cosmos 49* data by Zietz, Andreasen, and Cain (1970). However, those data are partially contaminated by fields originating in the spacecraft, and the map has not been widely used. Data from a series of satellites called Polar Orbiting Geophysical Observatories (POGOs), studied by Dr. Joseph Cain and his associates at Goddard Space Flight Center (GSFC), provided the main impetus for pursuing lithospheric studies. Those data clearly showed the effects of a current, called the equatorial electrojet, in the ionosphere below the satellites. Concentrated near noon local time, and presumably absent near midnight, that current resulted in a decrease in the field strength of about 5–12 nanoteslas (nT) as the satellites crossed equatorial latitudes. That decrease was observed at all longitudes, and the position of its peak fell almost exactly on the known location of the dip equator. An exception, however, was observed over the Central African Republic, where the decrease was still observed at midnight local time. Also in that region, the minimum was slightly displaced from the dip equator, even at noon local time. Regan, Cain, and Davis (1975) had found, with satellite data, the magnetic anomaly due to crustal rocks now known as the Bangui anomaly, named for the capital city, Bangui, of the Central African Republic, in which the anomaly is centered.

Thus began the science of mapping lithospheric magnetic fields, or anomalies, using satellite data. The mapped anomalies are spatially larger than those mapped by all but the most extensive aeromagnetic or shipborne surveys. Logistically, mapping these anomalies on a global scale can be accomplished most effectively by satellite. One consequence of the discovery of lithospheric magnetic anomalies in POGO data was the NASA mission *Magsat*, launched in November 1979, designed to extend the findings from the POGO data in two ways. First, *Magsat* measured the field direction as well as its magnitude. The technology existing at the time of POGO did not permit a full vector measurement. Second, *Magsat* orbited the earth at a lower altitude than the POGO satellites (closer to the sources of the anomaly fields), in order to increase the ability of the magnetic measurements to resolve individual anomalies. *Magsat* operated until mid-1980, when it reentered the earth's atmosphere. Satellite data have now been acquired from orbits as low as about 350 km and, in principle, can be recovered from altitudes as low as 150 km. Such data know no political boundaries, are of uniform precision and accuracy, have a uniform spatial distribution, and can be acquired in a time span short enough that the time variation of the earth's main field does not become a limiting factor.

1.1 THE MAGNETIC FIELD FROM THE LITHOSPHERE OF THE EARTH

1.1.1 DEFINITIONS AND RELATIONSHIPS

The convention adopted is that where a working definition of some term is first presented, the term will appear in italics. Système Internationale (SI) units are adopted throughout.

In the space exterior to the earth's surface, Maxwell's equations, assuming the absence of magnetic material, tell us that

$$\nabla \cdot \mathbf{B} = 0, \tag{1.1}$$

$$\nabla \times \mathbf{B} = \mu_0 \mathbf{J}, \tag{1.2}$$

where boldface symbols represent vector quantities in the usual three-dimensional space. In SI units, $\mathbf{B}$, called the *magnetic induction* or flux density, is measured in teslas (T); however, the strength of the geomagnetic field is such that it is convenient to work in nanoteslas (nT). Older literature commonly used a unit called the *gamma* (γ), where $1\gamma = 1$ nT. $\mathbf{J}$ is the *current density* in amperes per square meter (A/m^2), and μ_0, known as the *permeability of free space*, is a constant equal to $4\pi \times 10^{-7}$ henrys per meter (H/m) in the SI system.

If there is no current, $\mathbf{B}$ is curl-free (i.e., $\nabla \times \mathbf{B} = 0$), and it can be described by the gradient of a potential function Ψ,

$$\mathbf{B} = -\nabla \Psi, \tag{1.3}$$

because the curl of a gradient is zero by definition. Substitution of (1.3) into (1.1) gives

$$\nabla^2 \Psi = 0, \tag{1.4}$$

that is, the potential satisfies Laplace's equation. The quantity **B** is measured by satellite magnetic field experiments and hence is the quantity with which this book is concerned. It will be referred to simply as "the magnetic field."

Lithospheric magnetic fields are called *anomaly fields* because in practice what is studied is the residual field when estimates of fields from other sources have been subtracted from the measured field. If **r** is the position vector, **B**(**r**, t) is the measured magnetic field at location **r** and time t, and it is convenient to write

$$\mathbf{B}(\mathbf{r}, t) = \mathbf{B}_m(\mathbf{r}, t) + \mathbf{A}(\mathbf{r}) + \mathbf{D}(\mathbf{r}, t) + \mathbf{e}(t), \tag{1.5}$$

where $\mathbf{B}_m(\mathbf{r}, t)$ is the field from the earth's core (the *main field*), $\mathbf{A}(\mathbf{r})$ is the field from the earth's lithosphere, $\mathbf{D}(\mathbf{r}, t)$ is the *field from magnetospheric and ionospheric sources*, including portions induced in the earth, and **e** is the *measurement error*. The induced portion of **A** varies as $\mathbf{B}_m$ changes, but that effect is negligible; other variations of **A** are on a geologic time scale. The origins and characteristics of $\mathbf{B}_m(\mathbf{r}, t)$ and $\mathbf{D}(\mathbf{r}, t)$ are discussed in Chapter 2. Identification and analyses of both $\mathbf{A}(\mathbf{r})$ and $\mathbf{D}(\mathbf{r}, t)$ require subtracting an estimate of the main field $\mathbf{B}_m(\mathbf{r}, t)$, including its temporal (or secular) change, from $\mathbf{B}(\mathbf{r}, t)$. The difference, or *residual field*, is then

$$\Delta\mathbf{B}(\mathbf{r}, t) = \mathbf{B}(\mathbf{r}, t) - \underline{\mathbf{B}}_m(\mathbf{r}, t) = \mathbf{A}(\mathbf{r}) + \mathbf{D}(\mathbf{r}, t) + \boldsymbol{\eta}(t), \tag{1.6}$$

where $\underline{\mathbf{B}}_m(\mathbf{r}, t)$ is an estimate of $\mathbf{B}_m(\mathbf{r}, t)$, and $\boldsymbol{\eta}$ is a combination of **e** and the estimation error in $\underline{\mathbf{B}}_m$. The convention adopted is that estimated quantities are indicated by underlining. The *scalar residual field* (ΔB) is, by definition,

$$\Delta B(\mathbf{r}, t) = |\mathbf{B}(\mathbf{r}, t)| - |\underline{\mathbf{B}}_m(\mathbf{r}, t)|, \tag{1.7}$$

which is different from $|\Delta\mathbf{B}(\mathbf{r}, t)|$. Now,

$$|\mathbf{B}| = [(\Delta\mathbf{B} + \underline{\mathbf{B}}_m) \cdot (\Delta\mathbf{B} + \underline{\mathbf{B}}_m)]^{1/2} = [\Delta\mathbf{B} \cdot \Delta\mathbf{B} + 2\Delta\mathbf{B} \cdot \underline{\mathbf{B}}_m + \underline{\mathbf{B}}_m \cdot \underline{\mathbf{B}}_m]^{1/2}. \tag{1.8}$$

Under all circumstances of interest, $\Delta\mathbf{B} \cdot \Delta\mathbf{B}$ in (1.8) is much smaller than the other terms and can be neglected. Furthermore, the square root can be well approximated by the first two terms of its Taylor expansion. As a result,

$$\Delta B = \frac{\Delta\mathbf{B} \cdot \underline{\mathbf{B}}_m}{|\underline{\mathbf{B}}_m|} \tag{1.9}$$

is a good approximation.

Substituting (1.6) into (1.9) gives

$$\Delta B = A^s(\mathbf{r}) + D^s(\mathbf{r}, t) + \eta^s(\mathbf{r}, t), \tag{1.10}$$

where

$$A^s(\mathbf{r}) = \frac{\mathbf{A}(\mathbf{r}) \cdot \mathbf{B}_m}{|\mathbf{B}_m|}$$

$$D^s(\mathbf{r}, t) = \frac{\mathbf{D}(\mathbf{r}, t) \cdot \mathbf{B}_m}{|\mathbf{B}_m|} \qquad (1.11)$$

$$\eta^s(\mathbf{r}, t) = \frac{\eta(\mathbf{r}, t) \cdot \mathbf{B}_m}{|\mathbf{B}_m|}$$

are the scalar anomaly field, scalar external field, and scalar noise, respectively.

By definition, the quantities $\mathbf{A}$ and A^s are the *anomaly field*. Given a collection of data, one of the first and most difficult tasks is to isolate $\mathbf{A}(\mathbf{r})$ from $\mathbf{B}(\mathbf{r}, t)$ and $\mathbf{D}(\mathbf{r}, t)$, and $A^s(\mathbf{r})$ from $B_m(\mathbf{r}, t) = |\mathbf{B}_m|$ and $D^s(\mathbf{r}, t)$. Methods used to isolate the anomaly field are discussed in Chapter 4. In practice, application of these methods results in an estimate of $\mathbf{A}$ or A^s, when possible accompanied by some sort of estimate of its probable error. Maps of such estimates are called *anomaly maps*.

1.1.2 MAGNETISM OF ROCKS

Here the origin and properties of lithospheric fields are only introduced; for more details, see, for example, Irving (1964), Stacy and Banerjee (1974), O'Reilly (1984), Banerjee (1989), and Piper (1989). To define the magnetic properties of minerals, let $\mathbf{m}$ denote the magnetic moment of a dipole, from some form of atomic or otherwise localized current. For example, a stationary current loop of area $\mathbf{S}$ (written as a vector to indicate the associated direction normal to the surface, with the current circulating in a "right-hand" sense), with the current in the loop of magnitude I, has a *dipole moment* $\mathbf{m} = I\mathbf{S}$. The units for $\mathbf{m}$ are (amperes) (meters)^2 ($\mathrm{A} \cdot \mathrm{m}^2$). For a larger volume, with contributions from many individual dipoles, the quantity of interest is the *dipole moment per unit volume*, or *magnetization*:

$$\mathbf{M} = \frac{\sum_i \mathbf{m}_i}{V}, \qquad (1.12)$$

where the summation is taken over all dipoles in the volume V. $\mathbf{M}$ has SI units of amperes per meter (A/m). The curl of $\mathbf{M}$ enters into Maxwell's equations as an effective current, called the *magnetization current* (e.g., Panofsky and Phillips, 1962; Jackson, 1975) $\mathbf{J}_m$:

$$\mathbf{J}_m = \nabla \times \mathbf{M}. \qquad (1.13)$$

It is then convenient to write

$$\nabla \times (\mathbf{B} - \mu_0 \mathbf{M}) = \mu_0 \mathbf{J}, \qquad (1.14)$$

where **J** are the "true" currents, and

$$H = \frac{B - \mu_0 M}{\mu_0} \tag{1.15}$$

is called the *magnetic intensity*. Note that in some of the geophysical literature the symbol "**J**" is used to denote $\mu_0 M$ and is called the magnetic polarization or simply the magnetization. In principle, either **B** or **H** or both can be referred to as the magnetic field. As mentioned, we shall use that terminology for **B**. See, for example, Shive (1986) and Blakely (1995) for discussions of units.

It is commonly assumed (but see Section 5.1.2) that there is a linear relationship between **M** and **H**. Thus,

$$M = \kappa_m H, \tag{1.16}$$

where κ_m, a dimensionless scalar, is the ease with which a material is magnetized. Then

$$B = \mu_0(\kappa_m + 1)H = \mu H, \tag{1.17}$$

where κ_m is the volume magnetic susceptibility, and μ is the absolute permeability of the region. In this case the magnetization is said to be *induced* by the field **H**.

In rocks, the magnetization may be induced or, if **M** does not return to zero when **H** is removed, permanent or *remanent*. Permanent magnetism in a rock is generally referred to as *natural remanent magnetization* (NRM). Then

$$M = M_r + M_i, \tag{1.18}$$

where M_r is the NRM, and M_i is the induced magnetization. The ratio

$$Q = \frac{M_r}{M_i} = \frac{M_r}{\kappa_m|H|} \tag{1.19}$$

is called the Koenigsberger ratio.

A magnetic anomaly map reflects the physical and chemical properties of the rocks from which the lithospheric magnetic field originates. Rocks are assemblages of minerals, and though all minerals have magnetic characteristics, not all are important geophysically. In practice, the magnetic minerals of geophysical importance are the oxides of iron, often with some titanium, and some sulfides of iron. Beyond a critical temperature, called the *Curie temperature* (T_c) or *Neel temperature*, a mineral becomes paramagnetic, which for most geophysical purposes is nonmagnetic. Magnetite is of particular importance in local magnetization because it is a common accessory mineral that is strongly magnetized and has a relatively high T_c of 580°C. Depending on the crustal thickness, which can range from 10 to 60 km, and rarely to 80 km, and on the value of heat flow, the Curie temperature may be reached within the crust or may lie below the crust in the upper mantle. For oceanic regions, the Curie isotherm depth most likely occurs in the upper mantle rather than the crust. Magnetic mineralogy and petrology, including factors leading to magnetization variations in the lithosphere, are discussed in more detail in Chapter 7.

1.2   INTRODUCTION TO ANOMALY MAPS

Anomaly maps derived from data taken with airborne and shipborne instruments are often used as aids in locating subsurface features when making geologic and tectonic maps and when formulating geologic and geophysical models of the lithosphere. On a much larger scale, satellite anomaly maps can be used in the same way.

Marine magnetic anomaly profiles, derived from magnetometers towed by ships and carried by aircraft, and compiled into surface maps, revealed the striped anomalies formed at oceanic spreading centers. Interpretation of these anomalies was crucial in the development of the notions of seafloor spreading and plate tectonics (Heirtzler et al., 1968).

Aeromagnetic continental anomaly maps provide information regarding the geologic, tectonic, and thermal state of the earth's lithosphere. Such maps reflect not only magnetization but also anomaly fabric (i.e., trends or lineations in size and in amplitude). Often their analysis is able to estimate the depth to the top and, less frequently, the base of the magnetic source.

Satellite anomaly maps have some inherent differences from surface anomaly maps. One of the most obvious has to do with geometry. For most purposes, when analyzing surface anomaly data the earth can be treated as a plane surface, with the analysis being conducted in rectangular coordinates. In many cases, anomalies extend in one direction more than another, permitting a two-dimensional analysis. Although in some analyses of local regions a plane-surface geometry can be used, generally satellite data should be treated in three-dimensional spherical coordinates. For most surface surveys, the direction and strength of the main field do not vary greatly over the area of the survey. This greatly simplifies the procedures used for modeling the anomaly source. For satellite data, the strength and particularly the direction of the main field usually change significantly over any anomaly that is analyzed. Again in contrast to satellite data, surface data are generally acquired near the source, in the sense that the height from which the observations are made above the causative sources is smaller than or is of the same order as the vertical, and sometimes horizontal, dimensions of the anomaly source.

Satellite anomaly maps describe only anomalies of the very broadest scale. Figure 1.1 shows the first published lithospheric anomaly map derived from satellite data. Preparation of this map and other improved maps is discussed in

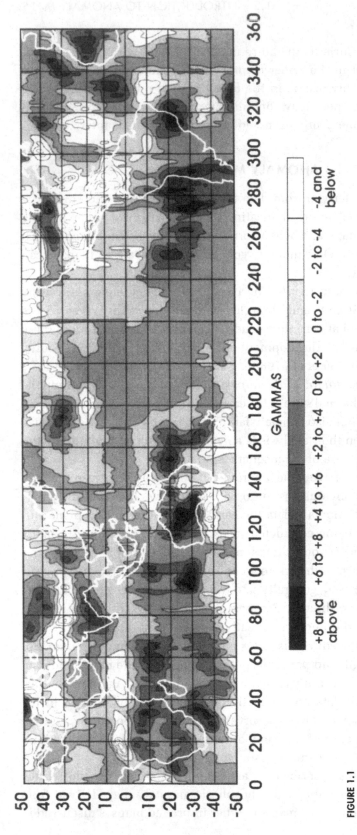

FIGURE 1.1

Magnetic anomaly map derived from data acquired by three POGO spacecraft. (Adapted from Regan et al., 1975.)

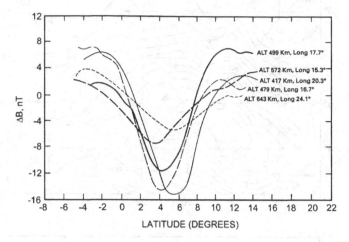

FIGURE 1.2
The Bangui anomaly as measured during several POGO satellite passes at midnight local time at various altitudes. (From Regan and Marsh, 1982, with permission.)

later chapters. The data used in Figure 1.1 were acquired at altitudes between 400 and 600 km. At these altitudes, anomaly magnitudes are generally within the range ±15 nT, with wavelengths between 700 and 3,000 km.

To illustrate the use of satellite magnetic anomaly data, consider the intense negative Bangui anomaly previously mentioned. As seen in Figure 1.1, this anomaly is located at about 5° N, 20° E, in central Africa. Regan and Marsh (1982) studied its characteristics and offered an interpretation. Figure 1.2 shows scalar residual values from the POGO satellite data plotted along five passes over the anomaly. From such profiles, Regan and Marsh (1982) concluded that the anomaly shape and amplitude are independent of the local time associated with the data and of the level of magnetic activity and that the anomaly amplitude decays with altitude in a manner consistent with a source in the lithosphere. Interpretation of such data is more nearly definitive when supplemented with data of other types. Figure 1.3 shows three maps of the Bangui region: (a) the total field anomaly from the POGO satellite data, (b) the total field anomaly based on a land survey, and (c) a Bouguer gravity anomaly map. A more recent Bouguer gravity map is available (Albouy and Godivier, 1981); in its main features it does not differ significantly from Figure 1.3c. Also available are two traverses of aeromagnetic data (project MAGNET) over the anomaly. Figure 1.4 shows a simplified regional geologic map. Regan and Marsh (1982) interpreted the anomalous magnetic fields as originating in a Precambrian mafic pluton that crops out only locally in the region.

A quantitative computer model, composed of a number of small prisms, reproduces many of the features of the data. Each prism is assigned values for magnetization and density, adjusted by trial and error to obtain what is regarded as an adequate fit to the data. Figure 1.5 shows cross-sectional and planar views of the model, and Figure 1.6 compares the resulting computed magnetic anomaly to the satellite magnetic and aeromagnetic data. This interpretation is plausible, but not unique, and alternatives are discussed in Chapter 9.

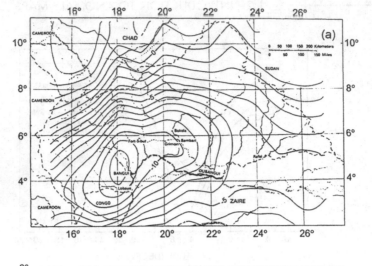

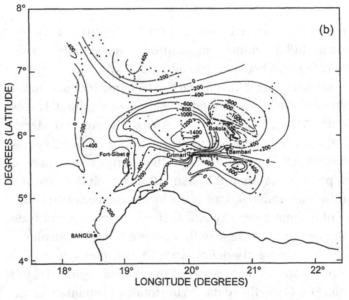

FIGURE 1.3
Magnetic and gravity data for the Bangui region in central Africa. Data from Regan and Marsh (1982). (a) Total field magnetic anomaly at an average altitude of 525 km; contour interval, 1 nT. (b) Total field magnetic anomaly at the earth's surface over central portion of anomaly; contour interval 100 nT (J. Vassal, unpublished data, 1978). (c) Simple Bouguer gravity anomaly map of the Central African Republic; contour interval 10 mGal (ORSTOM data, Albouy and Godivier, 1981).

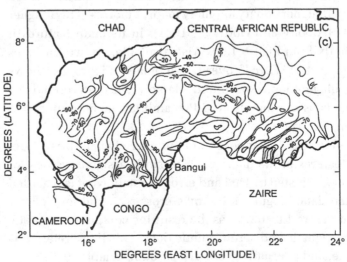

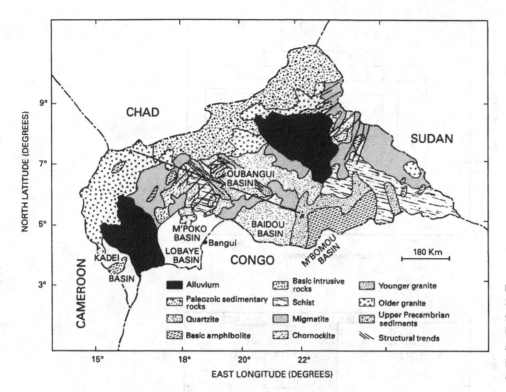

FIGURE 1.4
Simplified geologic map of the Central African Republic. (From Regan and Marsh, 1982, with permission; based on Mestraud, 1964.)

1.3   GENERAL CHARACTERISTICS OF SATELLITE DATA

It is appropriate here to take a brief look at some characteristics that set satellite data apart from surface data. One of the principal contributions of satellite magnetic field measurements to the study of geomagnetism has been to make available a global distribution of data from which a geographically uniform subset can be extracted. Furthermore, satellite data can be acquired within a relatively short time span. These data characteristics provide distinct advantages for the description of the main field from the core and for the mapping of long-wavelength anomalies. With satellite data, it becomes possible to model the core field with truly global data and without the compromise in accuracy imposed by the necessity of reducing data widely spaced in time to a common epoch. At the same time, regional anomalies can be studied without having to piece together data from surveys taken at different times, with different specifications, and reduced using differing models of the core field (e.g., Morley, MacLaren, and Charbonneau, 1968; Makarova, 1974; Hinze and Zietz, 1985).

It is important to understand the general nature of satellite orbits and their impact on satellite magnetic anomaly maps. Referring to Figure 1.7, an earth-orbiting satellite traverses an elliptical path, with the center of mass of the earth located at one focus of the ellipse. As the satellite traverses this path, its altitude

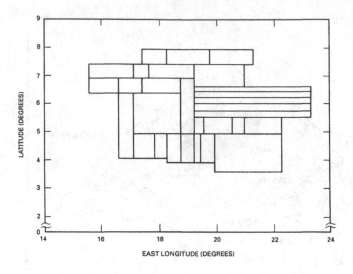

FIGURE 1.5
Computer block model of Regan and Marsh (1982), see Regan and Marsh for detailed susceptibility values. Top: Planar view. Bottom: Cross section (A–A' in Figure 1.4).

above the earth's surface varies because of the orbital ellipticity (or *eccentricity*); its closest and farthest points from the earth are respectively *perigee* and *apogee*. At near-earth altitudes, here defined as having apogee between 200 and 2,000 km, the time for one full orbit, the *orbital period*, varies from about 90 to 200 minutes (min). The angle of the plane of the satellite path with respect to the earth's equatorial plane is the *inclination*, i, and the intersection of the orbital plane and the equatorial plane, at the point where the satellite is going north, is the *ascending node*. To a first approximation, the orbital geometry remains fixed in inertial space, while the earth rotates beneath. The projection of the satellite position onto the surface of the earth is called the *ground track*, or subsatellite track, and this path over a period of time is a measure of the coverage of the globe from which data can be acquired.

Data are acquired between north and south latitudes only up to the value of the inclination of the orbit, with 100% coverage at $i = 90°$. An inclination greater than 90° means that the rotational component of the satellite direction

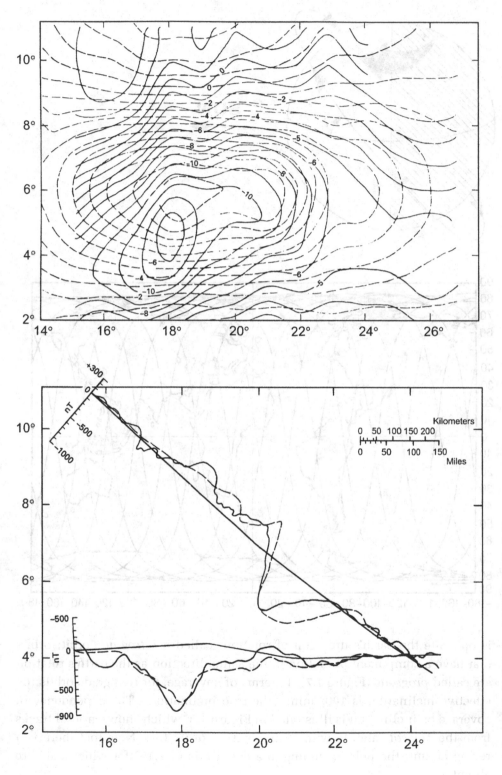

FIGURE 1.6
Comparison of magnetic anomaly fields from computer model and from satellite magnetic and aeromagnetic data. Dashed lines are computed values. Top: Measured and computed satellite altitude magnetic anomalies. Bottom: Measured and computed aeromagnetic values along project MAGNET flight lines. (From Regan and Marsh, 1982, with permission.)

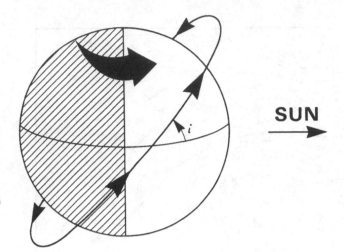

FIGURE 1.7
Illustration of the path of a spacecraft at inclination *i* in orbit around the rotating earth. Ruled area is the dark side of the earth.

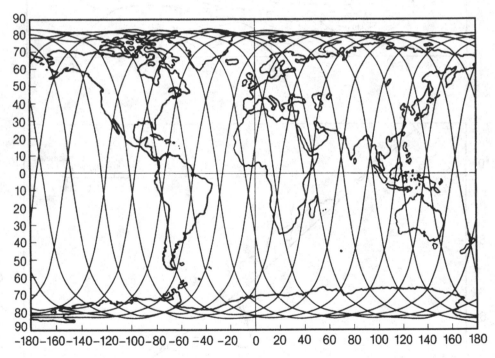

FIGURE 1.8
Ground track for the *Magsat* spacecraft for 24 hours; apogee was 550 km, perigee 325 km, and inclination 97.15°.

is opposite the earth's direction of rotation, called a *retrograde* orbit; orbits that have a component of motion in the same direction as the earth's rotation are called *prograde* (Figure 1.7). In terms of coverage, for retrograde orbits the effective inclination is 180° minus the true inclination. The dependence of coverage on inclination is illustrated in Figure 1.8, which shows ground tracks from the *Magsat* satellite with an inclination of 97.15°. Note that there is a region around the pole extending to a colatitude of 7.15° for which data are absent.

Two other aspects of data coverage are important, besides the latitude range of the satellite: data spacing in longitude and in local time. Data spacing in longitude depends on the way the subsatellite tracks vary in longitude. For example, it is possible to choose the orbital parameters so that the tracks will repeat after a fixed number of orbits, leaving gaps in longitude coverage, or so that the tracks will never repeat. Data spacing in longitude is a complicated function of satellite altitude, orbit ellipticity, and inclination.

Another factor in spacecraft surveys is that because the earth has an equatorial bulge, the plane of a satellite orbit precesses very slowly in inertial space. The rate of this precession depends on the orbital geometry (i.e., apogee, perigee, inclination, ellipticity). Because the earth's rotation brings each longitude under the orbit plane, the data are globally well distributed. However, all observations at one latitude may have nearly the same local time for an extended period of time. In fact, the orbital parameters can be chosen so that the orbital plane will remain fixed in its relation to the sun and hence fixed in local time. Such an orbit is called *sun-synchronous*. The *Magsat* spacecraft was in near-sun-synchronous orbit in the dawn–dusk meridian plane of the earth.

Of major importance to achieving optimal coverage is the capability to store and transmit the data obtained. In the absence of on-board data storage, data acquisition can take place only at those times when the satellite can transmit data directly to a ground receiving station, called real-time data acquisition. The total time available for data acquisition at near-earth altitudes as a satellite passes by a ground receiving station is generally from 3 to 15 min. Relying on real-time data acquisition severely limits the actual coverage capability. To overcome this limitation, a storage device (e.g., a tape recorder) must be carried on board the spacecraft.

CHAPTER 2

EARTH'S MAGNETIC FIELD

2.0 INTRODUCTION

To study the lithospheric component of the earth's magnetic field $\mathbf{A(r)}$, this anomaly field must be identified and isolated from the fields due to other sources. This chapter gives an overview of those other fields and of how they compare with $\mathbf{A(r)}$. Figure 2.1 schematically pictures the various source regions for the geomagnetic field. In the absence of outside currents and fields, the magnetic field of the earth would extend indefinitely into space. However, a plasma (ionized gas), called the *solar wind*, streams from the sun, enclosing the earth's magnetic field and confining it to a cavity called the magnetosphere. The outer boundary of the magnetosphere is called the magnetopause, and its inner boundary is the ionosphere. Except during magnetic storms, some 97–99% of the magnetic field at the earth's surface is produced by electric currents driven by a self-sustaining dynamo process in the earth's conducting liquid outer core. The resulting field is the main field, $\mathbf{B}_m$ in equation (1.5). The remainder of $\mathbf{B}$ is produced by electric currents induced in the earth by time variations in $\mathbf{B}$ (negligible in the present context), by remanent (permanent) and induced magnetization in the lithosphere [the $\mathbf{A}$ in equation (1.5)], by tidal currents excited by the ionospheric dynamo (driven mostly by the thermal solar tide), which are designated by $\mathbf{S}$, and by the effects of the solar-wind plasma in distorting the magnetopause (the boundary of the magnetosphere) and in producing the currents in the magnetosphere and ionosphere that generate magnetic storms and substorms. Fields from ionospheric and magnetospheric sources are the $\mathbf{D}$ in equation (1.5).

In the usual (r, θ, ϕ) spherical coordinate system, the vector field is specified by the three components B_r, B_θ, and B_ϕ. However, the earth is not a sphere, but an oblate spheroid. It is important to take this oblateness into account in models of the main field, but for satellite studies of the lithospheric field the distinction is unimportant.

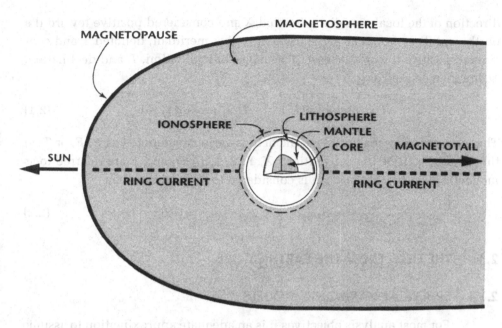

FIGURE 2.1
Schematic drawing showing the source regions for the near-earth geomagnetic field.

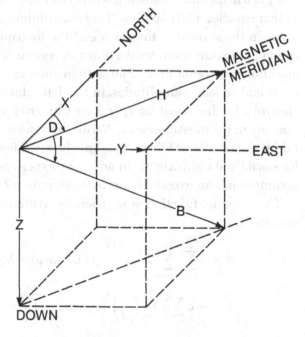

FIGURE 2.2
The magnetic elements in the local topocentric coordinate system. *D* and *I* are called declination and inclination, respectively; *H* is the horizontal magnetic field; *B* is the field magnitude.

Historically, measurements at the earth's surface, including data taken with shipborne and airborne instruments, have been made in a topocentric system. Figure 2.2 illustrates the magnetic elements in this system. The field is resolved into horizontal (H) and vertical (Z) components, with positive Z downward along the local vertical. H, in turn, is resolved into its component in the direction of the local meridian, denoted X and considered positive toward the north, and its component perpendicular to that meridian, denoted Y and considered positive toward the east. The angles of inclination, I, and declination, D, are then defined as

$$I = \arctan(Z/H), \qquad D = \arctan(Y/X). \tag{2.1}$$

The total, or scalar, magnitude of the field is variously denoted as B, F, or T; in this book, B will be the convention. X, Y, Z, I, D, H, and B are referred to as *magnetic elements*. If the earth is considered to be a sphere, then

$$X = -B_\theta, \qquad Y = B_\phi, \qquad Z = -B_r. \tag{2.2}$$

2.1   THE FIELD FROM THE EARTH'S CORE

2.1.1   SPHERICAL HARMONIC MODELS

For most analysis objectives it is an adequate approximation to assume that $\mathbf{J} = 0$ in the space outside the earth's surface extending to several earth radii, so that equation (1.3) applies. This assumption is certainly true for the region between the surface of the earth and the ionosphere. In and above the ionosphere, the assumption breaks down in regions where field-aligned currents or meridional currents flow. Fields from such currents occur in the auroral regions and, at lesser amplitudes, in low latitudes associated with the equatorial electrojet (as discussed later) and are primarily present in the horizontal components of the measurements. Various measures are taken to account for or to mitigate their effects when modeling the main field (Langel and Estes, 1985a,b; Kawasaki and Cain, 1992). In any case, experience indicates that the curl-free assumption is an excellent approximation for most purposes.

The potential function Ψ is generally written in the form of a spherical harmonic series:

$$\Psi = a \sum_{n=1}^{\infty} \sum_{m=0}^{n} \left(\frac{a}{r}\right)^{n+1} \left[g_n^m \cos(m\phi) + h_n^m \sin(m\phi)\right] P_n^m(\cos\theta)$$

$$+ a \sum_{n=1}^{\infty} \sum_{m=0}^{n} \left(\frac{r}{a}\right)^{n} \left[q_n^m \cos(m\phi) + s_n^m \sin(m\phi)\right] P_n^m(\cos\theta), \tag{2.3}$$

where a is the mean radius of the earth (taken to be 6,371.2 km), r, θ, and ϕ are the standard spherical polar coordinates referenced to the center of the earth, the $P_n^m(\cos\theta)$ are the Schmidt quasi-normalized form of associated Legendre functions of degree n and order m [sometimes abbreviated to $P_n^m(\theta)$ or simply P_n^m], and the g_n^m, h_n^m, q_n^m, and s_n^m are called the Gauss coefficients of $\mathbf{B}$ relative to P_n^m. These are functions of time alone. In equation (2.3), the g_n^m and h_n^m

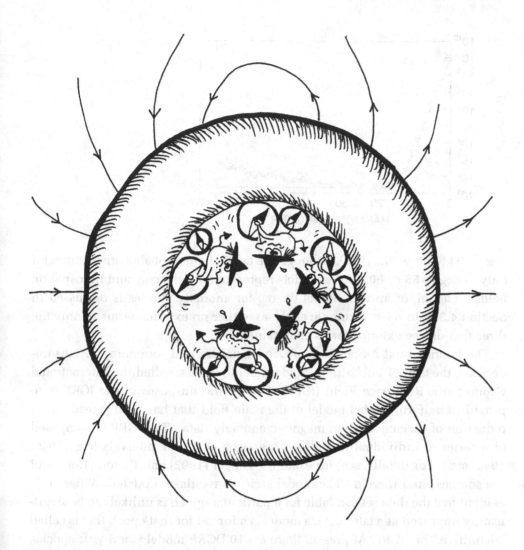

describe fields originating within the earth, and the q_n^m and s_n^m describe fields originating outside the earth. The internal **B** contains contributions from the earth's core, from its lithosphere, and, secondarily, from induced currents. Langel (1987) discusses questions of the existence and uniqueness of such solutions and reviews modeling methods, models, and characteristics of the main field.

Although equations (1.3) and (2.3), in principle, provide a complete description of the field, in practice it generally is not practical to represent the entire field in this way. Rather, the maximum degree of the internal terms, called n^*, is often chosen so that (2.3) represents the core field only, and alternative descriptions are chosen for the lithospheric field. The lithospheric field is shown (Figure 2.3) to be measurable for degrees of about 14 and above. It is generally considered that the magnetic field data available from surface and aircraft measurements are not adequate to determine coefficients beyond degree 10. However, high-quality satellite data have been used to derive models of up to

FIGURE 2.3
Geomagnetic field spectrum. $R_n(\langle|B_n|^2\rangle)$ is the total mean square contribution to the vector field at the earth's surface by all harmonics of degree n. The increase in the power at higher degrees is due to the presence of noise in the data. (From Cain et al., 1989; reprinted with kind permission of Blackwell Science.)

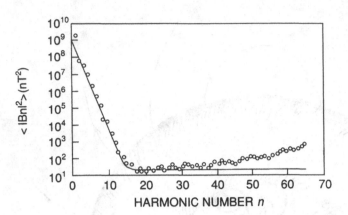

degree 63 (Cain et al., 1989), although the coefficients probably are meaningful only to degree 55 or 60. These models represent both the core and lithospheric fields. The importance of model degree for anomaly studies is discussed in Section 4.3.1. In most harmonic analyses, either no external terms or only the three first-degree external terms are retained.

The International Association of Geomagnetism and Aeronomy (IAGA) undertakes the task of publishing standard field models, called the International Geomagnetic Reference Field (IGRF). One of the purposes of the IGRF is to provide a uniform, global model of the main field that has been agreed on for reduction of surface and aeromagnetic anomaly data. The IGRF is composed of a series of individual models with epochs at 5-year intervals (e.g., 1940, 1945, etc.). For details, see, for example, Langel (1992) and Barton (1997) and the sources cited therein. The model series is regularly updated. Where it is evident that the data set available for a particular epoch is unlikely to be significantly improved at a later date, a model is adopted for that epoch that is called "definitive," or DGRF. At present there are 10 DGRF models for 5-year epochs, commencing at 1945 and ending at 1990 (Barton, 1997). For data between the model epochs, the DGRF is defined by linear interpolation of the model coefficients bracketing the time for which the field is to be computed. For years beyond 1990, the IGRF includes a model of epoch 1995, designated IGRF 1995, with secular variation coefficients for extrapolation to 2000. All DGRF models are of maximum degree 10.

The IGRF or even the DGRF may or may not be the most accurate model for a given epoch. For epochs when no satellite data are available, there is a good chance that it is indeed the most accurate model available. However, for epochs for which high-quality satellite magnetic data are available, permitting solution to degrees higher than 10, it is probable that the higher-degree models are more accurate than any version of the IGRF, even when the IGRF is based on satellite data. More to the point, the most recent IGRF is always a predictive model. The data used in the model predate the epoch of the model by 1–2 years and

predate the extended projection of the model by 6–7 years. This means that models that are derived after the latest IGRF and prior to the next IGRF, and that incorporate data acquired and analyzed after the latest IGRF, are likely to be more accurate than that IGRF.

IGRF models generally are not used in the reduction of satellite data. Models based on those satellite data are of greater accuracy than the DGRF and generally extend to degree 13. Residuals from the lower-degree IGRF models retain a significant portion of the main field.

A quantitative summary of the importance of different wavelengths at the earth's surface can be obtained by considering separately the fields for various spherical harmonic degrees n. The strength of the internal part of the "signal" in **B** at wavelengths corresponding to a given n in equation (2.3) can be measured by the square of the average over the earth's surface of

$$\mathbf{B}_n = -\nabla \Psi_n, \tag{2.4}$$

where

$$\Psi_n = a \sum_{m=0}^{n} \left(\frac{a}{r}\right)^{n+1} \left[g_n^m \cos(m\phi) + h_n^m \sin(m\phi)\right] P_n^m(\cos\theta). \tag{2.5}$$

The square of that average is given by

$$R_n(r) = (n+1) \sum_{m=0}^{n} \left(\frac{a}{r}\right)^{2n+4} \left[(g_n^m)^2 + (h_n^m)^2\right].$$ (2.6)

Usually $R_n(r)$ is evaluated at $r = a$; the notation R_n, *sans* (r), means $R_n(a)$. Figure 2.3 shows a plot of R_n as determined by Cain et al. (1989) using data from the *Magsat* spacecraft. There is a sharp break in the spectrum at about degree 15. Langel and Estes (1982) found a similar break at about degree 14 and interpreted it to mean that at the earth's surface, $\mathbf{B}_n$ comes mainly from the core for $1 \leq n \leq 12$, and mainly from the lithosphere for $n \geq 16$, with both sources contributing when $13 \leq n \leq 15$. The precise degree at which the break occurs, 14 or 15, remains to be definitively determined.

This interpretation of Figure 2.3 indicates that the low-degree lithospheric field and the high-degree core field can never be measured independently and can be separated only by using other information. For lithospheric field studies, the dominance of the core field in the Gauss coefficients for $n < 14$ means that the largest-scale components of the lithospheric field are unknown. As will be seen in a later chapter, this includes many important features, including the field resulting from the difference in magnetization between continents and oceans. A spherical harmonic model of degree 1–13 comprises an estimate of the main field, $\underline{\mathbf{B}}_m(\mathbf{r})$.

2.1.2 MAGNETIC COORDINATES

Examination of Figure 2.3 shows that the dipole term, R_1, stands significantly above the smooth line through the other main field terms and is more than an order of magnitude greater than R_2. At the earth's surface, approximately 90% of the magnetic field produced by the core is the field of a dipole at the earth's center. This dipole field, represented by the three first-degree terms in equation (2.3), is about 60,000 nanoteslas (nT) at the poles and 30,000 nT at the equator. The positions where the extension of the dipole axis intersects the earth's surface are called the north and south geomagnetic, or dipole, poles. In 1980, the position of the northern pole was 78.8° N latitude, 289.3° E longitude. The coordinate system with its polar axis along the tilted dipole axis is called the *dipole* or *geomagnetic coordinate system*. Thus, if (θ_0, ϕ_0) are the colatitude and longitude of the dipole pole,

$$\cos\theta_0 = \frac{g_1^0}{m_0}, \qquad \tan\phi_0 = \frac{h_1^1}{g_1^1},$$ (2.7)

where

$$m_0^2 = (g_1^0)^2 + (g_1^1)^2 + (h_1^1)^2.$$ (2.8)

If $\mathbf{B}_d$ is the field due to a dipole of moment $\mathbf{m}_d$, then $\mathbf{B}_d = -\nabla\Psi_d$, where $\Psi_d = \mu_0\mathbf{m}_d \cdot \hat{\mathbf{r}}/4\pi r^2$, and $\hat{\mathbf{r}}$ is the unit vector in the r direction. The dipole moment of the earth, m_d, then, is equal to $4\pi\, m_0 a^3/\mu_0$. Dipole, or geomagnetic, colatitude, θ_d, is defined by

$$\cos\theta_d = \cos\theta_0 \cos\theta + \sin\theta \sin\theta_0 \cos(\phi - \phi_0), \qquad (2.9)$$

and geomagnetic longitude, ϕ_d, by

$$\tan\phi_d = \frac{\sin\theta \sin(\phi - \phi_0)}{\cos\theta_0 \sin\theta \cos(\phi - \phi_0) - \sin\theta_0 \cos\theta}. \qquad (2.10)$$

Geomagnetic local time (MLT) is defined relative to the dipole system. Geomagnetic noon at a location P occurs when the subsolar point is on the geomagnetic meridian of P, and the MLT for P is defined, in degrees, as

$$\text{MLT} = 180° + \phi_d + \phi_{d,s}, \qquad (2.11)$$

where ϕ_d and $\phi_{d,s}$ are the geomagnetic longitudes of P and the subsolar point, respectively.

A magnetic *dip pole* is defined observationally as a location at which a freely suspended magnetic needle assumes a vertical position. Dip poles can occur at local maxima, minima, or saddle points of the potential, Ψ. The poles at which Ψ assumes its maximum and minimum values are called the principal poles. These are the north and south dip poles; they do not coincide with the geomagnetic poles. The *dip equator* is defined as the locus of points where the inclination I is zero (i.e., the field is horizontal). Again, the dip equator and the dipole equator do not coincide.

2.1.3   TIME SCALES

The time change of the main field, often referred to as *secular variation*, although that term properly refers only to the first time derivative, is easily observable in magnetic observatory records. It is of the order of 60 nT per year, but varies greatly with position and time. Models predicting its future variation have proven to be imprecise. On a more rapid time scale, a distinct global change in the rate of secular variation was detected during 1969–70 (e.g., Courtillot, Ducruix, and LeMouël, 1978; Ducruix, Courtillot, and LeMouël, 1980; LeMouël, Ducruix, and Duyen, 1982; Malin, Hodder, and Barraclough, 1983). This change is called a geomagnetic impulse, or *jerk*. Over very long time scales, the main magnetic field has been shown by paleomagnetic data to have reversed its direction many times. The importance of such reversals for lithospheric anomaly studies lies in the resulting remanent magnetism in rocks that has a direction other than that of the present-day main field and can be of sufficient intensity to contribute significantly to magnetic anomalies.

2.2   THE FIELD FROM SOURCES EXTERNAL TO THE EARTH

In addition to long-term variations of the main field, the geomagnetic field exhibits shorter-period variations that originate outside the earth, the $\mathbf{D}(\mathbf{r}, t)$ of equation (1.5). Because the earth is a conductor, these time-varying fields induce currents in the earth that in turn produce secondary fields. Here, the original and induced fields are considered together as "external" fields, although in equation (2.3), the original fields contribute to the external coefficients (q_n^m and s_n^m), whereas the induced field contributes to the internal coefficients (g_n^m and h_n^m). External variations divide into those that occur with a regular daily pattern (almost every day) and those that, interrupting the regular pattern, occur in a more random fashion, often initiated very abruptly. The character of each varies with location on the earth, particularly latitude. For a more extended treatment, the reader is referred to the contributions edited by Jacobs (1987–91).

2.2.1   THE MAGNETOSPHERE

Figure 2.4 depicts the magnetic environment of the earth, the magnetosphere, as mapped by spacecraft. Attention is called to the ubiquitous presence of currents: on the outer boundary, or magnetopause, across the tail, and, in the equatorial magnetosphere, the ring current. Currents also flow along magnetic field lines connecting the ionosphere with the ring current and tail current. For a more detailed summary, see, for example, Langel et al. (1996). Near the earth (i.e., for anomaly studies), the magnetic fields from these currents are indistinguishable. Their superposition results in a long-wavelength field that varies considerably with time. Examples of magnetospheric field signatures are given in Figures 2.6 and 2.11.

2.2.2   FIELDS FROM THE IONOSPHERE DURING MAGNETIC QUIET CONDITIONS

Diurnal currents (Figure 2.5) are present in the ionosphere primarily because of differential heating on the dayside and nightside of the earth and secondarily because of the differential attractions of the sun and moon. These currents are prominent in the sunlit hours and are stronger in the summer hemisphere than in the winter hemisphere. Ionospheric conductivity results from ultraviolet radiation from the sun. During periods of magnetic disturbance, the resulting fields may be masked by other fields. Consequently, they are designated *solar quiet-day variations* (*Sq*); see Campbell (1989a,b) for a discussion of *Sq*. Figure 2.6 shows a 30-day trace of the magnetic field recorded in San Juan, Puerto Rico, called a *magnetogram*. Note the basic pattern of variation, which repeats daily. This is the *Sq* variation.

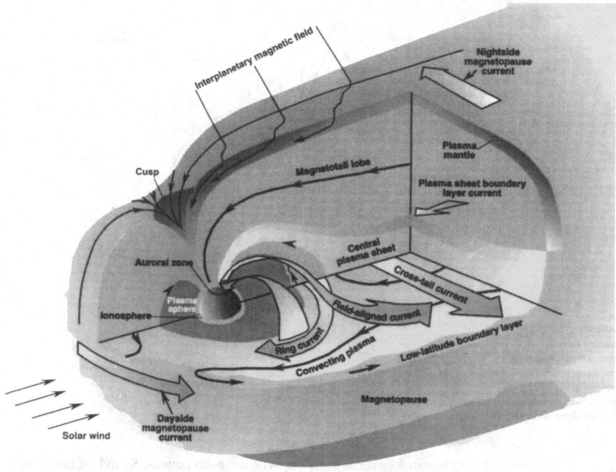

FIGURE 2.4
Dominant space plasma current systems shown in a "cutaway" view of the earth's magnetosphere. (From Fischbach et al., 1994, with permission.)

There are lunar tides as well as solar tides (Campbell, 1989b), and they also drive ionospheric currents. The resulting fields, called L fields, are nearly an order of magnitude smaller than the Sq fields.

The fact that the magnetic field is horizontal at the dip equator results in enhanced conductivity. As a result, within about 5° of the dip equator, the quiet-day variations at surface observatories are enhanced by factors of up to about 5 relative to observations at higher latitudes. The enhanced eastward current flowing along the dayside dip equator is called the *equatorial electrojet* (EE) (e.g., Forbes, 1981; Rastogi, 1989). Theoretical considerations, as summarized by Forbes (1981), indicate that meridional currents related to the EE must exist. Such currents were discovered by analysis of magnetic field data from the *Magsat* satellite by Maeda et al. (1982) and are discussed in more detail by Langel, Rajaram, and Purucker (1993) and Olsen (1997). Their signature in the *Magsat* satellite magnetic field data is discussed in Section 4.7 and illustrated in Figures 4.13 and 4.17.

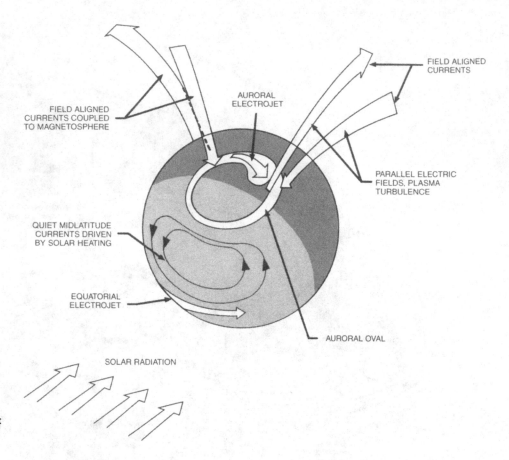

FIGURE 2.5
Dominant currents
associated with the earth's
ionosphere. (From
Lanzerotti et al., 1993;
reprinted with permission of
VCH Publishers; © 1993.)

Although they occur every day with a regular pattern, *Sq* and EE fields show considerable day-to-day variation. This can be seen quite clearly in Figure 2.6. Under some conditions the EE apparently reverses direction, resulting in a counter-electrojet. Although not shown in the figures, there are *Sq* effects that can extend into auroral regions when these are in sunlight.

Fields from quiet-time ionospheric currents are major sources of noise in satellite magnetic anomaly studies. Figure 2.7 shows the component and scalar residual fields at 500 km from a model of *Sq*. Note the major foci at local noon in ΔB, ΔB_r, and ΔB_ϕ and at and to either side of noon in ΔB_ϕ. Note also that fields whose amplitudes approach 6 nT are found well away from local noon. The pattern shifts with longitude (i.e., Universal Time) and season.

2.2.3 TEMPORAL MAGNETIC DISTURBANCES

Feldstein (1963) defined the notion of an *auroral oval* as the region in local time and geomagnetic latitude within which aurora are most likely to occur. Figure 2.5 depicts a current, the auroral electrojet, within the oval. A more

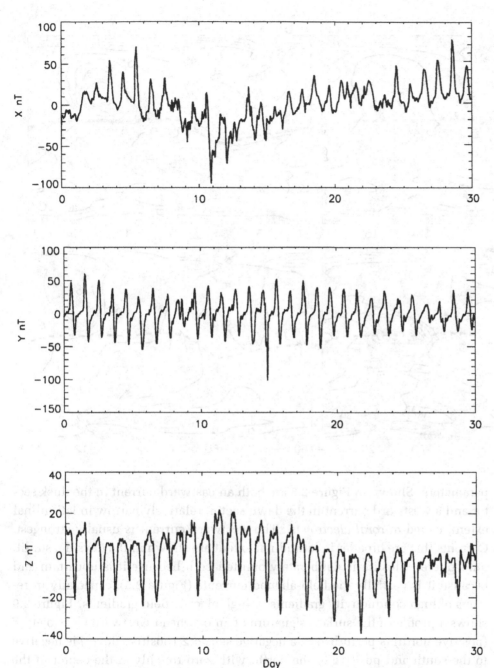

FIGURE 2.6
Trace of hourly values of the magnetic field from San Juan, Puerto Rico, for April 1980. Top: X after subtracting 26,900 nT. Middle: Y after adding 4,500 nT. Bottom: Z after subtracting 31,000 nT. Abscissa is day of the month.

representative, though still schematic, depiction, at least for the ionospheric portion of the current system, is given in the cartoon of Figure 2.8. The morphology of the auroral currents is relatively fixed in the coordinate system using magnetic local time and dipole latitude. However, variations from the nominal morphology are large. Some variations are systematic, depending upon the direction and strength of the interplanetary magnetic field and other solar-wind

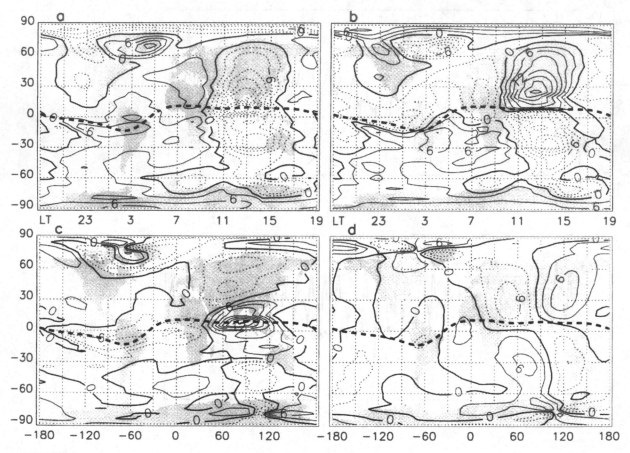

FIGURE 2.7

Magnetic fields at 500-km altitude from a model of the *Sq* current system at 7:00 UT on March 21: (a) ΔB, (b) ΔB_r, (c) ΔB_θ, (d) ΔB_ϕ. Abscissa for parts a and b is solar local time, and for parts c and d, longitude. Abscissa for part a applies also to part c, and vice versa. Abscissa for part b applies also to part d, and vice versa. A dashed line indicates the dip equator. Units are nanoteslas; contour interval is 3 nT.

parameters. Shown in Figure 2.8 are both an eastward current in the dusk sector and a westward current in the dawn sector, relatively narrow in latitudinal extent, called *auroral electrojets*. The westward current is usually strongest. Currents that are broader in latitudinal extent are shown in the dayside, sunlit, ionosphere. Closure is deliberately omitted both because it is uncertain and because it is partially via field-aligned currents (Figure 2.5), especially in regions of high conductivity gradients or high electric field gradients. Figure 2.9 shows a profile of the surface signature of an overhead westward electrojet: X (positive north) is perturbed in a negative sense; Z (positive down) is negative to the south and positive to the north, with zero roughly at the center of the latitudinal extent of the current. Above the current (e.g., at typical satellite altitudes) the sense of the disturbance in Z is the same as on the ground, whereas the sense of the disturbance in X is reversed. An eastward electrojet would result in disturbances of the opposite sense. As shown in Figure 2.5, closure is partly via current flowing into and out of the auroral oval along magnetic field lines. Although they are greatly enhanced during magnetic disturbances,

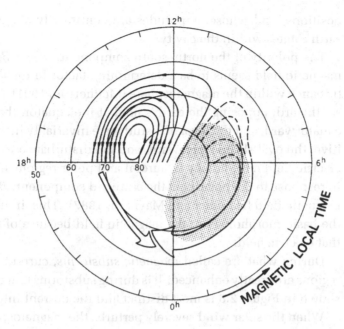

FIGURE 2.8
Schematic diagram of a proposed morphology for the auroral electrojet system. (From Langel, 1974, with permission.)

satellite data indicate that both the ionospheric currents and the field-aligned currents are always present in some form.

The magnetosphere and ionosphere are not static in the configurations depicted in Figures 2.4, 2.5, and 2.8. Rather, it is a dynamic system, constantly changing in response to varying conditions in the solar wind, particularly the direction of the interplanetary magnetic field. The magnetosphere should be visualized as a great cavity, with currents whose morphology is nominally fixed relative to the sun (i.e., in local time), but which move systematically in response to solar-wind stimuli, constantly varying around their nominal

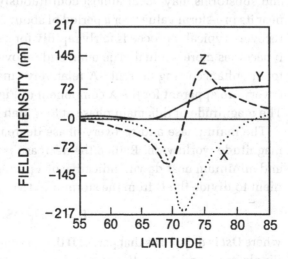

FIGURE 2.9
Latitude profile for hourly averaged ground magnetic perturbations taken near local dawn at the meridian chain of magnetometers of the University of Alberta, Canada. Note the level shift in Y. This is an indicator of the presence of field-aligned currents. (From Hughes and Rostoker, 1977, with permission.)

positions, and whose amplitudes are continually changing. Meanwhile, the earth rotates within the cavity.

The polarity of the north–south component, called B_z, of the interplanetary magnetic field seems to be a determining factor in the degree of magnetic disturbance within the magnetosphere (McPherron, 1991). When B_z is negative, or southward, opposing the earth's field at the equator, the auroral oval expands equatorward, and the level of disturbance invariably increases. When B_z is positive, the oval shrinks poleward, and the disturbance wanes. Additionally, the electric field morphology in auroral and polar regions has been shown to shift in response to the polarity of the eastward component, B_y, of the interplanetary magnetic field (Heppner and Maynard, 1987). This, in turn, causes a change in the basic morphology of the magnetic field because of the currents driven by that electric field.

During what are called *magnetic substorms*, current systems in the auroral regions are greatly enhanced. It is during substorms that the current morphology shown in Figure 2.8 is most distinct and the current intensities greatest.

When the solar wind severely perturbs the magnetosphere, a magnetic storm occurs. It is convenient to characterize a *magnetic storm* in terms of its effect on the horizontal component of the magnetic field at the equator. Typically such a storm has three phases, all global in extent: an initial phase, often with sudden commencement, a main phase, and a recovery phase. A sudden, global, increase in H of about 5–50 nT at equatorial latitudes is called a *sudden commencement* (SC). This is thought to be caused mainly by arrival of a shock wave in the solar wind that suddenly compresses the magnetosphere. The field then remains positive for a period of minutes, after which H rapidly decreases to a minimum value in half an hour to several hours. This decrease is termed the storm *main phase*. During the main phase, the auroral oval expands southward, and substorms may occur almost continuously. Then H gradually recovers to near its pre-storm value over a period of about 2–3 days, sometimes longer. This recovery typically proceeds fairly rapidly for several hours to a day, after which it becomes more gradual. The main and recovery phases usually are attributed to an enhanced ring current. A relatively small main-phase decrease and recovery are apparent for the X component of Figure 2.6, commencing on day 11. The magnitude of this main phase at San Juan was about 30 nT.

The main-phase and recovery-phase decrease in H does not have the same magnitude worldwide. Rather, the decrease is maximum near dusk local time, and minimum near dawn, indicative of a partial ring current. Thus, it is convenient to divide the **D** from the storm into two parts:

$$\mathbf{D} = \mathbf{D}st + \mathbf{DS}, \tag{2.12}$$

where **D**st is defined as that part of **D** that is symmetric with respect to the earth's dipole axis, and **DS** is the remainder. During the rapid part of the recovery phase, **DS** declines, so that **D** is nearly symmetric during the gradual phase of recovery. A commonly used measure of magnetic storm activity is the *Dst* index, to be discussed in the next section.

2.2.4   MAGNETIC ACTIVITY INDICES

Several indices have been devised to provide an indication of the magnitude and extent of **D**; for general reviews, see Mayaud (1980), Rangarajan (1989), and Siebert and Meyer (1996). Some of the indices are used in selecting satellite data for lithospheric studies.

Kp, ap, and Am. *Kp* is a worldwide index assigned to 3-hour intervals in Universal Time (UT) (00–03, 03–06, etc.). It is derived as a composite from what are called *K indices* at the 13 observatories listed in Table 2.1. Each observatory, whether among the 13 listed in Table 2.1 or not, assigns a *K* index to each 3-hour interval, based on the amplitude range of the most disturbed of the horizontally measured field elements, *X* or *Y*. The "amplitude range" is the difference between the maximum and minimum values for the element after an estimate of the effects of *Sq* and *L* has been removed. Given the measured range, the *K* index is then determined from a table customized for that observatory. Table 2.2 is that for the Niemegk observatory. Thus, if the range (R) at Niemegk is between zero and 5 nT, $K=0$; if R is 5–10 nT, $K=1$; and so forth. The *K* scales, which are quasi-logarithmic, are standardized between observatories by comparison with the *K* values at Niemegk, such that over a sufficiently long period of time the number of intervals at a particular *K* value should be the same as at Niemegk.

An index based on a single observatory is not satisfactory as a global index. The planetary index, *Kp*, is currently derived by combining the *K* indices from the 13 observatories listed in Table 2.1 using a procedure described, for example, by Siebert and Meyer (1996). *Kp* values are more finely divided than *K* values and are ordered as follows:

$$0, 0+, 1-, 1, 1+, 2-, 2, 2+, \ldots, 9-, 9,$$

with 0 corresponding to most quiet, and 9 to most disturbed.

A corresponding linear index, otherwise equivalent to *Kp*, is the *ap* index. Table 2.3 shows the relationship of *Kp* to *ap*.

The *Kp* index is published monthly, along with other indices, in the space science (blue) volume of the *Journal of Geophysical Research*. Table 2.4 reproduces the *Kp* index for April of 1980. The eight columns numbered 1–8 correspond to the eight 3-hour intervals of UT for each day. A sum value is given for each day, and the 10 quietest (Q1, . . . , Q0) and 5 most disturbed days (D1, . . . , D5) are indicated. April 1980 is the month for which the field variation

Table 2.1. *Magnetic observatories used for activity indices*

Station	IAGA code	Geographic Lat.	Geographic Long.	Geomagnetic Lat.	Geomagnetic Long.
Kp and ap indices (as of April 1, 1988)					
Lerwick	LER	60.13	−1.18	62.25	89.64
Lovö	LOV	59.35	17.83	57.67	106.72
Sitka	SIT	57.06	−135.33	60.25	277.08
Brorfelde	BFE	55.62	11.67	55.34	98.71
Eskdalemuir	ESK	55.32	−3.20	58.17	84.05
Meanook	MEA	54.62	−113.33	61.97	302.97
Wingst	WNG	53.75	9.07	54.29	95.15
Niemegk	NGK	52.07	12.68	51.78	97.71
Hartland	HAD	51.00	−4.48	54.31	80.20
Ottawa	OTT	45.40	−75.55	56.66	353.43
Fredericksburg	FRD	38.20	−77.37	49.42	351.58
Canberra	CAN	−35.32	149.36	−43.32	226.35
Eyrewell	EYR	−43.42	172.35	−47.67	253.52
Dst					
Honolulu	HON	21.32	−158.00	21.36	267.97
San Juan	SJG	18.12	−66.15	29.41	4.65
Hermanus	HER	−34.42	19.23	−33.59	81.99
Alibag	ABG	18.64	72.87	9.24	144.58
Kakioka	KAK	36.23	140.19	26.31	207.25
AE, AU, and AL					
Abisko	ABK	68.36	18.82	65.83	115.70
Dixon Island	DIK	73.54	80.56	63.10	162.32
Cape Chelyuskin	CCS	77.72	104.28	66.43	177.18
Tixie Bay	TIK	71.58	129.00	60.67	192.38
Cape Wellen	CWE	66.16	−169.84	62.13	238.48
Barrow	BRW	71.32	−156.62	68.88	242.60
College	CMO	64.86	−147.84	64.94	258.14
Yellowknife	YKC	62.48	−114.48	69.16	295.38
Fort Churchill	FCC	58.76	−94.09	68.72	325.00
Great Whale River	GWC	55.27	−77.78	66.46	349.46
Narssarssuaq	NAQ	61.10	−45.20	70.78	38.70
Leirvogur	LRV	64.18	−21.70	69.90	72.19

Table 2.2. *K scale for the Niemegk observatory*

K	0	1	2	3	4	5	6	7	8	9
R (nT)	5	10	20	40	70	120	200	330	500	

at San Juan is plotted in Figure 2.6. Because *Kp* is a global index, it does not completely reflect the disturbance level at San Juan. However, a general correspondence is evident. For example, April 11 and 12 are seen in Table 2.4 to be the most disturbed days, and examination of the *X* component in Figure 2.6 clearly shows a disturbed field.

Table 2.3. *Equivalent amplitude of ap for a given Kp*

Kp	0	0+	1−	1	1+	2−	2	2+	3−	3	3+	4−	4	4+	5−
ap	0	2	3	4	5	6	7	9	12	15	18	22	27	32	39
Kp	5	5+	6−	6	6+	7−	7	7+	8−	8	8+	9−	9		
ap	48	56	67	80	94	111	132	154	179	207	236	300	400		

Table 2.4. *Three-hourly Kp indices for April 1980*

Day		1	2	3	4	5	6	7	8	Sum
1	(Q2)	2	1−	1−	0+	1+	1+	1	0	7+
2	(Q1)	1−	1−	0+	1−	1−	1−	0+	2−	5−
3		1+	3−	2−	2+	2−	2+	2	1−	15−
4		0+	1+	3	3	3−	3+	3	1	18−
5		1	2−	1	1	2+	3+	2	1−	13
6	(D3)	0+	2+	1+	5−	5	5	5+	3+	27+
7		3+	3−	3+	2−	3−	3−	5−	4	25
8		3+	4−	3−	4	4	3−	2−	3−	25−
9		3	1+	1−	5−	4	3+	2−	1+	26
10	(D4)	4	5	4	5−	3−	2+	2	2+	27
11	(D1)	3	3−	4+	3	3	4+	6−	5	31
12	(D2)	5	4−	3	3+	3+	4+	3+	5+	31+
13		5−	3+	4−	3	3−	3+	3−	2−	25
14		1	2	2−	0+	1+	3−	3	4	16
15	(D5)	5−	4−	3+	2+	2+	3+	4−	4	27+
16		3	3	3−	1+	2+	3	2	3	20+
17		4−	4	2	1−	2	2+	2+	2	19
18	(Q3)	1	2	1−	1	1	1−	1−	2−	9−
19	(Q5)	1	0+	1	1	1+	2−	2−	2−	10−
20	(Q0)	2+	2	1−	1+	1	1	2	2−	12
21	(Q4)	2−	2−	1	1	1−	1	1+	1+	10−
22		3+	3−	1	1+	3−	2	2−	2−	16+
23		3−	2+	1−	1	2−	2−	1−	2	13−
24		3−	3−	1	1	1	2	1	2	13+
25		3−	2+	2	3−	3+	2−	1−	1	16+
26	(Q7)	1−	2	1+	0+	2	1+	1	1+	10
27	(Q8)	0+	1+	0+	1	2−	1+	1−	3−	9+
28	(Q6)	2−	1	0+	1+	2−	2	1−	1−	9+
29	(Q9)	1+	2−	2−	1+	2	2−	1	1+	12
30		2−	2+	1	3−	3	2+	2−	2−	16+

Source: From Lincoln (1980), with permission.

Also of interest is the global *Am* index introduced by Mayaud (1980). To derive *Am*, 24 subauroral observatories, roughly uniformly spaced in longitude, are subnetworked into nine longitude sectors, five in the Northern Hemisphere and four in the Southern Hemisphere. The *K* indices from the observatories in each sector are used to derive an index, *Aj*, for the *j*th sector. These in turn are used to derive two hemispheric indices, As (Southern Hemisphere) and An (Northern Hemisphere), and the global 3-hour index $Am = (An + As)/2$.

Kp and Am are global indices and may not accurately reflect a magnetic disturbance confined to a particular region of the earth. The K indices, on the other hand, apply only to the area near a single observatory. Some investigators have used the sector indices Aj when analyzing satellite magnetic data. The AE, AL, and AU indices, unrelated to Aj, are of particular value in the auroral region, as discussed later.

Dst. The horizontal component of the **D**st of equation (2.12) is regularly determined from a set of magnetic observatories (Table 2.1), all located at about the same dipole latitude. After correcting for secular variation and for Sq, the remaining variation at each observatory is averaged in 1-hour intervals and projected to the dipole equator by assuming that its amplitude varies as $\sin \theta_d$. Then the part of H that is symmetric is estimated by averaging over the set of observatories, giving the Dst index. The method is described by Sugiura (1964).

AE, AL, and AU. For each month, a quiet-time level is determined at each participating observatory. The variation of the H component at each of the observatories from its quiet level, at either 1-minute or 2.5-minute intervals, is superposed in UT with the variations from the other stations. At any UT, the largest value of the superposed variations is defined as the AU index, and the smallest value as the AL index. Then $AE = AU - AL$.

Table 2.1 shows the stations currently used to define these indices. However, in the past, a smaller selection of stations was used, especially for Dst and AE. Dst has sometimes been determined from as few as three stations, and often from four stations. When Davis and Sugiura (1966) first developed the concept of the AE index, they used seven stations.

In general, the meanings of all the indices are different. The Kp stations tend to be at mid-latitudes, and Kp is probably the best index of the general magnetic disturbance level for the entire planet. However, it can be quite low while a major substorm is occurring at auroral latitudes. The AE index is a better index of auroral electrojet activity. AU and AL are indicators of the contributions of the eastward and westward electrojets, respectively. Dst is generally considered an indicator of the strength of the ring current. However, it also reflects contributions from magnetopause and magnetotail currents. A Dst level of zero does not mean that the ring current is absent, but rather that it is at some relatively quiescent level.

As might be expected, the occurrence rate for a particular value of Kp or AE varies greatly from year to year. This is well illustrated by a comparison between years of minimal and maximal sunspot numbers. Figure 2.10 shows the number of occurrences of Kp for 1965, a year of a sunspot minimum, and

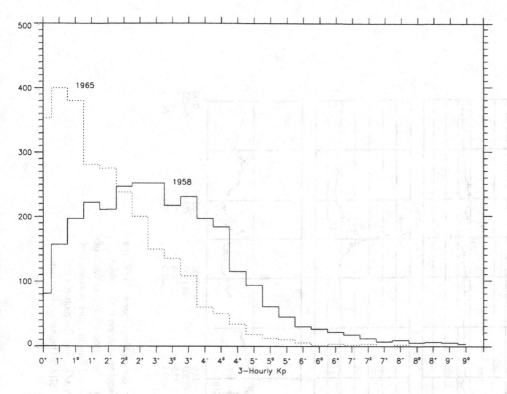

FIGURE 2.10
Frequency distribution of the *Kp* index for a year of sunspot minimum, 1965, and a year of sunspot maximum, 1958. Ordinate is number of occurrences.

1958, a year of a sunspot maximum. During sunspot minima, the occurrence is skewed toward lower index values, and vice versa during sunspot maxima.

All of these indices are available from the National Geophysical Data Center (NGDC) of the U.S. National Oceanographic and Atmospheric Administration, located in Boulder, Colorado, and from the World Data Centers in Edinburgh, Copenhagen, and Kyoto.

2.3   THE MEASURED FIELD AT SATELLITE ALTITUDE

Figure 2.11 shows one full orbit of data from the *Magsat* satellite for November 6, 1979. The field from a degree-13 spherical harmonic model of the main field, but no model of the external field, has been subtracted from these data. Four plots are shown: the residuals from the main-field model of the field magnitude, and the X, Y, and Z components, all in nanoteslas. For the time period shown, the Kp index was 1, and the Dst index was between -2.0 and 0.0 nT.

Although the data are from a period of magnetic quiet, there are noticeable effects from the ring current, magnetopause current, tail current, and field-aligned currents (FACs). The effect of the ring, magnetopause, and tail currents is seen in the value of the X residual, which is about -20 to -25 nT at the equator crossing in the middle of the plot. It is also seen in the variation of the Z

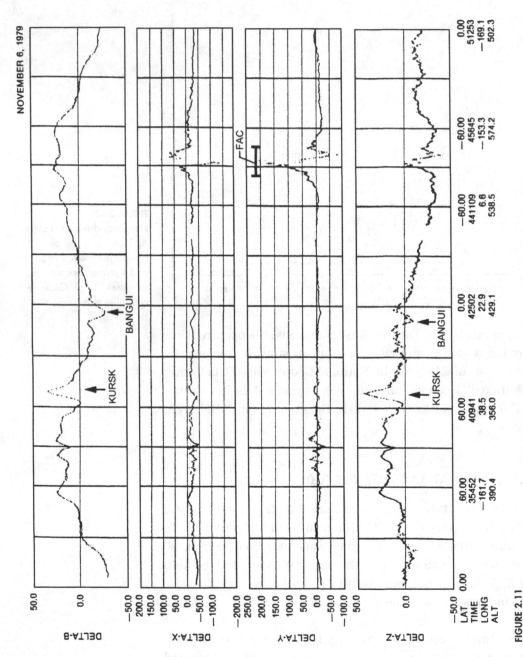

FIGURE 2.11

One orbit of data from the *Magsat* satellite during a period of magnetic quiet on November 6, 1979. The primary scale on the abscissa is latitude, extending from the northward equator crossing of the satellite, proceeding over the North Pole and returning, southbound, to the equator, then proceeding over the South Pole and back to the equator. Indicated in subsidiary scales are the Universal Time in hours, minutes, and seconds, the longitude in degrees, and the altitude in kilometers above 6,371.2 km. The plots are the measured field minus the field (i.e., the residual) from the degree-13 MGST(4/81) spherical harmonic analysis. Units along the ordinate are nanoteslas.

residual, from positive at the northern polar cap to negative at the southern polar cap. These are the signatures expected from a field directed primarily north to south, as produced by a westward ring of current, or a dawn-to-dusk cross-tail current relatively distant from the earth. The near-earth field from currents at the magnetopause is primarily south to north. Variations in the near-earth field from these sources seem to be dominated by variations in the ring current.

The effects of the FACs are seen most clearly in the X and Y residuals near the south pole, as indicated on the Y-component plot. Near the pole the increase becomes very rapid, after which there is a rapid decrease. The peak-to-peak magnitude of the variation in Y is about 300 nT. X shows a simultaneous variation with a smaller amplitude. The associated variation in the Z residual is only about 25 nT and is likely due to a current in the ionosphere. Close examination of the residuals in the northern polar region shows small, rapid variations in the X and Y residuals that are due to FACs. The large difference between northern and southern polar cap disturbances is attributed to the fact that the southern polar cap is sunlit and the northern polar cap is in darkness.

Field variations due to lithospheric sources are also evident in this plot. The Bangui anomaly in central Africa is the cause of the indicated negative Z excursion, with variations in the other components also, and the Kursk anomaly in Russia is the cause of the indicated large positive fluctuation in the Z and B residuals. Variations are also present in X and Y, but the difference in scale renders them hard to discern.

CHAPTER 3

MEASUREMENT OF THE NEAR-EARTH MAGNETIC FIELD FROM SPACE

3.0 INTRODUCTION TO SATELLITE DATA

Unlike a geomagnetic observatory, a satellite does not acquire a time series of measurements at a fixed location. Unlike a ship or aircraft, a satellite cannot be maneuvered to control the locations where data are acquired. As discussed in Section 1.3, a satellite flies in a fixed great-circle trajectory specified by its inclination, apogee, and perigee. Further, its orbit moves only slowly, if at all, in local time, because it is nearly fixed in inertial space. Being in orbit, a satellite is not accessible for repair, testing of equipment, or calibration, unless these functions are built in for remote performance.

An additional complication for magnetic field data is that **B** is a vector. This means that its direction must be measured on an inaccessible moving platform. The accuracy of vector measurements depends on how well the directional axes of the magnetometers are known, and so the measurements contain an inherent

source of error beyond those in total intensity data measured by instruments with no directional sensitivity.

All of the error sources inherent in surface magnetic field data are also present in satellite data (e.g., instrument error, fields from contaminating sources, knowledge of position, and orientation problems). The measurements are more complicated in space because the ability to mitigate effects due to magnetic contamination by spatial separation is limited and because the position and attitude of the spacecraft are constantly changing.

In this chapter, the acquisition of satellite magnetic data, the errors in such data, and other characteristics of these data are discussed in detail. Specific characteristics of those satellites that have furnished data for lithospheric anomaly studies are presented, and issues concerned with processing of the observations are pointed out, but not treated exhaustively.

3.1 MAGNETOMETERS

This section provides a summary of some characteristics of magnetometers flown on near-earth satellites. A comprehensive review of magnetometers flown on spacecraft has been provided by Ness (1970), and there is a more recent, although less specialized, review of magnetic field measurements by Forbes (1987).

Because the geomagnetic field is a vector field, a complete measurement comprises three vector components or, equivalently, the field magnitude and direction. Component measurements on near-earth satellites have generally been made with fluxgate magnetometers. The heart of a fluxgate magnetometer is one or more fluxgate sensors that measure the field along a single axis. Such sensors have a very wide dynamic range but, to improve the noise performance of the instruments, the range of measurements is sometimes limited to $\pm1,000$ to $\pm2,500$ nT. Extension of the total instrument range to the $\pm64,000$ nT needed to measure the earth's field is accomplished by injecting precisely known currents into coils surrounding each sensor, resulting in uniform-bias fields along each sensor axis. Current strength is varied stepwise so that the bias magnetic field from the coil cancels most of the ambient field, allowing the sensor to function in the desired range. The sum of the bias field and the field measured by the fluxgate sensor equals the total ambient field.

To find the complete vector of **B**, measurements are required along three axes. Such fluxgate instruments are called "triaxial." In most cases, each axis has a separate bias coil. Another approach is to place the entire triaxial fluxgate inside a three-axis coil system. One mode of operation for such a system is to adjust the currents in the coil system so that the field at the fluxgate is zero (nulled). The ambient field is then determined by measuring the currents in the coil system and applying suitable geometric constants. In this approach, the accuracy depends upon the mechanical and thermal stability of the coil system and on the accuracy to which the current in that system is measured.

Measurement of the field magnitude only, called a scalar measurement, has been accomplished with two types of magnetometers: the proton-precession magnetometer and the alkali-vapor magnetometers. One might ask why scalar measurements are needed if it is possible to measure the three components, because the scalar magnitude can be determined by taking the square root of the sum of the squared components. The reason is that *all* measurements by vector instruments flown on spacecraft are intrinsically relative in nature. Scale and zero-offset variations can occur because of effects such as sensor and electronic drift. For the accuracies required in solid-earth studies, it is necessary to turn to scalar instruments that are able to achieve absolute accuracies of better than 10 parts per million (≤ 0.6 nT in the earth's field) to calibrate the vector instruments. Proton-precession and alkali-vapor magnetometers are quantum-mechanical devices whose accuracies depend upon the fundamental properties of the atom. The principal disadvantages of proton-precession instruments are that large (several milliteslas) polarizing fields must be frequently applied and that the instrument is unable to make continuous measurements. Proton magnetometers using the Overhauser effect are able to overcome both of these disadvantages and are now coming into use in the space environment (Abragam, 1961; Pake, 1962; Slichter, 1963; Schumacher, 1970; Hrvoic, 1973). In these magnetometers, polarization is accomplished continuously by a weak excitation magnetic field.

Inherent in the operation of most scalar instruments is the existence of "dead" or "null" zones. These are conical regions relative to the center of the instrument such that if the magnetic vector is within the cone, instrument oscillation ceases. In the case of the alkali-vapor magnetometers, to minimize the chances of this happening, it is usual to incorporate two separate magnetometers rotated with respect to one another, leaving only a small dead zone oriented away from those directions expected of the measured field vectors.

Magnetometers on board spacecraft measure *in situ* fields at the sensor location. Because the spacecraft is moving [at about 7 kilometers per second (km/s) for near-earth orbits], these fields change with position and time; so when a finite counting time is necessary, as for the proton-precession and alkali-vapor magnetometers, the field readout from the instrument is an average of the ambient field over the counting time. It is desirable to minimize this time, because nonlinearities in the variation of the field over the counting time result in the average value differing from the field value at the middle of the counting interval.

3.2   ERROR SOURCES

Errors in satellite geomagnetic data are of three types. First are the usual errors that affect any magnetic field measurement whether in space or on earth. These include instrument errors, contaminating fields, and digitization and sampling-interval errors. Second are those errors that are peculiar to satellites and are encountered during every satellite mission. These include errors due to imprecise knowledge of time and position and of the magnetometer attitude (and the stability of that knowledge). Finally, there are errors that are unique to a given experiment.

Any magnetic fields from the spacecraft that are above the noise level of the instrument constitute an error source. Ideally, magnetometer sensors are separated sufficiently from magnetic sources to eliminate spacecraft fields as a significant noise source. Typically, this is accomplished by placing the magnetometer sensors at the end of a nonmagnetic boom, while most of the noise-producing electronics are placed in the spacecraft proper (Figure 3.1). Obviously, careful

FIGURE 3.1
Pictorial diagram of the *Magsat* spacecraft. (From Langel et al., 1982b, with permission.)

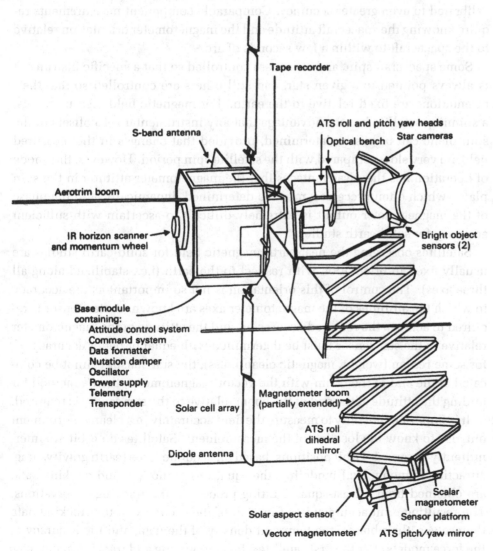

Tape recorder

ATS roll and pitch yaw heads

Optical bench   Star cameras

S-band antenna

Aerotrim boom

IR horizon scanner and momentum wheel

Bright object sensors (2)

Base module containing:
Attitude control system
Command system
Data formatter
Nutation damper
Oscillator
Power supply
Telemetry
Transponder

Solar cell array

Magnetometer boom (partially extended)

ATS roll dihedral mirror

Dipole antenna

Scalar magnetometer

Sensor platform

Solar aspect sensor

Vector magnetometer   ATS pitch/yaw mirror

spacecraft design to minimize interfering magnetic fields decreases the required length of the boom. These steps have not been possible on all spacecraft carrying magnetometers, nor have all such spacecraft been equipped with booms of sufficient length. For these, the noise level in the resulting data is dominated by the spacecraft fields. This is particularly troublesome when spacecraft subsystems that generate magnetic fields are turned off and on, resulting in a changing spacecraft field.

To measure the direction of the geomagnetic field, one must know the directions of the magnetometer axes in an earth-centered coordinate system. If the main field along one axis is of magnitude 50,000 nT, then an error of 20 seconds (s) of arc in the magnetometer orientation will result in a 5-nT error in measuring the field perpendicular to that axis. Scalar spacecraft magnetometers have attained absolute accuracies of 1–2 nT and, in principle, can be constructed and calibrated to even greater accuracy. Comparable component measurements require knowing the spacecraft attitude and the magnetometer orientation relative to the spacecraft to within a few seconds of arc.

Some spacecraft spin, and others are controlled so that a specific instrument is always pointed to a given star, and still others are controlled so that their orientations are fixed relative to the earth. For magnetic field measurements, a spinning satellite has the advantage that any instrumental zero offsets in the spin plane can be easily determined, provided that changes in the measured field are very slow compared with the satellite spin period. However, this mode of operation has the disadvantages that the magnetometer attitude in the spin plane, which often precesses, must be determined dynamically, and the phase of the magnetometer output is extremely difficult to ascertain with sufficient accuracy for solid-earth studies.

Satellites observing the near-earth magnetic field for solid-earth studies are usually fixed in orientation with respect to the earth (i.e., stabilized along all three axes). The control of this orientation is not so important as the accuracy to which the attitudes of the magnetometer axes are known. A star sensor is required to achieve the required accuracies, and the attitude of the magnetometer relative to the star sensor must be determined with equal or better accuracy. If, for some reason (weight, magnetic cleanliness), the star sensor cannot be colocated at the end of the boom with the vector magnetometer, then a method for finding the attitude at the end of the boom relative to the star sensor is required.

It is important not only to measure the field accurately in a clean environment but also to know the location of the measurement. Satellites on orbit are intermittently observed. Orbit positions, based on accurate force (earth gravity, drag, attraction of moon, etc.) modeling, the equations of motion, and tracking data, are obtained from a least-squares fitting process to the tracking observations. The resulting orbit accuracy is a function of the accuracy of the tracking data themselves, their biases, the temporal density of the data, and the accuracy of the force models. For the first satellites, force modeling and tracking accuracies

Table 3.1. *Gradient of magnetic field: equivalent measurement error for 1 km of orbit error*

Component	Maximum gradient (nT/km)		
	Vertical	Along track	Across track
B_r	−18.0	−13.3	6.8
B_θ	18.0	−6.5	23.4
B_ϕ	8.4	−2.0	23.3
B	−28.0	−6.1	5.7

Table 3.2. *Translation of position error to equivalent field error*

Position error	Equivalent maximum field error (nT)			
	B_r	B_θ	B_ϕ	B
100 m vertical	2.8	1.8	0.8	2.8
100 m along track	1.3	0.7	0.2	0.6
100 m across track	0.6	2.3	2.3	0.5

were both of great concern, and orbit accuracies were not very good. The situation had improved considerably by the time of the POGO satellites. A progression in position accuracy for magnetic field missions can be discerned in Table 3.4, to be discussed later. With modern tracking technologies and the latest force models, decimeter-level or better orbit modeling is now routinely achieved. With the use of the Global Positioning System (GPS), direct three-dimensional navigation of the satellite position is achieved using reduction strategies that are nearly completely free of concerns about force modeling.

One way of looking at the effect of satellite mispositioning is to ask what the field difference is between the actual position and the estimated position. By far the largest differences are due to the gradients in the main geomagnetic field, which can be estimated. Table 3.1 summarizes the global maxima for the various field gradients, and Table 3.2 gives the equivalent field error for a position error of 100 meters (m). Note, for example, that if one desires that the maximum error be no more than 1 nT, then the spacecraft's radial position must be known to within 36 m.

Any error in assigning time to a data point translates directly into a position error. A near-earth satellite has an orbital velocity near 7 km/s. Thus, an error of 10 milliseconds (ms) in time will result in a 70-m position error along the track direction, or a maximum field error of nearly 2 nT. Early spacecraft clocks and their calibration could easily be in error by 10–100 ms. That has now been

Table 3.3. *Spacecraft obtaining near-earth magnetic field measurements*

Satellite	Inclination	Altitude range (km)	Dates	Instruments	Approximate accuracy (nT)	Coverage
Sputnik 3	65°	226–1,881	5/58–6/58	fluxgates	100	USSR
Vanguard 3	33°	510–3,750	9/59–12/59	proton	10	near ground station
1963-38C	polar	1,100	9/63–1/74	fluxgate	unknown	near ground station
Cosmos 26	49°	270–403	3/64	proton	unknown	whole orbit
Cosmos 49	50°	261–488	10/64–11/64	proton	22	whole orbit
1964-83C	90°	1,040–1,089	12/64–6/65	rubidium	22	near ground station
OGO-2	87°	413–1,510	10/65–9/67	rubidium	6	whole orbit
OGO-4	86°	412–908	7/67–1/69	rubidium	6	whole orbit
OGO-6	82	397–1,098	6/69–7/71	rubidium	6	whole orbit
Cosmos 321	72°	270–403	1/70–3/70	cesium	unknown	whole orbit
Triad	polar	750–832	9/72–1/84	fluxgate	about 200	near ground station
S3-2	97°	230–900	10/72–5/1/78	fluxgate	>300	whole orbit
Magsat	97°	325–550	11/79–5/80	fluxgate and cesium	6 3	whole orbit
DE-2	89.97°	309–1,012	8/81–2/83	fluxgate	about 100 per axis, 28 for scalar	whole orbit
DE-1	89.91°	570–3.6 RE	8/81–3/91	fluxgate	about 30 for scalar	whole orbit
ICB-1300	81°	825–906	10/81–8/7/83	fluxgate	>75	part orbit
AUREOL-3	82.5°	408–2,012	9/81–?	fluxgate	>150	part orbit
Hilat	82°	800	6/83–7/18/89	fluxgate	about 200	near ground station
DMSP F-7	polar	835	11/83–1/88	fluxgate	>1,000	whole orbit
Polar Bear	polar	1,000	11/86–12/89	fluxgate	about 200	near ground station
POGS	polar	800	7/90–8/93	fluxgate	50(?)	whole orbit
Ørsted	96.1°	450–850	launch mid-1998	fluxgate and Overhauser	<5 2	whole orbit

improved so that it is routinely possible to determine the time for each data point to about 1 ms.

3.3 SATELLITES MEASURING THE NEAR-EARTH MAGNETIC FIELD

Spacecraft that have made significant contributions to our understanding of the near-earth geomagnetic field are listed in Table 3.3. The first satellite magnetic field measurement was accomplished by a triaxial fluxgate magnetometer on board *Sputnik 3*. The instrument was mounted in a gimballed fashion so that it could be reoriented in flight. One axis was maintained along the ambient field by reorienting the instrument until the fields measured by the other axes were zero. The axis parallel to the field gave the field magnitude, and the position of the gimbal gave the orientation of the spacecraft relative to the field; that is, the magnetometer was used to measure the spacecraft attitude, a common practice on many spacecraft. Spacecraft magnetic fields were high on *Sputnik 3*, and the coverage was limited to the Soviet Union.

The U.S. Navy satellite *1963-38C* was magnetically stabilized to within about 6° of the ambient magnetic field by a permanent magnet. Its fluxgate magnetometers gave useful data only for the field transverse to the permanent magnet. Those data provided the first evidence for the presence of transverse magnetic fields due to field-aligned currents in the auroral belt; see Potemra (1982) for a review. The U.S. Navy *Triad* satellite, operative from late 1972 to early 1984, carried a triaxial fluxgate that obtained higher-quality data than *1963-38C* and mapped the characteristics of the fields due to the field-aligned currents.

Missions to study ionosphere–magnetosphere coupling have included *DE-2*, *ICB-1300*, *AUREOL-3*, *Hilat*, *DMSP F-7*, and *Polar Bear*. The *DMSP F-7* spacecraft was intended primarily to provide optical-image information for weather monitoring. Use of data from all of those satellites for solid-earth studies was minimal, for several reasons. Except for *DMSP F-7*, none had really adequate attitude determination; none carried an absolute magnetometer to calibrate the fluxgate; and, except for *DE-2*, none had on-board data storage for extended coverage. *ICB-1300* and *AUREOL-3* had limited on-board data storage, so data could be obtained for extended periods, but the orbital coverage was not 100%. Although the *DE-2* spacecraft was intended to measure primarily fields from field-aligned and auroral currents, the scalar magnitude of the field is used in studies of the earth's main magnetic field.

Until the *Magsat* mission, other surveys suitable for main-field or lithospheric-field studies were performed with either proton-precession or alkalivapor magnetometers, which measure only the field magnitude. Lack of onboard recording devices limited the coverage of *Vanguard 3* and of *1964-83C*, whereas the Cosmos, POGO, *Magsat*, and *DE-2* satellites all carried tape recorders and achieved full-orbit coverage. Spacecraft fields were well above the noise level for *Sputnik 3*, *1963-38C*, *Cosmos 49*, *1964-83C*, *S3-2*, *ICB-1300*, *AUREOL-3*, and *DMSP F-7*. There are additional errors in the *Cosmos 49* data because the times assigned to the data are uncertain to ±0.5 s.

The first survey to combine near-polar inclination, on-board data storage, and high measurement accuracy was conducted by the *OGO-2*, *-4*, and *-6* (POGO) satellites that operated between 1965 and 1971. OGO stands for Orbiting Geophysical Observatory, of which there were six. Three were placed in highly eccentric orbits for exploration of the magnetosphere. Three, called Polar Orbiting Geophysical Observatories (POGOs), were placed in near-earth orbits.

Magsat, launched in October of 1979, was the first satellite (and to date the only satellite) to survey the vector components of the field with high accuracy.

A cooperative venture between Danish research institutes and space-related industry is under way to build and launch a small satellite for acquisition of near-earth magnetic field data and charged-particle data. Called *Ørsted*, its instrument complement includes triaxial fluxgate and Overhauser scalar magnetometers at the end of an 8-m deployable boom. A low-magnetic-field star imager is located near the vector magnetometer on the boom. The fluxgate magnetometer is enclosed in a spherical coil that nulls the field at the sensor (Nielsen et al., 1995). The French Centre National d'Etudes Spatiales (CNES, the French space agency) will furnish the Overhauser magnetometer (Kernevez and Glenat, 1991; Kernevez et al., 1992), and the satellite position will be determined by the Global Positioning System (GPS).

Table 3.4 summarizes the error budgets for the satellites contributing to modeling the main field and/or mapping fields originating in the lithosphere. The position errors attributed to *Vanguard 3* and *Cosmos 49* are estimated maximum errors. In both cases the usual position error probably contributes less than 50 nT. Two sets of statistics are compiled in Table 3.4. The first set comprises the mean difference, and the standard deviation (σ) about that mean, of the data from each satellite with respect to a field model derived from those data alone. The σ values from such models give an estimate of the scatter or internal consistency in the data set. The second set of statistics comprises the mean and σ about the mean for the relevant IGRF. In most cases this is a DGRF. Note that the higher means and σ values for these models may, in some cases, reflect the truncation level of the model, which is degree 10.

3.4   SATELLITES CONTRIBUTING TO LITHOSPHERIC MAGNETIC FIELD STUDIES

To date, five of the spacecraft included in Tables 3.3 and 3.4 have contributed to studies of the field from the lithosphere: *Cosmos 49*, the POGO satellites *OGO-2*, *OGO-4*, and *OGO-6*, and *Magsat*. The measurements taken on board the *Cosmos 49* and the POGO spacecraft are of the field magnitude, but not its direction; instruments on board *Magsat* measured both the direction and magnitude.

Cosmos 49 (Benkova and Dolginov, 1971) provided the first data set that could be called global in that data were obtained worldwide at latitudes below 49°, the satellite inclination. Because of the low perigee, 260 km, the satellite had a very short lifetime: October 24 to November 6, 1964. Measurements were made every 32.76 s and were stored in an on-board tape recorder. Two orthogonally mounted proton-precession magnetometers were located 3.3 m from the spacecraft center at the end of a boom. At that distance, the spacecraft

Table 3.4. *Error budget summary by satellite*

Satellite	Vanguard 3	Cosmos 49	POGO			Magsat		DE-2	DE-1 (perigee)	POGS
			2	4	6	Scalar	Vector			
Instrument (nT)	<1	2.0	0.9	0.9	0.9	1.5	3.0 (calibrated)	?	?	?
Digitization resolution (nT)	negligible	?	0.44	0.44	0.6	0.6	0.5	1.5	1.5	2.0
Instrument particular (nT)	spin frequency shift 6.7 nT (2 nT avg.); signal noise: 4 nT									
Time ms	100	500	30–70	30–70	10	1	1	5(?)	5(?)	>2.5
nT	4.3	21.4	1.3–3	1.3–3	0.43	0.04	0.04	0.21	0.21	1.1
Spacecraft field	<1.0	2 (compensated)	<1.0	<1.0	<1.0	<1.0	<1.0	<10.0	<10.0	<5.0
Position Vertical km	1	1	0.25	0.25	0.25	0.06	0.06	0.12	0.12	0.06
nT	4(?)	28	7	7	7	1.7	0.8	3.4	3.4	1.7
Horizontal km	9(?)	3	1	1	1	0.2	0.2	0.6	0.6	0.2
nT	207(?)	69	6	6	6	5.6	1.4	16.8	16.8	5.6
Attitude arc-seconds							20			
nT							4.8			
Statistics from model fit to the data Mean	1.2	?	0.3	0.1	−0.6	0.5–1.0	1.0–2.0	−3.5		0.0
σ	12.0	22.0	5.2	6.8	6.3	7.0	6–8.2	23.0		23.2
Statistics from IGRF model Mean	−1.1	−7.8	1.4	−1.1	−4.6	−5.1	−7.8	−8.0	−25.2	21.7
σ	12.1	69.3	14.0	18.2	17.6	15.6	12.0	31.7	26.9	40.7

field was not negligible. It was compensated to a claimed accuracy of 2 nT by a set of permanent magnets; however, the data indicate a much higher noise level than 2 nT. A total of 18,000 usable scalar intensity values were acquired and published in a catalog.

OGO-2 acquired data from launch until October 2, 1967. Data were, however, limited to twilight local times (when the orbit was in full sunlight) because of an early failure in the attitude control system. *OGO-4* operated almost continuously from launch until January 19, 1969, and *OGO-6* operated almost continuously until August 29, 1970, and sporadically from then until June 1971.

Measurements on board these satellites were taken with optically pumped self-oscillating rubidium-vapor magnetometers (Farthing and Folz, 1967). These instruments measured the absolute scalar field to better than 2 nT. Extraneous magnetic fields from the spacecraft were measured prior to launch and were found to be below 1 nT at the magnetometer sensor, which was mounted at the end of a 6-m boom. Data were stored by an on-board tape recorder. Satellite characteristics and data accuracy have been discussed in some detail by Langel (1967, 1973), Cain et al. (1967), and Cain and Langel (1971). The standard deviation of the error estimates from all known sources of errors, including inaccuracies in orbital position, is 5.63 nT.

Magsat (Langel et al., 1982b) was operational from November 4, 1979, through June 11, 1980. However, acquisition of data was severely curtailed toward the end of that period because of battery overheating. Until April 12, the satellite was in full sunlight. After April 12, eclipse by the earth occurred on every orbit, with increasingly lengthy periods in shadow. Battery capacity was not sufficient to maintain continuous operation. Nevertheless, substantial data were acquired until mid-May, and some data up to reentry. Also, as the satellite approached reentry, the accuracy of orbit determination and attitude determination deteriorated. Scalar magnitude is unaffected by deterioration of attitude accuracy.

The *Magsat* scalar magnetometer (Farthing, 1980) was of the cesium-vapor type, accurate to about 1.5 nT. Vector data were acquired with fluxgate magnetometers that performed flawlessly except for a gradual drift (about 20 nT over the mission lifetime), which was corrected by calibration against the cesium-vapor magnetometer (Lancaster et al., 1980). The vector magnetometer (Acuna et al., 1978; Acuna, 1980) consisted of three fluxgate sensors mounted on a ceramic structure; its construction was designed to minimize the effects of thermal expansion and contraction. Its accuracy, after in-flight calibration, was estimated to be within 3 nT for each axis.

Attitude determination data were acquired from two star trackers and a gyro on board the spacecraft, a precision sun sensor mounted at the end of the boom near the vector magnetometer, and an optical attitude transfer system (ATS) for determining the vector magnetometer axis relative to the spacecraft. Figure 3.1 shows a diagram of the spacecraft. The two star cameras and the ATS optics were mounted on a temperature-controlled graphite-epoxy optical bench. As part of the ATS, two mirrors were mounted on the back of the vector magnetometer, which in turn was located on the end of a boom 6 m in length. The mirrors provided reflected beams of collimated light for accurate magnetometer axis determination by the ATS. In addition, measurements were made by a precision sun sensor mounted on the vector magnetometer and by a pitch gyro mounted in the spacecraft. The star camera measurements were combined with the sun sensor and gyro measurements to provide a cross check and to allow for interpolation of attitude between star camera tracks.

3.5   DATA-PROCESSING ISSUES

In any experiment, the integrity of the results is dependent on the quality of the data. In the case of satellite data, the instrument is in a remote location and is not available for testing, calibration, or repair. To ensure the highest possible quality, every test and verification study that is practical should be employed. This includes careful verification of the determination and assignment of the time and position of each data point. Often, the data themselves carry information about their quality and their magnetic environment. That was the case for *Magsat*, which is used as an example in the following subsections.

3.5.1   CALIBRATION OF A VECTOR MAGNETOMETER BY A SCALAR MAGNETOMETER

As previously noted, vector magnetometers are inherently relative instruments. However, it is, in principle, possible to calibrate a vector fluxgate magnetometer using simultaneous data from an absolute scalar magnetometer, provided (1) that the magnetic field gradient between the instruments is small or known and (2) that during the period of calibration the parameters of the fluxgate instrument determined by the calibration do not change significantly. This calibration does not depend upon determination of satellite position or attitude. The second condition is self-evident. The first means that if the measurement of field magnitude for the scalar instrument is known, it can be extrapolated to the position of the fluxgate instrument.

Calibration of a vector magnetometer consists of determining those instrument parameters that, if known, allow one to convert the magnetometer output into magnetic field values along three known axes. Those parameters vary between instruments, but the basic process is similar for all instruments. Calibration of the fluxgate instrument on board *Magsat* has been described by Lancaster et al. (1980) and Langel et al. (1981), and a more generalized approach has been described by Langel et al. (1997). For *Magsat*, a typical calibration solution

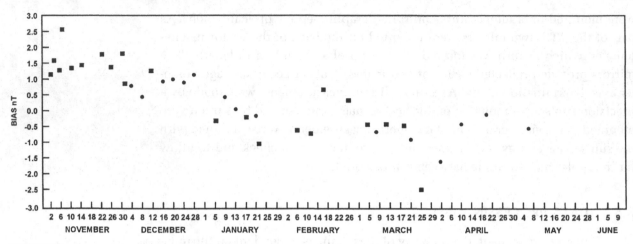

FIGURE 3.2
Variation of the bias on one axis of the *Magsat* fluxgate magnetometer. (From Langel et al., 1981, with permission.)

used data acquired over several days. Figure 3.2 shows the variation of the bias, or zero offset, for one axis of the *Magsat* fluxgate magnetometer.

3.5.2 DETERMINATION OF SPACECRAFT FIELD AND ATTITUDE BIAS

An earth-oriented spacecraft rotates with respect to the earth's magnetic field. Because of this rotation, field characteristics fixed relative to the spacecraft or instrument are separable from those fixed to earth. This separation can be effected naturally during the field modeling process. Suppose that

$\mathbf{B}$ is the measured field, in (orthogonal) instrument coordinates,
$\mathbf{B}_s$ is the contribution to $\mathbf{B}$ from spacecraft-generated fields,
$\mathbf{B}_m$ is the contribution to $\mathbf{B}$ from the earth's field,
Ψ_m is the spherical harmonic potential representation of $\mathbf{B}_m$ in earth-fixed geocentric coordinates [i.e., following equation (2.3)],
T_r is the transformation matrix from instrument coordinates to earth-fixed geocentric coordinates, and
T_ε is a matrix accounting for errors in knowledge of the orientation of the instrument coordinate system relative to the determined spacecraft celestial attitude.

The matrix T_r is known. It is desired to solve for the coefficients in Ψ_m while simultaneously solving for the elements of T_ε and $\mathbf{B}_s$. The elements of T_ε are conveniently expressed in Euler angles along the spacecraft pitch, roll, and yaw axes [i.e., (γ, α, β) respectively]. Typically, the range of these angles will permit a small-angle approximation to be made. In particular,

$$
T_\varepsilon =
\begin{bmatrix}
\cos\gamma\cos\beta & \cos\gamma\sin\beta\cos\alpha + \sin\gamma\sin\alpha & -\cos\gamma\sin\beta\sin\alpha + \sin\gamma\cos\alpha \\
-\sin\beta & \cos\beta\cos\alpha & -\cos\beta\sin\alpha \\
-\sin\gamma\cos\beta & -\sin\gamma\sin\beta\cos\alpha + \cos\gamma\sin\alpha & \sin\gamma\sin\beta\sin\alpha + \cos\gamma\cos\alpha
\end{bmatrix}
$$

$$
\doteq
\begin{bmatrix}
1 & \beta & \gamma \\
-\beta & 1 & -\alpha \\
-\gamma & \alpha & 1
\end{bmatrix},
\tag{3.1}
$$

where $\doteq$ means "approximately equal to."
Then

$$
\mathbf{B} = -(T_\varepsilon)^{-1}(T_r)^{-1}\nabla\Psi_m + \mathbf{B}_s
$$

$$
= -(I + \Delta)(T_r)^{-1}\nabla\Psi_m + \mathbf{B}_s,
\tag{3.2}
$$

$$
\mathbf{B} \doteq -(T_r)^{-1}\nabla\Psi_m + \mathbf{B}_s - \Delta(T_r)^{-1}\underline{\mathbf{B}}_m,
\tag{3.3}
$$

where

$$
\Delta =
\begin{bmatrix}
0 & -\beta & -\gamma \\
\beta & 0 & \alpha \\
\gamma & -\alpha & 0
\end{bmatrix}.
\tag{3.4}
$$

$\underline{\mathbf{B}}_m$ is a previous estimate of $\mathbf{B}_m$. Equation (3.3) is linear in the unknowns and can be solved iteratively, updating the estimate $\underline{\mathbf{B}}_m$ at each iteration, until the solution converges.

Using this formulation, it is possible to solve for any errors in the attitude of the magnetometer and for any residual spacecraft field. For *Magsat* there was no residual spacecraft field, but the roll, pitch, and yaw alignments changed as shown in Figure 3.3.

3.5.3   DEALING WITH DISCONTINUITIES IN THE ATTITUDE SOLUTION

The attitude determination system on *Magsat* incorporated three sensors: two star trackers and a precision sun sensor. Finding the spacecraft attitude at any particular time involved a process of fitting data from whichever sensors were contributing measurements at that time. If data were available from any two of the three sensors, an attitude solution was possible. However, the alignment between the three sensors was not known perfectly. As a result, whenever the available combination of sensors changed, a small discontinuity was introduced into the attitude solution. Many of the resulting field discontinuities are small, within 1–3 nT. Significant numbers, however, are larger and affect attempts to derive anomaly maps.

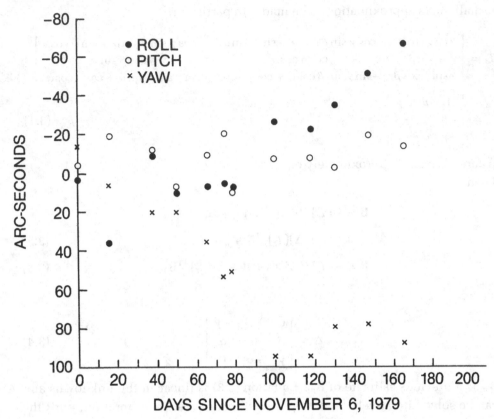

FIGURE 3.3
Changes in magnetometer alignment in terms of pitch, roll, and yaw Euler angles for the *Magsat* mission. Fine attitude date. (From Langel et al., 1981, with permission.)

Purucker (1991) developed a procedure to remove the resulting discontinuities in the magnetic component data. The method is totally empirical and works in the following manner, using the vertical component, Z, for illustration. The difference between adjacent field measurements, say $Z(I)$ and $Z(I+1)$, is computed. If the absolute value of the difference exceeds 5 nT, the observed difference is compared to nearby differences. If all of those differences are less than 50% of the difference between $Z(I)$ and $Z(I+1)$, then the algorithm considers that a discontinuity has been identified. This successfully identifies discontinuities flanked by three or more measurements without another discontinuity. Because discontinuities frequently are paired, are of opposite signs, and are close together, the algorithm includes identification of cases where a discontinuity is followed by a second discontinuity within one or two measurements. In these cases, the data between the discontinuities are discarded.

After discarding the single-, double-, and triple-point discontinuities, what are left are identified discontinuities with four or more points between. Corrections are then applied as follows. The first discontinuity in the pass is identified, say at $\Delta Z(I)$. Then all data between the first point in the pass and the first discontinuity are adjusted linearly (i.e., tilted), with the first point in the pass held

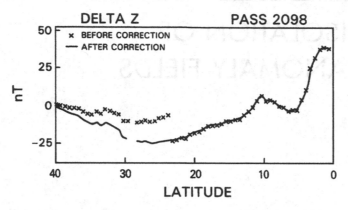

FIGURE 3.4
Illustration of correction of attitude jump in *Magsat* data; xxx indicates original ΔZ data; solid line indicates corrected data. (From Purucker, 1991, with permission.)

fixed, as illustrated in Figure 3.4. In the next step, $Z(I)$ is held fixed; that is, if the next discontinuity is at $\Delta Z(J)$, $J > I + 4$, then all the points between I and J are linearly adjusted. The procedure is repeated for other discontinuities in the pass. This procedure is straightforward in its implementation and preserves $B = (B_r^2 + B_\theta^2 + B_\phi^2)^{1/2}$ at the points of discontinuity.

An examination of 51 passes with this new algorithm yielded a total of 396 discontinuities. Visual inspection confirmed those discontinuities and identified an additional 15 discontinuities, for a 96% success rate. In total, nearly 12,000 discontinuities were identified in the entire *Magsat* ΔZ data set acquired at dawn, encompassing some 1,186 passes. About 25% of the discontinuities in that worldwide data set were followed by a second discontinuity within one or two measurements. In the remainder of the cases, the discontinuity was flanked by three or more measurements without another discontinuity. Lithospheric anomaly maps of the field components and accompanying maps for the standard error of the mean, produced before and after correction for discontinuities, are very similar. However, examination of such figures shows a reduction in overall noise level.

CHAPTER 4

ISOLATION OF ANOMALY FIELDS

4.0   LONG-WAVELENGTH MAGNETIC ANOMALIES

Magnetic anomalies derived from the lithosphere cover a broad band of wavelengths (or wavenumbers). In the 1960s the availability of the fast Fourier transform and rapid increases in computational power and speed led to studies of the spectra of both theoretical and observed magnetic anomalies and to their interpretation (e.g., Bhattacharyya, 1966, 1967; Gudmundsson, 1967; Spector and Grant, 1970). One of the more important relations that came out of those studies was the anomaly wavelength amplitude attenuation as a function of source depth (Dean, 1958). This relationship (the *upward continuation relationship*) states that

$$AA_\lambda = A_\lambda e^{-2\pi \Delta h/\lambda}, \qquad (4.1)$$

where λ is a specific wavelength, A_λ is the amplitude of the λ wavelength component at the original source depth, and AA_λ is the attenuated wavelength component at an elevation Δh above the surface on which A_λ is given. It shows the profound change in spectra with increasing depth of the source or increasing height of the observations above the earth's surface. In either case, higher-wavenumber (shorter-wavelength) components of an anomaly will decrease with increasing observation elevation or source depth. Thus the spectrum of a magnetic anomaly from a specific source is a function not only of body size and configuration but also, very prominently, of the distance from the source to the point of observation.

The term "long wavelength" is not precisely defined, but herein is assumed to apply to all anomalies resolvable in satellite data (i.e., half-wavelengths greater than about 200 km). The term was originally introduced as a way of referring to near-surface anomalies with horizontal dimensions on the order of tens to thousands of kilometers (e.g., Zietz et al., 1970; Hall, 1974). This is in contrast to the smaller horizontal dimensions typically of interest in geophysical exploration.

Anomalies of exploration surveys have been separated into *noise fields*, *residual fields*, and *regional fields*, where the residual field is of interest for the exploration objective. Its wavelength is then defined by the context of the survey, and so has no generally agreed-upon definition. Residual fields are superimposed on the longer-wavelength regional fields, which are removed because they tend to obscure the anomalies sought in the survey. Regional fields, or anomalies, are derived from deeper, larger sources. These, at least in part, constitute the long-wavelength features found in satellite data. An international agreement on definitions is needed, as shown by the possible confusion inherent in referring to anomalies of extent 300–2,600 km as "intermediate" (Arkani-Hamed and Hinze, 1990). Taylor, Hinze, and Ravat (1992) have suggested guidelines to be followed in developing definitions.

4.1 NOTATION AND PROCEDURE

To study the field originating in the earth's lithosphere, that field must be identified and isolated from the fields due to other sources. It is common when interpreting aeromagnetic data to use the word "isolate" in the sense of identifying the field from a particular geologic source, or suite of sources, in the crust (e.g., to isolate the field from a dike or from a seamount). That is not the meaning used here. Here, the word "isolation," or "identification," means to identify an estimate of the field that originates in the lithosphere, as opposed to the core, ionosphere, or magnetosphere. Implicit in carrying out such identification is the prior, or simultaneous, identification of the fields from other sources.

Each of the techniques that we know to have been used to facilitate the identification of lithospheric fields in satellite data is described in this chapter. Although these methods generally have been developed individually, the presentation is made in as near a logical sequence as possible. Section 4.10 discusses a recommended combination of techniques.

Consider the chart depicting data flow in Figure 4.1. The problem to be addressed is the identification of $\mathbf{A}(\mathbf{r})$ in equation (1.5). As a first step (Section 4.2), one eliminates data from periods of time for which it is known that the fields from ionospheric and magnetospheric sources were large (i.e., during magnetically disturbed periods). Aside from the main field, data from such periods are generally dominated by fields from external sources. The remaining data are termed "quiet."

The next step (Section 4.3) is to compute and subtract an estimate of the main field, $\underline{\mathbf{B}}_m(\mathbf{r}, t)$, including its temporal change, as in equations (1.6) and (1.7). For anomaly studies, the remaining $\mathbf{D}(\mathbf{r}, t)$ and η constitute "noise." It is necessary to either minimize or model $\mathbf{D}(\mathbf{r}, t)$ in order to isolate $\mathbf{A}(\mathbf{r})$, as discussed in Sections 4.4–4.8.

Some sections in this chapter contain detailed information and mathematical development. For the casual reader seeking only an overview, the preceding paragraphs and Figure 4.1 should suffice. The reader who seeks more information regarding the methods recommended by the authors, but does not want to read all the details, should turn to Section 4.10 and use it as a guide to what are probably the topics of main interest. Many readers can pass over part of Section 4.5.4, which explores the relationship between two types of spherical harmonic representation. Its main points can be discerned from the first and last paragraphs. Also, Section 4.6, which explores analytically the possibilities for separating fields originating in the lithosphere from those originating in the ionosphere in the case where an ideal or nearly ideal data distribution is available, is not required for understanding of subsequent sections. Note, however, that Section 4.6.3 discusses the pitfalls of an averaging procedure that is in common use with existing data.

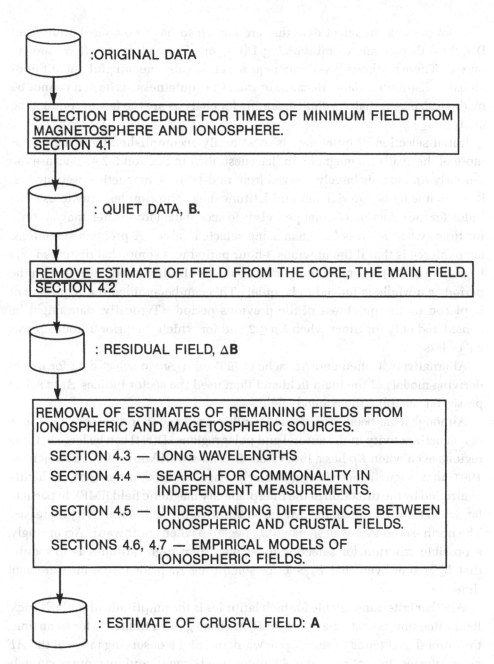

:ORIGINAL DATA

SELECTION PROCEDURE FOR TIMES OF MINIMUM FIELD FROM
MAGNETOSPHERE AND IONOSPHERE.
SECTION 4.1

: QUIET DATA, **B.**

REMOVE ESTIMATE OF FIELD FROM THE CORE, THE MAIN FIELD.
SECTION 4.2

: RESIDUAL FIELD, Δ**B**

REMOVAL OF ESTIMATES OF REMAINING FIELDS FROM
IONOSPHERIC AND MAGETOSPHERIC SOURCES.

SECTION 4.3 — LONG WAVELENGTHS

SECTION 4.4 — SEARCH FOR COMMONALITY IN
                        INDEPENDENT MEASUREMENTS.

SECTION 4.5 — UNDERSTANDING DIFFERENCES BETWEEN
                        IONOSPHERIC AND CRUSTAL FIELDS.

SECTIONS 4.6, 4.7 — EMPIRICAL MODELS OF
                        IONOSPHERIC FIELDS.

: ESTIMATE OF CRUSTAL FIELD: **A**

FIGURE 4.1
Flow chart for crustal-field
identification process.

4.2   SELECTION OF QUIET DATA

When deriving maps of **A(r)**, it is desirable to begin with data from times
of magnetically quiet conditions. However, this involves a trade-off. Because
of the ubiquitous presence of Sq and EE fields during daylight hours, and because field-aligned currents and auroral E-region currents are always present,

it is not possible to select data that are known to have no contribution from $D(r, t)$. Other means of minimizing $D(r, t)$ or of modeling its effects are required. These methods invariably require statistically meaningful quantities of globally distributed data. Because of this, the "quietness" criterion cannot be made so stringent that the data remaining for analysis are too few to conduct the analysis.

Initial selection of quiet data is most easily accomplished by using one or more of the available magnetic indices described in Section 2.2.4. Because the 3-hourly Kp index is largely derived from mid-latitude magnetic observatories, it is suitable for equatorial and mid-latitude data. The simplest way to use this index for the selection of quiet periods is to accept data for further analysis only for times when Kp was less than some selected value. A problem sometimes encountered is that if the previous 3-hour period was somewhat disturbed, the beginning of the current 3-hour period could also be disturbed, even though the period as a whole is judged to be quiet. This can be avoided if a requirement is placed on the quietness of the previous period. Typically, data might be considered only for times when $Kp \leq 2$ and for which the prior 3-hour Kp was 2+ or less.

Alternatively, Cohen and Achache (1990) used Am to select data for use in deriving models of the main field and then used the sector indices Aj to select passes for mapping the anomaly field.

Although it has been so used, the Kp index is not a good indicator of the degree of magnetic activity in the auroral and polar regions. $D(r, t)$ can be large in these regions even when Kp has a low value. It is now well known (e.g., McPherron, 1991) that magnetic activity, particularly magnetic substorm activity, is highly controlled by the direction of the interplanetary magnetic field (IMF). In particular, when the direction of the IMF turns southward, magnetic activity increases. The north–south IMF is designated B_z, negative when southward. Accordingly, a possible criterion for selecting quiet data from high latitudes is to require that $B_z > 0$ or even that $B_z > 0$ for some time Δt prior to the measurement time.

Another criterion suitable for high latitudes is the amplitude of the AE index. Its shortcoming is that when the IMF B_z has been greater than zero for some time, the auroral oval tends to shrink poleward, out of the observing range of the AE observatories. In that case, the AE index can be small and may not accurately reflect the presence of disturbance in the polar cap. Nevertheless, AE is a very useful criterion for magnetic quiet conditions at high latitude. For example, Ravat et al. (1995) retained high-latitude data only when $AE \leq 50$ nT.

Still another criterion is to use the variance of ΔB along a pass to judge the magnetic quietness of that pass (e.g., Alsdorf, 1991; Alsdorf et al., 1994, 1997). The premise is that $A(r)$ is stationary and always present, whereas $D(r, t)$ is dynamic, with significant variations in amplitude from pass to pass, and it always adds energy. Profiles in the same location with smallest variance are least likely to be contaminated by $\mathbf{D}(\mathbf{r}, t)$. For *Magsat* data Ravat et al. (1995) rejected any pass for which the variance exceeded 80 nT^2. A similar criterion was adopted for the analysis of POGO high-latitude data, where Langel (1990a) rejected passes during which any value of $|\Delta B(\mathbf{r}, t)|$ exceeded 20 nT.

Data from the POGO satellites include all local times. Because there were three spacecraft acquiring data for about 6 years, the amount of data is large enough to permit application of relatively stringent selection criteria. At low latitudes, data between local times of 0900 and 1500 were discarded by Langel (1990a) to minimize the effects of the EE and *Sq* fields. In addition, adjacent passes were compared visually in an attempt to identify the time-invariant features, the lithospheric field, from the time-varying external field. Passes believed to have had significant contribution from the latter were discarded.

For best results, a combination of these criteria should be used. At low latitudes, the combination might be an initial selection based on the *Kp* or *Aj* indices, followed by selection on the basis of pass variance. At high latitudes, three steps are appropriate, namely, selection on interplanetary B_z, the *AE* index, and the pass variance.

4.3 ISOLATION FROM THE MAIN FIELD

Of all the contributions to the measured geomagnetic field, the main field has by far the largest magnitude and wavelengths. At the satellite altitudes considered here, its strength is not greatly diminished from that at the earth's surface: 30,000–60,000 nT. In contrast, at the altitudes at which the existing satellite data were taken, the magnitude of the lithospheric field is no greater than about 30 nT. If, in the future, data should be acquired at altitudes as low as 250 km, that maximum amplitude might rise to 60 nT. Thus, accuracy in modeling the main field, $\mathbf{B}_m(\mathbf{r}, t)$, including its variation with time, is crucial both to the accuracy of estimating $\mathbf{A}(\mathbf{r})$ and to the study of $\mathbf{D}(\mathbf{r}, t)$. Inaccurate representation of the main field can introduce systematic variations in data (i.e., the η). Models $\underline{\mathbf{B}}_m(\mathbf{r}, t)$ of $\mathbf{B}_m(\mathbf{r}, t)$ consist of the Gauss coefficients $\{g, h\}$ in equation (2.3).

Model error occurs when $\underline{\mathbf{B}}_m(\mathbf{r}, t)$ is not equal to $\mathbf{B}_m(\mathbf{r}, t)$; the model degree and/or order may be too large, so that $\underline{\mathbf{B}}_m(\mathbf{r}, t)$ includes some lithospheric field, or too small, so that $\underline{\mathbf{B}}_m(\mathbf{r}, t)$ does not represent all of $\mathbf{B}_m(\mathbf{r}, t)$, or the model may be in error because of contamination or inadequacy of the data upon which it is based. From the discussion of Section 2.1.1, there is some degree, say n_c, at which the spectrum of R_n shows a distinct break. For $n < n_c$ the main field is dominant; for $n > n_c$ the lithospheric field is dominant. However, this description is idealized. More likely, for n just below and just above n_c, neither the main nor the lithospheric field is sufficiently dominant that the other can be ignored. Furthermore, from Section 2.1.1, the value of n_c is not known definitively, but is in the range of 13–15. Further discussion of the overlap region $(12 < n < 16)$ and the effects of including terms of a particular degree on the resulting anomaly map and its interpretation is given in Chapter 6.

Model error is a long-wavelength effect, confined to spatial variations that can be described by spherical harmonics of degree less than 16, or wavelengths longer than 2,500 km. In the following paragraphs, three common sources of model error are discussed: (1) choosing $n^* \neq n_c$, so that $\underline{\mathbf{B}}_m(\mathbf{r}, t)$ does not represent all of the main field or includes part of the field from the lithosphere; (2) error in, or failure to account for, temporal change in $\underline{\mathbf{B}}_m(\mathbf{r}, t)$; (3) model error due to contamination by fields originating in the ionosphere.

4.3.1 MODEL DEGREE

The choice of the maximum degree for the model $\underline{\mathbf{B}}_m(\mathbf{r}, t)$ used to represent the main field, the n^* of Chapter 1, will affect the long-wavelength nature of the resulting anomaly map after the main-field estimate is removed. However, the lithospheric field does contain fields of degree $n \leq n_c$. Thus, for any choice of n^*, the lithospheric field at $n \leq n^*$ is deleted from the analysis. The goal is to choose $n^* = n_c$ so that the residual field after subtraction of $\underline{\mathbf{B}}_m$ retains the longest wavelengths of $\mathbf{A}$ that can be reliably identified in the data. If $n^* > n_c$, part of the distinguishable long-wavelength portion of $\mathbf{A}$ (i.e., for $n_c < n \leq n^*$) would be incorporated into $\underline{\mathbf{B}}_m(\mathbf{r}, t)$. Long-wavelength features of $\mathbf{A}$ would be lost needlessly in the data reduction process. Because the exact value of n_c is still debated, and the values of n^* currently being used are all 13 or less, which is the lowest value proposed for n_c, it seems unlikely that long-wavelength features of $\mathbf{A}$ are needlessly lost.

If $n^* < n_c$, $\underline{\mathbf{B}}_m(\mathbf{r}, t)$ will not represent all of the main field, and the resulting estimates of $\mathbf{A}$ will contain a spurious long-wavelength contribution. Choosing n^* to be 13 or 15 is quite feasible with data of the quality available from the *Magsat* satellite. Based on Langel and Estes (1982), most models of $\underline{\mathbf{B}}_m(\mathbf{r}, t)$ from *Magsat* are of degree 13, including those used in the analyses presented herein. But it is not necessarily possible to derive models of degree 13 and above from other magnetic data sets. Using scalar data from the POGO satellites, models were derived with $n^* = 13$ (e.g., Langel, Coles, and Mayhew, 1980). However, with only scalar data, the sectorial $(n = m)$ terms in equation (2.3) are not well resolved (Backus, 1970, 1974; Lowes, 1975; Stern and Bredekamp, 1975; Stern, Langel, and Mead, 1980). Reliable estimates of $\underline{\mathbf{B}}_m(\mathbf{r}, t)$ from *Cosmos 49* data could be obtained only up to degree 11 (Langel, 1990b). IGRF models, even the definitive versions, extend only to degree 10.

The consequence of analyzing $\Delta\mathbf{B}(\mathbf{r}, t)$ relative to a model of degree n^* less than n_c is that unless additional steps are taken, the resulting $\Delta\mathbf{B}(\mathbf{r}, t)$ will retain

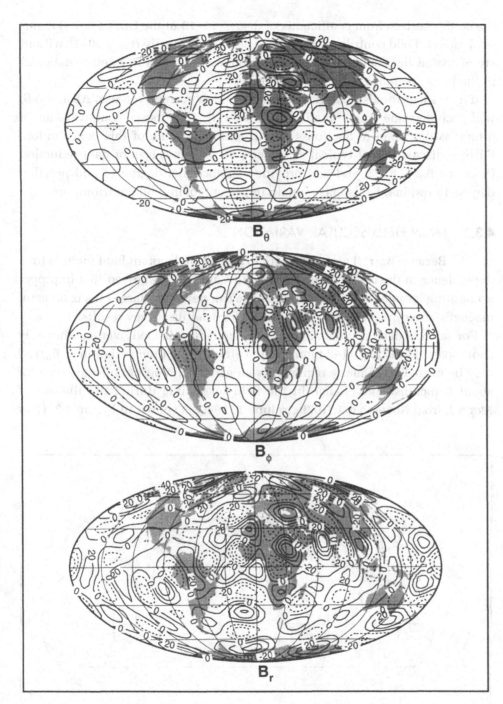

B_θ

B_ϕ

B_r

FIGURE 4.2
Contribution to the main
field at 1980 from spherical
harmonic degrees 11
through 13. From the
GSFC(12/83) field model
(Langel and Estes, 1985b).
Contour interval 10 nT.
Mollweide projection.

long-wavelength features originating in the earth's core, rather than the lithosphere. This is particularly true for aeromagnetic and marine magnetic surveys,
which typically are analyzed using the IGRF models. To illustrate the scale and
amplitude of the fields involved, Figure 4.2 shows contours for B_r, B_θ, and B_ϕ at

the earth's surface from coefficients of degree 11–13 of the GSFC(12/83) main-field model. Field contributions at these degrees can easily reach 10–20 nT and are of spatial sizes equivalent to some anomalies mapped in satellite data and in the larger, regional aeromagnetic surveys.

If $n_c > 13$, residuals from the main-field models currently in use, $\Delta \mathbf{B}$ and ΔB, will include some contamination from $\mathbf{B}_m(\mathbf{r}, t)$. The use of long-wavelength filters, as described in Section 4.4, will eliminate some of this contribution. Further discussions of typical R_n spectra at various stages of data reduction (Sections 6.2.1, 6.2.3, and 6.3) provide evidence that degree-14 and possibly degree-15 residual fields may contain important main-field contributions.

4.3.2   MAIN-FIELD SECULAR VARIATION

Because $\mathbf{B}_m(\mathbf{r}, t)$ varies with time, models of the main field include time dependence in the coefficients. Regan and Cain (1975) showed that improper accounting for main-field temporal change can have very large effects on aeromagnetic survey results. The same is true for satellite survey results.

For satellite data, if the data to be analyzed are collected over a sufficiently short span of time (e.g., 1–2 months), the effect of the time variation of $\mathbf{B}_m(\mathbf{r}, t)$ may be negligible in many regions. As an example, *Magsat* collected data for about 6 months. The change in the field magnitude during the lifetime of *Magsat*, from November 1, 1979, to June 1, 1980, is shown in Figure 4.3. Over

FIGURE 4.3
Change in main-field magnitude from the start to the end of the *Magsat* mission (November 1, 1979, to June 1, 1980). From the GSFC(12/83) field model (Langel and Estes, 1985b). Contour interval 10 nT/yr. Mollweide projection.

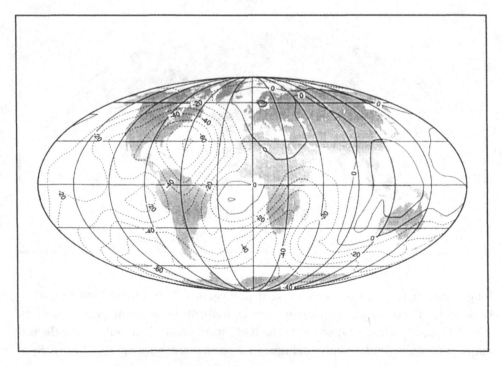

much of Europe, Asia, India, and Africa the changes were quite small. However, over the North Atlantic and Antarctic regions the changes were substantial. In contrast to the short lifetime of *Magsat*, the POGO satellites were in operation for about 6 years. During such a long time span large, changes will occur at all spatial scales. Attempts at prediction of main-field temporal changes for periods beyond 1–2 years have not been successful. This means that it is not possible to anticipate with any confidence the length of time over which one can ignore temporal changes. In general, it is prudent to take account of the temporal variation of $\underline{\mathbf{B}}_m(\mathbf{r}, t)$ in all cases.

4.3.3 COORDINATE SYSTEMS

Equation (2.3) gives the geomagnetic potential in geocentric spherical coordinates. Normally these are the same coordinates used in specifying the satellite data. However, as noted in Section 2.0, the earth is not a sphere, and the usual latitude is geodetic, not geocentric. On the surface of the earth, observations are usually labeled with geodetic rather than geocentric latitude, and the usual geographic locations, including continental boundaries, and so forth, are expressed in geodetic coordinates. Spherical harmonic models come in the form of software that often is able to compute in either geodetic or geocentric position. Whereas the distinction does not matter when considering $\mathbf{A}$, it does matter for $\underline{\mathbf{B}}_m$, and one must be certain to use the correct system. Transformations between the geodetic and geocentric systems are discussed elsewhere (Langel, 1987, 1992).

4.3.4 MODEL CONTAMINATION BY IONOSPHERIC FIELDS

Satellite magnetic data are acquired above the ionospheric region where substantial currents flow; thus the fields from such currents are curl-free and can be represented by an internal potential. Hence there is the possibility that $\underline{\mathbf{B}}_m(\mathbf{r}, t)$ represents some portion of the ionospheric field as well as the main field. In particular, the *Sq* current system, say **S**, and the equatorial electrojet (EE) are part of $\mathbf{D}(\mathbf{r}, t)$. Models of **S** in terms of potential functions require terms of degree n up to about 23 and of order m up to about 6 (W. H. Campbell, personal communication, 1992). Because **S** and EE originate internal to the satellite orbit, those terms in their potentials that are of degree 13 or less may be at least partly subsumed into $\underline{\mathbf{B}}_m(\mathbf{r}, t)$.

Of particular importance for anomaly studies are situations in which such contamination of $\underline{\mathbf{B}}_m(\mathbf{r}, t)$ leads to spurious signals that can be interpreted as part of $\mathbf{A}(\mathbf{r})$. Such a situation arose in the analysis of *Magsat* data. As part of the derivation of the GSFC(12/83) field model Langel and Estes (1985a,b) compared models based solely on data from either dawn or dusk local time, designated

DAWN(6/83-6) and DUSK(6/83-6), respectively. Differences between the two models of up to 15 nT clearly reflected the presence of fields from the EE in the dusk data. That was accounted for in the derivation of the GSFC(12/83) model, so that the effects were minimized. However, earlier models, such as MGST(6/80) (Langel et al., 1980b) and MGST(4/81-2) (Langel et al., 1981), were contaminated at the 10-nT level by fields from the EE. Those models were widely distributed before the problem became evident. In particular, MGST(4/81-2) was used to derive $\Delta \mathbf{B}(\mathbf{r}, t)$ values for data on computer tapes that were, and are, distributed to *Magsat* investigators and other interested scientists. Langel et al. (1993) described the effects of that contamination in some detail.

Because the GSFC(12/83) model has become the "standard" *Magsat* model and was the basis for the 1980 DGRF, Langel et al. (1993) investigated the extent to which it might be contaminated by the dusk fields associated with the EE. They concluded that such effects may be of the order of 1–4 nT, which can be significant in studies of **A**.

4.3.5 SPHERICAL HARMONIC DEGREE AND WAVELENGTH

The notion of wavelength is associated with the circular functions (e.g., sine and cosine). It is also common to associate the notion of wavelength with spherical harmonics. This can be misleading unless care is taken. Consider the surface harmonic of degree n and order m from equation (2.3):

$$S_n^m(\theta, \phi) = \left[g_n^m \cos(m\phi) + h_n^m \sin(m\phi) \right] P_n^m(\cos \theta). \tag{4.2}$$

At a fixed value of θ, $S_n^m(\theta, \phi)$ is clearly periodic, with a wavelength $\lambda_m = 2\pi / m$. Note that for a given degree n, equation (2.3) contains values of m from zero to n. That is, the shortest "wavelength" associated with degree n is $\lambda_n = 2\pi/n$. But λ_n is only one of $n + 1$ wavelengths associated with degree n; the other n wavelengths are all longer than λ_n. Spherical harmonics of degree n represent wavelengths from *zero* to λ_n and should not be thought of in terms of λ_n only.

Furthermore, the periodicity in $S_n^m(\theta, \phi)$ is explicit only in the ϕ coordinate. To be sure, there is periodicity in $P_n^m(\cos\theta)$, but it is not in the form of a Fourier series. To be explicit, Schmidt-normalized associated Legendre functions up to degree and order 4 are shown in Table 4.1, and Figure 4.4 shows the seventh-degree associated Legendre functions as functions of colatitude θ. The functions $P_n^m(\cos \theta) \cos(m\phi)$ and $P_n^m(\cos \theta) \sin(m\phi)$ are zero along $(n-m)$ circles of latitude (the zeros of P_n^m) and along $2m$ meridians [the zeros of $\cos(m\phi)$ or $\sin(m\phi)$], dividing the sphere into regions, called tesserae, in which the sign of the spherical harmonic is either positive or negative. Because of this, the spherical harmonics are also called tesseral surface harmonics. When $m = 0$, they are called zonal surface harmonics because they divide the sphere into zones along lines

Table 4.1. *Schmidt-normalized Legendre functions up to degree and order 4*

n	m	$P_n^m(\cos\theta)$
0	0	1
1	0	$\cos\theta$
1	1	$\sin\theta$
2	0	$(1/2)(3\cos^2\theta - 1)$
2	1	$\sqrt{3}(\cos\theta\sin\theta)$
2	2	$(1/2)\sqrt{3}\sin^2\theta$
3	0	$(1/2)(5\cos^3\theta - 3\cos\theta)$
3	1	$(\sqrt{3}/2\sqrt{2})\sin\theta(5\cos^2\theta - 1)$
3	2	$(1/2)\sqrt{15}\sin^2\theta\cos\theta$
3	3	$(\sqrt{5}/2\sqrt{2})\sin^3\theta$
4	0	$(1/8)(35\cos^4\theta - 30\cos^2\theta + 3)$
4	1	$(\sqrt{5}/2\sqrt{2})\sin\theta(7\cos^3\theta - 3\cos\theta)$
4	2	$(\sqrt{5}/4)\sin^2\theta(7\cos^2\theta - 1)$
4	3	$(\sqrt{35}/2\sqrt{2})\sin^3\theta\cos\theta$
4	4	$(\sqrt{35}/8)\sin^4\theta$

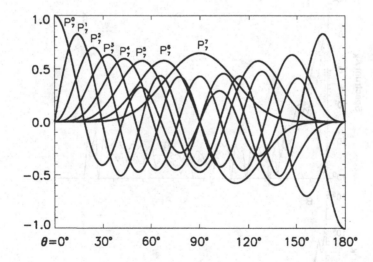

FIGURE 4.4
Associated Legendre functions of degree 7 as functions of colatitude θ.

of latitude; when $m = n$, they are called sectorial surface harmonics because they divide the sphere into sectors along lines of equal longitude and there are no circles of latitude along which the function vanishes. The regions of constant sign are illustrated in Figure 4.5.

In view of this discussion, it is useful to consider the Fourier spectra of the X, Y, and Z geomagnetic components along lines of constant latitude and along a longitudinal profile. These have been derived by Wang (1987). Fourier amplitude spectra for latitudinal profiles for degrees 1–64 are shown in Figure 4.6.

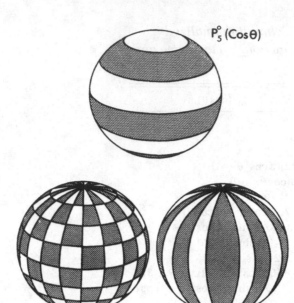

$P_5^0 (\cos\theta)$

$P_{16}^9 (\cos\theta) \cos 9\phi$

$P_9^9 (\cos\theta) \cos 9\phi$

FIGURE 4.5
Map of zero lines and tesserae in which the sign of the spherical harmonic is constant for a zonal harmonic (P_5^0), a sectorial harmonic ($P_9^9 \cos 9\phi$), and a tesseral harmonic ($P_{16}^9 \cos 9\phi$). (Courtesy of D. Barraclough.)

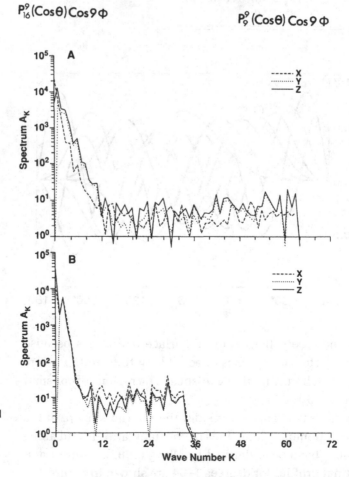

FIGURE 4.6
Fourier amplitude spectrum derived from degree-1–64 terms of a spherical harmonic model: (A) equatorial profile, (B) 60° N latitudinal profile. (From Wang, 1987, with kind permission of TERRAPUB.)

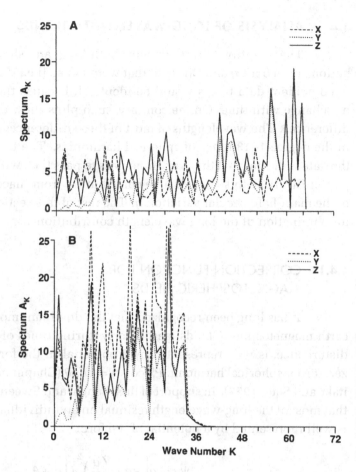

FIGURE 4.7
Fourier amplitude spectrum
derived from degree-14–64
terms of a spherical
harmonic model: (A)
equatorial profile, (B) 60° N
latitudinal profile. (From
Wang, 1987, with kind
permission of TERRAPUB.)

A large amount of power is concentrated at low wavenumbers, as would be expected, where wavenumber is defined as $k = 2\pi R/\lambda$, with R the radius of the latitudinal circle. The range of k is thus smaller at higher latitudes.

All of the spectra in Figure 4.6 show clear breaks, at about $k = 12$ at the equator, and $k = 8$ at 60° latitude. The break represents the dividing point between dominance of the main field and dominance of the lithospheric field. Now, it might be thought that deleting the wavenumbers below the break would remove the main field and leave the lithospheric field. That is not true; some lithospheric field would also be removed. Figure 4.7 shows the same spectra as in Figure 4.6, except that only contributions from degrees 14–64 are included (i.e., the lithospheric field). Note that significant power remains at lower wavenumbers – power contributed by *lower-order* terms.

In using analyses, such as double Fourier representations, that filter on a wavenumber basis, one must consider the nature of spherical harmonics and note that it is easy to unintentionally discard low-order anomaly fields.

4.4 ANALYSIS OF LONG-WAVELENGTH TRENDS

The quantity $\Delta\mathbf{B}(\mathbf{r}, t)$ contains both long- and short-wavelength contributions from both η and $\mathbf{D}(\mathbf{r}, t)$. If that were not so [i.e., if $\Delta\mathbf{B}(\mathbf{r}, t) = \mathbf{A}(\mathbf{r})$], plots of coincident data tracks would be identical, because the anomaly field does not change with time. On the contrary, such plots show clear long-wavelength differences. The wavelengths of most of these differences are such that they lie in the degree-1–13 range of spherical harmonics. The fact that they remain in the data indicates that they vary from pass to pass (i.e., with time) and likely are part of $\mathbf{D}(\mathbf{r}, t)$, with some possible contribution from inaccurate representation of the main-field secular variation η. The rest of this section describes methods for elimination of the long-wavelength contributions.

4.4.1 CORRECTION FUNCTION FOR MAGNETOSPHERIC FIELDS

It has long been recognized that the dominant morphology in the near-earth magnetospheric fields, particularly during times of worldwide magnetic disturbance, is well represented by a potential in the form of the first degree, zero-order spherical harmonic $P_1^0(\cos\theta) = \cos\theta$ (Chapman and Price, 1930; Rikitake and Sato, 1957). In support of this, Langel and Sweeney (1971) have shown that most of the long-wavelength residual on an individual pass of satellite data is well represented by a potential of the form

$$\Psi_d = a\left[\frac{r}{a}e + \left(\frac{a}{r}\right)^2 i\right]\cos\theta. \qquad (4.3)$$

This function (i.e., the coefficients e and i) can be determined for any individual satellite pass by a least-squares process (Langel and Sweeney, 1971; Davis and Cain, 1973). The variable θ in equation (4.3) is a colatitude. But there are several definitions of latitude (i.e., geographic, geocentric, dipole or geomagnetic, and dip latitudes). Langel and Sweeney (1971) used geocentric colatitude. Values for e and i on the distributed *Magsat* Investigator tapes were determined using θ equal to the colatitude relative to the magnetic dip latitude (Langel et al., 1981). The use of geocentric, geodetic, or dip colatitude, however, leaves long-wavelength trends in the data (e.g., Singh, Agarwal, and Rastogi, 1986), which led Langel et al. (1982c,d) to adopt the colatitude relative to the magnetic dipole latitude and led Zaaiman and Kuhn (1986) to recommend taking the colatitude relative to the magnetic dip latitude computed at three earth radii above the surface of the earth. These latter two approaches give very similar results.

However, regardless of the choice of θ, use of the function of equation (4.3) leaves long-wavelength trends in the data, although these are less when the dipole colatitude is used. One reason for the remaining long-wavelength trends is that the near-earth morphology of the magnetospheric fields changes slightly with time, particularly with season, as described in some detail by Langel et al. (1996).

One solution to the problem of the shifting magnetosphere fields is that of Regan, Handschumacher, and Sugiura (1981), who recommended using an alternative correction to equation (4.3) in the form of a four-term cosine series. Other alternatives are described in the next two sections.

4.4.2   AD HOC TREND CORRECTIONS

After application of the correction for magnetospheric fields, there still remain significant differences, with wavelengths of more than about 2,000–3,000 km, between nearby passes. Figure 4.8 illustrates one method adopted for additional filtering. Three sets of three passes are shown; each of the three passes in a set follows very nearly the same track in longitude, and they are at similar altitudes, so that the lithospheric field should be nearly identical

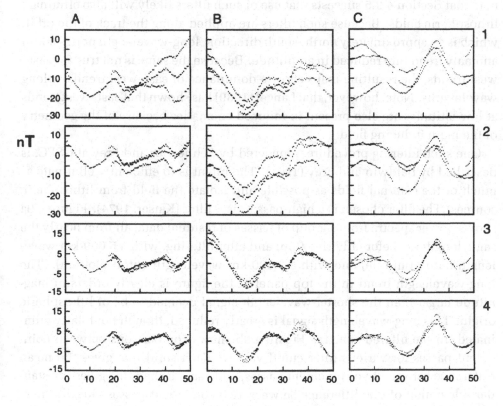

DEGREES LATITUDE

FIGURE 4.8

Satellite magnetic profiles from POGO data: row 1, raw residual data; row 2, residual data with correction from equation (4.3) applied; row 3, data from row 2 with additional linear trend removed; row 4, data from row 2 with additional parabolic trend removed. Ordinate is degrees of latitude; abscissa is nanoteslas. (From Mayhew, 1977, with permission.)

for all. The top row shows the raw residual after removal of a field model. If the residual were purely of lithospheric origin, the three passes should be identical. Clearly they are not. The second row shows the same sets of passes with the magnetospheric correction of equation (4.3) removed. The agreement is better, but not satisfactory. The third and fourth rows show the results when an additional linear trend or parabolic trend, respectively, is removed. In each case the agreement among the three passes is now satisfactory. Because there will be an edge effect at the segment end, the segments are chosen to overlap significantly. Data used in deriving average maps and inversion solutions are then taken from the interior of each segment.

4.4.3   HIGH-PASS FILTERING

Removing linear or parabolic trends is an ad hoc procedure, with no physical basis. An alternative procedure is to eliminate trends with a high-pass or band-pass filter (Ridgway and Hinze, 1986; Baldwin and Frey, 1991). Such a filter can be applied after the correction of equation (4.3) or can replace both the functional correction of equation (4.3) and the linear-trend removal. But note that Section 4.3.5 suggests that use of such filters likely will also eliminate lithospheric fields. Because such filters are applied along the track of the orbit, which is an approximately north–south direction, long-wavelength north–south anomaly trends are reduced in amplitude. Because the same is not true for east–west trends, the resulting maps show a dominance of east–west trends at long wavelengths. Note, however, that Langel (1989) has shown that east–west trends at low latitudes are also present in induced anomalies because of the geometry of the main inducing field.

One such filtering procedure, pioneered by Schnetzler and Frey at GSFC, is described by Baldwin and Frey (1991). The intent is to efficiently eliminate as much of the external fields as possible and isolate the field from lithospheric sources. The filter chosen is a high-pass Kaiser filter (Kaiser, 1974). Figure 4.9a shows power spectra from a group of passes of residual data, all from nearly the same longitude, before filtering (top) and after filtering, with a 5,000-km wavelength cutoff (middle) and with a 3,000-km wavelength cutoff (bottom). The long-wavelength trend in the top panel of the figure is clearly orders of magnitude larger than the shorter-wavelength signal deemed to be of lithospheric origin. This long-wavelength signal is greatly reduced, though not totally eliminated, in the filtered spectra. Figure 4.9b shows comparisons between coincident passes for various filter cutoff values. Each solid line gives the mean difference between two coincident passes, and each dashed line gives the standard deviation of the difference between two coincident passes. Ideally there would be a sharp break in the curve at some wavelength, which would then be chosen as the optimal filter cutoff. That clearly is not the case, and the decision

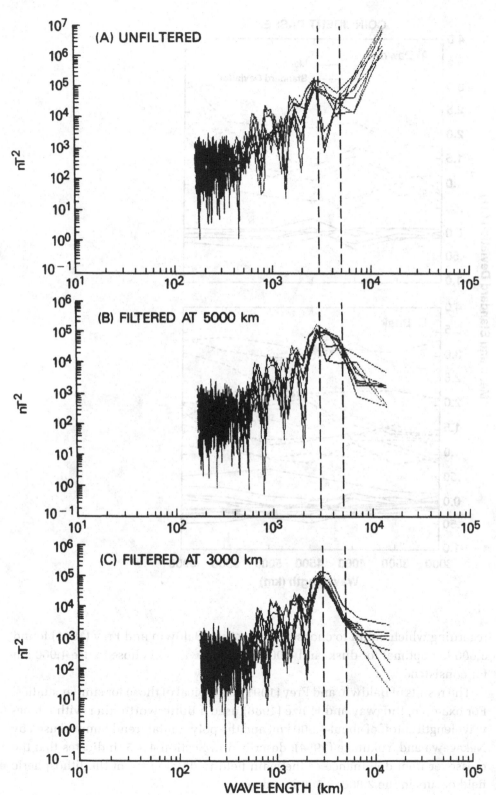

FIGURE 4.9a
Power spectra for eight
dawn passes: (A) unfiltered
passes, (B) passes filtered
with a Kaiser filter having
long-wavelength cutoff of
5,000 km, (C) same as part
B but with a 3,000-km
cutoff. (Reprinted from R.
Baldwin and H. Frey,
*Magsat crustal anomalies
for Africa: dawn and dusk
data differences and a
combined data set. Physics
of the Earth and Planetary
Interiors, 67, 237–50,
1991,* with kind permission
of Elsevier Science NL, Sara
Burgerhartstraat 25, 1055
KV Amsterdam, The
Netherlands.)

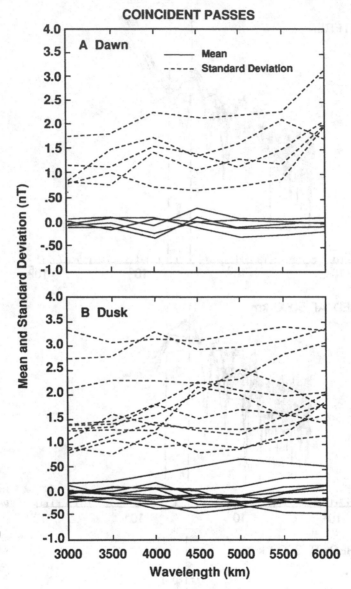

FIGURE 4.9b
Standard deviation of the mean difference (dashed) and difference of the mean of coincident pass pairs (solid) as a function of the long-wavelength cutoff of the Kaiser filter: (A) 5 dawn pass pairs, (B) 11 dusk pass pairs. (Reprinted from R. Baldwin and H. Frey, *Magsat* crustal anomalies for Africa: dawn and dusk data differences and a combined data set. *Physics of the Earth and Planetary Interiors, 67*, 237–50, 1991, with kind permission of Elsevier Science NL, Sara Burgerhartstraat 25, 1055 KV Amsterdam, The Netherlands.)

regarding which cutoff to choose is subjective. Baldwin and Frey (1991) found 3,000 km optimal for dusk, and 4,000 km for dawn, and chose to use 4,000 km for consistency.

The results of Baldwin and Frey (1991) are typical of those for similar studies. For example, Ridgway and Hinze (1986) used a Butterworth filter with a long-wavelength cutoff of about 4,900 km, and the polynomial trend removal used by Nakagawa and Yukutake (1984), described in Section 4.4.5, indicates that the crossover from dominance of the main field to dominance of the lithospheric field occurs in the 2,800–3,500-km range.

4.4.4 CROSSOVER ANALYSIS

When the orbital data available include tracks crossing one another, then another method can be used for removing, or adjusting, long-wavelength trends. The basic assumption is that the signal, or lithospheric field, corrected to a common altitude and measured at the crossover point, is identical for each track. Suppose there are K tracks of data, ordered as $k = 1, 2, \ldots, K$, and there are $n = 1, 2, \ldots, N$ locations, where two of the K tracks cross one another. Let f be the measured quantity. It could be the scalar residual, ΔB, or one of the components of the vector residual, $\Delta \mathbf{B}$. Assume that f is made up of the real lithospheric signal, say A, plus a nonlithospheric trend that can be represented by a polynomial, p:

$$f = A - p. \tag{4.4}$$

By definition, for the kth track,

$$p_k(d_k) = \sum_{j=0}^{J} a_{kj} d_k^j = \mathbf{d}_k^T \mathbf{a}_k, \tag{4.5}$$

where $\mathbf{a}_k$ and $\mathbf{d} = (1, d, d^2, \ldots, d^J)$ are $(J+1)$-component vectors of coefficients and of powers of the distance along the kth track. For satellite analysis, J is generally 2 or less, often 0 or 1. Superscript T denotes the transpose. If y_n is the difference in measured values for the two crossing tracks, say track i and track k, and if $\mathbf{y}$ is the matrix of values of y_n, then the objective is to find the values of the a_{kj} that will minimize the quantity $\mathbf{y}^T \mathbf{y}$ under the assumption that $A^i = A^k$ at the nth crossover. Then

$$y_n = f_n^i - f_n^k = A_n^i - A_n^k + \mathbf{d}_{k,n}^T \mathbf{a}_k - \mathbf{d}_{i,n}^T \mathbf{a}_i = \mathbf{d}_{k,n}^T \mathbf{a}_k - \mathbf{d}_{i,n}^T \mathbf{a}_i. \tag{4.6}$$

If $\mathbf{a} = (\mathbf{a}_1, \mathbf{a}_2, \ldots, \mathbf{a}_k)^T$ is the $k \cdot J$-dimensional vector of all polynomial parameters, and if $\mathbf{d}_n = (0, \ldots, -\mathbf{d}_{i,n}, 0, \ldots, 0, \mathbf{d}_{k,n}, 0, \ldots)^T$ is a $k \cdot J$-dimensional vector with nonzero values only for the entries corresponding to the distances of the two tracks at the nth crossover, and if D is the matrix with its nth row equal to $\mathbf{d}_n^T$, then

$$\mathbf{y} = D\mathbf{a}, \tag{4.7}$$

with least-squares solution

$$\underline{\mathbf{a}} = [D^T D]^{-1} D^T \mathbf{y}. \tag{4.8}$$

The inverse in equation (4.8) may not be well-behaved. Taylor and Frawley (1987) found that imposing the condition that $a_{10} = 0$ results in a stable solution for data in the Kursk region of Ukraine. This condition means that the bias on the first track is assumed to be zero. It establishes the arbitrary zero level for the crossover analysis.

Equation (4.8) can also be stabilized by adding an a priori condition that $\underline{a}^T\underline{a}$ be "small" in some sense, that is, by using ridge regression (Hoerl and Kennard, 1970a,b). In that case, equation (4.8) becomes

$$\underline{a} = [D^T D + \lambda^d I]^{-1} D^T \mathbf{y}, \tag{4.9}$$

where λ^d is a damping factor to be determined in the analysis. In practice, there is a trade-off between the goodness of fit and the magnitude of $\underline{a}^T\underline{a}$ that must be explored as part of the analysis. A more extended discussion of ridge regression is given in Section 5.6.2.

An advantage of this method is that it removes a set of self-consistent long-wavelength biases without resorting to filtering, which can introduce artificial effects into the data.

4.4.5 POLYNOMIAL FIT ALONG ORBITAL TRACKS

Nakagawa and Yukutake (1984, 1985) and Nakagawa, Yukutake, and Fukushima (1985) described an alternative method for removing long-wavelength fields. Instead of subtracting an estimate of the main field from a spherical harmonic analysis, their method removes *all* long-wavelength fields together. The essence of the procedure is to fit each component of the data along each pass with a polynomial expansion whose degree is chosen to remove long-wavelength fields, while leaving the shorter-wavelength lithospheric fields.

Suppose the region to be studied lies between latitudes λ_1 and λ_2. Let $P_n(z)$ be completely normalized Legendre functions; that is,

$$\int_{-1}^{1} P_n(z) P_k(z)\, dz = \delta_{n,k}. \tag{4.10}$$

If the scaled distance between λ_1 and λ_2 is given by $z(\lambda)$, where z is chosen so that $z(\lambda_1) = -1$ and $z(\lambda_2) = 1$, then any component B_i of the field can be represented in the form

$$B_i(z) = \sum_{n=0}^{n^*} \mathcal{A}_n^i P_n(z), \tag{4.11}$$

with the $\mathcal{A}_n^i$ determined for each component of each pass by least-squares fitting. Figure 4.10 shows the magnitude of the $\mathcal{A}_n$ and the root-mean-square (rms) residual of the fit for the vertical component from a series of passes in the range of $\lambda_1 = 8°$ N to $\lambda_2 = 68°$ N near Japan. In Figure 4.10a the coefficient amplitude drops sharply up to n values of about 6, after which it is relatively constant. In Figure 4.10b, the goodness of fit improves dramatically as n increases, up to n

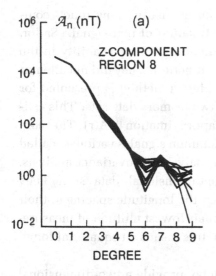

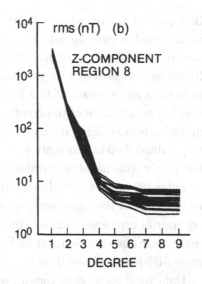

FIGURE 4.10
Results of application of the analysis of equation (4.11): (a) absolute values of the coefficients of the polynomials determined for the Z component data, (b) root-mean-square (rms) residuals of the component data after removal of terms up to and including degree n^* of the polynomial series. (From Nakagawa et al., 1985, with permission.)

of 5 or 6, after which the change is negligible. Figure 4.10a is the single-pass analog of Figure 2.3, which shows a spectrum for the global potential function. In Figure 2.3, a slope change occurs at about degree 13 or 14, or at a wavelength of about 3,040–3,270 km. In the polynomial spectrum of Figure 4.10a the break occurs between degrees 5 and 6, corresponding to 2,800–3,500 km. The obvious interpretation is that the change in slope of the spectrum corresponds to the change from dominance of the main field to dominance of the lithospheric field, with important caveats raised by the discussion in Section 4.3.5.

Nakagawa and Yukutake (1984, 1985) and Nakagawa et al. (1985) analyzed all of the data components from *Magsat* data for several subregions near Japan. Data from dawn and dusk local times were considered separately. In all cases, the spectral break occurred at or near degree 5 or 6. They subtracted a sixth-degree representation from all the data, taking the residual to be the anomalous field.

The basic premise of this analysis does not depend upon the type of polynomial chosen. In principle, other choices would work as well. Nor is the analysis necessarily limited to a short latitude range. However, the method does seem particularly well suited to smaller areas and to analyses seeking for finer resolution. As the length of the latitude range is decreased, the resolution of the series in equation (4.11) will increase. Removal of long-wavelength external fields [with the potential function of equation (4.3), with ad hoc trend corrections, or with high-pass filtering] is unnecessary in this method.

4.5 WAVELENGTH DOMAIN ESTIMATION FOR THE COMMON SIGNAL IN MULTIPLE SOURCES

When two independent data sets are available for a given geographic region [i.e., with a "common signal," $\mathbf{A}(\mathbf{r})$, but with different interfering signals,

$\mathbf{D}(\mathbf{r}) + \eta]$, it is convenient to have a method for estimating how much common signal is present and for extracting an estimation of that signal. Sailor, Lazarewicz, and Brammer (1982) quantified the notion of repeatability in the frequency domain by using a spectral coherence function. They did not attempt to extract an estimate of that common signal. Here a method is presented for making an estimate of the common portion of two or more data sets. This estimated common signal is then assumed to be an approximation to $\mathbf{A}(\mathbf{r})$. The technique described in this section to estimate the common signal is variously called harmonic wavenumber correlation, correlation analysis, or covariance analysis.

Satellite passes provide independent one-dimensional data sets. They roughly cover the same geographic region when the longitude spacing at their equatorial crossings is closer than about half their lowest altitude of measurement. For measurements at and above 400 km, this means an equatorial longitude difference of less than 2°.

Data from an area or region, or from the globe, provide a two-dimensional data set. Two-dimensional independent data sets can be extracted from the overall data set by subdividing the data according to some suitable parameter (e.g., altitude or local time). For example, the *Magsat* data divide naturally into two data sets with different local times, dawn and dusk.

It is assumed that each set of independent measurements, signal and noise together, can be sufficiently extracted from any accompanying random noise to permit representation by spectral decomposition or a harmonic series, either one- or two-dimensional. Harmonic analysis can be accomplished by a one-dimensional Fourier series for individual passes or by either a two-dimensional Fourier series or a spherical harmonic analysis for two-dimensional data. After each data set is harmonically decomposed, the correlation coefficients and amplitude ratios between the two decompositions are computed *for each harmonic*. Only harmonics that satisfy predetermined selection criteria are retained. Those criteria, for example, may be in the form of requiring the correlation coefficient to exceed some magnitude, say ρ_0, and the ratio of largest to smallest amplitude not to exceed some value, say R_0. Summation of the retained harmonics gives an estimate of the common signal, separately for each of the original data sets. The resulting pair of retained signals typically are then very highly correlated, and the final estimate of the common signal is made by averaging the two sets of selected harmonics. Differences between the two data sets, both before and after the correlation analysis, can provide estimates of the level of contamination or noise.

4.5.1 FOURIER HARMONIC CORRELATION ANALYSIS

The simplest example is a one-dimensional Fourier series (e.g., Jenkins and Watts, 1968; Hildebrand, 1976; Walker, 1988). Suppose the data are represented as a function of x in the interval $-L \leq x \leq L$, subdivided into K

equal-length segments of length $\Delta x = 2L/K$, with segment end points denoted by x_p, $p = 0, \ldots, K - 1$. Assume that there are K data points, say $\{f_p, p = 0, \ldots, K - 1; f_p = f(x_p)\}$. Then the *discrete Fourier series* representing the data is

$$f(p) = f(x_p) = \sum_{k=-K/2}^{K/2} F(k)e^{-ik(2\pi/K)p}, \tag{4.12}$$

where

$$F(k) = \frac{1}{K} \sum_{p=0}^{K-1} f(p)e^{+ik(2\pi/K)p}, \tag{4.13}$$

where $i = \sqrt{-1}$, $F(k)$ is complex, p is real, $F(-k) = F^*(k)$, and where the asterisk denotes the complex conjugate. $F(k)$ is the discrete Fourier transform of $f(x)$. Because $F(-k) = F(K - k)$, (4.12) can be rewritten as

$$f(p) = f(x_p) = \sum_{k=0}^{K-1} F(k)e^{-ik(2\pi/K)p}. \tag{4.14}$$

If there are two signals $f(x)$ and $g(x)$, each measured at the points $\{x_p : p = 0, \ldots, K - 1\}$, then the correlation between $f(x)$ and $g(x)$ is estimated by

$$\rho_{fg} = \frac{\sum_{p=0}^{K-1} f_p g_p^*}{\{[\sum_{p=0}^{K-1} f_p f_p^*][\sum_{p=0}^{K-1} g_p g_p^*]\}^{1/2}}, \tag{4.15}$$

$$\rho_{fg} = \frac{\sum_{k=-K/2}^{K/2} F(k)G^*(k)}{\{[\sum_{k=-K/2}^{K/2} F(k)F^*(k)][\sum_{k=-K/2}^{K/2} G(k)G^*(k)]\}^{1/2}}. \tag{4.16}$$

Suppose f and g are simple sinusoidal functions,

$$f^k(x) = A_f^k \cos(k\phi) + B_f^k \sin(k\phi) = \mathcal{X}_f^k \cos(k\phi - \zeta_f^k),$$
$$g^k(x) = A_g^k \cos(k\phi) + B_g^k \sin(k\phi) = \mathcal{X}_g^k \cos(k\phi - \zeta_g^k), \tag{4.17}$$

where $\phi = (\pi/L)x$, $(\mathcal{X}_f^k)^2 = (A_f^k)^2 + (B_f^k)^2$, $\tan \zeta_f^k = B_f^k/A_f^k$, $A_f^k = \mathcal{X}_f^k \cos \zeta_f^k$, and $B_f^k = \mathcal{X}_f^k \sin \zeta_f^k$, with similar relationships for A_g^k, B_g^k, $\mathcal{X}_g^k$, and ϕ_g^k. Defining

$$\Delta \zeta^k = \zeta_f^k - \zeta_g^k, \tag{4.18}$$

then

$$\rho_{fg}^k = \cos(\Delta \zeta^k). \tag{4.19}$$

According to (4.19), the correlation coefficient between two sinusoidal signals is equal to the cosine of the difference between their phase angles and is independent of their amplitudes (e.g., Alsdorf, 1991; Alsdorf et al., 1994). The overall correlation between two Fourier-analyzed signals is a weighted average of the cosines of the phase differences; that is,

$$\rho_{fg} = \frac{\sum_{k=0}^{K/2} \mathcal{X}_f^k \mathcal{X}_g^k \cos(\zeta_f^k - \zeta_g^k)}{\{[\sum_{k=0}^{K/2} (\mathcal{X}_f^k)^2][\sum_{k=0}^{K/2} (\mathcal{X}_g^k)^2]\}^{1/2}}. \tag{4.20}$$

One scheme for estimating the common signal in adjacent passes of satellite magnetic field data is that of Alsdorf (1991) and Alsdorf et al. (1994). A particular pass is chosen for study, in this case to be compared with the adjacent passes to the east and to the west. The southern and northern end points (latitudes) must be the same for a Fourier comparison, so any necessary truncation is applied. Each of the three passes is then Fourier-analyzed. The pass chosen for study is then compared with, for example, the pass to the west, wavenumber by wavenumber. For each wavenumber, the correlation coefficient, say ρ_w^k, is computed using equation (4.19). If $\rho_w^k < \rho_0$, where ρ_0 is the selected correlation criterion, then the contribution from that wavenumber is considered suspect and is rejected. The Fourier sum of the nonrejected wavenumbers, denoted w, constitutes the signal "correlated" with the pass to the west. The procedure is repeated with the pass to the east, giving the correlated signal e (i.e., w and e are the estimated parts of the selected pass that correlate with the adjacent west and east passes). These two signals are then correlated with each other. If the portion of w that is correlated with e is denoted ww, and the portion of e that is correlated with w is denoted ee, then the adopted signal is taken to be $(ee + ww)/2$. Note that ee and ww are not identical, because $\rho_0 < 1.0$.

Alsdorf (1991) and Alsdorf et al. (1994) selected correlated signals only on the basis of the ρ_w^k. A more effective selection criterion would also compare the amplitudes of the wavenumbers and would not accept wavenumbers with amplitudes different by more than a preselected ratio (Langel, 1995). Other schemes of comparison are possible.

The method is readily adaptable to two dimensions for map comparison, and such analyses have been carried out by Arkani-Hamed, Urquhart, and Strangway (1984), Arkani-Hamed, Zhao, and Strangway (1988), Arkani-Hamed and Strangway (1986b), Alsdorf (1991), and Alsdorf et al. (1994, 1997). As in the case of inter-pass comparisons, there is more than one way to go about selecting the common signal.

There are two shortcomings in the application of the two-dimensional version of this method to satellite data: First, the underlying coordinate system is rectangular, which deviates from the actual spherical geometry. This means that one must consider the earth to be locally flat, or else project the data onto a flat plane, presumably tangent at its center to the sphere at the mean altitude of the satellite data. This shortcoming can be remedied by the analysis described in the next section. The second shortcoming is the deletion of the lithospheric field of low spherical harmonic order and long wavelength, as discussed in Section 4.3.5.

4.5.2 SPHERICAL HARMONIC CORRELATION ANALYSIS

Spherical harmonic correlation analysis (SHCA) was developed by Arkani-Hamed and Strangway (1985a,b, 1986a) for analysis of scalar anomaly data, but it is applicable to analysis of any component of $\mathbf{A}(\mathbf{r})$ at a fixed altitude. That component is expressed in terms of spherical harmonics as

$$A^i(\theta, \phi) = \sum_{n=1}^{N} \sum_{m=0}^{n} \left[C_{nm}^i \cos(m\phi) + S_{nm}^i \sin(m\phi) \right] P_n^m(\cos\theta)$$

$$= \sum_{n=1}^{N} \sum_{m=0}^{n} \left[A_{nm}^i \cos\left(m\phi + \zeta_{nm}^i\right) \right] P_n^m(\cos\theta), \tag{4.21}$$

where A^i is the ith component of $\mathbf{A}(\mathbf{r})$, or is the scalar anomaly field, A^s, and C_{nm} and S_{nm} are the spherical harmonic coefficients of the magnetic anomalies. For A^s, the summation over n begins with zero. Section 5.4.3 gives an efficient method for deriving the coefficients in (4.21). $A_{nm} = [C_{nm}^2 + S_{nm}^2]^{1/2}$ is an amplitude factor, and $\zeta_{nm} = \tan^{-1}(-S_{nm}/C_{nm})$ is a phase factor. As will be seen in Chapter 6, for existing data, a realistic value for N, the limit of summation in (4.21), would be between 60 and 70. Note that the expansion of equation (4.21) is not the usual potential function as given in equation (2.3). The relationship between the two is the topic of Section 4.5.4.

By analogy with Fourier analysis, the corresponding power spectrum is defined by

$$P_n = \frac{1}{2n+1} \sum_{m=0}^{n} [(C_{nm})^2 + (S_{nm})^2]. \tag{4.22}$$

The degree correlation between two such spherical harmonic analyses is estimated by

$$\rho_n = \frac{\sum_{m=0}^{n} \left(C_{nm} C'_{nm} + S_{nm} S'_{nm} \right)}{\left\{ \left[\sum_{m=0}^{n} (C_{nm})^2 + (S_{nm})^2 \right] \left[\sum_{m=0}^{n} \left(C'_{nm} \right)^2 + (S'_{nm})^2 \right] \right\}^{1/2}}. \tag{4.23}$$

As in Fourier correlation, the correlation coefficient for an individual degree and order between the two analyses can be estimated as

$$\rho_{nm} = \cos\left(\zeta_{nm} - \zeta'_{nm} \right). \tag{4.24}$$

A criterion is needed to identify "common" signals between two harmonically analyzed maps. As an example, one could require C_{nm} and C'_{nm} (and S_{nm} and S'_{nm}) to be of the same sign and to have magnitudes within some specified

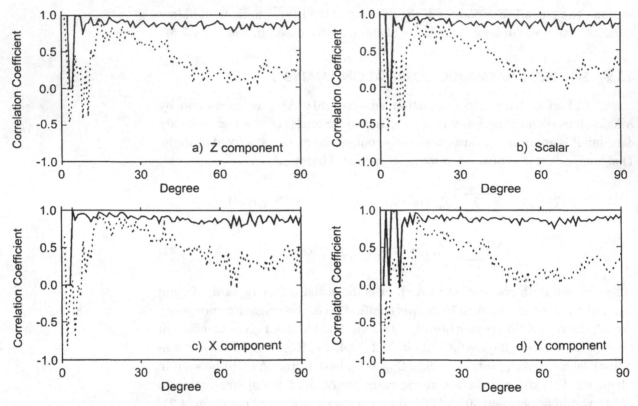

FIGURE 4.11

Degree correlation between final dawn and dusk *Magsat* maps: (a) ΔZ, (b) ΔB, (c) ΔX, (d) ΔY. Dotted line is before SHCA; solid line is after. (From Ravat et al., 1995, with permission.)

ratio (Arkani-Hamed and Strangway, 1985a,b, 1986a). Alternatively, one may require the ρ_{nm} of equation (4.24) to be greater than some specified number ρ_0, and the amplitudes $A_{nm} = [(C_{nm})^2 + (S_{nm})^2]^{1/2}$ and $A'_{nm} = [(C'_{nm})^2 + (S'_{nm})^2]^{1/2}$ to have an appropriately bounded ratio, say $R_{nm} \leq R_0$ (Arkani-Hamed, Langel, and Purucker, 1994; Ravat et al., 1995).

SHCA has been used by Ravat et al. (1995) to estimate the common signal between dawn and dusk *Magsat* data after each has had an estimate of the contribution from external fields removed. Figure 4.11 shows the degree correlation, ρ_n, for each component before and after application of SHCA. The improvement in correlation is dramatic.

4.5.3 TRADE-OFFS IN THE CORRELATION ANALYSIS

Correlative harmonic analysis has both positive value and shortcomings, and it must be used carefully, with its limitations recognized and accounted for. In the course of applying the analysis, one rejects individual harmonics that do not agree between two sets of measurements, with the result that the modified sets of measurements are more highly correlated than the original measurements. This high correlation does not guarantee that the selected signal will be a good representation of the anomaly field, because the deleted harmonics may carry some, or even a significant portion, of the anomaly signal, and retained harmonics may contain highly correlated disturbance fields. It is important to recognize the inevitable distortion of the estimate because of the elimination of wavenumbers important to the anomaly field. Also, there are several circumstances in which the interfering signals may be dominant in the estimate, and yet be undetectable (Langel, 1995).

The choice of ρ_0 and R_0 is subjective. It is desired to make a choice that will preserve most of what are regarded as "common" features, while rejecting most of what is regarded as interference. This involves a trade-off between the amount of power retained versus the degree of correlation of the retained signal. One would like ρ_k (or ρ_{nm}) and R_k (or R_{nm}) both to be near 1 (i.e., for the retained signal to be very highly correlated between the two sources). At the same time, one would like to retain as much power as possible. Both are feasible only if contamination of the original signals is low. In practice, a subjective choice is made, and the larger the contamination, the more difficult the trade-off.

For each of ρ_0 and R_0 there is a range of values for which the analysis gives satisfactory results under most conditions. Langel (1995) examined the effectiveness of these criteria and suggested that appropriate values ρ_0 lie between 0.25 and 0.6, and for R_0 between 3.0 and 6.0, depending on the relative amplitudes of the primary and interfering signals. Most published analyses using the correlation method have adopted values of ρ_0 within this range. Some have not used the amplitude criteria. Arkani-Hamed et al. (1994) adopted $\rho_0 = 0.7$ and $R_0 = 2.0$ for their "stringent" criteria. Based on the analysis of Langel (1995), those values reject too many wavenumbers. Reasonable general-purpose values are $\rho_0 = 0.3$ and $R_0 = 3.0$ for most levels of disturbance, and $\rho_0 = 0.6$ and $R_0 = 3.0$ when the level of disturbance is known to be high.

Of the two selection criteria, the correlation criterion usually is most important, but the amplitude criterion becomes important when the disturbances are nearly in phase. Jones (1988) used the method to compare aeromagnetic data with first-vertical-derivative gravity data, and von Frese et al. (1997) discussed correlation analysis in a broader geophysical context.

4.5.4 SPHERICAL HARMONIC RELATIONSHIPS

Representation of the magnetic anomaly components, the A^i and A^s, in terms of spherical harmonics, as in equation (4.21), does not result in the potential function as given by

$$\Psi_a(r, \theta, \phi) = a \sum_{n=1}^{N} \left(\frac{a}{r}\right)^{n+1} \sum_{m=0}^{n} \left[G^a_{nm} \cos(m\phi) + H^a_{nm} \sin(m\phi)\right] P^m_n(\cos\theta), \qquad (4.25)$$

where all quantities are as in equation (2.3), and where

$$\mathbf{A}(\mathbf{r}) = -\nabla\Psi_a. \qquad (4.26)$$

However, the coefficients in (4.21) can be related to those in (4.25).

In the case where the A^i of equation (4.21) is the radial component of the anomaly field (i.e., A_r), the coefficients of (4.21) and (4.25) are related by

$$G_{nm}^a = \frac{1}{n+1}\left(\frac{r}{a}\right)^{n+2} C_{nm},$$

$$H_{nm}^a = \frac{1}{n+1}\left(\frac{r}{a}\right)^{n+2} S_{nm},$$

(4.27)

where the r in (4.27) corresponds to the altitude for which (4.21) is derived.

In the case where equation (4.21) is used to represent the scalar anomaly field A^s, the relationship between the coefficients of (4.21) and (4.25) is more complex. Here, an approximation is developed.

First, assume that the main field can be approximated by an axial dipole so that

$$\underline{\mathbf{B}}_m = -2\frac{(\cos\theta)\hat{\mathbf{r}} + (\sin\theta)\hat{\boldsymbol{\theta}}}{(1 + 3\cos^2\theta)^{1/2}}.$$

(4.28)

Substituting (4.27), (4.26), and (4.25) into (1.10), and taking $r = c = a + h$, where $a = 6{,}371.2$ km, gives

$$\Delta B^a = \frac{1}{(1 + 3\cos^2\theta)^{1/2}} \sum_{n=1} \left(\frac{a}{c}\right)^{n+2} \sum_{m=0}^{n} \left[G_{nm}^a \cos(m\phi) + H_{nm}^a \sin(m\phi)\right]$$

$$\times \left[\sin\theta \frac{\partial P_n^m}{\partial\theta} - 2(n+1)P_n^m(\theta)\right].$$

(4.29)

Use of the orthogonality relationships for spherical harmonics leads to

$$\begin{bmatrix} C_{nm} \\ S_{nm} \end{bmatrix} = \frac{2n+1}{4\pi} \int_0^\pi \int_0^{2\pi} \Delta B^a(\theta, \phi) \begin{bmatrix} \cos(m\phi) \\ \sin(m\phi) \end{bmatrix} P_n^m \sin\theta \, d\phi \, d\theta.$$

(4.30)

To find the C_{nm} and S_{nm} in terms of the G_{nm}^a and H_{nm}^a, substitute (4.29) into (4.30) and perform the integration. Begin by noting the well-known recursion relations (e.g., Langel, 1987)

$$\sin\theta(\partial P_n^m / \partial\phi) = K1_{nm}P_{n+1}^m - K2_{nm}P_{n-1}^m,$$

(4.31)

$$\cos\theta\left(P_n^m\right) = K3_{nm}P_{n+1}^m + K4_{nm}P_{n-1}^m,$$

(4.32)

where

$$K1_{nm} = \frac{n[(n+1)^2 - m^2]^{1/2}}{2n+1}, \qquad K2_{nm} = \frac{(n+1)[n^2 - m^2]^{1/2}}{2n+1},$$

$$K3_{nm} = \frac{[(n+1)^2 - m^2]^{1/2}}{2n+1}, \qquad K4_{nm} = \frac{[n^2 - m^2]^{1/2}}{2n+1}.$$

(4.33)

Let

$$K5_{nm} = K1_{nm} - 2(n+1)K3_{nm} = -\frac{(n+2)[(n+1)^2 - m^2]^{1/2}}{2n+1},$$

$$K6_{nm} = -K2_{nm} - 2(n+1)K4_{nm} = -\frac{3(n+1)(n^2 - m^2)^{1/2}}{2n+1},$$

so equation (4.29) becomes

$$\Delta B^a = \frac{1}{(1 + 3\cos^2\theta)^{1/2}} \sum_{l=1} \left(\frac{a}{c}\right)^{l+2} \sum_{k=0}^{l} [G_{lk}^a \cos(k\phi) + H_{lk}^a \sin(k\phi)]$$

$$\times [K5_{lk} P_{l+1}^k + K6_{lk} P_{l-1}^k]. \tag{4.34}$$

When (4.34) is substituted into (4.30) and the integral over ϕ is taken, only terms with $k = m$ are nonzero, and we obtain, using $\mu = \cos\theta$,

$$\begin{bmatrix} C_{nm} \\ S_{nm} \end{bmatrix} = \frac{2n+1}{4} \sum_{l=1} \left(\frac{a}{c}\right)^{l+2} \begin{bmatrix} G_{lm}^a \\ H_{lm}^a \end{bmatrix} \int_{-1}^{1} \frac{P_n^m(\mu)}{(1 + 3\mu^2)^{1/2}} [K5_{lm} P_{l+1}^m + K6_{lm} P_{l-1}^m] \, d\mu. \tag{4.35}$$

To estimate the integral in (4.35), use the further approximation

$$(1 + 3\mu^2)^{-1/2} \doteq 1 - \mu^2/2, \tag{4.36}$$

which is exact at $\mu = 0$ and ± 1, and apply (4.34). The result is

$$C_{nm} = K7_{nm} G_{n+3,m}^a + K8_{nm} G_{n-3,m}^a + K9_{nm} G_{n+1,m}^a + K10_{nm} G_{n-1,m}^a, \tag{4.37}$$

with a similar expression for S_{nm}, where

$$K7_{nm} = -\frac{K6_{n+3,m} K4_{n+2,m} K4_{n+1,m}}{4} \left(\frac{a}{c}\right)^{n+5},$$

$$K8_{nm} = -\frac{K5_{n-3,m} K3_{n-2,m} K3_{n-1,m}}{4} \left(\frac{a}{c}\right)^{n-1},$$

$$K9_{nm} = \frac{K6_{n+1,m} KK_{nm} - K5_{n+1,m} K4_{n+2,m} K4_{n+1,m}}{4} \left(\frac{a}{c}\right)^{n+3},$$

$$K10_{nm} = \frac{K5_{n-1,m} KK_{nm} - K6_{n-1,m} K3_{n-2,m} K3_{n-1,m}}{4} \left(\frac{a}{c}\right)^{n+1},$$

$$KK_{nm} = 2 - K3_{nm} K4_{n+1,m} - K4_{nm} K3_{n-1,m}.$$

Table 4.2 shows a tabulation of $K7$, $K8$, $K9$, and $K10$ for degrees 13 through 15 for $h = 400$ km. Note that the dominant contribution comes from $K9$. This means that the $(n+1)$th degree in G_{nm}^a and H_{nm}^a corresponds nearly to the nth degree in C_{nm} and S_{nm}.

Table 4.2. *Computed values for coefficients in equation (4.37)*

n	m	K7	K8	K9	K10
13	1	0.4793	0.2031	−2.8903	−0.6261
13	2	0.4697	0.1967	−2.8872	−0.6380
13	3	0.4537	0.1861	−2.8810	−0.6570
13	4	0.4316	0.1716	−2.8699	−0.6817
13	5	0.4037	0.1535	−2.8516	−0.7104
13	6	0.3705	0.1322	−2.8226	−0.7403
13	7	0.3324	0.1084	−2.7783	−0.7678
13	8	0.2901	0.0828	−2.7124	−0.7880
13	9	0.2445	0.0563	−2.6159	−0.7937
13	10	0.1965	0.0302	−2.4760	−0.7745
13	11	0.1474	0.0000	−2.2725	−0.7120
13	12	0.0986	0.0000	−1.9688	−0.5660
13	13	0.0521	0.0000	−1.4787	0.0000
14	1	0.4809	0.2053	−2.9086	−0.6263
14	2	0.4724	0.1998	−2.9059	−0.6366
14	3	0.4582	0.1906	−2.9004	−0.6531
14	4	0.4387	0.1781	−2.8911	−0.6750
14	5	0.4141	0.1624	−2.8761	−0.7008
14	6	0.3845	0.1438	−2.8527	−0.7286
14	7	0.3506	0.1228	−2.8177	−0.7557
14	8	0.3127	0.0999	−2.7666	−0.7789
14	9	0.2714	0.0757	−2.6932	−0.7935
14	10	0.2276	0.0512	−2.5891	−0.7928
14	11	0.1821	0.0273	−2.4422	−0.7670
14	12	0.1360	0.0000	−2.2331	−0.6990
14	13	0.0907	0.0000	−1.9272	−0.5510
14	14	0.0477	0.0000	−1.4415	0.0000
15	1	0.4805	0.2065	−2.9146	−0.6243
15	2	0.4730	0.2017	−2.9122	−0.6333
15	3	0.4605	0.1938	−2.9074	−0.6478
15	4	0.4432	0.1829	−2.8995	−0.6672
15	5	0.4213	0.1692	−2.8869	−0.6903
15	6	0.3950	0.1529	−2.8678	−0.7158
15	7	0.3647	0.1343	−2.8396	−0.7417
15	8	0.3307	0.1139	−2.7991	−0.7657
15	9	0.2934	0.0920	−2.7419	−0.7845
15	10	0.2535	0.0694	−2.6621	−0.7938
15	11	0.2117	0.0466	−2.5519	−0.7874
15	12	0.1687	0.0247	−2.3997	−0.7561
15	13	0.1255	0.0000	−2.1871	−0.6841
15	14	0.0834	0.0000	−1.8810	−0.5353
15	15	0.0437	0.0000	−1.4020	0.0000

4.6 ANALYTICS OF LITHOSPHERIC AND IONOSPHERIC FIELDS

4.6.1 INTRODUCTION

Given different, at least partially independent, data sets (e.g., *Magsat* data from dawn and dusk local times), the common lithospheric fields can, in principle, be identified and separated from ionospheric fields to the extent that the ionospheric fields are uncorrelated between those data sets. This is accomplished by the use of (1) filtering and crossover analysis at long wavelengths and (2) correlation analysis for all wavelengths. However, when the amplitude of the ionospheric fields is more than about 20% greater than that of the lithospheric fields (Langel, 1995), the trade-offs in correlation analysis become difficult, and the method may fail to identify a good approximation to the common lithospheric signal. Any reduction of the effects of ionospheric fields in the data prior to correlation analysis is not only useful in itself but also renders the correlation analysis more effective. In this section, the spatial and temporal characteristics of both lithospheric and ionospheric fields will be described in order to explore ways to more definitively separate the two and to evaluate strategies already in use.

Lithospheric fields are conceptually simple. They do not vary with time and can be represented by the usual spherical harmonic potential analysis, equation (4.25). Ionospheric current systems, on the other hand, organize in a complex combination of geomagnetic and geographic coordinates. Within that coordinate-system combination, the ionospheric current systems generally have a consistent morphology, as described in Chapter 2. However, it is not a static morphology, but rather one with current amplitudes and flow patterns that constantly vary from the statistical average in complex ways.

For clarity, the development begins with simplifying assumptions regarding the behavior of ionospheric currents and regarding the spatial and temporal distributions of available data. In particular, it is assumed that the ionospheric currents are fixed in dipole latitude–magnetic local time coordinates and that data are available at all positions at all local times. It is shown that for these conditions it is possible to uniquely identify lithospheric and ionospheric fields, except for those parts that are symmetric around the axis of some coordinate system.

Actual data are not ideally distributed. Because of the long time span over which data were acquired, the POGO satellite data distribution probably approximates the ideal as well as data from any single satellite is able to do. *Magsat* data, however, have a very biased distribution in local time, with data only from dawn and dusk. The implications of typical averaging techniques in this biased situation are explored in Section 4.6.3.

Finally, in Section 4.6.4, the restriction of ionospheric currents in a fixed morphology is lifted, and a crude statistical estimate of the effects of temporal variations is developed.

At satellite altitude, fields from both ionospheric currents and lithospheric sources can be represented by internal spherical harmonic potential functions. The approach followed is to examine the potential function representations to explore possible ways to take advantage of the different properties of lithospheric and ionospheric signals in order to isolate each. The focus for ionospheric fields will be on the known properties of the Sq current system. However, the results are easily generalized to other ionospheric sources.

Suppose that $\mathbf{S} = \mathbf{S}(\mathbf{r}_d, t)$ is the ionospheric field from Sq currents, where $\mathbf{r}_d = (r, \theta_d, \lambda_d)$ is the geomagnetic system in which θ_d is the dipole colatitude and λ_d is magnetic local time. Then, for the present discussion,

$$\Delta \mathbf{B} = \mathbf{A} + \mathbf{S} + \mathbf{e}, \tag{4.38}$$

where $\mathbf{e}$ is measurement noise, assumed gaussian, with variance σ^2 and a mean of zero. The field is assumed to be curl-free and thus representable by potential functions

$$\mathbf{A} = -\nabla \Psi_a, \tag{4.39}$$

with Ψ_a given by equation (4.25),

$$\mathbf{S} = -\nabla \Psi_s, \tag{4.40}$$

$$\Psi_s = a \sum_{n=1}^{\infty} \left(\frac{a}{r}\right)^{n+1} \sum_{m=0}^{n} \left[G_{nm}^{d,s} \cos(m\lambda_d) + H_{nm}^{d,s} \sin(m\lambda_d)\right] P_n^m(\cos \theta_d) \tag{4.41}$$

$$\Delta \mathbf{B} = -\nabla \Psi = -\nabla(\Psi_a + \Psi_s), \tag{4.42}$$

$$\Psi = a \sum_{n=1}^{\infty} \left(\frac{a}{r}\right)^{n+1} \sum_{m=0}^{n} \left[G_{nm} \cos(m\phi) + H_{nm} \sin(m\phi)\right] P_n^m(\cos \theta). \tag{4.43}$$

The superscript d in (4.41) indicates that the Gauss coefficients are in the dipole coordinate system; the superscript and subscript s indicates the Sq field. This formulation assumes that the measurements are taken above the currents in the ionosphere. Note that the G_{nm} and H_{nm} are functions of time.

The potentials ψ_a and ψ_s, equations (4.25) and (4.41), are expressed in different coordinate systems. Accordingly, the potential for the field from the lithosphere, equation (4.25), is transformed to the dipole system:

$$\Psi_a = a \sum_{n=1}^{\infty} \left(\frac{a}{r}\right)^{n+1} \sum_{m=0}^{n} \left[G_{nm}^{d,a} \cos(m\phi_d) + H_{nm}^{d,a} \sin(m\phi_d)\right] P_n^m(\cos \theta_d). \tag{4.44}$$

The relationship between $G_{nm}^{d,a}$ and $H_{nm}^{d,a}$ and G_{nm}^{a} and H_{nm}^{a} is given by Bernard et al. (1969).

If τ_d is defined as

$$\tau_d = \pi + \phi_{d,s}, \tag{4.45}$$

where $\phi_{d,s}$ is the dipole longitude of the subsolar point [see equation (2.11)], then

$$\lambda_d = \tau_d + \phi_d, \tag{4.46}$$

where τ_d is the counterpart of Greenwich time, and λ_d the counterpart of local time in the dipole coordinate system. It is helpful to think of λ_d and τ_d (also of λ, local time, and τ, Greenwich time) as positional, or geographic, variables, with a range of 0–2π, rather than as time variables. In addition, in Section 4.6.4, consideration will be given to proper variations with time t, where t is a continuous variable, not limited to 0–2π.

It is convenient to deal with the radial component of $\Delta \mathbf{B}$. Using equations (4.46) and (4.44), the radial component of the lithospheric field in the magnetic coordinate system is given by

$$A_r = a \sum_{n=1}^{\infty} \left(\frac{a}{r}\right)^{n+2} (n+1) \sum_{m=0}^{n} \left\{ \left[G_{nm}^{d,a} \cos(m\tau_d) - H_{nm}^{d,a} \sin(m\tau_d) \right] \cos(m\lambda_d) \right.$$

$$\left. + \left[G_{nm}^{d,a} \cos(m\tau_d) + H_{nm}^{d,a} \sin(m\tau_d) \right] \sin(m\lambda_d) \right\} P_n^m(\cos\theta_d). \tag{4.47}$$

4.6.2 ESTIMATION OF THE COEFFICIENTS

The objective is to determine the two sets of coefficients $\{G_{nm}^{d,s},\ H_{nm}^{d,s}\}$ and $\{G_{nm}^{d,a},\ H_{nm}^{d,a}\}$ that completely define the S and A fields. This is possible under the somewhat idealized circumstance that the coefficients $G_{nm}^{d,s}$ and $H_{nm}^{d,s}$ in equation (4.41) are fixed in τ_d ("dipole" Universal Time), that is, are independent of ϕ_d, and also are fixed in time, t. Also assume that $r = c = a + h$, where h, the altitude, is fixed.

Using (4.47) and (4.41), equation (4.42) gives

$$\Delta B_r = a \sum_{n=1}^{\infty} \left(\frac{a}{r}\right)^{n+2} (n+1) \sum_{m=0}^{n} \left\{ \left[G_{nm}^{d,s} + G_{nm}^{d,a} \cos(m\tau_d) - H_{nm}^{d,a} \sin(m\tau_d) \right] \cos(m\lambda_d) \right.$$

$$\left. + \left[H_{nm}^{d,s} + H_{nm}^{d,a} \cos(m\tau_d) + G_{nm}^{d,a} \sin(m\tau_d) \right] \sin(m\lambda_d) \right\} P_n^m(\cos\theta_d), \tag{4.48}$$

$$\Delta B_r = a \sum_{n=1}^{\infty} \left(\frac{a}{r}\right)^{n+2} (n+1) \sum_{m=0}^{n} \left[G_{nm}^{d,\lambda} \cos(m\lambda_d) + H_{nm}^{d,\lambda} \sin(m\lambda_d) \right] P_n^m \cos\theta_d). \tag{4.49}$$

With some manipulation, equation (4.49) can be written in the form

$$\Delta B_r = a \sum_{n=1}^{\infty} \left(\frac{a}{r}\right)^{n+2} (n+1) \sum_{m=0}^{n} \left\{ \left[G_{nm}^{d,a} + G_{nm}^{d,s} \cos(m\tau_d) + H_{nm}^{d,s} \sin(m\tau_d) \right] \cos(m\phi_d) \right.$$

$$\left. + \left[H_{nm}^{d,a} + H_{nm}^{d,s} \cos(m\tau_d) - G_{nm}^{d,s} \sin(m\tau_d) \right] \sin(m\phi_d) \right\} P_n^m(\cos\theta_d), \qquad (4.50)$$

$$\Delta B_r = a \sum_{n=1}^{\infty} \left(\frac{a}{r}\right)^{n+2} (n+1) \sum_{m=0}^{n} \left[G_{nm}^{d} \cos(m\phi_d) + H_{nm}^{d} \sin(m\phi_d) \right] P_n^m(\cos\theta_d).$$

$$(4.51)$$

Equations (4.49) and (4.51) are equivalent representations of ΔB_r, both in geomagnetic coordinates. The first, (4.49), uses λ_d, magnetic local time, as the longitudinal variable, whereas (4.51) uses ϕ_d, dipole longitude. If the available data are sufficient to determine the potential completely in both forms, then appropriate averaging can distinguish the anomaly and ionospheric fields. Define the average over τ_d as

$$\langle f(\tau_d) \rangle_{\tau_d} = \frac{1}{2\pi} \int_0^{2\pi} f(\tau_d) \, d\tau_d. \qquad (4.52)$$

Then

$$\langle G_{nm}^{d} \rangle_{\tau_d} = G_{nm}^{d,a} + G_{nm}^{d,s} \delta_{m,0}, \qquad \langle H_{nm}^{d} \rangle_{\tau_d} = H_{nm}^{d,a} + H_{nm}^{d,s} \delta_{m,0}. \qquad (4.53)$$

Alternatively, one can estimate the $G_{nm}^{d,s}$ and $H_{nm}^{d,s}$ as $\langle G_{nm}^{d,\lambda} \rangle_{\tau_d}$ and $\langle H_{nm}^{d,\lambda} \rangle_{\tau_d}$, where

$$\langle G_{nm}^{d,\lambda} \rangle_{\tau_d} = G_{nm}^{d,s} + G_{nm}^{d,a} \delta_{m,0}, \qquad \langle H_{nm}^{d,\lambda} \rangle_{\tau_d} = H_{nm}^{d,s} + H_{nm}^{d,a} \delta_{m,0}. \qquad (4.54)$$

The preceding formalism requires not only that the coefficients be independent of τ_d and t but also that the data distribution be near enough to ideal to allow estimation of the $\{G_{nm}^{d,\lambda}, H_{nm}^{d,\lambda}\}$ and $\{G_{nm}^{d}, H_{nm}^{d}\}$ in equations (4.49) and (4.51) as functions of τ_d. Note that the $m=0$ estimates are identical for $\mathbf{S}$ and $\mathbf{A}$, an ambiguity that can be resolved by assigning all of the $m=0$ averages to the anomaly field. In principle, $\langle G_{nm}^{d} \rangle_{\tau_d}$ and $\langle H_{nm}^{d} \rangle_{\tau_d}$ are estimates of $G_{nm}^{d,a}$ and $H_{nm}^{d,a}$, and $\langle G_{nm}^{d,\lambda} \rangle_{\tau_d}$ and $\langle H_{nm}^{d,\lambda} \rangle_{\tau_d}$ are estimates of $G_{nm}^{d,s}$ and $H_{nm}^{d,s}$. Appropriate transformations can then be made to the geographic system.

4.6.3 DATA FROM A SINGLE LOCAL TIME

For the restricted local-time coverage of *Magsat* (a situation that may pertain to future data acquired by satellites), the data needed to estimate the averages of equations (4.53) and (4.54) are not available. In this situation, alternative approaches are attempted. One such alternative is to attempt to estimate $\mathbf{S}$ by averaging the measurements $\Delta \mathbf{B}$ for one specific local time over some longitude interval. The assumption is that $\mathbf{A}$ will have sufficient variability that it will contribute little or nothing to the average. To investigate the validity of this approach, assume that data are available from only one local time, that is, λ_d is fixed at the value $\lambda_d = \lambda_d^1$. Then the radial ionospheric field is given by

$$\Delta B_r^s = a \sum_{n=1}^{\infty} \left(\frac{a}{c}\right)^{n+2} (n+1) \sum_{m=0}^{n} \left[G_{nm}^{d,s} \cos\left(m\lambda_d^1\right) + H_{nm}^{d,s} \sin\left(m\lambda_d^1\right)\right] P_n^m(\cos\theta_d)$$

(4.55)

$$\Delta B_r^s = \sum_{n=1}^{\infty} \sum_{m=0}^{n} A_{nm}^s P_n^m(\cos\theta_d),$$

(4.56)

where

$$A_{nm}^s = \left(\frac{a}{c}\right)^{n+2} (n+1)\left[G_{nm}^{d,s} \cos\left(m\lambda_d^1\right) + H_{nm}^{d,s} \sin\left(m\lambda_d^1\right)\right]$$

(4.57)

$$= \text{constant}.$$

Because ΔB_r^s is a function of θ_d alone, one can write

$$\Delta B_r^s = \sum_{k=1}^{\infty} A_k P_k(\cos\theta_d),$$

(4.58)

where the $P_k(\cos\theta_d) = P_k^0(\cos\theta_d)$ are the usual Legendre polynomials.

It is desired to explore what happens when data are averaged over a range of longitudes, for a fixed local time, under the assumption that the contribution of the anomaly field will be reduced or eliminated and the contribution of the ionospheric field will be unaffected. Consider the radial field, ΔB_r, which is the sum of ΔB_r^s, from (4.58), and A_r, from the gradient of (4.25) and from (4.47). The procedure is to average ΔB_r over the geomagnetic longitude range ϕ_1 to ϕ_2, that is, with $\phi_2 - \phi_1 = \Delta\phi$ (the d subscript has been dropped from ϕ for convenience):

$$\langle \Delta B_r \rangle_{\phi_1 - \phi_2} = \frac{1}{\Delta\phi} \int_{\phi_1}^{\phi_2} \Delta B_r \, d\phi.$$

(4.59)

It is instructive to ask what one intuitively expects from carrying out such an averaging procedure. Suppose at $\theta = \theta'$, and for $\phi_a \le \phi \le \phi_b$, $\phi_b - \phi_a = 5°$, there is an isolated positive anomaly of constant amplitude 5 nT. Then when carrying out the average, whenever both ϕ_b and ϕ_a are within the $\Delta\phi$ averaging window, the average is expected to be equal to $5 \cdot (5°/\Delta\phi°)$ nT, that is, a positive value lowered in amplitude by the ratio of the longitudinal anomaly width to the width of the $\Delta\phi$ window. If only a part of the anomaly width is within the averaging window, the average will be proportionately lower. So the average will be an "anomaly" elongated along lines of latitude, but of reduced amplitude; that is, isolated anomalies will be smeared along bands of constant latitude. If, on the other hand, there is also a negative anomaly of -5 nT of width $5°$ within the averaging window, the average will be zero. Such logic is easily extended to the expectation that if at a particular latitude the positive and negative anomaly spacings in longitude are fairly regular, with comparable amplitudes, then the average will indeed be near zero. But if there is a particularly large amplitude anomaly at a specific latitude, with no or few anomalies of the opposite sign, then the average will be a streak, or elongated anomaly, of the same sign as the large-amplitude anomaly, but reduced in amplitude.

To proceed analytically, computing the average gives

$$\langle \Delta B_r \rangle_{\phi_1 - \phi_2} = \Delta B_r^s + \sum_{n=1}^{\infty} \frac{n+1}{\Delta\phi} \left(\frac{a}{c}\right)^{n+2} \sum_{m=0}^{n} P_n^m(\cos\theta_d)$$

$$\times \left[G_{nm}^{d,a} \int_{\phi_1}^{\phi_2} \cos(m\phi)\, d\phi + H_{nm}^{d,a} \int_{\phi_1}^{\phi_2} \sin(m\phi)\, d\phi \right], \qquad (4.60)$$

$$\langle \Delta B_r \rangle_{\phi_1 - \phi_2} = \Delta B_r^s + \sum_{n=1}^{\infty} (n+1) \left(\frac{a}{c}\right)^{n+2} \left\{ P_n^0(\cos\theta_d) G_{n0}^{d,a} + \frac{2}{\Delta\phi} \sum_{m=1}^{n} m^{-1} \sin(m\Delta\phi/2) \right.$$

$$\times \left. \left[G_{nm}^{d,a} \cos(m\underline{\phi}) + H_{nm}^{d,a} \sin(m\underline{\phi}) \right] P_n^m(\cos\theta) \right\}, \qquad (4.61)$$

where $\underline{\phi} = (\phi_2 - \phi_1)/2$. If $\Delta\phi = 2\pi$, all of the $\sin(m\Delta\phi/2)$ factors are zero, and (4.61) reduces to

$$\langle \Delta B_r \rangle_{0-2\pi} = B_r^s + \sum_{n=1}^{\infty} (n+1) \left(\frac{a}{c}\right)^{n+2} G_{n0}^{d,a} P_n^0(\cos\theta_d). \qquad (4.62)$$

In this case the average gives an estimate of the ionospheric field, except for the symmetric portion of the anomaly field, which is expected to be quite small. In principle, this type of analysis should be effective in estimating ΔB^s. In practice, variations of the Sq field with longitude, discussed in Section 4.6.4, make it necessary to average over smaller $\Delta\phi$ intervals.

Because of the m^{-1} factor, higher-order terms in (4.61) experience the largest reduction in amplitude. To visualize the effect of the averaging, $\langle \Delta B_r \rangle_{\Delta\phi}$ was computed from the potential function of a typical *Magsat* anomaly map. For this case, $B_r^s = 0$, so the quantity computed corresponds to $\langle \Delta B_r \rangle_{\Delta\phi} - B_r^s$, or the error in using $\langle \Delta B_r \rangle_{\Delta\phi}$ to estimate B_r^s. The results for $\Delta\phi = 45°$ are shown in Figure 4.12 and confirm the previous discussion about the expected results. In particular, the figure is dominated by east–west streaks resulting from smearing of the fields over a range of longitudes because of large-amplitude anomalies. In regions where positive and negative anomalies of comparable amplitudes are adjacent in longitude, the result is near zero. This occurs, for example (Figure 6.11) in southern Alaska and adjacent Canada, in the region extending east and west of the southern tip of Greenland, in a narrow band just to the south of Norway and Sweden, at and near South Africa, and in northern Australia.

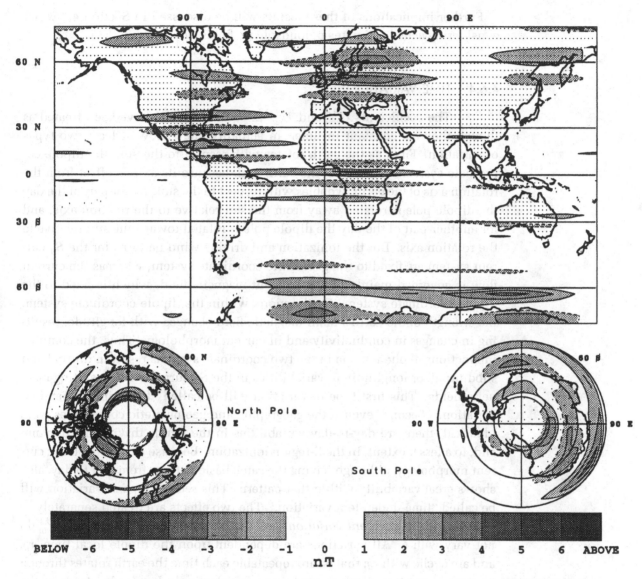

FIGURE 4.12
Averages over longitude of the radial crustal field, A_r, as computed using equation (4.61). The averaging window, $\Delta\phi$, was 45°.

It is apparent that the success of this method for reducing the effect of lithospheric fields and allowing estimates of **S** (or of the equatorial electrojet) depends on the local distribution of anomalies. However, the typical ionospheric fields to be estimated generally have strong trends in a nearly east–west direction. This means that the smeared-out anomalies might easily be mistaken for ionospheric fields. In Figure 4.12, equatorward of about 60° latitude, the smeared amplitudes are mostly in the 1–3-nT range, so that the error in the estimated ionospheric fields will usually be less than about 3 nT. At higher latitudes the amplitudes are higher, and the method probably will give very spurious results.

Further implications of these results will be discussed in Section 4.7 when considering actual estimates of ionospheric fields from *Magsat* data.

4.6.4 TIME VARIATIONS

The coefficients $G_{nm}^{d,s}$ and $H_{nm}^{d,s}$ in equation (4.41) have been treated as though they are constant in time. In actuality, they show at least two types of variation. First, as the earth rotates with respect to the sun, the dipole coordinate system rotates with the earth and, because its axis is offset from the rotation axis of the earth, wobbles with respect to the sun. So for part of the day the dipole pole is rotated away from the sun relative to the rotation axis, and for another part of the day the dipole pole is rotated toward the sun relative to the rotation axis. But the ionization and driving wind patterns for the *Sq* current system are fixed to the geocentric coordinate system, whereas the current flow is governed mainly by the ambient magnetic field, which is fixed to the dipole coordinate system. Furthermore, within the dipole coordinate system, the strength of the earth's main magnetic field changes with longitude, resulting in changes in conductivity and in current morphology. Thus, the complex interactions of phenomena in the two coordinate systems result in diurnal and subdiurnal, or longitudinal, variabilities in the *Sq* currents and resulting magnetic fields. This first type of variation will be called "local-time-dependent variation." Second, even if the geographic and geomagnetic coordinates were identical, there are day-to-day variabilities in the winds driving the dynamo and, to a lesser extent, in the E-region ionization. Because of this, the basic current morphology, although having the same basic pattern from day to day, also shows great variability within that pattern. This second type of variation will be called "time-dependent variation." The two effects are treated separately.

Local-time-dependent variations are treated assuming that $G_{nm}^{d,s}$ and $H_{nm}^{d,s}$ do not vary with t. All variations are dependent upon the dipole local time, τ_d, and are cyclic with τ_d; that is, are repeatable each time the earth rotates through 24 hours, and thus they depend upon which dipole longitude, ϕ_d, is at the subsolar point. Note that these variations are not in terms of magnetic local time, λ_d. Magnetic local-time variations are explicit in the sine and cosine terms in equation (4.41). Rather, in view of equation (4.46), the variations with τ_d can be viewed as equivalent to variations with ϕ_d. It is therefore possible to write

$$G_{nm}^{d,s} = CG_{nm0}^{d,s} + \sum_{k=1}^{\infty} \left[CG_{nmk}^{d,s} \cos(k\tau_d) + SG_{nmk}^{d,s} \sin(k\tau_d) \right],$$

$$H_{nm}^{d,s} = CH_{nm0}^{d,s} + \sum_{k=1}^{\infty} \left[CH_{nmk}^{d,s} \cos(k\tau_d) + SH_{nmk}^{d,s} \sin(k\tau_d) \right],$$

(4.63)

where all of the coefficients (CG, SG, CH, SH) are independent of τ_d and t. The decomposition of equation (4.63) separates the **S** field into two parts: (1) the part that is independent of τ_d (or ϕ_d) (i.e., is independent of variations caused by the complex interactions between processes in the geographic and geomagnetic coordinate systems), the $CG_{nm0}^{d,s}$ and $CH_{nm0}^{d,s}$, and (2) the part that depends on those interactions, the $CG_{nmk}^{d,s}$ and $CG_{nmk}^{d,s}$, $k > 0$. Data from a single satellite inherently are unable to determine the latter. The question to be answered is this: To what accuracy can the spherical harmonic coefficients describing the lithospheric fields and/or the non-τ_d-dependent **S** fields be determined?

Estimation of the terms in (4.63) that are dependent on τ_d requires a sophisticated analysis with a more comprehensive data set. In particular, data from a constellation of three or more near-earth satellites in orbits well spaced in local time probably will be required.

To estimate the error due to the local-time-dependent variations, the orthogonality relationships for circular functions are used, giving

$$\langle G_{nm}^{d,s}\rangle_{\tau_d} = CG_{nm0}^{d,s}, \qquad \langle H_{nm}^{d,s}\rangle_{\tau_d} = CH_{nm0}^{d,s}, \tag{4.64}$$

and

$$\langle G_{nm}^{d,s}\cos(m\tau_d)\rangle_{\tau_d} = \frac{CG_{nmm}^{d,s}}{2}, \qquad \langle H_{nm}^{d,s}\cos(m\tau_d)\rangle_{\tau_d} = \frac{CH_{nmm}^{d,s}}{2}. \tag{4.65}$$

Then, taking averages of the coefficients in equations (4.49) and (4.51) with respect to τ_d gives

$$\langle G_{n0}^{d}\rangle_{\tau_d} = G_{n0}^{d,a} + CG_{n00}^{d,s}, \qquad \langle G_{nm}^{d}\rangle_{\tau_d} = G_{nm}^{d,a} + \frac{CG_{nmm}^{d,s}}{2}, \qquad m \neq 0,$$
$$\langle H_{n0}^{d}\rangle_{\tau_d} = H_{n0}^{d,a} + CH_{n00}^{d,s}, \qquad \langle H_{nm}^{d}\rangle_{\tau_d} = H_{nm}^{d,a} + \frac{CH_{nmm}^{d,s}}{2}, \qquad m \neq 0, \tag{4.66}$$

and

$$\langle G_{n0}^{d,\lambda}\rangle_{\tau_d} = G_{n0}^{d,s} + CG_{n00}^{d,a}, \qquad \langle G_{nm}^{d,\lambda}\rangle_{\tau_d} = CG_{nm}^{d,s}, \qquad m \neq 0,$$
$$\langle H_{n0}^{d,\lambda}\rangle_{\tau_d} = H_{n0}^{d,a} + CH_{n00}^{d,s}, \qquad \langle H_{nm}^{d,\lambda}\rangle_{\tau_d} = CH_{nm}^{d,s}, \qquad m \neq 0. \tag{4.67}$$

Equations (4.66) and (4.67) indicate that if, having an ideal data distribution, one attempts to estimate $G_{nm}^{d,a}$ by taking the average $\langle G_{nm}^{d}\rangle_{\tau_d}$ and $G_{nm}^{d,s}$ by taking the average $\langle G_{nm}^{d,\lambda}\rangle_{\tau_d}$, then the following problems arise: (1) As for equations (4.56) and (4.55), it is not possible to distinguish the $m = 0$ terms; they are identical. (2) The estimate of $G_{nm}^{d,s}$, $m = 0$, includes only the terms in (4.63) that are independent of τ_d. (3) The estimate of $G_{nm}^{d,a}$, $m > 0$, is now in error by half the magnitude of the mth order term in (4.63). At present, the distribution of magnitudes for the coefficients in (4.63) has not been explicitly investigated, so quantitative error estimates are not possible. However, some qualitative comments can be made. First, from the nature of the problem, the diurnal and semidiurnal terms in (4.63) are likely to be the most significant. This means that the terms of order $m = 1$ and $m = 2$ in $\{G_{nm}^{d,a}, H_{nm}^{d,a}\}$ will be most affected. Second, most of **S** is represented by spherical harmonics of degree less than 13, so that the resulting contamination of **A** will be at longer wavelengths, that is, lower order in the spherical harmonic expansion.

Time-dependent variations are considered assuming that the variation with τ_d is zero. The superscript d will be dropped for convenience. The notation $\langle\,\rangle_t$ indicates taking the time average of the quantity within the brackets. Define the expectation value with respect to time averaging as

$$E_t[\,f(t)] = \langle\,f(t)\rangle_t. \tag{4.68}$$

Further define

$$\underline{G}_{nm}^s = \langle G_{nm}^s\rangle_t, \qquad \underline{H}_{nm}^s = \langle\underline{H}_{nm}^s\rangle_t, \qquad \tilde{G}^s = G^s - \underline{G}^s, \qquad \tilde{H}^s = H^s - \underline{H}^s. \tag{4.69}$$

If it is assumed that

$$E_t(\tilde{G}^s) = E_t(\tilde{H}^s) = E_t(\tilde{G}^s\tilde{H}^s) = 0,$$

$$E_t(\tilde{G}_{nm}^s\tilde{G}_{lk}^s) = E_t(\tilde{H}_{nm}^s\tilde{H}_{lk}^s) = (\tilde{\sigma}_{nm}^s)^2\delta_{nl}\delta_{mk}, \tag{4.70}$$

then it is straightforward to show that

$$(\tilde{\sigma}_{B_r}^s)^2 = E_t(B_r - \underline{B}_r)^2 = \sum_{n=1}^{\infty}\left(\frac{a}{r}\right)^{2n+4}(n+1)^2\sum_{m=0}^{n}(\tilde{\sigma}_{nm}^s)^2\left[P_n^m(\cos\theta)\right]^2. \tag{4.71}$$

Equation (4.71) expresses the variability in B_r (e.g., from orbit to orbit) as a function of the variability of the spherical harmonic coefficients describing **S**. Extrapolating from the residuals of the data to the models of Campbell and Schiffmacher (1985; see also Campbell, 1989b), one gets the impression that $\sigma_{B_r}^s$ is at least 20% of the value of B_r itself, perhaps higher. At dipole latitudes poleward of 50° it is certainly higher, perhaps reaching 50%. This variability is probably the chief difficulty in applying the analyses presented in the preceding paragraphs.

4.6.5   SUMMARY AND DISCUSSION

Can **A** and **S** be definitively untangled? Probably not, given the existing data. With ideal data, Section 4.6.2 indicates that, yes, it is in principle possible, with difficulties due to complexities and time variations in **S**. A formalism describing the effects of those complexities and time variations is given in Section 4.6.4. Some key statistics in the parameters of that formalism are unknown, but would be accessible quickly from a more nearly ideal distribution of surface and satellite data. Without simultaneous surface and satellite data well distributed in local time, a definitive simultaneous characterization of **A** and **S** is not likely.

It is often assumed that averages of residual data from a sufficiently large longitude band will average out contributions from **A** and permit determination of **S**. In principle, that is true only for averages over 360° of longitude. Section 4.6.3 shows that averaging anomaly fields over the usual longitude bands of 30° to 90° in width can result in 2–3-nT field features elongated along bands constant latitude – similar to some features due to the equatorial electrojet. Such averages should thus be regarded with caution.

In addition to the considerations already given, it is important to note that part of $G_{nm}^{d,s}$ will be absorbed into $\underline{\mathbf{B}}_m$ during derivation of spherical harmonic models of the main field. This contributes to the errors in those models. When data from a single local time are used to find $\underline{\mathbf{B}}_m$, then τ_d in (4.63) is nearly constant, and $\underline{\mathbf{B}}_m$ can, in principle, incorporate most of **S**, at that local time and up to the degree of the model, usually $n = 13$. For nightside and near-dawn local times it is thought that **S** is small, so that the effects on $\underline{\mathbf{B}}_m$ are also small.

If $\underline{\mathbf{B}}_m$ includes a range of local times, then at least partial averaging over τ_d occurs. In this case the averaged coefficients of equation (4.67) may be approximately absorbed by $\underline{\mathbf{B}}_m$, again up to the degree of the model. The greater part of **S** will not become part of $\underline{\mathbf{B}}_m$, but will remain to contribute to $\Delta\mathbf{B}$.

The effect of averaging over only nighttime hours of τ, as was done for part of the data from the POGO satellites (Langel, 1990a), remains to be explored.

4.7 MODELS OF THE EQUATORIAL ELECTROJET FROM *MAGSAT* DATA

It is useful to divide *Magsat* data into independent dawn and dusk subsets (Arkani-Hamed, Urquhart, and Strangway, 1985b; Yanagisawa and Kono, 1985). Figures 4.13 and 4.14 are residual maps of the three components of *Magsat* data at dawn and at dusk. Data used in these maps were selected from magnetically quiet times and were low-pass-filtered with a Kaiser filter with cutoff at 12,000 km. It is immediately obvious from these maps that the dusk data are highly contaminated. The ΔY map shows distinct positive residuals to the north and negative residuals to the south of the dip equator. Similarly, the ΔZ map shows negative residuals to the north and positive residuals to the south of the dip equator, and the ΔX map shows a negative residual along the dip equator. Maeda et al. (1982, 1985) showed that this distinctive ΔY pattern is due to a meridional current associated with the equatorial electrojet (EE). The ΔX and ΔZ patterns are consistent with an eastward-flowing EE below the satellite. Data from dawn, especially for ΔY, are relatively free from these

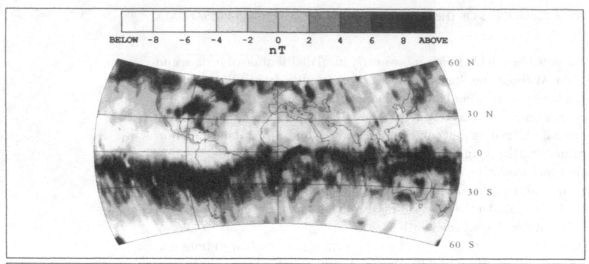

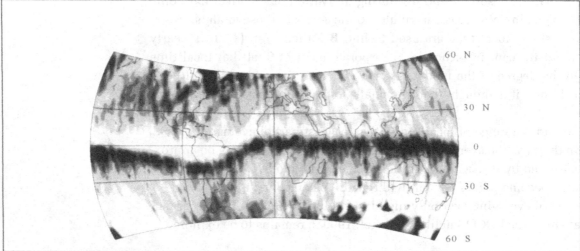

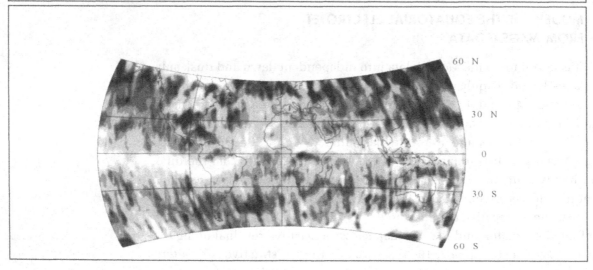

FIGURE 4.13
Uncorrected residual maps of the three components of *Magsat* data at dusk local time: top, ΔZ; middle, ΔY; bottom, ΔX.

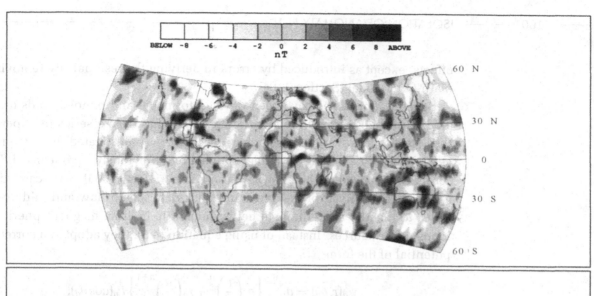

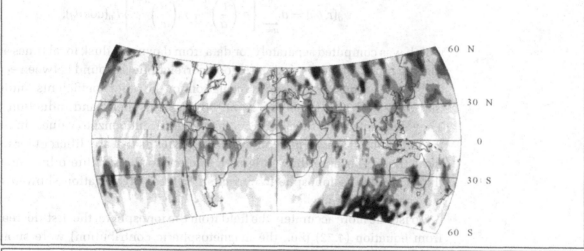

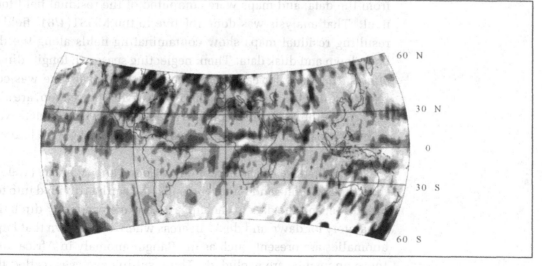

FIGURE 4.14
Uncorrected residual maps of the three components of *Magsat* data at dawn local time: top, ΔZ; middle, ΔY; bottom, ΔX.

effects, except as introduced by errors in deriving the residuals by removing a main-field model (Section 4.3.4).

Several approaches have been devised to model and remove fields associated with the EE from the *Magsat* data. There has been a series of papers on these topics from various researchers, each more sophisticated than its predecessors, and each taking advantage of what had previously been learned. That series commenced with Yanagisawa and Kono (1984, 1985), who carried out a systematic analysis of the dawn and dusk data sets for low and middle latitudes to identify and eliminate fields thought to be from magnetospheric and ionospheric sources. Instead of using equation (4.3), they adopted a correction potential of the form

$$\psi_d(r, \theta_d) = a \sum_{n=1}^{6} \left\{ e_n \left(\frac{r}{a} \right)^n + i_n \left(\frac{a}{r} \right)^{n+1} \right\} P_n(\cos \theta_d), \qquad (4.72)$$

which was computed separately for data from dawn and dusk local times and for varying ranges of the *Dst* index. A good correlation was found between e_1 and i_1 and the *Dst* index, but no such correlation for the other coefficients, indicating that the first-degree terms represent the magnetospheric and induction fields. They then noted that the i_n for $n > 1$ have significant nonzero values, indicating the existence of systematic, long-wavelength fields that are different for the two local times. Because those originate internal to the satellite orbit, they most likely are due to ionospheric currents. Further investigation showed that i_n varies with longitude.

To isolate more accurately the field from the ionosphere, the first-degree terms from equation (4.72) (i.e., the magnetospheric contribution) were subtracted from the data, and maps were computed of the residual field for each component. That analysis was done relative to the MGST(4/81) field model, so the resulting residual maps show contaminating fields along the dip equator for both dawn and dusk data. Then, neglecting seasonal, longitudinal, and altitude effects, the mean for each component at each local time was computed. The results, called the mean ionospheric field (MIF) correction, are shown in Figure 4.15. After the MIF correction was applied, the corrected data were averaged in 5°-by-5° bins, and the resulting dawn and dusk maps were found to be in better agreement.

That procedure has been refined by Cohen and Achache (1990): After one selects the quiet data and removes trends, the earth is divided into four quarters by longitude, and the data in each quarter are averaged in 1° dip latitude intervals, separately for dawn and dusk. In areas where it is known that large lithospheric anomalies are present, such as the Bangui anomaly in Africa, profiles crossing those anomalies are excluded. The resulting averages, called the mean equatorial anomaly (MEA), are very similar from quarter to quarter, but are nearly antisymmetric between dawn and dusk. What is termed an average electrojet

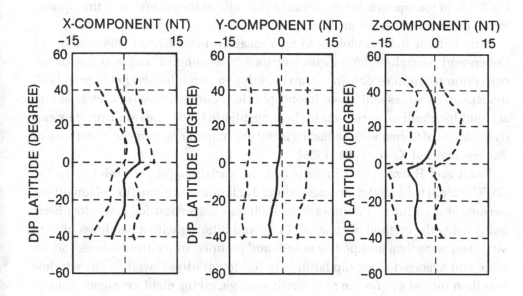

MEAN IONOSPHERIC FIELD : 0600
(DIP LATITUDE)

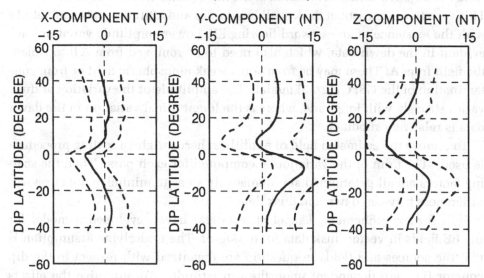

MEAN IONOSPHERIC FIELD : 1800
(DIP LATITUDE)

FIGURE 4.15
Mean ionospheric field (MIF) for 0600 (top) and 1800 (bottom) local time meridians. The solid and dashed lines indicate the mean value and plus or minus one standard deviation, respectively. Horizontal components (*X* and *Y*) in this figure are in the direction of magnetic north and east defined by the model field, MGST(4/81). (From Yanagisawa and Kono, 1985, with permission.)

effect (AEE) is then computed under the assumption that the latitudinal distribution of the EE is the same at dawn and dusk within a given quarter, except for amplitude and sign. Thus the AEE in each sector is taken to be the mean of the difference between the dusk and dawn MEAs. Each individual pass is then corrected by removal of the AEE, scaled to that pass by a least-squares procedure.

The main-field model of Cohen and Achache uses uncorrected dusk data. From Section 4.3.4, it is expected that their model partly includes the effects of the fields of ionospheric origin, resulting in antisymmetric effects at the equator in the dawn and dusk anomaly maps.

Although the field models used by Yanagisawa and Kono (1984, 1985) and Cohen and Achache (1990) cause spurious "ionospheric" fields at dawn, the correction techniques developed are effective in removing the fields from both ionospheric sources and "pseudoionospheric" sources. Because they take into account longitudinal variations in the ionospheric fields, the resulting dusk and dawn maps of Cohen and Achache (1990) are in better agreement than those of Yanagisawa and Kono (1984, 1985).

Ravat and Hinze (1993) investigated the variations of ΔB, relative to the GSFC(12/83) field model, averaged along dip latitude as a function of longitude, season, and altitude. Pre-processing included correction for magnetospheric fields using the method of equation (4.3), and filtering with a bandpass of 5° to 40°. Data were then grouped by month and grouped into altitude ranges 50 km wide and averaged along dip latitude in 90° longitudinal swaths. The window was then moved 45° for the next swath average, giving eight averaged profiles. Data in the vicinity of the prominent Bangui anomaly were excluded from the analysis. Figure 4.16 shows results for the swath from −90° to 0° longitude. Note the negative variation centered at the dip equator with positive side lobes. This variation is present at dusk at all longitudes and altitudes. It is compatible with the existence of an eastward-flowing EE. Lower-amplitude variations are evident in the dawn data, which also need to be removed from ΔB to isolate the field from **A**. These may be from some weak ionospheric field or from contamination of the GSFC(12/83) model. The amplitude of the variation at dusk varies strongly with longitude, whereas the longitudinal variation in the dawn data is relatively subdued.

To remove the estimated field of nonlithospheric origin, a scaling procedure is used. The form of the variation is computed for each pass. Then the scaling factors are all determined simultaneously so as to minimize the crossover differences between dawn and dusk data.

In a further refinement, Langel et al. (1993) have developed a model for the EE fields in vector dusk data from *Magsat*. The underlying assumption is that the sources and fields in question are organized with respect to the dip equator (i.e., are dependent upon the dip latitude). To minimize the effects of lithospheric fields in modeling EE fields, an estimate of $\mathbf{A(r)}$, derived from *Magsat* dawn data, is removed from the dusk $\Delta\mathbf{B(r}, t)$. The data are then filtered to remove magnetospheric fields and most of the *Sq* fields, and the results are analyzed relative to dip latitude. A moving-window approach in longitude is adopted, with window size equal to 45°. EE-associated fields are found to have variations with respect to season, daily sunspot number, and longitude.

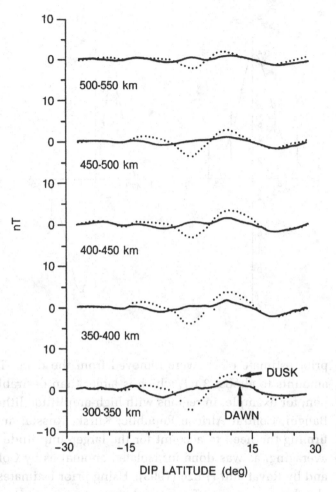

FIGURE 4.16
Averages along dip latitude over eastern South America at longitudes −90° to 0°. (From Ravat and Hinze, 1993; reprinted with kind permission of Blackwell Science.)

Figure 4.17 shows ΔB_r, ΔB_θ, and ΔB_ϕ from the model of Langel et al. (1993), as based on data from March and April 1980 in the longitude band 240°−285°. Error bars are based on the data scatter to the model.

To "correct" $\Delta \mathbf{B}(\mathbf{r}, t)$ for the effects of $\mathbf{D}(\mathbf{r}, t)$ remaining after filtering, the appropriate season, sunspot number, and longitude bins from the model are identified, and the estimated $\mathbf{D}(\mathbf{r}, t)$ is computed along each pass. That $\mathbf{D}(\mathbf{r}, t)$ is then modified by keeping its form, but scaling its amplitude by least-squares fit to the measured $\Delta \mathbf{B}(\mathbf{r}, t)$. The resulting scaled $\mathbf{D}(\mathbf{r}, t)$ is subtracted from $\Delta \mathbf{B}(\mathbf{r}, t)$. Figure 4.18 shows the X and Z components of the resulting dusk map, to be compared with Figures 4.13 and 4.14. It is apparent that this scheme has been successful.

The subdivision of the data into 45° blocks by Langel et al. (1993) corresponds to $\phi_2 = \phi_1 + 45°$, or $\Delta\phi = 45° = 0.785$ radian in equation (4.61). The previously discussed Figure 4.12 shows the errors that would result in estimating the radial component of the EE within a 45° averaging window if no

FIGURE 4.17
Variation of each component organized by dip latitude, based on Langel et al. (1993). Data are from 240–285° longitude. The data are analyzed relative to the DAWN(6/83-6) field model and filtered with a Kaiser filter with cutoff of 12,000 km. The abscissa is degrees dip latitude, and the ordinate is nanoteslas.

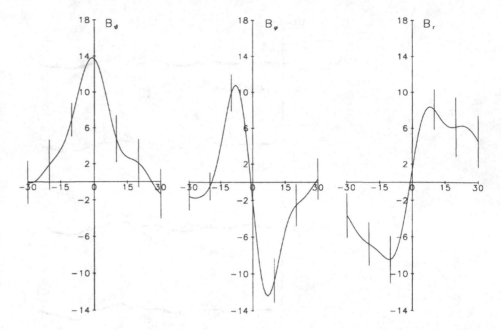

prior estimate of A_r were removed from the data. In some places the error amounts to almost 3 nT, which is larger than desirable. This is likely to happen, for example, in regions with high-amplitude lithospheric anomalies (e.g., Bangui, Central African Republic; Kursk, Russia; and Kentucky, USA), confirming the need to account for the larger-amplitude anomalies prior to such averaging, as was done for isolated anomalies by Cohen and Achache (1990) and by Ravat and Hinze (1993). Using prior estimates for all of the anomalies should prove more effective and, even if such an estimate is in error by 2–3 nT, is likely to reduce the error due to anomaly fields to 1 nT or less when estimating EE fields. This is an error level below the natural variability of the EE fields.

4.8   A HIGH-LATITUDE DISTURBANCE MODEL

At high latitudes (i.e., especially the auroral belt), $D(r, t)$ is much more complex than at low and middle latitudes. In particular, ΔX and ΔY show extremely large variations in the auroral belt because of the presence of field-aligned currents. Although attempts have been made to identify high-latitude anomaly fields in these components, it is not clear that they yield reliable estimates. Accordingly, the high-latitude analyses to be described apply only to ΔB and to ΔZ. It is recommended, on the basis of Section 4.6.3, and the discussion in the preceding section, that a preliminary estimate of A first be removed from ΔB before trying to estimate D.

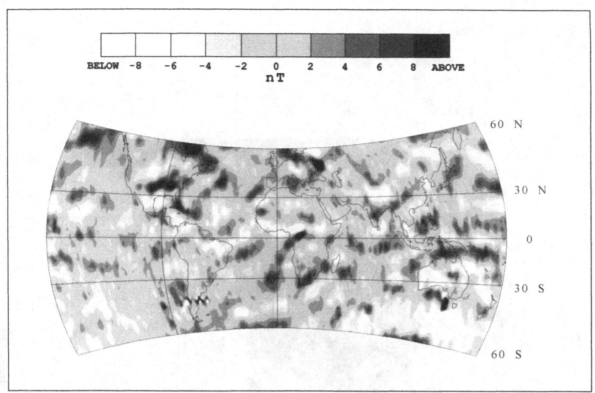

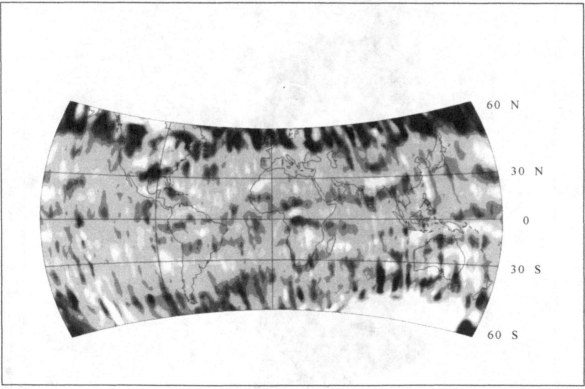

FIGURE 4.18
Residual maps of the *Z* (top) and *X* (bottom) components of *Magsat* data at dusk local time after correction for fields from the equatorial electrojet.

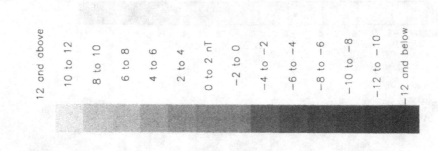

12 and above

10 to 12

8 to 10

6 to 8

4 to 6

2 to 4

0 to 2 nT

−2 to 0

−4 to −2

−6 to −4

−8 to −6

−10 to −8

−12 to −10

−12 and below

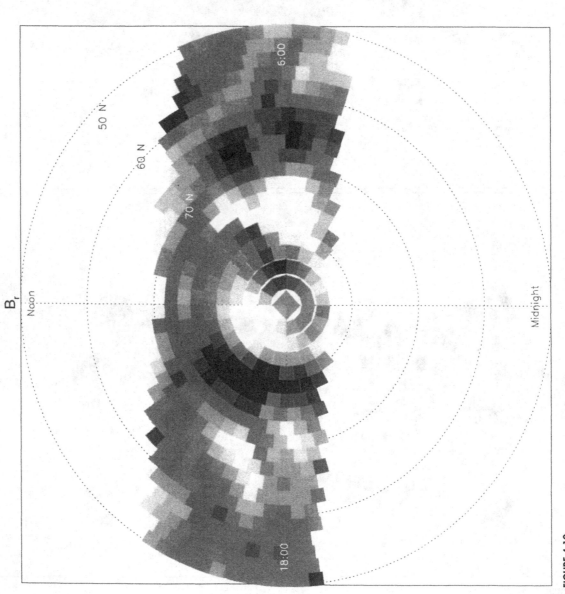

B_r

Noon

50 N

60 N

70 N

6:00

18:00

Midnight

FIGURE 4.19

Average polar ΔB_r relative to the GSFC(12/83) field model in dipole colatitude–MLT coordinates. Data were selected such that the variance of ΔB_r along each pass was less than 80 (nT)². An estimate of the

The morphology and amplitude of **D** at auroral and polar latitudes depend heavily on the IMF components, B_z and B_y. Further, **D** is known to "organize" in magnetic coordinates. Several candidate magnetic coordinate systems exist for the polar regions. Here, dipole latitude (or colatitude) and magnetic local time are adopted because of their simplicity. Other systems involve a nonlinear transformation from the usual geocentric system. A logical procedure, then, would be to divide the data into subsets according to the values of the IMF B_z and B_y, and perhaps also according to the level of a disturbance index, preferably AE, and to average the resulting data sets in a suitable geomagnetic coordinate system.

Such a procedure may yet prove to be optimal, but it has not been implemented. However, two simpler, though related, models have been shown to be useful, though selection of the data by IMF is not carried out in these models. The first, due to Takenaka et al. (1991), is called the mean polar disturbance field (MPDF). It depends on invariant latitude, on magnetic local time, and on the Kp disturbance index. For various levels of Kp, the polar data are divided into 1-hour MLT intervals and $1°$ intervals of invariant latitude. This results in 1,200 bins into which the data are distributed and then averaged. No prior correction for estimated lithospheric fields is made, and it is assumed that those fields are random in the chosen coordinate system and are averaged out. The resulting estimated field is considered consistent with the expected morphology of quiet-time fields at high latitudes. These estimates do not change drastically for lower Kp indices, so the final correction is derived for Kp from 0 to 2.

The second model is due to Ravat et al. (1995). Here, data are selected strictly on the basis of the variance criterion already described. In this case the variance of a pass extending from $50°$ latitude, over the pole, to $50°$ latitude is required to be less than or equal to 80 nT^2. Then, after the initial estimate of **A(r)** is subtracted from $\Delta\mathbf{B}(\mathbf{r}, t)$, the data averages are compiled into bins in the dipole latitude–MLT coordinate system, roughly equal in area to $2°$ by $2°$ at the equator. The results for ΔZ for the Northern Hemisphere dawn and dusk data are shown in Figure 4.19. The premise is that these averages provide an estimate of the average ionospheric field [i.e., constitute a $\underline{\mathbf{D}}(\mathbf{r})$] and can be used at least partially to correct $\Delta\mathbf{B}(\mathbf{r}, t)$ for the effects of $\mathbf{D}(\mathbf{r}, t)$.

The correction of $\Delta\mathbf{B}(\mathbf{r}, t)$ by $\underline{\mathbf{D}}(\mathbf{r})$ is accomplished in a manner similar to the procedure for equatorial and low latitudes. Along each pass the values of $\underline{\mathbf{D}}(\mathbf{r})$ are collected, and a common scaling factor, say s, is determined for that pass such that the scaled $\underline{\mathbf{D}}(\mathbf{r})$ best fits $\Delta\mathbf{B}(\mathbf{r}, t)$ in a least-squares sense. It is assumed that altitude and seasonal variations are accounted for by the scaling procedure. Then the estimate of the lithospheric field is $\underline{\mathbf{A}}(\mathbf{r}) = \Delta\mathbf{B}(\mathbf{r}, t) - \underline{\mathbf{D}}(\mathbf{r})$.

4.9   SPHERICAL HARMONIC MODELS

In principle, the core field and lithospheric field can be jointly represented by a spherical harmonic analysis. For *Magsat* data, this probably requires an n^* value between 60 and 65 in equation (2.3) before the coefficients become dominated by noise. Cain, Schmitz, and Muth (1984) derived a spherical harmonic model of degree/order 29. They show global maps, at 400 km, representing the contribution due to terms in the expansion above degree 13, that are similar to the average maps derived in the usual way, except that their resolution is less. In principle, if the expansion were carried to a sufficiently high degree, such a map would have the same resolution as the conventional maps, with the additional advantage of being reduced to a common altitude. Schmitz, Meyer, and Cain (1989) and Cain, Holter, and Sandee (1990) presented a technique for deriving models of still higher degree ($n^* \geq 50$), and anomaly maps for selected regions are included in the publications of Cain et al. (1989, 1990).

As a technique for isolating lithospheric fields, the difficulty with spherical harmonic analysis is the problem of contamination, as discussed in Section 4.3.4. Any ionospheric field whose spherical harmonic representation is similar to that of the lithospheric field will be absorbed into the analysis. Either some means must be devised to separately distinguish ionospheric effects in the model, or to identify the two effects in the spherical harmonics after modeling, or else the ionospheric contributions must be removed before fitting the spherical harmonic model. Some modeling efforts are proceeding in the direction of including ionospheric effects into main-field models (Langel et al., 1996). Such efforts may prove to be optimal for isolating the lithospheric field (Purucker et al., 1997), but the procedures are not yet sufficiently advanced for that purpose. In spite of the current limitations, anomaly maps produced by this method can be of high quality.

4.10   A RECOMMENDED PROCEDURE

It should be obvious from the preceding paragraphs that there has been an ongoing process of development and refinement of techniques to isolate the field originating in the lithosphere. This problem has, in fact, dominated satellite studies of the lithospheric field. The reason is clear: At satellite altitudes the lithospheric signal is extremely small compared with the signals from other sources. Fields from external sources can be of the same amplitude as the lithospheric field or larger, and they often occupy the same wavelength band. If doubt remains as to the accuracy of how well the anomaly field is isolated, that doubt will extend to any subsequent interpretation in terms of, for example, lithospheric structure or tectonic activity.

The techniques presented in this chapter have been used individually and in combination according to the experience and bias of the investigator. Used together they furnish the means to derive credible and interpretable maps of the lithospheric field. One such combination is presented in the following paragraphs.

Central to gaining real confidence in any map of the lithospheric field is agreement between one or more independent data sets. One method for obtaining, and measuring, such agreement quantitatively is the method of spherical harmonic correlation analysis (SHCA) described in Section 4.5.2. However, this analysis is apt to encounter problems in regions where ionospheric fields are very dominant. In such regions, the correlation analysis is likely to eliminate both ionospheric and lithospheric fields. A way to optimize the use of SHCA is to first use a combination of other methods to reduce the larger external-field contributions. Another condition for application of SHCA is that it be kept global in nature, meaning that whatever prior analyses are necessary must be accomplished for the entire globe.

The conclusion is that the place of SHCA comes at the end of the suggested procedure. Other techniques are first used to separate as much as possible of the external fields from the lithospheric field.

For low latitudes, say 50° dipole latitude or less, a reasonable procedure is as follows:

1. Select data according to appropriate criteria (Section 4.2). For low latitudes, the choice of $Kp \leq 2-$ in the selected period and $Kp < 2+$ in the preceding 3-hour period seems adequate.
2. Compute data residuals, corrected for attitude jumps (Section 3.5.3) if they are *Magsat* vector data, relative to a model with as little contamination from ionospheric fields as possible (Section 4.3); for example, the DAWN(6/83-6) model.
3. High-pass-filter the residuals (Section 4.4.3) using, for example, the Kaiser filter with a 12,000-km cutoff.
4. Correct the component and scalar data for ionospheric fields according to season and longitude using, for example, the models of Ravat and Hinze (1993) and Langel et al. (1993) scaled to best fit each individual pass (Section 4.7).
5. Carry out pass-by-pass Fourier correlation (Section 4.5.1) for sets of adjacent passes.
6. Apply a quadratic trend to all passes simultaneously, with the parameters of each trend determined so as to minimize dawn–dusk crossover differences (Section 4.4.4).

The specification of the 12,000-km filter in step 3 is to be consistent with the EE model of Langel et al. (1993). If a different model of the EE is chosen, the filtering scheme may require adjustment.

For high latitudes, the situation is more difficult. The selection of quiet data in step 1 should be modified to use the AE index, say $AE < 50$ nT, and one should limit passes to those with variance below some value, say 80 nT^2. Other criteria might also be applied. The EE model of step 4 should be replaced by an estimate of the high-latitude disturbance fields like that of Section 4.8. If the model of Section 4.8 is chosen, then the filter cutoff in step 3 must be changed to 4,000 km. In this or any such procedure, the B_θ and B_ϕ residuals may be dominated by fields from field-aligned currents and thus not be useful for anomaly studies. However, for B_r and scalar residuals, the method applies.

The resulting corrected passes, high- and low-latitude, are used to derive equivalent source solutions (to be described in Section 5.1.3), separately, for dawn and dusk, from which spherical harmonic analyses can be derived. These, in turn, can be used in SHCA (Section 4.5.2) to isolate the common features of the two data sets.

This discussion is not meant to imply that the development of techniques for isolation of lithospheric fields is a closed issue. On the contrary, additional research effort is clearly needed for high-latitude data. It also is probable that future extensions of main-field models to include Sq, EE, and other fields will increase the reliability of the estimates of those fields and permit a more definitive isolation of the lithospheric field. Progress in this direction is being made by Purucker et al. (1997).

CHAPTER 5

REDUCTION AND INVERSION

5.0 OVERVIEW

The ultimate goal of acquisition and study of magnetic field data from lithospheric sources is to understand the nature of the source regions and, as far as possible, the geologic history and physical processes that led to the present state of the lithosphere. Procedures described in earlier chapters are used to bring the data to the point where such interpretation can be attempted. In particular, Chapter 4 described ways of isolating anomaly fields from other fields contributing to the measurements. Regardless of the method used, the result will be one or more "cleaned-up" data sets, each as free from contaminating

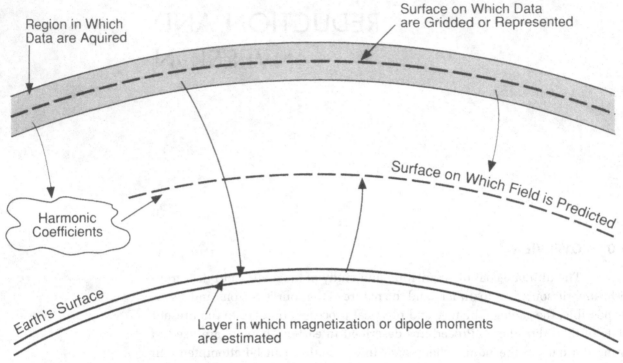

Model Parameters: $\underline{x}$ = ?

A. Harmonic Coefficients
1. Spherical Harmonics (Global), $\underline{x} = \underline{p} = \{g_n{}^m, h_n{}^m\}$ or $\underline{x} = \underline{h} = \{C_n{}^m, S_n{}^m\}$
2. Spherical Cap Harmonics (Local), $\underline{x} = \underline{q} = \{g_k{}^m, h_k{}^m\}$
3. Rectangular Harmonics (Local), $\underline{x} = \underline{z} = \{X_0, Y_0, Z_0, \varkappa_k\}$

B. Layer of Dipole Moments (Local), $\underline{x} = \underline{m}$
-can be expanded to Global

C. Field values, $\underline{x} = \underline{b}$
1. At any location (Minimum norm)
2. Gridded at satellite altitude (Collocation).

FIGURE 5.1
Schematic of reduction and inversion methods. Data are acquired in an altitude band, the shaded region. Either a global or a local selection of data is inverted to give a set of model parameters. A generic form for that inversion is
$\underline{x} = (H^T W H)^{-1} H^T W \mathbf{c}$,
where $\mathbf{c}$ is the data vector, H is the matrix of derivatives of the model (*continued*)

fields as possible, presumably with some sort of error estimate or covariance specifying the accuracy to which the data are thought to represent the crustal field. Accompanying results may be in the form of a map of averaged data, a gridded data set, a spherical harmonic model, or similar representations.

Averaging, map presentation, and profiling of the anomaly data are the first steps toward interpretation. However, ultimately the goal is to determine the distribution of magnetization, $\mathbf{M(r)}$, within the earth that gives rise to the anomaly field, $\mathbf{A(r)}$, and to explain that distribution in terms of tectonic and geologic processes. That is not a trivial task. Toward that goal, the schematic of Figure 5.1 depicts various procedures used to reduce or invert the data. Anomaly data, in as pure a form as possible, are collected within a layer surrounding the earth whose upper and lower altitudes are determined by some subset of the altitude range defined by the satellite apogee and perigee. In the ideal case, those data are used to determine $\mathbf{M}(\mathbf{r})$ in that portion of the earth's lithosphere in which the temperature is below the Curie point. This ideal case will be discussed in Section 5.1, in which it will be clear, first, that the parameter space of the solution is infinite, whereas the data are finite, and, second, that even with ideal data the solution for the distribution of $\mathbf{M}(\mathbf{r})$ is not unique. As a result, viable approaches involve compromises that lead, first, to alternative methods for reduction and initial inversion, which is the topic of this chapter, and, second, to the application of constraints from other geoscience data (e.g., surface geology, gravity, and heat flow), as discussed in later chapters.

Methods of reduction and inversion, as depicted in Figure 5.1, may be as simple as gridding the data or as complex as deriving an estimate of the magnetization in a layer. It turns out that there is an underlying formalism common to most such models, as discussed in Section 5.2. Detailed treatments of the individual procedures, in the context of that formalism, are given in the remainder of the chapter.

By its nature, the material in this chapter is mathematical. The reader seeking a partial overview should read Sections 5.0–5.2, 5.4.1, and 5.4.2, and if concerned with the stability of inversion, Section 5.5. Section 5.6 is highly mathematical and can be passed over by readers unconcerned with minimum-norm methods. Few (but important) published inversions use this method.

5.1 THE FUNDAMENTAL INVERSION PROBLEM

5.1.1 THE MAGNETIC FIELD OF A DIPOLE

5.1.1.1 Basic Principles, the Dipole Potential

Maxwell's equations tell us that magnetic fields are caused by electric currents. Anomaly fields result from magnetized rocks. The currents involved occur at the atomic and molecular scale, and their study is outside the scope of this book. Rather, it is convenient to consider the magnetic sources in terms of the sum of individual dipoles. Discussion relating microscopic currents to macroscopic dipoles and magnetization can be found in Jackson (1975), Panofsky and Phillips (1962), and Blakely (1995).

Consider the magnetic potential at location $\mathbf{r}_i$ from a dipole at point $\mathbf{r}_j$ with magnetic moment $\mathbf{m}_j$, as in Figure 5.2. Let r_i and r_j be the magnitudes of the position vectors. Then

$$\mathbf{r}_{ij} = \mathbf{r}_i - \mathbf{r}_j, \qquad r_{ij} = |\mathbf{r}_{ij}|. \tag{5.1}$$

Also,

$$\mu_{ij} = \frac{\mathbf{r}_i \cdot \mathbf{r}_j}{|\mathbf{r}_i||\mathbf{r}_j|} \tag{5.2}$$

FIGURE 5.1 (*continued*) relative to the model parameters, W is a weight matrix, and $\underline{x}$ is a vector of estimated parameters. The parameters and the elements of H depend upon the chosen model, as indicated in the figure.

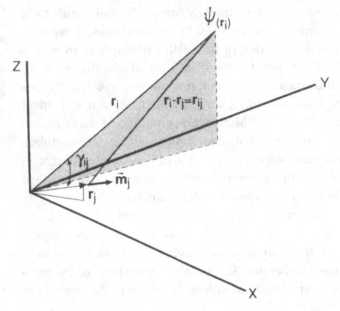

FIGURE 5.2
Geometry for dipole
potential: $\mathbf{m}_j$ is the vector
dipole moment at location
$\mathbf{r}_j$; $\Psi(\mathbf{r}_i)$ is the resulting
potential at point $\mathbf{r}_i$
according to equation (5.8);
$\mathbf{r}_{ij} = \mathbf{r}_i - \mathbf{r}_j$; γ_{ij} is the angle
between $\mathbf{r}_i$ and $\mathbf{r}_j$.

is the cosine of γ_{ij}, the angle between $\mathbf{r}_i$ and $\mathbf{r}_j$. It is easy to show that

$$\mu_{ij} = \cos\gamma_{ij} = \cos\theta_i \cos\theta_j + \sin\theta_i \sin\theta_j \cos(\phi_i - \phi_j) \tag{5.3}$$

and that

$$(r_{ij})^2 = (r_i)^2 + (r_j)^2 - 2r_i r_j \mu_{ij}. \tag{5.4}$$

Unit vectors will be indicated by a circumflex (^):

$$\hat{\mathbf{r}}_i = \frac{\mathbf{r}_i}{|\mathbf{r}_i|}. \tag{5.5}$$

The magnetic potential at point $\mathbf{r}_i$ is

$$\Psi_j(\mathbf{r}_i) = -\mu_0 \frac{\mathbf{m}_j}{4\pi} \cdot \nabla_i \frac{1}{r_{ij}} = \mu_0 \frac{\mathbf{m}_j}{4\pi} \cdot \nabla_j \frac{1}{r_{ij}}, \tag{5.6}$$

where ∇_i and ∇_j are taken with respect to $\mathbf{r}_i$ and $\mathbf{r}_j$, respectively, and μ_0 is the magnetic permeability of free space. Note that m_j is the magnitude of the vector $\mathbf{m}_j$. Taking note that

$$\nabla_i \frac{1}{r_{ij}} = \frac{\mathbf{r}_{ij}}{(r_{ij})^3} = -\nabla_j \frac{1}{r_{ij}}, \tag{5.7}$$

the potential becomes

$$\Psi_j(\mathbf{r}_i) = \frac{\mu_0}{4\pi} \frac{\mathbf{m}_j \cdot \mathbf{r}_{ij}}{(r_{ij})^3}. \tag{5.8}$$

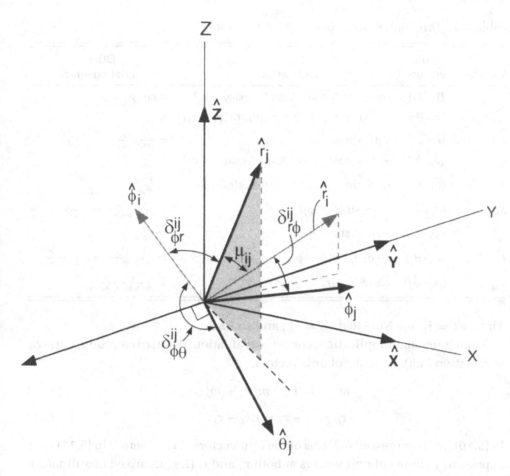

FIGURE 5.3
Partial geometry illustrating relationships among unit vectors. In the x–y–z cartesian system, the unit vectors are $\hat{\mathbf{x}}, \hat{\mathbf{y}}, \hat{\mathbf{z}}$; in the r–θ–ϕ system at position $\mathbf{r}_j$ the unit vectors are $\hat{\mathbf{r}}_j, \hat{\boldsymbol{\theta}}_j, \hat{\boldsymbol{\phi}}_j$; in the r–θ–ϕ system at position $\mathbf{r}_i$ the unit vectors are $\hat{\mathbf{r}}_i, \hat{\boldsymbol{\theta}}_i, \hat{\boldsymbol{\theta}}_i, \hat{\boldsymbol{\theta}}_i$ is not shown. The notation $\delta_{r\phi}^{ij}$ indicates the direction cosine between the radial unit vector at $\mathbf{r}_i$ and the ϕ-direction unit vector at $\mathbf{r}_j$, i.e., $\delta_{r\phi}^{ij} = \hat{\mathbf{r}}_i \cdot \hat{\boldsymbol{\phi}}_j$.

5.1.1.2 Geometric Relationships and Transformations

The local unit coordinate vectors, $\hat{\mathbf{r}}$, $\hat{\boldsymbol{\theta}}$, and $\hat{\boldsymbol{\phi}}$, change direction with $\mathbf{r}$. For transformation purposes, it is useful to have the relationships between the unit vectors at positions $\mathbf{r}_i$ and $\mathbf{r}_j$ (Figure 5.3). At any position $\mathbf{r}$, the unit vectors in the spherical coordinate system are related to the fixed unit vectors $\hat{\mathbf{x}}, \hat{\mathbf{y}}$, and $\hat{\mathbf{z}}$ in the usual rectangular cartesian system by

$$\hat{\mathbf{r}} = \hat{\mathbf{z}}\cos\theta + \sin\theta[\hat{\mathbf{x}}\cos\phi + \hat{\mathbf{y}}\sin\phi],$$

$$\hat{\boldsymbol{\theta}} = -\hat{\mathbf{z}}\sin\theta + \cos\theta[\hat{\mathbf{x}}\cos\phi + \hat{\mathbf{y}}\sin\phi], \qquad (5.9)$$

$$\hat{\boldsymbol{\phi}} = \hat{\mathbf{y}}\cos\phi - \hat{\mathbf{x}}\sin\phi.$$

Using equations (5.9), it is easy to derive the direction cosines between the various unit vectors, as given in Table 5.1. The notation adopted is, for example, $\delta_{r\theta}^{ij}$, where the superscripts i and j indicate the two position vectors, $\mathbf{r}_i$ and $\mathbf{r}_j$, and the subscripts indicate which unit vectors are considered (i.e., for $\delta_{r\theta}^{ij}$ the r-direction unit vector at position $\mathbf{r}_i$ and the θ-direction position vector at $\mathbf{r}_j$).

Table 5.1. *Direction cosines between unit vectors*

Variable	Unit vectors	Expression	Other relationships
δ_{rr}^{ij}	$(\hat{\mathbf{r}}_i \cdot \hat{\mathbf{r}}_j)$	$\cos\theta_i \cos\theta_j + \sin\theta_i \sin\theta_j \cos(\phi_i - \phi_j)$	$= \cos\gamma_{ij} = \mu_{ij}$
$\delta_{r\theta}^{ij}$	$(\hat{\mathbf{r}}_i \cdot \hat{\boldsymbol{\theta}}_j)$	$-\cos\theta_i \sin\theta_j + \sin\theta_i \cos\theta_j \cos(\phi_i - \phi_j)$	$= \frac{\partial \mu_{ij}}{\partial \theta_j}$
$\delta_{r\phi}^{ij}$	$(\hat{\mathbf{r}}_i \cdot \hat{\boldsymbol{\phi}}_j)$	$\sin\theta_i \sin(\phi_i - \phi_j)$	$= \frac{1}{\sin\theta_j} \frac{\partial \mu_{ij}}{\partial \phi_j}$
$\delta_{\theta r}^{ij}$	$(\hat{\boldsymbol{\theta}}_i \cdot \hat{\mathbf{r}}_j)$	$-\sin\theta_i \cos\theta_j + \cos\theta_i \sin\theta_j \cos(\phi_i - \phi_j)$	$= \frac{\partial \mu_{ij}}{\partial \theta_i}$
$\delta_{\theta\theta}^{ij}$	$(\hat{\boldsymbol{\theta}}_i \cdot \hat{\boldsymbol{\theta}}_j)$	$\sin\theta_i \sin\theta_j + \cos\theta_i \cos\theta_j \cos(\phi_i - \phi_j)$	$= \frac{\partial^2 \mu_{ij}}{\partial \theta_j \partial \theta_i}$
$\delta_{\theta\phi}^{ij}$	$(\hat{\boldsymbol{\theta}}_i \cdot \hat{\boldsymbol{\phi}}_j)$	$\cos\theta_i \sin(\phi_i - \phi_j)$	$= \frac{1}{\sin\theta_j} \frac{\partial^2 \mu_{ij}}{\partial \theta_i \partial \phi_j} = -\frac{1}{\sin\theta_j} \frac{\partial^2 \mu_{ij}}{\partial \theta_i \partial \phi_i}$
$\delta_{\phi r}^{ij}$	$(\hat{\boldsymbol{\phi}}_i \cdot \hat{\mathbf{r}}_j)$	$-\sin\theta_j \sin(\phi_i - \phi_j)$	$= \frac{1}{\sin\theta_i} \frac{\partial \mu_{ij}}{\partial \phi_i}$
$\delta_{\phi\theta}^{ij}$	$(\hat{\boldsymbol{\phi}}_i \cdot \hat{\boldsymbol{\theta}}_j)$	$-\cos\theta_j \sin(\phi_i - \phi_j)$	$= \frac{1}{\sin\theta_i} \frac{\partial^2 \mu_{ij}}{\partial \theta_j \partial \phi_i} = -\frac{1}{\sin\theta_i} \frac{\partial^2 \mu_{ij}}{\partial \theta_j \partial \phi_j}$
$\delta_{\phi\phi}^{ij}$	$(\hat{\boldsymbol{\phi}}_i \cdot \hat{\boldsymbol{\phi}}_j)$	$\cos(\phi_i - \phi_j)$	$= \frac{1}{\sin\theta_i \sin\theta_j} \frac{\partial^2 \mu_{ij}}{\partial \phi_j \partial \phi_i}$

Thus, $\delta_{r\theta}^{ij} = \hat{\mathbf{r}}_i \cdot \hat{\theta}_j$. Note that $\delta_{r\theta}^{ij} = \delta_{\theta r}^{ji}$, and so forth.

As an immediate application, consider evaluation of the scalar product $\mathbf{m}_j \cdot \mathbf{r}_{ij}$ in equation (5.8). In terms of unit vectors,

$$\mathbf{m}_j = m_r^j \hat{\mathbf{r}}_j + m_\theta^j \hat{\boldsymbol{\theta}}_j + m_\phi^j \hat{\boldsymbol{\phi}}_j, \tag{5.10}$$

$$\mathbf{r}_{ij} = \mathbf{r}_i - \mathbf{r}_j = r_i \hat{\mathbf{r}}_i - r_j \hat{\mathbf{r}}_j. \tag{5.11}$$

In (5.10), $\mathbf{m}_j$ is expressed in terms of the unit vectors at $\mathbf{r}_j$, whereas in (5.11) $\mathbf{r}_{ij}$ is expressed in terms of unit vectors at both $\mathbf{r}_j$ and $\mathbf{r}_i$ (i.e., in mixed coordinates). To evaluate the scalar product, it is necessary to express $\mathbf{r}_{ij}$ in terms of unit vectors at $\mathbf{r}_j$. But

$$\hat{\mathbf{r}}_j = (\hat{\mathbf{r}}_i \cdot \hat{\mathbf{r}}_j)\hat{\mathbf{r}}_i + (\hat{\boldsymbol{\theta}}_i \cdot \hat{\mathbf{r}}_j)\hat{\boldsymbol{\theta}}_i + (\hat{\boldsymbol{\phi}}_i \cdot \hat{\mathbf{r}}_j)\hat{\boldsymbol{\phi}}_i$$
$$= \delta_{rr}^{ij}\hat{\mathbf{r}}_i + \delta_{\theta r}^{ij}\hat{\boldsymbol{\theta}}_i + \delta_{\phi r}^{ij}\hat{\boldsymbol{\phi}}_i, \tag{5.12}$$

and similarly,

$$\hat{\mathbf{r}}_i = \delta_{rr}^{ij}\hat{\mathbf{r}}_j + \delta_{r\theta}^{ij}\hat{\boldsymbol{\theta}}_j + \delta_{r\phi}^{ij}\hat{\boldsymbol{\phi}}_j, \tag{5.13}$$

with corresponding expressions for the other unit vectors. Combining equations (5.10), (5.11), and (5.13), it is easy to show that

$$\mathbf{m}_j \cdot \mathbf{r}_{ij} = m_r^j[r_i \mu_{ij} - r_j] + m_\theta^j \delta_{r\theta}^{ij} r_i + m_\phi^j \delta_{r\phi}^{ij} r_i. \tag{5.14}$$

Suppose the vector $\mathbf{m}_j$ is needed in terms of the spherical coordinates at position $\mathbf{r}_i$. Using the transformations for unit vectors, (5.10) is transformed to

$$\mathbf{m}_j = \left[m_r^j \delta_{rr}^{ji} + m_\theta^j \delta_{\theta r}^{ji} + m_\phi^j \delta_{\phi r}^{ji} \right]\hat{\mathbf{r}}_i + \left[m_r^j \delta_{r\theta}^{ji} + m_\theta^j \delta_{\theta\theta}^{ji} + m_\phi^j \delta_{\phi\theta}^{ji} \right]\hat{\boldsymbol{\theta}}_i$$
$$+ \left[m_r^j \delta_{r\phi}^{ji} + m_\theta^j \delta_{\theta\phi}^{ji} + m_\phi^j \delta_{\phi\phi}^{ji} \right]\hat{\boldsymbol{\phi}}_i. \tag{5.15}$$

Table 5.2. *Elements of the matrix* $H'_{pq} = (H'_{pq})_i^{m_j}$ *for computing field from magnetization [H' in equations (5.17) and (5.30)]*

Element	Expression
H'_{11}	$\left(\frac{\mu_0}{4\pi}\right)\left(\frac{1}{r_{ij}}\right)^3 \left[3\left(\frac{(r_i - r_j \mu_{ij})(r_i \mu_{ij} - r_j)}{(r_{ij})^2}\right) - \mu_{ij}\right]$
H'_{12}	$\left(\frac{\mu_0}{4\pi}\right)\left(\frac{1}{r_{ij}}\right)^3 \left[3\left(\frac{(r_i - r_j \mu_{ij})[r_i(\hat{\boldsymbol{\theta}}_j \cdot \hat{\mathbf{r}}_i)]}{(r_{ij})^2}\right) - (\hat{\boldsymbol{\theta}}_j \cdot \hat{\mathbf{r}}_i)\right]$
H'_{13}	$\left(\frac{\mu_0}{4\pi}\right)\left(\frac{1}{r_{ij}}\right)^3 \left[3\left(\frac{(r_i - r_j \mu_{ij})[r_i(\hat{\boldsymbol{\phi}}_j \cdot \hat{\mathbf{r}}_i)]}{(r_{ij})^2}\right) - (\hat{\boldsymbol{\phi}}_j \cdot \hat{\mathbf{r}}_i)\right]$
H'_{21}	$-\left(\frac{\mu_0}{4\pi}\right)\left(\frac{1}{r_{ij}}\right)^3 \left[3\left(\frac{r_j(\hat{\mathbf{r}}_j \cdot \hat{\boldsymbol{\theta}}_i)(r_i \mu_{ij} - r_j)}{(r_{ij})^2}\right) + (\hat{\mathbf{r}}_j \cdot \hat{\boldsymbol{\theta}}_i)\right]$
H'_{22}	$-\left(\frac{\mu_0}{4\pi}\right)\left(\frac{1}{r_{ij}}\right)^3 \left[3\left(\frac{r_j(\hat{\mathbf{r}}_j \cdot \hat{\boldsymbol{\theta}}_i)[r_i(\hat{\boldsymbol{\theta}}_j \cdot \hat{\mathbf{r}}_i)]}{(r_{ij})^2}\right) + (\hat{\boldsymbol{\theta}}_j \cdot \hat{\boldsymbol{\theta}}_i)\right]$
H'_{23}	$-\left(\frac{\mu_0}{4\pi}\right)\left(\frac{1}{r_{ij}}\right)^3 \left[3\left(\frac{[r_j(\hat{\mathbf{r}}_j \cdot \hat{\boldsymbol{\theta}}_i)][r_i(\hat{\boldsymbol{\phi}}_j \cdot \hat{\mathbf{r}}_i)]}{(r_{ij})^2}\right) + (\hat{\boldsymbol{\phi}}_j \cdot \hat{\boldsymbol{\theta}}_i)\right]$
H'_{31}	$-\left(\frac{\mu_0}{4\pi}\right)\left(\frac{1}{r_{ij}}\right)^3 \left[3\left(\frac{r_j(\hat{\mathbf{r}}_j \cdot \hat{\boldsymbol{\phi}}_i)(r_i \mu_{ij} - r_j)}{(r_{ij})^2}\right) + (\hat{\mathbf{r}}_j \cdot \hat{\boldsymbol{\phi}}_i)\right]$
H'_{32}	$-\left(\frac{\mu_0}{4\pi}\right)\left(\frac{1}{r_{ij}}\right)^3 \left[3\left(\frac{r_j(\hat{\mathbf{r}}_j \cdot \hat{\boldsymbol{\phi}}_i)[r_i(\hat{\boldsymbol{\theta}}_j \cdot \hat{\mathbf{r}}_i)]}{(r_{ij})^2}\right) + (\hat{\boldsymbol{\theta}}_j \cdot \hat{\boldsymbol{\phi}}_i)\right]$
H'_{33}	$-\left(\frac{\mu_0}{4\pi}\right)\left(\frac{1}{r_{ij}}\right)^3 \left[3\left(\frac{[r_j(\hat{\mathbf{r}}_j \cdot \hat{\boldsymbol{\phi}}_i)][r_i(\hat{\boldsymbol{\phi}}_j \cdot \hat{\mathbf{r}}_i)]}{(r_{ij})^2}\right) + (\hat{\boldsymbol{\phi}}_j \cdot \hat{\boldsymbol{\phi}}_i)\right]$

Note: Expressions for the scalar products of unit vectors are found in Table 5.1.

5.1.1.3 Dipole Field Expressions

The anomaly field, $\mathbf{A}(\mathbf{r}_i)$, from a dipole at $\mathbf{r}_j$ is then found by taking the negative gradient of equation (5.8):

$$\mathbf{A}(\mathbf{r}_i) = -\nabla \Psi_j(\mathbf{r}_i) = \frac{\mu_0}{4\pi} \frac{m_j}{(r_{ij})^3}[3(\hat{\mathbf{m}}_j \cdot \hat{\mathbf{r}}_{ij})\hat{\mathbf{r}}_{ij} - \hat{\mathbf{m}}_j]. \tag{5.16}$$

$$\mathbf{A}(\mathbf{r}_i) = (H')_i^{m_j}\mathbf{m}_j, \tag{5.17}$$

where H' is the matrix whose elements are given in Table 5.2. The notation is that there is a 3×3 matrix H' for each pair of positions $\mathbf{r}_i$ and $\mathbf{r}_j$, with elements, say, H'_{pq} ($p, q = 1, 2, 3$). The superscript m_j and subscript i indicate that the matrix is for computing the field at $\mathbf{r}_i$ from a dipole $\mathbf{m}_j$. The full notation for a matrix element would be $(H'_{pq})_i^{m_j}$. For shorthand notation, in Table 5.2, $H'_{pq} = (H'_{pq})_i^{m_j}$. The equations in this section assume that the earth is spherical, which is an acceptable approximation in this context.

5.1.2   MAGNETIZATION AND SUSCEPTIBILITY

Given a dipole with magnetic moment $\mathbf{m}_j$ representing a particular volume in space, say δV_j at $\mathbf{r}_j$, then the dipole moment per unit volume, $\mathbf{M}_j$, or *magnetization*, is given by

$$\mathbf{M}(\mathbf{r}_j) = \frac{\mathbf{m}_j}{\delta \mathsf{V}_j}. \tag{5.18}$$

Now suppose that there are many dipoles (i.e., that $j = 1, 2, \ldots$). Then the potential at $\mathbf{r}_i$ becomes

$$\Psi(\mathbf{r}_i) = \sum_j \Psi_j(\mathbf{r}_i) = \frac{\mu_0}{4\pi} \sum_j \mathbf{m}_j \cdot \nabla_j \frac{1}{r_{ij}}. \tag{5.19}$$

There are lots of dipoles in the lithosphere of the earth. Whether the number is large and finite or infinite is a matter of philosophy. For all practical purposes the number is infinite, and the volume represented by each dipole is microscopically small. We then rewrite (5.19), using (5.18), and pass to the limit of an integral:

$$\Psi(\mathbf{r}_i) = \frac{\mu_0}{4\pi} \sum_j \frac{\mathbf{m}_j}{\delta \mathsf{V}_j} \cdot \nabla_j \frac{1}{r_{ij}} \delta \mathsf{V}_j \rightarrow \frac{\mu_0}{4\pi} \int_\mathsf{V} \mathbf{M}(\mathbf{r}_j) \cdot \nabla_j \frac{1}{r_{ij}} d^3 r_j. \tag{5.20}$$

Using the identity

$$\nabla \cdot (\psi \mathbf{M}) = \mathbf{M} \cdot \nabla \psi + \psi \nabla \cdot \mathbf{M}, \tag{5.21}$$

where ψ is any scalar, in this case $1/|\mathbf{r} - \mathbf{r}'|$, and the divergence theorem, the integral in (5.20) is transformed as follows:

$$\Psi_a(\mathbf{r}) = \frac{\mu_0}{4\pi} \int_\mathsf{V} \mathbf{M}(\mathbf{r}') \cdot \nabla' \frac{1}{R} d^3 r'$$

$$= \frac{\mu_0}{4\pi} \int_S \frac{\mathbf{M}(\mathbf{r}') \cdot dS'}{R} - \frac{\mu_0}{4\pi} \int_\mathsf{V} \frac{\nabla' \cdot \mathbf{M}(\mathbf{r}')}{R} d^3 r'. \tag{5.22}$$

where $R = |\mathbf{r} - \mathbf{r}'|$, and S is the surface of the volume V.

It is often assumed that the magnetization sources for the satellite-measured anomaly fields are strictly due to induction by the ambient main field. Then, if the ambient magnetic field is $\mathbf{B}(\mathbf{r})$, with $\mathbf{B}(\mathbf{r}) = \mu_0 \mathbf{H}(\mathbf{r})$,

$$\mathbf{M}(\mathbf{r}) = \mathcal{K}(\mathbf{r})\mathbf{H}(\mathbf{r}) = \frac{\mathcal{K}(\mathbf{r})}{\mu_0 \mathbf{B}(\mathbf{r})}, \tag{5.23}$$

where $\mathcal{K}$ is called the magnetic susceptibility. $\mathcal{K}$ is a second-order symmetric tensor (matrix) that relates the magnetization induced in the rock to the applied magnetic field. In the analysis of satellite data, it is often assumed, as in equation (1.16), that

$$\mathcal{K}(\mathbf{r}) = \kappa(\mathbf{r})I, \tag{5.24}$$

where κ is a scalar susceptibility, and I is the identity matrix. If, in (5.18), the volume is given by

$$\delta \mathsf{V} = \Delta h \Delta W, \tag{5.25}$$

where ΔW is the area at the earth's surface, and Δh is the thickness of the magnetized layer, then, taking

$$\sigma(\mathbf{r}) = \kappa(\mathbf{r})\Delta W(\mathbf{r}), \tag{5.26}$$

$$\mathbf{m}(\mathbf{r}_j) = \frac{\sigma(\mathbf{r}_j)}{\mu_0}\Delta h\mathbf{B}(\mathbf{r}_j). \tag{5.27}$$

If the earth's main field is approximated by an axial dipole with moment p_0, then

$$m_r(\mathbf{r}_j) = 2(\sigma\Delta h)\frac{p_0}{4\pi a^3}\cos\theta_j,$$

$$m_\theta(\mathbf{r}_j) = \sigma\Delta h\frac{p_0}{4\pi a^3}\sin\theta_j, \tag{5.28}$$

$$m_\phi = 0,$$

where a is the radius of the earth. The susceptibility is assumed to be nonzero only in the region between the Curie isotherm and the earth's surface. With satellite data, it is difficult, if not impossible, to distinguish variations of κ with depth, so the effective κ is the average value over the volume δV_i. Taking ΔW as a unit area, equation (5.28) makes explicit the dependence of $\mathbf{m}$ on (1) the geometry of the inducing field, (2) the thickness of the magnetized layer, and (3) the bulk susceptibility. It is variation in the product of σ and Δh that causes variations in satellite magnetic anomaly data.

5.1.3 THE EQUIVALENT SOURCE METHOD, A FINITE-PARAMETER SOLUTION

Equation (5.20) and the discussion preceding it highlight one of the problems of inverting magnetic field data to find a magnetization distribution. The magnetization distribution is essentially infinite-dimensional, but the numbers of data are finite. In mathematical terms, the problem is underdetermined. However, at the altitude of satellite data, which at the lowest is not likely to be less than 150 km, the measured magnetic field can be regarded as the combined effect of the dipoles from a relatively large area. Accordingly, it is convenient to revert to the form

$$\Psi(\mathbf{r}_i) = \frac{\mu_0}{4\pi}\sum_j \mathbf{m}_j \cdot \nabla_j \frac{1}{r_{ij}} \tag{5.29}$$

with

$$\mathbf{A}(\mathbf{r}_i) = \sum_{j=1}^{k}(H')_i^{m_i}\mathbf{m}_j, \tag{5.30}$$

where k is the number of dipoles needed to represent the field. Typically, the dipoles are chosen and located so that each dipole represents the same area,

$\Delta W_j = \Delta W$, on the surface of the earth. If the thickness, Δh_j, of the magnetized layer is taken to be constant, then each dipole represents the same volume.

To this point, nothing has been assumed about magnetization direction. It is usual to assume that the orientation of each $\mathbf{m}_j$ is along the present-day ambient main field. Then, if D and I are the declination and inclination of the ambient field, and m is the dipole moment,

$$m_r^j = -m_j \sin I_j, \qquad m_\theta^j = -m_j \cos I_j \cos D_j, \qquad m_\phi^j = m_j \cos I_j \sin D_j.$$

$$(5.31)$$

The elements D and I usually are estimated from a spherical harmonic model of the main field. Let

$$\mathbf{w}_b(\mathbf{r}_j) = (-\sin I_j, -\cos I_j \cos D_j, \cos I_j \sin D_j)^T, \qquad (5.32)$$

so that

$$\mathbf{m}_j = m_j \mathbf{w}_b(\mathbf{r}_j), \qquad (5.33)$$

and

$$\mathbf{A}_j(\mathbf{r}_i) = m_j (H')_i^{m_j} \mathbf{w}_b(\mathbf{r}_j), \qquad (5.34)$$

where the subscript j indicates that $\mathbf{A}_j(\mathbf{r}_i)$ is the field from the jth dipole, remembering that $(H')_i^{m_j}$ is a 3-by-3 matrix. Consider a single field component, designated p (for r, θ, or ϕ), and let the pth component

$$\left[(H')_i^{m_j} \mathbf{w}_b(\mathbf{r}_j) \right]_p = H_{ij}^{m,p}. \qquad (5.35)$$

Then the p component of $\mathbf{A}$ due to all dipoles is given by

$$A_p(\mathbf{r}_i) = \sum_j H_{ij}^{m,p} m_j. \qquad (5.36)$$

Expressions for $H_{ij}^{m,p}$ are found in Table 5.3. The assumption that $\mathbf{m}_j$ is oriented along the ambient field ignores the possibility of remanent magnetization in other directions. Although that is a common assumption, it is not required by the formalism. For example, Whaler and Langel (1996) derived a model with no assumption regarding the direction of magnetization.

The preceding formulation assumes that the anomaly data, $\mathbf{A}$, are linear field components. However, the data from the POGO satellites are scalar field magnitude. Further, many studies with *Magsat* data use only the scalar data because, not having to contend with inaccuracies in attitude (direction), they are more accurate. Scalar measurements are inherently nonlinear functions of the model parameters, of the m_j in the present case. However, because the magnitude of the anomaly field, $|\mathbf{A}|$, is small compared with the main field, $B_m = |\mathbf{B}_m|$, the approximation of equation (1.9) is accurate and can be used to linearize the

Table 5.3. *Matrix elements for derivation of magnetic moments $H_{ij}^{m,p}$ in equation (5.36)*[a]

Element	Expression
$H_{ij}^{m,r}$	$(H'_{11})^{m_j}(\sin I_j) - (H'_{12})^{m_j}(\cos I_j \cos D_j) + (H'_{13})^{m_j}(\cos I_j \sin D_j)$
$H_{ij}^{m,\theta}$	$(H'_{21})^{m_j}(\sin I_j) - (H'_{22})^{m_j}(\cos I_j \cos D_j) + (H'_{23})^{m_j}(\cos I_j \sin D_j)$
$H_{ij}^{m,\phi}$	$(H'_{31})^{m_j}(\sin I_j) - (H'_{32})^{m_j}(\cos I_j \cos D_j) + (H'_{33})^{m_j}(\cos I_j \sin D_j)$
$H_{ij}^{m,s}$	$-\sin I_i[(H'_{11})^{m_j}(\sin I_j) - (H'_{12})^{m_j}(\cos I_j \cos D_j) + (H'_{13})^{m_j}(\cos I_j \sin D_j)]$
	$\quad - \cos I_i \cos D_i[(H'_{21})^{m_j}(\sin I_j) - (H'_{22})^{m_j}(\cos I_j \cos D_j)$
	$\quad + (H'_{23})^{m_j}(\cos I_j \sin D_j)] + \cos I_i \sin D_i[(H'_{31})^{m_j}(\sin I_j)$
	$\quad - (H'_{32})^{m_j}(\cos I_j \cos D_j) + (H'_{33})^{m_j}(\cos I_j \sin D_j)]$
or	
$H_{ij}^{m,s}$	$-\sin I_j[(H'_{11})^{m_j}(\sin I_i) + (H'_{21})^{m_j}(\cos I_i \cos D_i) - (H'_{31})^{m_j}(\cos I_i \sin D_i)]$
	$\quad + \cos I_j \cos D_j[(H'_{12})^{m_j}(\sin I_i) + (H'_{22})^{m_j}(\cos I_i \cos D_i)$
	$\quad - (H'_{32})^{m_j}(\cos I_i \sin D_i)] - \cos I_j \sin D_j[(H'_{13})^{m_j}(\sin I_i)$
	$\quad + (H'_{23})^{m_j}(\cos I_i \cos D_i) - (H'_{33})^{m_j}(\cos I_i \sin D_i)]$

[a]The indices i and j correspond to the point at which the field is computed and to the dipole location, respectively; p is the component, and m indicates magnetic moment model.
Note: Expressions for $(H'_{ij})^{m_j}$ are found in Table 5.2.

problem. In particular, if the scalar anomaly field is denoted by $A^s(\mathbf{r})$, then

$$A^s(\mathbf{r}) = \mathbf{A}(\mathbf{r}) \cdot \hat{\mathbf{b}}(\mathbf{r}), \qquad (5.37)$$

where

$$\hat{\mathbf{b}}(\mathbf{r}) = \frac{1}{B_m(\mathbf{r})}(B_r \hat{\mathbf{r}} + B_\theta \hat{\boldsymbol{\theta}} + B_\phi \hat{\boldsymbol{\phi}}) \qquad (5.38)$$

is the unit vector along the direction of the main field at location $\mathbf{r}$. Equation (5.37) states that the scalar anomaly field is always computed relative to the field directions from some model of the earth's main field. In practice, the calculation is not sensitive to the model used. Combining equations (5.37), (5.38), and (5.36),

$$A^s(\mathbf{r}_i) = \frac{1}{B_m(\mathbf{r}_i)} \sum_{p=1}^{3} A_p(\mathbf{r}_i) B_p(\mathbf{r}_i), \qquad (5.39)$$

$$A^s(\mathbf{r}_i) = \sum_j H_{ij}^{m,s} m_j, \qquad (5.40)$$

where

$$H_{ij}^{m,s} = \frac{1}{B_m(\mathbf{r}_i)} \sum_{p=1}^{3} B_p(\mathbf{r}_i) H_{ij}^{m,p}. \qquad (5.41)$$

Assuming a spherical earth, and recalling equations (2.1) and (2.2), leads to the expression for $H_{ij}^{m,s}$ given in Table 5.3.

Now let $\mathbf{a}$ be the vector of all, say N, magnetic component measurements (the three components measured at a single location will furnish three elements of $\mathbf{a}$), and $\mathbf{m}$ the vector of all, say k^*, magnetic moment magnitudes to be determined. Then the general statement of the finite-dimensional inverse problem is given by

$$\mathbf{a} = H^m \mathbf{m}, \tag{5.42}$$

which has the usual least-squares solution

$$\underline{\mathbf{m}} = [(H^m)^T H^m]^{-1} (H^m)^T \mathbf{a}, \tag{5.43}$$

where an underbar will generally indicate an estimate of the quantity, and the superscript T indicates the matrix transpose. Formally, H^m is the matrix of source functions of partial derivatives of the field with respect to the m_j. For a realistic solution covariance, the data must be weighted properly. If V_d is the data covariance matrix, then the weight matrix is $W = (V_d)^{-1}$, and the solution becomes

$$\underline{\mathbf{m}} = [(H^m)^T W H^m]^{-1} (H^m)^T W \mathbf{a}, \tag{5.44}$$

with covariance

$$V_m = [(H^m)^T W H^m]^{-1}. \tag{5.45}$$

Note that the superscript m is carried to distinguish the H associated with magnetic moments from other H matrices to be discussed in following paragraphs. A more extensive development of the least squares formalism is given in Section 5.2. In practice, solutions of the form (5.44) usually are for a limited area and may suffer from instability. Stabilization is discussed in Sections 5.5 and 5.6.

Representation of a data set in terms of a fixed distribution of elementary sources is known as an *equivalent source* (ES) *model*. It is called "equivalent" because, although it is nonunique, it can be transformed into other, equally nonunique source distributions and because it can be used to calculate the same field, or equivalent field, that results from the true source distribution.

The formalism described was adapted by Mayhew (1979) and von Frese, Hinze, and Braile (1981a) from a method of Dampney (1969) for use with gravity measurements. Dampney seems to have coined the term "equivalent source model." Although other choices of alignment are possible, as described earlier, the dipoles typically are assumed to be aligned along the direction of the earth's main field. The question of dipole alignment will be encountered again in a later chapter.

5.1.4 REDUCTION TO COMMON INCLINATION (TO THE POLE) AND MAGNITUDE

An ES solution is a tool for analysis. Among its uses is the transformation of a map of the observed field into maps of the field that would arise from the same sources (rocks) if the inducing magnetic field were different. Such a transformation involves some critical assumptions regarding the nature of the magnetization of the sources. If it is assumed that all of the magnetic moment inferred from the data is induced by the present-day ambient main field [i.e., following equations (5.31)], then the following transformations are meaningful.

Reduction to common inclination consists of transforming the solution to the geometry where the declination D of the inducing field is zero and where the inclination I of the inducing field is constant, $I = I_0$. This is accomplished by taking $m = |\underline{\mathbf{m}}|$, from the solution of (5.44), and computing m_r, m_θ, and m_ϕ from (5.31) using $D = 0$ and $I = I_0$. The resulting, transformed, field components are then found from (5.30), (5.36), or (5.40), with $I = I_0$ and $D = 0$ in the main field [i.e., in the $\mathbf{w}_b$ of equation (5.32)].

The case $I_0 = 90°$ corresponds to an inducing field at the geomagnetic north pole and is therefore referred to as *reduced to the pole*. In this case the calculated anomalies are centered over their sources, facilitating interpretation. Reduction to the pole can be an unstable transformation, particularly for regions near the geomagnetic equator. Sections 5.5 and 5.6 describe ways of dealing with instability.

Reduction to common magnitude consists of transforming the solution to that which corresponds to the same magnitude of inducing field for the entire area. As noted in (5.27), if the magnetic moments $\mathbf{m}_j$ are induced by the ambient field $\mathbf{B}$, then the magnitude of $\mathbf{m}_j$, m_j, is directly proportional to B, the magnitude of $\mathbf{B}$. To transform a solution of dipoles $\{\mathbf{m}_j\}$ to a common magnitude of inducing field, say B_0, the transformation is accomplished by taking

$$\mathbf{m}'_j = \mathbf{m}_j \frac{B_0}{B_j}. \tag{5.46}$$

Again, the resulting fields are computed from (5.30) or (5.36).

In practice, reduction to the pole and reduction to common magnitude generally are carried out together. In this case, not only are the calculated anomalies centered over their sources but also the anomaly magnitudes are proportional to the strength of those sources.

5.1.5   UNIQUENESS CONSIDERATIONS

There are at least two fundamental sources of nonuniqueness to be considered. The first was mentioned in Section 2.1.1; that is, the presence of the large field from the core effectively masks all crustal anomalies below about spherical harmonic degree 14.

The second source of nonuniqueness is inherent in the nature of the inverse problem. If the volume in equation (5.22) extends over all of space, so that the surface integral is over $\mathbf{R} = \infty$, then the surface integral vanishes, leaving

$$\Psi_a = -\frac{\mu_0}{4\pi} \int \frac{\nabla' \cdot [\mathbf{M}(\mathbf{r}')]}{|\mathbf{r} - \mathbf{r}'|} d^3\mathbf{r}', \tag{5.47}$$

where the integral is taken over all space. Determining $\mathbf{M}(\mathbf{r}')$ from measurements of $\Psi_a(\mathbf{r})$ [i.e., from measurement of $\mathbf{A}(\mathbf{r}) = -\nabla\Psi_a(\mathbf{r})$] is a linear inverse problem of the form $A = L(\mathbf{M})$, where L is a linear operator. If a magnetization $\mathbf{M}$ exists such that $\mathbf{M}$ is nonzero and $L(\mathbf{M}) = 0$, then $\mathbf{M}$ is called an annihilator for the operator L (Parker, 1977; Parker and Huestis, 1974), or a member of the null space of L. If such an $\mathbf{M}$ exists and if $\mathbf{M}$ is a solution to the inverse problem, then $\mathbf{M} + \alpha\mathbf{M}$, for any finite α, is also a solution to the inverse problem (i.e., the solution $\mathbf{M}$ is not unique). In the magnetic field case, it is immediately clear from equation (5.47) that any magnetization $\mathbf{M}$, such that $\nabla \cdot \mathbf{M} = 0$ everywhere, including boundaries, is a member of the null space of the problem. More practically, Runcorn (1975) pointed out that for a spherical shell, any magnetization that is everywhere in the direction of an internally generated field and whose magnitude is proportional to the strength of that internal field is such an annihilator.

In addition to defining a class of annihilators for equation (5.22), this result means that if, for a spherical earth, the magnetic layer is of uniform thickness and susceptibility, then its external magnetic field is zero. More practically, if the susceptibility of the earth's magnetic layer can be decomposed into a part that corresponds to a uniform thickness and average susceptibility and a remainder, then the external magnetic field of the first part will be zero.

Harrison, Carle, and Hayling (1986), Harrison (1987), and Hayling and Harrison (1986) computed magnetizations and then added the magnetization due to some annihilator so as to obtain solutions with positive magnetization everywhere. This is considered more realistic than the ES solution alone, because the ES magnetizations are both positive and negative, and only positive magnetizations should result from induction (see Section 8.5).

The existence of one or more annihilators means that there is a basic ambiguity in magnetization solutions that can be removed only by using other geophysical experience or data to restrict the solution.

5.2 BASICS OF ESTIMATION

5.2.1 OVERVIEW

Return to Figure 5.1 and its depiction of various procedures used to reduce or invert the data. Assume that anomaly data, suitably cleansed of the effects of the main field, ionospheric fields, and magnetospheric fields, are distributed within a spherical shell, generally within an altitude range of 200–600 km. The equivalent source method described in Section 5.1.3 belongs to a class of inverse methods that have many features in common. For example, data can simply be gridded, typically on some surface within or at the lower boundary of the data shell. In a more sophisticated approach, the parameters in some analytic model can be determined, and the model can be used to compute, or predict, the magnetic field at any altitude. In order to correspond to the physics of magnetic fields, such a model should be a potential function. Or one might invert the data to determine some magnetization distribution.

All of these models can be formulated in terms of a least-squares or minimum-variance problem in a vector space. A summary of vector-space notions is presented in Appendix 5.2. Let $\{c_i, i = 1, \ldots, N\}$ be N measurements of the magnetic anomaly field; that is,

$$c_i = A_j(\mathbf{r}) + e_i, \tag{5.48}$$

where A_j is one component of the anomaly field (A_r, A_θ, or A_ϕ) measured at some location $\mathbf{r}$, and e_i is the measurement noise. Then the vector $\mathbf{c}$ of all measurements is

$$\mathbf{c} = \mathbf{a} + \mathbf{e}. \tag{5.49}$$

The vectors $\mathbf{c}$, $\mathbf{a}$, and $\mathbf{e}$ are members of the real vector space $\mathbf{V}_N$ formed by all N-dimensional vectors with N real-valued components. The usual inner product on $\mathbf{V}_N$ is

$$\langle \mathbf{v}, \mathbf{w} \rangle = \sum_{i=1}^{N} v_i w_i, \tag{5.50}$$

where the v_i and w_i are the components of $\mathbf{v}$ and $\mathbf{w}$. Sometimes it is convenient to let the probability distribution for the random errors, e_i, induce the inner product, in which case

$$\langle \mathbf{v}, \mathbf{w} \rangle_\sigma = \sigma^{-2} \sum_{i=1}^{N} v_i w_i, \tag{5.51}$$

where the errors are assumed gaussian and independent, with zero mean and variance σ^2.

Suppose the model space $\mathbf{X}$ is a real, K-dimensional linear vector space. The model is defined by

$$\mathbf{a} = H\mathbf{x}, \qquad \mathbf{c} = H\mathbf{x} + \mathbf{e}, \tag{5.52}$$

with $\mathbf{a}$, $\mathbf{c}$, and $\mathbf{e}$ as before, and H the geometric matrix defining the model (i.e., the matrix of partial derivatives of measurements with respect to the $\mathbf{x}$). The dimension of the matrix H is $N \times K$, with H_{ij} relating the ith data point to the jth model parameter. The explicit form of H depends upon the model.

Table 5.4. *List of inversion models and parameters*

Model type/ parameters	Symbol for $\mathbf{x}$	H matrix			
		Symbol	Section where defined	Table of expressions	Model equations
Equivalent source/ magnetization	$\mathbf{m}$	H^m	5.1.3	5.3	(5.42), (5.43)
Spherical cap harmonic analysis	$\mathbf{q} = \{g_k^m, h_k^m\}$	H^q	5.3.1	5.5	(5.76)
Rectangular harmonic analysis	$\mathfrak{z} = \{\mathcal{X}_0, \mathcal{Y}_0, \mathcal{Z}_0, \mathcal{X}_{kj}\}$	$H^{\mathfrak{z}}$	5.3.2	5.6	(5.77), (5.79)
Collocation	$\mathbf{b}$ = predicted anomaly values	H^b	5.3.3		(5.84), (5.89)
Spherical harmonic analysis (global, potential)	$\mathbf{p} = \{g_n^m, h_n^m\}$	H^p	5.4.1	5.8	(1.3), (2.3), (5.101)
Spherical harmonic analysis (global, component only, i.e., not potential)	$\mathbf{h} = \{C_n^m, S_n^m\}$	H^h	5.4.2	5.8	(5.102), (5.104)
Generalized inverse methods			5.6.4		(5.152), (5.153), (5.167), (5.172)
ψ-norm	$\mathbf{t}$	Λ		5.10	
B_r-norm	$\mathbf{t}$	Λ		5.11	
Minimum-norm magnetization	$\mathbf{t}$	K	5.6.5		(5.186), (5.189)

One example is given by the ES model of Section 5.1.3, in which $\mathbf{x} = \mathbf{m}$ and $H = H^m$. Table 5.4 lists the various models considered, the sections in which they are discussed, the equations defining the models, and the tables in which the matrix elements of H are given explicitly.

5.2.2 MINIMUM-VARIANCE ESTIMATION WITH A PRIORI INFORMATION

It is convenient, following Jackson (1979), to discuss model solutions in the context of minimum-variance estimation with a priori information. Suppose there is prior information about the solution available for the problem to be solved that is expressed in terms of an a priori estimate of the model parameters, called $\underline{\mathbf{x}}_0$. Assume that the new estimate of $\mathbf{x}$, $\underline{\mathbf{x}}$, can be written as a linear combination of the data $\mathbf{c}$ and of $\underline{\mathbf{x}}_0$:

$$\underline{\mathbf{x}} = Q\mathbf{c} + G\underline{\mathbf{x}}_0, \tag{5.53}$$

where Q and G are matrix operators to be determined. It is assumed that $\underline{x}_0$ was obtained independently of the data $\mathbf{c}$. Combining (5.52) and (5.53) gives

$$\underline{x} = QH\mathbf{x} + Q\mathbf{e} + G\underline{x}_0,$$
$$\underline{x} = R\mathbf{x} + Q\mathbf{e} + G\underline{x}_0, \tag{5.54}$$

where $R = QH$. The estimation error, $\underline{x} - \mathbf{x}$, is then

$$\varepsilon_x = \underline{x} - \mathbf{x} = (R - I)\mathbf{x} + Q\mathbf{e} + G\underline{x}_0. \tag{5.55}$$

If the estimate is to be unbiased, the expected value of $\underline{x} - \mathbf{x}$ must be zero. Assume $E(\mathbf{e}) = 0$ and $E(\underline{x}_0) = \mathbf{x}$. Then

$$G = I - R, \tag{5.56}$$

and

$$\varepsilon_x = \underline{x} - \mathbf{x} = (R - I)(\mathbf{x} - \underline{x}_0) + Q\mathbf{e}. \tag{5.57}$$

The interpretation of (5.57) is that the first term is the resolving error, and the second the random error. Resolving error is dependent on two factors. First, the difference between R and I depends upon how well the data resolves, or determines, the parameters $\mathbf{x}$. If $\mathbf{x}$ is completely determined by the data, then $QH = I$. If QH is not the identity matrix, then the solution will depend upon the a priori information, and the error will depend both on the amount of that dependence, measured by $R - I$, and on the amount that $\underline{x}_0$ deviates from the true value $\mathbf{x}$.

For a good estimate $\underline{x}$, both error contributions should be small. Taking the covariance of (5.57) gives

$$E(\underline{x} - \mathbf{x})(\underline{x} - \mathbf{x})^T = V_\varepsilon = (R - I)X_0(R - I)^T + QV_dQ^T, \tag{5.58}$$

where V_d is the data covariance matrix, and X_0 is the a priori covariance matrix of $\mathbf{x} - \underline{x}_0$.

Q is specified by the estimation procedure. Least squares, or minimum variance, minimizes the variances in (5.58), that is, minimizes the trace of V_ε with respect to Q (see Appendix 5.1). The result is

$$Q = X_0H^T(HX_0H^T + V_d)^{-1}. \tag{5.59}$$

If (5.59) is premultiplied by $[H^T(V_d)^{-1}H + (X_0)^{-1}]^{-1}[H^T(V_d)^{-1}H + (X_0)^{-1}] = I$ and rearranged,

$$Q = [H^T(V_d)^{-1}H + (X_0)^{-1}]^{-1}H^T(V_d)^{-1}. \tag{5.60}$$

Substituting (5.60) and (5.59) appropriately into (5.58) and manipulating gives

$$V_{\hat{\varepsilon}} = [H^T(V_d)^{-1}H + (X_0)^{-1}]^{-1}. \tag{5.61}$$

Then

$$\underline{x} = [H^T(V_d)^{-1}H + (X_0)^{-1}]^{-1}[H^T(V_d)^{-1}\mathbf{c} + (X_0)^{-1}\underline{\mathbf{x}}_0]. \tag{5.62}$$

If $(X_0)^{-1}$ is zero, (5.62) reduces to the usual least-squares solution.

Equation (5.52) assumes that the model parameters and the data have a linear relationship; that is, H is a linear transformation from parameter space to data space. Consider briefly the situation where the relationship is nonlinear. For example,

$$\mathbf{c} = \mathbf{F}(\mathbf{x}) + \mathbf{e}, \tag{5.63}$$

where $\mathbf{F}$ is an N-vector-valued nonlinear function of the parameters $\mathbf{x}$. The standard procedure in this case is to linearize the problem and iterate toward a solution. Let $\underline{\mathbf{x}}_0$ be an initial estimate of $\mathbf{x}$, X_0 the covariance matrix of $\mathbf{x} - \underline{\mathbf{x}}_0$, and

$$\underline{\mathbf{c}}_0 = \mathbf{F}(\underline{\mathbf{x}}_0). \tag{5.64}$$

$\mathbf{F}(\mathbf{x})$ is expanded in a Taylor series around $\underline{\mathbf{x}}_0$:

$$\underline{\mathbf{c}} = \mathbf{F}(\underline{\mathbf{x}}_0) + \frac{\partial \mathbf{F}}{\partial \mathbf{x}}[\mathbf{x} - \underline{\mathbf{x}}_0] + \text{higher-order terms} + \mathbf{e}, \tag{5.65}$$

$$\underline{\mathbf{c}} = \underline{\mathbf{c}}_0 + H[\mathbf{x} - \underline{\mathbf{x}}_0] + \mathbf{e}, \tag{5.66}$$

where the matrix $H = \partial \mathbf{F}/\partial \mathbf{x}$ has elements

$$H_{ij} = \frac{\partial F_i}{\partial x_j}. \tag{5.67}$$

An iterative process is then defined by taking

$$\Delta \mathbf{c}^k = \mathbf{c} - \underline{\mathbf{c}}^k, \qquad \Delta \mathbf{x}^k = \mathbf{x} - \underline{\mathbf{x}}^k, \tag{5.68}$$

$$H^k = \frac{\partial \mathbf{F}(\mathbf{x}^k)}{\partial \mathbf{x}}, \tag{5.69}$$

$$\Delta \mathbf{c}^k = H^k \Delta \mathbf{x}^k + \mathbf{e}, \tag{5.70}$$

$$\Delta \underline{\mathbf{x}}^k = Q^k \Delta \mathbf{c}^k + G^k[\underline{\mathbf{x}}^k - \underline{\mathbf{x}}_0]. \tag{5.71}$$

Following the same steps as in the linear case leads again to equations (5.56), (5.59), and (5.60), with appropriate k dependence, and to

$$V^{k+1} = [(H^k)^T V_d H^k + (X_0)^{-1}]^{-1}, \tag{5.72}$$

$$\Delta \underline{\mathbf{x}}^{k+1} = V^{k+1}[(H^k)^T(V_d)^{-1}\Delta \mathbf{c}^k + (X_0)^{-1}(\underline{\mathbf{x}}^k - \underline{\mathbf{x}}_0)]. \tag{5.73}$$

The process is iterated until convergence.

5.3   LOCAL MODELS

Many inverse models are restricted to a specific, limited region under study. Also of importance, smaller-scale features can be more readily extracted from the data on a regional basis than on a global basis. Further, in a region where the effects of ionospheric currents are limited, a local analysis often can result in a higher signal-to-noise ratio for that region than would be obtained in a global approach. On the other hand, local models have the disadvantage of unavoidable distortion at and near the edge of the region modeled. That can foster misleading interpretations, as well as difficulty when piecing together models from adjacent regions. Section 5.5.3 discusses these disadvantages in more detail for local equivalent source models.

The first and most widely used local model is the ES model described in Section 5.1.3. Other approaches to local modeling are summarized in the following sections.

5.3.1   SPHERICAL CAP HARMONIC ANALYSIS

One reason that global spherical harmonic analysis (SHA) has not been widely used for representation of anomaly fields is that very large numbers of coefficients are necessary to achieve the desired resolution. Spherical cap harmonic analysis (SCHA) is a formulation suitable for a local region that retains some of the advantages of SHA, but does not require excessive numbers of coefficients to achieve the desired resolution.

The global potential function, equation (2.3), is made up of the sum of solutions of the form

$$\Psi_k^m(\mathbf{r}) = a\left(\frac{a}{r}\right)^{x_k+1} \left[g_{x_k}^m \cos(m\phi) + h_{x_k}^m \sin(m\phi)\right] P_{x_k}^m(\cos\theta), \qquad (5.74)$$

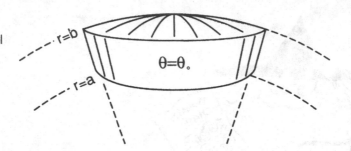

where external fields are ignored in the present discussion, and h_x^0 is zero. Continuity of Ψ_k^m and its derivatives in the ϕ variable requires that m be an integer. When analyzing the field over the entire earth, continuity arguments at the poles eliminate Legendre functions of the second kind and dictate that $x_k = n$, an integer, leading to $P_n^m(\cos\theta)$, the familiar associated Legendre polynomials.

Suppose that $\Psi_k^m(\mathbf{r})$ is to be applicable for a spherical cap, $0 \leq \theta \leq \theta_0$, (Figure 5.4), and that the boundary conditions are given by

$$\Psi(r, \theta_0, \phi) = \mathfrak{f}(r, \phi), \tag{5.75a}$$

$$\frac{\partial \Psi(r, \theta_0, \phi)}{\partial \theta} = \mathfrak{g}(r, \phi), \tag{5.75b}$$

where $\mathfrak{f}$ and $\mathfrak{g}$ are arbitrary continuous functions imposed by the particular problem to be solved. To satisfy these boundary conditions, Haines (1985a) divided the $\{\Psi_k^m(\mathbf{r})\}$ into two sets and considered nonintegral values of x_k. The members of the first set are all $\Psi_k^m(\mathbf{r})$, with $x_k = n_k$ chosen such that $P_{n_k}^m(\theta_0) = 0$ [i.e., θ_0 is a zero of $P_{n_k}^m(\theta)$]. Note that n_k is dependent on the value of m [i.e., $n_k = n_k(m)$]. For this set of $x_k = n_k(m)$,

$$\Psi_{x_k}^m(\theta_0) = 0,$$

but

$$\left.\frac{\partial \Psi_{x_k}^m(\theta)}{\partial\theta}\right|_{\theta=\theta_0}$$

will not, in general, be zero, so that the corresponding coefficients, $\{g_{n_k(m)}^m, h_{n_k(m)}^m\}$, can be chosen to satisfy (5.75b). Notice that the value of k in (5.74) is simply a label; its actual value can be assigned by any rule that proves convenient. Adopting the rule suggested by Haines (1985a), for this first set of $\{\Psi_k^m(\mathbf{r})\}$ the values of k are chosen so that $k - m$ is odd [i.e., for $k - m$ odd, $n_k(m)$ is chosen so that $P_{n_k}^m(\theta_0) = 0$].

Members of the second set of $\{\Psi_k^m(\mathbf{r})\}$ are those with $x_k = n_k$ chosen such that

$$\left.\frac{\partial P_{n_k(m)}^m(\theta)}{\partial\theta}\right|_{\theta=\theta_0} = 0, \quad \text{i.e., } \theta_0 \text{ is a zero of } \frac{\partial P_{n_k}^m(\theta)}{\partial\theta}.$$

Table 5.5. *Matrix elements for derivation of spherical harmonic coefficients H_{ij}^q*

Element	Expression	
H_{ij}^r	$[n_k(m) + 1] \cos(m\phi_i) P_{n_k}^m(\cos\theta_i) \left[\frac{a}{r_i}\right]^{n_k+2}$	(for $q_j = g_k^m$)
	$[n_k(m) + 1] \sin(m\phi_i) P_n^m(\cos\theta_i) \left[\frac{a}{r_i}\right]^{n_k+2}$	(for $q_j = h_k^m$)
H_{ij}^θ	$-\cos(m\phi_i) \frac{dP_{n_k}^m(\cos\theta_i)}{d\theta} \left[\frac{a}{r_i}\right]^{n_k+2}$	(for $q_j = g_k^m$)
	$-\sin(m\phi_i) \frac{dP_{n_k}^m(\cos\theta_i)}{d\theta} \left[\frac{a}{r_i}\right]^{n_k+2}$	(for $q_j = h_k^m$)
H_{ij}^ϕ	$\frac{m}{\sin\theta_i} \sin(m\phi_i) P_{n_k}^m(\cos\theta_i) \left[\frac{a}{r_i}\right]^{n_k+2}$	(for $q_j = g_k^m$)
	$-\frac{m}{\sin\theta_i} \cos(m\phi_i) P_{n_k}^m(\cos\theta_i) \left[\frac{a}{r_i}\right]^{n_k+2}$	(for $q_j = h_k^m$)

Note: Indices i and j correspond, respectively, to the point where the field is computed and to the jth parameter. q_j, either a g_k^m or an h_k^m.

For this second set of $x_k = n_k(m)$, $\Psi_{x_k}^m(\theta_0)$ will not, in general, be zero, so that the corresponding coefficients, $\{g_{n_k(m)}^m, h_{n_k(m)}^m\}$, can be chosen to satisfy (5.75a). The labeling rule of Haines (1985a) is that for this second set of $\{\Psi_k^m(\mathbf{r})\}$, $k - m$ is even. That is, for $k - m$ even, $n_k(m)$ is chosen so that

$$\left. \frac{\partial P_{n_k(m)}^m(\theta)}{\partial\theta} \right|_{\theta=\theta_0} = 0.$$

A combination of the two sets of $\Psi_k^m(\mathbf{r})$,

$$\Psi = a \sum_{m=0}^{\infty} \sum_{k=m}^{\infty} \left(\frac{a}{r}\right)^{n_k(m)+1} P_{n_k(m)}^m(\cos\theta)[g_k^m \cos(m\phi) + h_k^m \sin(m\phi)], \quad (5.76)$$

will satisfy both boundary conditions. In the notation of Section 5.2, $\mathbf{h} = \mathbf{q}$ is the vector of coefficients $\{g_k^m, h_k^m\}$, and H is equal to H^q, with elements as given in Table 5.5.

SCHA has been used for local analysis of anomaly data by Haines (1985b), DeSantis, Kerridge, and Barraclough (1989), and DeSantis, Battelli, and Kerridge (1990) and, in a modified form, by DeSantis (1991). It has also been used in other applications by, for example, Haines (1985c), Haines and Newitt (1986), Garcia et al. (1991), and Torta, Garcia, and DeSantis (1992, 1993).

Lowes and Haines (1993) and F. J. Lowes (personal communication, 1994) have pointed out that there can be problems in using SCHA to derive a potential function. First, the longest wavelength that can be represented is θ_0, so the method is not efficient for data with longer wavelengths. This should not be a problem when dealing with **A**. The second problem is more serious. The potential of the geomagnetic field of internal origin is a three-dimensional function applicable in all space outside the source region. Whereas the potential determined by SCHA is equal to (or a near estimate of) the real potential on the spherical cap, it is intrinsically different from the real potential elsewhere. The SCHA potential has infinite discontinuities at the opposite pole, and its radial dependence involves non-integer powers of n. Thus, for example, the actual radial dependence of an $n = 20$ harmonic of B_r is $1/r^{22}$, which must be approximated by non-integer inverse powers of r in SCHA. Lowes (personal communication, 1997) points out that one consequence is that theoretically it is not possible simultaneously to represent exactly the potential for the horizontal field components and the potential for the radial component. Thus, fitting all components of **A** simultaneously gives an approximation that does its best to fit both, but it is not as accurate as fitting A_r separately from A_θ and A_ϕ. Also, using SCHA to predict (continue) at a different altitude than the measurements will introduce errors that will increase as the range of altitude increases. At this time, no estimate of the magnitude of such errors is available, and so it is not known under what conditions the approximation is adequate. It is recommended that SCHA be used only for interpolation and continuation of B, of B_r, or of B_θ and B_ϕ together, and only within the limited altitude range of the data.

5.3.2 RECTANGULAR HARMONIC ANALYSIS

Another method of retaining high resolution in data or model representation, without the requirement of a large number of model parameters, is to use a local rectangular harmonic analysis (RHA) (Alldredge, 1981). This method involves an approximation that becomes less accurate as the area analyzed becomes larger. On the other hand, it is simpler to implement than is SCHA, and for sufficiently small regions it is an accurate approximation. Its use for analysis of satellite magnetic anomaly data was implemented by Nakagawa and Yukutake (1985) and Nakagawa et al. (1985), on which the following development is based.

Let θ_0 and ϕ_0 be the colatitude and longitude at the center of the region under consideration. As illustrated in Figure 5.5a, a local cartesian coordinate system is defined with origin at (θ_0, ϕ_0) at the earth's surface. In the local system, the x axis is northward in the meridian plane, the z axis is down, and the y axis is eastward. Data locations and component directions must be transformed to the local system.

The form of a potential function in rectangular coordinates is given by

$$\Psi(x, y, z) = \mathcal{X}_0 x + \mathcal{Y}_0 y + \mathcal{Z}_0 z + \sum_{k=0}^{K-1} \sum_{j=0}^{K-1} \mathcal{X}_{kj} e^{-\pi i [kx + jy]/L} e^{d_{kj} z}, \tag{5.77}$$

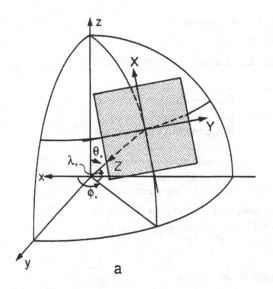

a

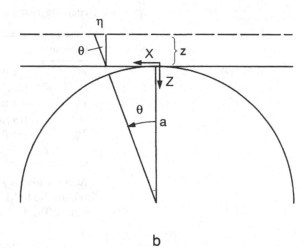

b

FIGURE 5.5
Geometry for rectangular harmonic analysis. (a) X is north directed, Y east directed, and Z downward; (b) geometry of the flat-earth approximation.

where the dimensions of the rectangular region are defined by $-L \le x \le L$, $-L \le y \le L$, where $i = \sqrt{-1}$, and where

$$d_{kj} = \frac{\pi}{L}[k^2 + j^2]^{1/2}. \tag{5.78}$$

The field in the local system is given by

$$\mathbf{A}_x = -\nabla\Psi. \tag{5.79}$$

Nakagawa and Yukutake (1985) used a normalized form of (5.77). Also, in order to minimize edge effects in the analysis, they applied a weighting function that was equal to unity except near the boundaries: $x \ge 0.8L$, $x \le -0.8L$, $y \ge 0.8L$, and $y \le -0.8L$. In the boundary regions, the weight is a cosine function that tapers to zero at the boundary.

The procedure is to transform the data into the local cartesian system and find the coefficients $\mathcal{X}_0$, $\mathcal{Y}_0$, $\mathcal{Z}_0$, and $\mathcal{X}_{kj}$ in (5.77) that best fit the data in a least-squares sense. This representation can then be used to reduce the data to any desired altitude, after which the results can be easily transformed back to the spherical-earth coordinates for display and analysis. However, Haines (1990) noted that the terms in $\mathcal{X}_0$, $\mathcal{Y}_0$, and $\mathcal{Z}_0$ do not go to zero as z approaches infinity and so cannot represent purely internal sources. In an evaluation of the method as applied to European magnetic observatory annual means in the data of Alldredge (1981), Malin, Düzgît, and Baydemîr (1996) suggested addition of terms varying as xy, yz, and xyz to remove trends, and they concluded that the method is suitable for interpolation but not extrapolation.

In the notation of Section 5.2, $\mathbf{h} = \hat{\gamma}$ is the vector of coefficients $\{\mathcal{X}_0, \mathcal{Y}_0, \mathcal{Z}_0, \mathcal{X}_{kj};$ $k, j = 0, \ldots, K - 1\}$, and H is equal to H^δ, with elements as given in Table 5.6.

Table 5.6. *Matrix elements for derivation of rectangular harmonic coefficients* $H_{ip}^{\tilde{\jmath}}$

Element	Expression	
H_{ip}^{x}	-1.0	(for constant coefficients)
	$(i\pi k)e^{-\pi i[kx+jy]/L}e^{\gamma_{kj}^{z}}$	(for series coefficients)
H_{ip}^{y}	-1.0	(for constant coefficients)
	$(i\pi j)e^{-\pi i[kx+jy]/L}e^{\gamma_{kj}^{z}}$	(for series coefficients)
H_{ip}^{z}	-1.0	(for constant coefficients)
	$(-\gamma_{kj})e^{-\pi i[kx+jy]/L}e^{\gamma_{kj}^{z}}$	(for series coefficients)

Note: Indices i and p correspond, respectively, to the point where the field is computed and to the pth parameter, $\tilde{\jmath}_p$, variously X_0, y_0, Z_0, or X_{kj} ($k, j = 0, \ldots, K-1$).

Table 5.7. *Altitude deviation in flat-earth approximation*

Central angle θ	$(h+a)(1-\cos\theta)$ (km)
5.0°	25.0
7.5°	57.9
10.0°	102.9
12.5°	160.5
15.0°	230.7

As indicated in Figure 5.5b, this method involves an approximation. In particular, the z axis of the local coordinate system is taken to represent the altitude. This is a flat-earth approximation. To consider the possible error involved, begin with the transformation of altitude in the spherical coordinate system to its z value in the local system. If the surface of the earth is taken to be $r = a = 6{,}371.2$ km, and altitude above that surface is denoted by h, then the transformation to z is equivalent to

$$z = -h + (h + a)(1 - \cos\theta), \tag{5.80}$$

where θ is the central angle between the location of the point of interest and the origin of the local coordinate system. The second term on the right is the difference between the actual altitude h and the flat-earth altitude. Table 5.7 gives the difference as a function of θ. Another aspect to the approximation is that the direction of z in the local, flat-earth coordinate system is perpendicular to the plane tangent to the earth's surface at (θ_0, ϕ_0). This direction is deflected from the true vertical to the (spherical) earth by the angle θ. Clearly the approximation deteriorates quickly as the distance from (θ_0, ϕ_0) increases.

Two areas were considered by Nakagawa and Yukutake (1985), with $L = 2,000$ and $L = 1,000$ [note that the definition of L used in this section is half of that used by Nakagawa and Yukutake (1985)]. They note that at an altitude of 430 km, roughly a mean altitude for the *Magsat* satellite, z in equation (5.80) becomes zero at the boundary of the $L = 2,000$ region. The success of their analysis indicates that the method is useful, but its use for continuation outside the altitude range of the data is not recommended.

5.3.3 GRIDDING BY COLLOCATION

5.3.3.1 The Basic Formalism

Suppose, as in Section 5.2, that a vector of measurements, $\mathbf{c}$, is available, where $\mathbf{c} = \mathbf{a} + \mathbf{e}$, with $\mathbf{a}$ the vector of anomaly fields, and $\mathbf{e}$ the vector of noise, equation (5.49). It is desired to estimate the anomaly field, say $\mathbf{b}$, at a set of locations different from those of $\mathbf{a}$. In this section, a method is discussed in which an estimate, $\underline{\mathbf{b}}$, can be made from estimates of the statistical properties of $\mathbf{c}$.

It is assumed, following Moritz (1980), that the expected value of $\mathbf{e}$ is zero and that covariance matrices are defined as

$$V_a = E(\mathbf{aa}^T), \qquad V_b = E(\mathbf{bb}^T), \qquad V_{ba} = E(\mathbf{ba}^T), \qquad V_d = E(\mathbf{ee}^T), \qquad (5.81)$$

where E indicates "expected value." Note that $V_{ab} = (V_{ba})^T$. It is also assumed that $\mathbf{e}$ and $\mathbf{a}$ are uncorrelated, so that $V_{ae} = V_{ea} = 0$. Then

$$V_c = E(\mathbf{cc}^T) = V_a + V_d. \qquad (5.82)$$

V_d is the usual a priori covariance matrix of measurement noise.

In the collocation model, the anomaly field to be estimated, $\mathbf{b}$, is expressed as a linear function of the anomaly field, $\mathbf{a}$. That is,

$$\mathbf{b} = H^b \mathbf{a}. \qquad (5.83)$$

To estimate $\mathbf{b}$, write

$$\underline{\mathbf{b}} = H^b \mathbf{c} = H^b \mathbf{a} + H^b \mathbf{e}, \qquad (5.84)$$

where H^b is chosen to minimize the covariance of

$$\varepsilon = \underline{\mathbf{b}} - \mathbf{b}, \qquad (5.85)$$

that is, to minimize

$$V_\varepsilon = E(\varepsilon\varepsilon^T). \qquad (5.86)$$

From (5.84), $E(\underline{\mathbf{b}}) = \mathbf{b}$, so the estimator is unbiased. Now

$$\varepsilon\varepsilon^T = (\underline{\mathbf{b}} - \mathbf{b})(\underline{\mathbf{b}} - \mathbf{b})^T = (H^b\mathbf{a} + H^b\mathbf{e} - \mathbf{b})(H^b\mathbf{a} + H^b\mathbf{e} - \mathbf{b})^T. \qquad (5.87)$$

Taking expectation values,

$$V_\varepsilon = H^b V_a (H^b)^T + H^b V_d (H^b)^T - H^b V_{ab} - V_{ba}(H^b)^T + V_b$$
$$= V_b - V_{ba}(V_a + V_d)^{-1} V_{ab} + [H^b - V_{ba}(V_a + V_d)^{-1}](V_a + V_d)$$
$$\times [H^b - V_{ba}(V_a + V_d)^{-1}]^T. \tag{5.88}$$

The first two terms (matrices) on the right are independent of H^b. The third is zero if

$$H^b = V_{ba}(V_a + V_d)^{-1} = V_{ba}(V_c)^{-1} = (V_{ab})^T (V_c)^{-1}. \tag{5.89}$$

Moritz (1980) shows that equation (5.89) gives the best minimum-variance estimate of H^b. In this case, (5.88) reduces to

$$V_\varepsilon = V_b - V_{ba}(V_c)^{-1} V_{ab} = V_b - H^b V_{ab}. \tag{5.90}$$

To apply the method,

1. the covariance matrices V_d, V_c, V_b, and V_{ab} are estimated,
2. H^b is computed from equation (5.89),
3. V_ε is computed from equation (5.90), and
4. **b** is computed from equation (5.84).

The desired solution is **b**, with error estimate given by V_ε.

In essence, collocation is a scheme for interpolation by the weighted average of the surrounding data. It is equivalent to the method called Kriging, as described, for example, by Isaaks and Srivastava (1989). The weighted averages are optimal in a minimum-variance sense provided that the covariances used are reasonable approximations. Clearly, the heart of the method lies in computation of the covariance matrices V_d, V_a, V_b, and V_{ab}. V_d must be estimated from properties of the noise in measuring **c**. V_a, V_b, and V_{ab} can be estimated from known properties of the geomagnetic field or, more efficiently, from properties of the data **c**.

5.3.3.2 Local Covariance Functions

Forms for covariance matrices can be carried over from gravity analyses. For data on a plane (flat-earth approximation); such as the x–y plane, as in Figure 5.5, Moritz (1980) suggested approximate covariance functions between points at (x, y) and (x', y') of the form

$$V_1(r) = V_0 e^{-\mathcal{B}^2 r^2}, \tag{5.91}$$

$$V_2(r) = \frac{V_0}{(1 + \mathcal{B}^2 r^2)^{1/2}}, \tag{5.92}$$

$$V_3(r) = \frac{V_0}{(1 + \mathcal{B}^2 r^2)^{3/2}}, \tag{5.93}$$

where

$$r = [(x' - x)^2 + (y' - y)^2]^{1/2}, \tag{5.94}$$

and

$$\mathcal{B} = (\ln 2)^{1/2}/r_c \quad \text{for (5.91),} \tag{5.95}$$

$$\mathcal{B} = \sqrt{3}/r_c \quad \text{for (5.92),} \tag{5.96}$$

$$\mathcal{B} = 0.766\ldots/r_c \quad \text{for (5.93),} \tag{5.97}$$

where the *correlation length*, r_c, is the value of r for which $V(r) = V(0)/2$.

Equations (5.91)–(5.93) apply where the data are at a common altitude. To extend the approximation to three dimensions, let z be the perpendicular distance above the plane. In order to ensure that the extension of $V(r)$, say $V(r, z, z')$, will be a harmonic function, a requirement for both gravity and magnetic field covariance functions, z and z' can enter $V(r, z, z')$ only through their sum, $z + z'$ (Moritz, 1980). It turns out that V_1 has no harmonic extension. However, taking $\ell = 1/\mathcal{B}$, V_2 and V_3 can be extended as

$$V_2(r, z + z') = \frac{V_0 \ell}{[r^2 + (z + z' + \ell)^2]^{1/2}} \tag{5.98}$$

and

$$V_3(r, z + z') = \frac{V_0 \ell^2 (z + z' + \ell)}{[r^2 + (z + z' + \ell)^2]^{3/2}}. \tag{5.99}$$

In the case of magnetic field anomaly data, covariance functions of both the potential and its radial derivative, A_r, depend only on r_i, r_j, and γ_{ij}, and the results for the gravity field extend immediately to the magnetic potential and A_r. That is not the case for the A_θ and A_ϕ components nor for the scalar magnitude, A^s, although for limited distances dependence only on r_i, r_j, and γ_{ij} is a good approximation. Covariance functions V_2 and V_3 are most appropriate for use with satellite magnetic anomaly data. The correlation distance is in the range 350–600 km (Langel, Estes, and Sabaka, 1989).

In a typical application of collocation to magnetic anomaly data, a collection of data points in a particular three-dimensional bin is used to predict the anomaly field at some location within the bin (Figure 5.6). Data **a** in the bin are to be used to compute the field, say b_k, at some specified location within the kth bin. In this case the "vector" **b** is of one dimension (i.e., one point is predicted). If a set of such bins were located at, say, $1°$ intervals in θ and ϕ, and if the b_k were all computed at $r = a + h$, then the resulting set of $\{b_k\}$ would comprise a gridded anomaly map at a constant altitude of h, relative to a spherical earth. It is this sort of application that defines what is meant by "local" (i.e., the area of a bin is usually less than $5°$ square, and the altitude range Δh less than 200 km).

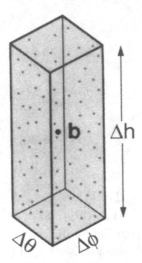

FIGURE 5.6
In collocation, a predicted
or interpolated value **b** is
found from the set of
measurements in a defined
volume.

5.3.3.3 Use of Collocation in Magnetic Anomaly Studies

To our knowledge, the only published result giving details of the use of collocation in magnetic anomaly studies is that of Goyal et al. (1990), in which the covariance function was V_1, with the correlation distance taken to be 300 km. The data were taken at different altitudes, so that was not the most appropriate covariance function. Even so, they found that the method gave good predictive results. This would seem to indicate that for the highly local applications considered, the procedure is not critically sensitive to the correlation length or the form of the correlation function.

We (R. A. Langel and D. N. Ravat, unpublished study) conducted a simulation study for the region surrounding Bangui, Africa, namely, from −5° to 50° W longitude and 15° N to 15° S latitude. A lithospheric magnetization model was derived that reproduced *Magsat* data in that region. Using that model, computed anomaly fields were derived at random positions in the specified latitude-longitude area and at altitudes between 300 and 400 km. Collocation models were then used to predict the field at gridded locations at 300-, 400-, and 500-km altitudes using the V_1 and V_2 potential functions, with the reference plane taken at 300 km. At each of the three altitudes, the results of the collocation prediction were compared with the exact results computed from the magnetization model by computing the rms difference. No data noise was included. At 400 km the results were excellent, with rms difference values of 0.0239 nT and 0.00346 nT for the V_1 and V_2 covariance models, respectively. The optimal scale lengths for the V_1 and V_2 covariances were 400 and 1,000 km, respectively.

At 300 and 500 km, none of the models was really satisfactory, with poorest results at 300 km. Those altitudes were at the upper and lower edges of the data volume. Using scale lengths of 100 km and more for V_1 and V_2, the rms differences were all about 5.2 nT and 2.38 nT at 300-km and 500-km altitudes, respectively. The conclusion is that collocation is a very good gridding method when the grid points are near the center of the altitude range of the data, but otherwise it is to be used with caution.

5.3.4 SUMMARY OF LOCAL MODELS

In principle, local models are capable of retaining more detail from the data than are global models, a topic that will be revisited in Section 6.9. Equivalent source inversion, spherical cap harmonic analysis, rectangular harmonic analysis, and collocation all have been used for preparation of local anomaly maps. To these should be added the use of crossover analysis, as described in Section 4.4.4. For gridding at satellite altitudes, collocation is the simplest and is recommended. For combining *Magsat* data from dawn and dusk local times in a consistent way, the method of choice is crossover analysis. However, it requires data within a limited altitude range. Although it does not seem to have been attempted, a useful procedure might be to use collocation to extrapolate data from individual satellite orbits to a common altitude and then apply a crossover analysis.

All of the local methods introduce inaccuracies when used for downward continuation. Spherical cap harmonic analysis, rectangular harmonic analysis, and collocation are not recommended for this application. The equivalent source method is suitable for continuation. However, it is subject to edge effects and to instability. These can be minimized, as discussed in Sections 5.5 and 5.6.

5.4 GLOBAL MODELS

Several models use data from the entire globe and apply to the entire globe. Most of these take the form of a spherical harmonic analysis, but recently approximate numerical techniques have been applied to obtain global equivalent source models.

5.4.1 SPHERICAL HARMONIC POTENTIAL

For continuous global representation of a quantity on a sphere, or when the coordinates of choice are spherical, spherical harmonics, as defined by equation (2.3), are the most natural functions to use. Consequently, spherical harmonic analysis (SHA) has become the generally agreed method for representing the earth's main field. In principle, SHA is also suitable for representation of the lithospheric field. In practice, it has not been widely used for this purpose, for at least two reasons. First, prior to the advent of satellites, anomaly data were collected and analyzed on a local or regional basis, not globally. Hence localized analyses rather than global SHA were employed. The second reason SHA has not been widely used for anomaly representation stems from its computational burden. For a model of the earth's main field, coefficients are required only up to about degree 13 (i.e., 195 coefficients). For adequate representation of satellite anomaly data, SHA of at least degree 60 is required (i.e., 3,720 coefficients). If higher-resolution global data become available, the computational burden will increase further. The availability of faster and larger computers and the use of efficient algorithms (Schmitz et al., 1989), including look-up tables and approximations (Arkani-Hamed and Strangway, 1986a), have led to new applications of SHA.

The most obvious approach to representing magnetic anomaly fields by SHA is to expand the derivation of a main field representation to a sufficiently high degree to include the fields of lithospheric origin (Cain et al., 1984, 1989), as described in Section 4.9. Alternatively, one can perform SHA on anomaly data isolated as in Chapter 4. In either case, the infinite series of equation (2.3) is truncated. Ideally, the level of truncation is chosen high enough to include all the parameters that are resolved by the data. If necessary, smoothing may be applied via a minimum-norm solution, as described in Section 5.6.

Equations (1.3) and (2.3) define spherical harmonic potential models. Reverting to the notation of Section 5.2, let $\mathbf{p}$ be the matrix of spherical harmonic coefficients; that is,

$$\{p_i\} = \{g_n^m, h_n^m\}. \tag{5.100}$$

The matrix H becomes H^p, so that

$$\mathbf{c} = H^p \mathbf{p} + \mathbf{e}. \tag{5.101}$$

In the formalism of Section 5.2, $\mathbf{x}$ is replaced by $\mathbf{p}$, $\underline{\mathbf{x}}_0$ by $\underline{\mathbf{p}}_0$ (an a priori estimate of $\mathbf{p}$), X_0 by P_0 (the a priori covariance matrix of $\mathbf{p} - \underline{\mathbf{p}}$), and H by H^p, where the elements of H^p are as summarized in Table 5.8.

5.4.2 NON-POTENTIAL SPHERICAL HARMONIC REPRESENTATION

Solutions for the coefficients $\{g_n^m, h_n^m\}$ of equation (2.3), via equation (1.3), represent all components of the field (i.e., X, Y, and Z). Accordingly, all of these components can be input data (i.e., elements of $\mathbf{c}$). However, it is also possible to represent the individual components, or the scalar anomaly field, by a spherical harmonic series (Arkani-Hamed and Strangway, 1986a). In that case the resulting spherical harmonic series is not a potential function and applies directly only at a single altitude.

Table 5.8. *Matrix elements for derivation of spherical harmonic coefficients H_{ij}^p and H_{ij}^h*

Element	Expression	
H_{ij}^r	$(n+1)\cos(m\phi_i)P_n^m(\cos\theta_i)\left[\frac{a}{r_i}\right]^{n+2}$	(for $p_j = g_n^m$)
	$(n+1)\sin(m\phi_i)P_n^m(\cos\theta_i)\left[\frac{a}{r_i}\right]^{n+2}$	(for $p_j = h_n^m$)
H_{ij}^θ	$-\cos(m\phi_i)\frac{dP_n^m(\cos\theta_i)}{d\theta}\left[\frac{a}{r_i}\right]^{n+2}$	(for $p_j = g_n^m$)
	$-\sin(m\phi_i)\frac{dP_n^m(\cos\theta_i)}{d\theta}\left[\frac{a}{r_i}\right]^{n+2}$	(for $p_j = h_n^m$)
H_{ij}^ϕ	$\frac{m}{\sin\theta_i}\sin(m\phi_i)P_n^m(\cos\theta_i)\left[\frac{a}{r_i}\right]^{n+2}$	(for $p_j = g_n^m$)
	$-\frac{m}{\sin\theta_i}\cos(m\phi_i)P_n^m(\cos\theta_i)\left[\frac{a}{r_i}\right]^{n+2}$	(for $p_j = h_n^m$)
H_{ij}^h	$\cos(m\phi_i)P_n^m(\cos\theta_i)$	(for $p_j = C_n^m$)
	$\sin(m\phi_i)P_n^m(\cos\theta_i)$	(for $p_j = S_n^m$)

Note: The indices i and j correspond, respectively, to the point where the field is computed and the jth parameter, p_j (g_n^m, h_n^m, C_n^m, or S_n^m).

Suppose the quantity measured is $A = A(\theta, \phi)$; that is, A is known at a constant altitude as a function of position. The SHA of A is given by

$$A(\theta, \phi) = \sum_{n=0}^{N}\sum_{m=0}^{n}\left[C_n^m\cos(m\phi) + S_n^m\sin(m\phi)\right]P_n^m(\cos\theta), \qquad (5.102)$$

where the C_n^m and S_n^m are coefficients to be determined, and N is taken high enough to represent the data.

Equation (5.102) is the defining model equation. In the notation of Section 5.2, **h** is the vector of spherical harmonic coefficients; that is,

$$\{h_i\} = \{C_n^m, S_n^m\}. \qquad (5.103)$$

The vector **a** is now restricted to values of A at a constant altitude. As usual, $\mathbf{c} = \mathbf{a} + \mathbf{e}$, and H is replaced by H^h, with elements as summarized in Table 5.8, with the $[a/r_i]$ factors deleted. The solution is

$$\underline{\mathbf{h}} = [(H^h)^T(H^h)]^{-1}(H^h)^T\mathbf{c}. \qquad (5.104)$$

Typically, no a priori is utilized in this method, although it is not excluded. Implementation of the Arkani-Hamed and Strangway (1986a) method is described in the next section.

5.4.3 EFFICIENT COMPUTATION TECHNIQUES WITH SPHERICAL HARMONICS

5.4.3.1 Formulation of Solution by Integration

Computation of SHA of high degree can make use of algorithms that avoid accumulation of the normal matrix $H^T H$ (or $H^T X_0 H$). In general, this is accomplished by use of an approximation based on the orthogonality properties of spherical harmonics. To be specific, let the starting point be the spherical harmonic representation of equation (5.102), with **h**, **a**, and H as defined in the preceding section. If equation (5.102) is multiplied by, for example, $\cos(p\phi)P_k^p(\theta)$ and integrated over the sphere,

$$C_k^p = \frac{2k+1}{4\pi} \int_0^\pi \int_0^{2\pi} A(\theta, \phi) \left[\cos(p\phi)P_k^p(\theta) \right] \sin\theta \, d\theta \, d\phi. \tag{5.105}$$

If $A(\theta, \phi)$ is known everywhere, the parameters can be determined by integration. In practice, the number of points is finite, and the integral must be estimated. The estimates to be discussed require data sets gridded in different ways.

5.4.3.2 Data Gridded Along Circles of Equal Latitude

Suppose the data are gridded along circles of latitude. Then, at each latitude circle, the coefficients $G_m(\theta)$ and $H_m(\theta)$ in

$$A(\theta, \phi) = \sum_{m=0}^N [G_m(\theta)\cos(m\phi) + H_m(\theta)\sin(m\phi)] \tag{5.106}$$

can be determined by a suitable quadrature method. Arkani-Hamed and Strangway (1986a) used a fast fourier transform (FFT) technique. Comparison of (5.106) and (5.102) shows that

$$\begin{bmatrix} G_m(\theta_i) \\ H_m(\theta_i) \end{bmatrix} = \sum_{n=m}^N \begin{bmatrix} C_n^m \\ S_n^m \end{bmatrix} P_n^m(\cos\theta_i). \tag{5.107}$$

If (5.107) is multiplied by $(\sin\theta)P_n^m(\cos\theta)$ and the result is integrated over θ from 0 to π, then using the orthogonality relationships of spherical harmonics gives

$$\begin{bmatrix} C_n^m \\ S_n^m \end{bmatrix} = \frac{(2n+1)}{4} \int_0^\pi d\theta \sin\theta \begin{bmatrix} G_m(\theta) \\ H_m(\theta) \end{bmatrix} P_n^m(\cos\theta). \tag{5.108}$$

To find C_{nm} and S_{nm} from (5.108), a numerical integration scheme is required. Arkani-Hamed and Strangway (1986a) used "data" gridded at 0.5° intervals. Equation (5.108) is then transformed into a finite sum. Numerical integration is facilitated by use of a pre-prepared table of the values for the $P_n^m(\cos\theta_i)$ at each half-degree of latitude.

Schmitz et al. (1989) used a Gauss-Legendre quadrature scheme in which

$$\int_{-1}^1 f(x) \, dx = \sum_{k=1}^N f(x_k)w_k, \tag{5.109}$$

where the weights w_k and node points x_k can be chosen so that the quadrature is exact whenever $f(x)$ is a polynomial of degree $2N - 1$ or less. The x_k are the roots of the Legendre polynomial of degree N. For tables of x_k and w_k, see Abramowitz and Stegun (1964).

5.4.3.3 Data Gridded According to Equal Area

If there are N data points equally spaced over a spherical surface, then each data point represents an angular area equal to $4\pi/N$. Then equation (5.105) can be estimated by a sum in which $[\sin\theta\, d\theta\, d\phi]$ is replaced by $4\pi/N$ to give

$$C_k^p = \frac{2k+1}{N} \sum_{i=1}^{N} A(\theta_i. \phi_i)[\cos(p\phi_i) P_k^p(\theta_i)]. \tag{5.110}$$

Equations (5.110) and (5.104) are equivalent provided that

$$[H^T H]_{ij} = \delta_{ij}\frac{N}{2k+1}, \tag{5.111}$$

the approximation corresponding to a perfect data distribution.

The approximation of equation (5.110), and its equivalent for S_k^p, can be improved by iterating on the results, as follows. Given the jth estimate of $\{C_k^p, S_k^p\}_j$, compute estimates for the $A(\theta_i, \phi_i)$ and the corresponding residuals $\Delta A(\theta_i, \phi_i)$, substitute the residuals into (5.110) in place of $A(\theta_i, \phi_i)$ and solve for the corrections $\{\Delta C_k^p, \Delta S_k^p\}_{j+1}$.

5.4.3.4 Interpolation for Missing Data

To carry out the numerical integrations properly, data are required from pole to pole. Data sets limited in latitude require artificial extension, as, for example, in the work of Arkani-Hamed and Strangway (1986a). Schmitz et al. (1989) discussed estimation errors resulting when the extension is omitted.

5.4.4 EQUIVALENT SOURCE MODELS

Global equivalent source models using dipole spacing of $2-3°$ are computationally formidable. Two approaches have been taken. In the first, local solutions are patched together; in the second, a computational approximation is employed.

Langel (1990a) used the first approach to derive a global solution based on POGO data. The final model comprises 92 local solutions, with dipole spacing of $3°$ at the equator, or 330 km. The number of dipoles in each solution varies from 104 to 144. To minimize edge effects, adjacent regions are heavily overlapped, and only the interior portion of each solution is utilized, as discussed in Section 5.5.3. Advances in computer technology now permit larger parameter spaces, and 400 dipoles per solution is now routine on a work station.

Using the second approach, Purucker, Sabaka, and Langel (1996) showed that the matrix H^m in equation (5.42) is well approximated by a sparse martix (i.e., a matrix with most elements equal to zero). Linear problems with such matrices are efficiently solved by conjugate gradient methods (e.g., Golub and Van Loan, 1989; Press et al., 1992); see also Fletcher (1976), Björck and Elfving (1979), and van der Sluis and van der Vorst (1987). Advantages of this approach over the patching together of local solutions include greater efficiency, elimination of edge effects, and generation of a globally self-consistent solution.

5.4.5 SUMMARY OF GLOBAL MODELS

Either a spherical harmonic potential function or a global equivalent source model will satisfy Laplace's equation and be suitable for downward continuation and other analysis purposes. To date, spherical harmonic analyses combining representations of both main and lithospheric fields have not dealt adequately with the presence of fields from the ionosphere and have not attained the resolution achieved with equivalent source models. Work is under way at GSFC to derive models that can simultaneously represent the main field, magnetospheric field, and quiet-day ionospheric field. A progress report is available (Langel et al., 1996). If this effort is successful, incorporation of higher-degree terms may also lead to optimum representation of the lithospheric field (Purucker et al., 1997). Section 5.7 discusses a method whereby a spherical harmonic potential and a spherical harmonic representation of susceptibility can be derived from processed anomaly values.

The non-potential spherical harmonic representation is useful for interpolation and features ease of computation. It is not capable of continuation. Because of this limitation, it is best applied to data already reduced to common altitude.

5.5 DEALING WITH INSTABILITY: PRINCIPAL-COMPONENT ANALYSIS

5.5.1 CAUSES AND EXAMPLES OF INSTABILITY

All of the model types that have been described are subject to instability, wherein it is possible to have large changes in the model parameters, $\underline{x}$, without corresponding large changes in the computed anomaly field $\mathbf{a}$. Such instability has characteristic effects on magnetization or downward-continuation solutions, although representation of the measured data at satellite altitude remains satisfactory. For example, when the model is a global spherical harmonic analysis, the magnitudes of higher-degree terms often become excessively large, as measured by the R_n of equation (2.6). Then, as the value of r decreases, the value of $(a/r)^{n+2}$ increases substantially at high n. The result is that the predicted downward-continued anomaly fields exhibit large-amplitude short-wavelength variations. This is a common problem in downward-continuation algorithms,

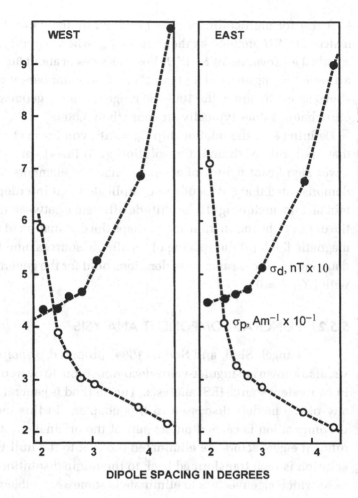

FIGURE 5.7
Trade-off between the standard deviation of the fit between the equivalent source magnetic anomaly field and the observed field, σ_d, versus the "stability" of inversion as indicated by standard deviation of magnetization solution parameters, σ_p. Optimal dipole spacing is taken to be about 2.7°. (Reprinted from M. A. Mayhew, B. D. Johnson, and R. A. Langel, An equivalent source model of the satellite-altitude magnetic anomaly field over Australia. *Earth and Planetary Science Letters*, 51, 189–98, 1980, with kind permission of Elsevier Science NL, Sara Burgerhartstraat 25, 1055 KV Amsterdam, The Netherlands.)

and all of the models discussed in this chapter can be used for downward continuation.

Mayhew, Johnson, and Langel (1980) examined the stability of the equivalent source dipole solution as a function of dipole spacing. Figure 5.7 is a plot of the standard deviation of the dipole parameters, σ_p, as a function of dipole spacing, for analysis of a portion of the POGO data over North America. Also shown is the standard deviation of the fit of the model to the data, σ_d. The parameter σ_p increases slowly as the dipole spacing is reduced until at about 2.7° its increase becomes very rapid as the dipole spacing is further reduced. Model goodness of fit, σ_d, shows the opposite behavior. At this same spacing, plots of the dipole moments begin to exhibit an oscillatory instability in which adjacent sources take on large alternating positive and negative values. Contours of the magnetic moments exhibit an alternating bull's-eye pattern with no physical meaning. It is concluded that 2.7° is the optimal dipole spacing for the data over the conterminous United States.

A test for multicollinearity is provided by the condition number C_n of the matrix $H^T W H$, defined as the ratio λ_1/λ_m, where λ_1 and λ_m are the largest and smallest eigenvalues of $H^T W H$. For area sizes of about 30° to 45° on a side, with a dipole spacing of about 3°, typical condition numbers at geomagnetic latitudes above about 30° are in the 100–600 range. Near the geomagnetic equator, on the other hand, values typically are more than 6,000.

Defining ξ by the relationship $C_n = 10^\xi$, von Frese et al. (1988) pointed out that, as a rule of thumb, the solution $\underline{\mathbf{m}}$ to (5.44) can be expected to have ξ fewer significant figures of accuracy than the elements of $H^T W H$. They also demonstrated that ξ depends in complicated and interdependent ways on certain factors, including (1) the altitude difference between the dipole sources and the data, (2) the inclination and perhaps the declination of the inducing ambient magnetic field, (3) the spacing of the dipole sources, and (4) the spacing of the data points. These same considerations hold for the general solution $\underline{\mathbf{x}}$ to (5.62), with $(X_0)^{-1} = 0$.

5.5.2   PRINCIPAL-COMPONENT ANALYSIS

Langel, Slud, and Smith (1984) proposed *principal-component analysis*, also known as singular-value decomposition, to overcome stability problems in equivalent source (ES) analysis. The method is general and can be used with any of the models discussed in this chapter. In this method, an eigenvalue decomposition is carried out as part of the original solution, and the less significant eigenvectors are eliminated (i.e., set to the null vector). The modified solution is then transformed back to the original solution space. The decision as to which eigenvectors to eliminate is somewhat subjective.

One begins with the singular-value decomposition (Stewart, 1973; Lawson and Hanson, 1974) of the matrix H of Section 5.2:

$$H = U \Lambda V^T \quad \text{or} \quad HV = U\Lambda, \tag{5.112}$$

where U is an $n \times n$ matrix whose columns are eigenvectors of HH^T, V is an $m \times m$ matrix whose columns are eigenvectors of $H^T H$, and Λ is an $n \times m$ matrix with nonnegative entries $\lambda_1 \geq \cdots \geq \lambda_m \geq 0$ along the main diagonal and zeros elsewhere. The values $(\lambda_i)^2$ are eigenvalues of both $H^T H$ and HH^T. For simplicity, the weight matrix is supposed to be $1/\sigma^2$ times the identity matrix, where σ^2 is the data variance.

The orthogonal matrices U and V are used to transform the observations and parameters as follows. Let

$$\mathbf{z} = U^T \mathbf{c}, \qquad \mathbf{q} = V^T \mathbf{x}, \qquad \mathbf{t} = U^T \mathbf{e}. \tag{5.113}$$

Then equation (5.52) transforms to

$$\mathbf{z} = \Lambda \mathbf{q} + \mathbf{t} \tag{5.114}$$

and

$$H^T H \mathbf{x} = V \Lambda^T \Lambda \mathbf{q}, \qquad H^T \mathbf{c} = V \Lambda^T \mathbf{z}, \tag{5.115}$$

so that the estimate of $\mathbf{q}$ is

$$\underline{q}_j = (\Lambda_j)^{-1} z_j, \tag{5.116}$$

with

$$\text{Var } \underline{q}_j = (\lambda_j)^{-2} \sigma^2 \quad \text{and} \quad \text{Cov}(q_j, q_k) = 0. \tag{5.117}$$

From (5.117) it is easily seen that when λ_j is small, $\underline{q}_j$ has a high variance.

In the original coordinate system, one can write

$$\left| \sum_{j=1}^{m} h_{ij} v_{jk} \right| = |u_{ik} \lambda_k| \le \lambda_k, \tag{5.118}$$

because U is orthogonal. Therefore, small λ values correspond to eigenvectors representing sources of multicollinearity. If λ_k is very small, q_k is estimated imprecisely. But through the transformation $\mathbf{x} = V^T \mathbf{q}$, that imprecision is propagated throughout $\mathbf{x}$.

Now the rationale for the principal-component method can be seen. If λ_k is small, the corresponding $\underline{q}_k$ is noisy (i.e., an imprecise estimate whose variability makes the estimate $\mathbf{x}$ unreliable). Furthermore, the corresponding quantity $\sum_j h_{ij} v_{jk}$ is uniformly small and can safely be replaced by zero. This process is equivalent to replacing $\underline{q}_k$ by zero. An algorithm is

1. Compute $H^T H$ and solve for $\mathbf{x} = [H^T H]^{-1} H^T \mathbf{c}$.
2. Compute the eigenvalues $(\lambda_1)^2, \dots, (\lambda_m)^2$ and V, the corresponding matrix of eigenvectors.
3. Transform to obtain $\mathbf{q} = V \mathbf{x}$.
4. Define $\mathbf{q}^*$ by

$$\underline{q}_j^* = \underline{q}_j \quad \text{if } j \le k^*$$
$$= 0 \quad \text{if } j > k^*,$$

   where

$$\frac{\sum_{j=1}^{k^*} (\lambda_j)^2}{\sum_{j=1}^{k} (\lambda_j)^2} = P$$

   where P and k^* are to be determined.
5. Compute $\mathbf{x}^* = V^T \mathbf{q}^*$.

The large matrix U is not computed, so the computation is manageable.

The determination of k^*, the number of eigenvectors retained, and P, the percentage of the trace of Λ, in step 4 is the crucial decision in applying the method, and it is somewhat subjective. A trade-off occurs between goodness of fit and solution "smoothness." This trade-off is seen, for example, in Figure 5.7, where σ_p measures smoothness of fit, and how well the model fits the data is measured by σ_d. A certain solution smoothness is required for the solution to be considered "well-behaved" or stable. This occurs at the expense of a worse fit of the model to the data.

5.5.3   EXAMPLE FROM EQUIVALENT SOURCE MODELING

To illustrate application of the principal-component analysis, consider the ES problem and, in particular, the global set of ES solutions of Langel (1990a) to the POGO data. Figure 5.8 shows an example of the variations of some of the crucial parameters as functions of the number of eigenvectors retained. C_n is the condition number of the normal matrix, σ_p is the standard deviation of the resulting dipole moments, and P and k^* are as defined in step 4. From the plot it is apparent that a large number of eigenparameters can be discarded before any significant change in P occurs. The full set of parameters k numbers 126; for $k^* = 94$, P is reduced only to 99%, whereas large improvements are evident in condition number and σ_p. Typical values for σ_p at higher latitudes, where the instability is negligible, are between 30 and 40. It might be expected, then, that k^* should be chosen so that σ_p would be in that range. In Figure 5.8 this gives k^* from 60 to 106, P from 93.5% to 99.7%, and C_n from 20 to 189.5. Unfortunately, none of the curves in the figure shows a sharp break or any other indication, in this k^* range, that can be used objectively to choose k^*.

A procedure for choosing k^* can be based on comparison of ES solutions from adjacent regions. Maps for larger regions are often derived by piecing together solutions from adjacent areas. Because the solution areas are finite and have edges, each solution has a certain amount of "edge effect"; the magnetization values and reduced maps are more accurate or realistic in the interior of the solution region. One practice is to overlap adjacent regions by about 50%. Final selection of k^* is then made by trying to minimize the disjointedness of adjacent solutions, while retaining as much solution detail as possible. The measure of disjointedness is taken to be the rms difference of magnetization in the region of overlap. Figure 5.9 shows a typical contour plot of this difference as a function of k^* for two overlapping solutions in the equatorial African region. For high-latitude regions, acceptable discontinuities over boundaries are obtained for rms magnetization differences lower than about 25 A·m^{-1}. Langel et al. (1984) suggested the choice of a value of about 20 A·m^{-1} for "safety," which gives $k^* < 87$ for the 0–40° longitude region, and $k^* < 90$ for the 20–60° longitude region. Each region has multiple sets of constraints, one from each adjacent region. For the example of Figures 5.8 and 5.9, the values chosen were $k^* = 74$ for the 0–40° longitude region and $k^* = 86$ for the 20–60° region. The result is a somewhat smoothed (at the lower latitudes) but stable solution.

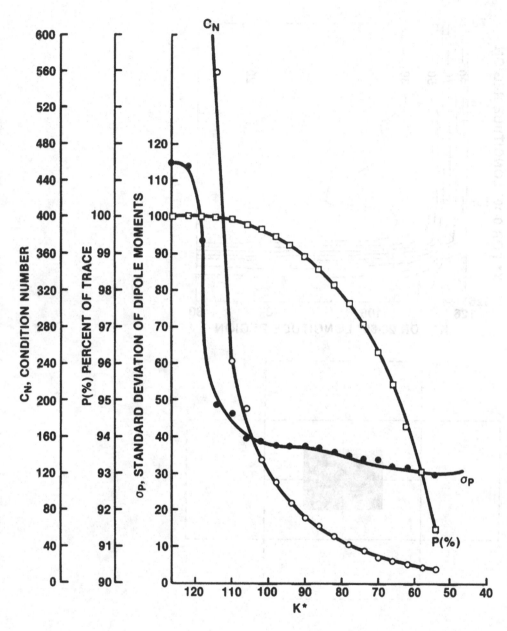

FIGURE 5.8
Variations of solution parameters (C_n, condition number; P, percentage of trace; σ_P, standard deviation of dipole moments) as functions of k^*, the number of retained eigenparameters. (From Langel et al., 1984; reprinted with kind permission of Springer-Verlag; © 1984.)

Figure 5.10 illustrates how the individual solutions are arranged and combined in the scheme of Langel (1990a). Suppose the ES solution under consideration covers the rectangular area outlined by the solid lines. Four adjacent solutions are outlined by dashed lines. Overlap in this case is taken to be 50%, and there are four regions of overlap. A plot such as Figure 5.9 is produced for each of these overlapping regions. The final selection of k^* for the primary region must attempt to minimize the rms difference of magnetization in each of the four regions, while not resulting in an overly smooth solution.

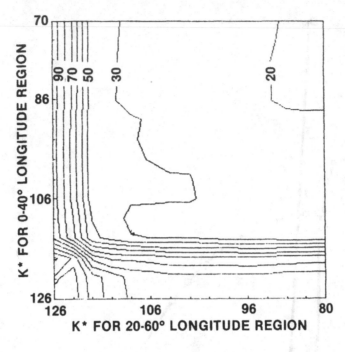

FIGURE 5.9
The rms difference in magnetization (A · m⁻¹) in a typical overlap region. (From Langel et al., 1984; reprinted with kind permission of Springer-Verlag; © 1984.)

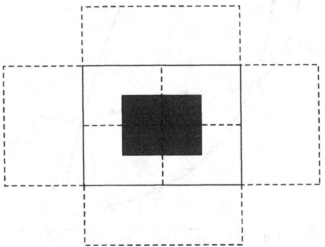

FIGURE 5.10
Schematic diagram of how equivalent source solutions are overlapped and combined. The solid rectangle is the equivalent source region under consideration. The dashed rectangles are adjacent equivalent source regions. Overlap is 50% for this figure. The shaded region is the "interior" region for this solution; the remainder of the solid rectangle is the "exterior" region.

5.6 DEALING WITH INSTABILITY: MINIMUM-NORM SOLUTIONS

5.6.1 MINIMUM-NORM CONCEPTS

Recalling that instability occurs when large changes in the model parameters $\underline{x}$ are possible without corresponding changes in the computed anomaly field leads to the notion of imposing a constraint on $\underline{x}$. For example, one can seek a solution in which $\underline{x}^T\underline{x}$ is minimized while still fitting the data, a method known as "ridge regression." It is convenient to develop this notion of constraint in the context of vector and Hilbert spaces. The vector of solution parameters, $\underline{x}$, is a member of a finite-dimensional vector space, say $\mathbf{X}$. By definition, an inner product is defined on $\mathbf{X}$ such that, in the terminology of Appendix 5.3, if $\mathbf{u}$ and $\mathbf{v}$ are in $\mathbf{X}$, then their inner product is called $\langle \mathbf{u}, \mathbf{v} \rangle$, and their norms $\langle \mathbf{u}, \mathbf{u} \rangle$ and $\langle \mathbf{v}, \mathbf{v} \rangle$. The function

$$\mathcal{G}(\mathbf{x}_a, \mathbf{x}_b) = (\mathbf{x}_a)^T F \mathbf{x}_b, \tag{5.119}$$

where F is a positive-definite diagonal matrix, is an inner product, and the corresponding function $\vartheta(\underline{x}) = \mathcal{G}(\underline{x}, \underline{x})$ is a norm. If $F = I$, the identity matrix, $\vartheta(\underline{x}) = \underline{x}^T \underline{x}$.

In some contexts it is useful to define inner products and norms in terms of integrals, as follows: If $\mathcal{F}_1(\mathbf{r})$ and $\mathcal{F}_2(\mathbf{r})$ are continuous integrable functions, then

$$\langle \mathcal{F}_1(\mathbf{r}), \mathcal{F}_2(\mathbf{r}) \rangle = \int_V \mathcal{F}_1(\mathbf{r}') \mathcal{F}_2(\mathbf{r}') \, d^3 r' \tag{5.120}$$

constitutes an inner product, and $\langle \mathcal{F}_1(\mathbf{r}), \mathcal{F}_1(\mathbf{r}) \rangle$ is a norm for $\mathcal{F}_1(\mathbf{r})$. These definitions are expanded to include the case where $\mathcal{F}_1(\mathbf{r}) \to \mathbf{F}_1(\mathbf{r})$, a vector function, by taking the usual vector dot product under the integral in (5.120). In the present context, $\mathcal{F}_1(\mathbf{r})$ is chosen to be a suitable function of the solution parameters; for example, $\mathcal{F}_1(\mathbf{r}) = A_r(\mathbf{r})$, the radial component of the anomaly field.

Minimization of $\vartheta(\underline{x}) = \mathcal{G}(\underline{x}, \underline{x})$, or of the norm of $\mathcal{F}_1(\mathbf{r})$, imposes a smoothness constraint on the solution. As in principal-component analysis, there is a trade-off between goodness of fit and smoothness of solution.

A convenient way of introducing constraints is to use the notion of Lagrange multipliers from the calculus of variations. Recall that if it is desired to find the minimum of $\mathfrak{f}(\mathbf{x})$ with respect to the parameters $\mathbf{x}$, the conditions are that $\partial \mathfrak{f}/\partial x_i = 0$ for all i. But if we wish to find the minimum of $\mathfrak{f}(\mathbf{x})$ subject to the condition $\vartheta(\mathbf{x}) = \vartheta_0$, where ϑ_0 is some constant, then the function to be minimized is

$$\mathfrak{k}(\mathbf{x}) = \mathfrak{f}(\mathbf{x}) + \lambda^d \vartheta(\mathbf{x}), \tag{5.121}$$

where λ^d is a constant, called a Lagrange multiplier. Minimization then requires that

$$\frac{\partial \mathfrak{f}}{\partial x_i} = \lambda^d \frac{\partial \vartheta}{\partial x_i}. \tag{5.122}$$

The solution depends upon λ^d, which is adjusted so that $\vartheta(\mathbf{x}) = \vartheta_0$.

Turning to the geomagnetic problem, suppose it is desired to minimize

$$\mathfrak{f}(\mathbf{x}) = (\mathbf{a} - H\underline{x})^T W (\mathbf{a} - H\underline{x}) \tag{5.123}$$

subject to the condition that

$$\vartheta(\underline{x}) = \underline{x}^T F \underline{x} \leq \vartheta_0. \tag{5.124}$$

Applying the method of Lagrange multipliers requires that the quantity

$$\kappa(\mathbf{x}) = \mathfrak{f}(\mathbf{x}) + \lambda^d \vartheta(\mathbf{x}) \tag{5.125}$$

be minimized. Expanding and minimizing with respect to $\underline{x}$ gives

$$\underline{x} = [H^T W H + \lambda^d F]^{-1} H^T W\mathbf{c}. \tag{5.126}$$

Suppose an inner product is defined by the form

$$\mathcal{G}(\mathbf{x}_a, \mathbf{x}_b) = \mathcal{C} \sum_{k=1}^{k^*} f_k x_k^a x_k^b, \tag{5.127}$$

where $\{x_k^a\}$ and $\{x_k^b\}$ are the components of $\mathbf{x}_a$ and $\mathbf{x}_b$, $\mathcal{C}$ is an arbitrary constant, and the factors of f_k are to be chosen to provide a suitable meaning to the inner product or norm. Comparison of (5.127) and (5.124) shows that

$$F_{ik} = \delta_{i,k} \mathcal{C} f_k \tag{5.128}$$

where $\delta_{i,k}$ is the Kronecker delta.

The norm of $\mathbf{x}$ is a measure of the smoothness, complexity, or regularity of $\mathbf{x}$, and the function $\mathfrak{f}(\underline{x}, \lambda^d)$ is a measure of the goodness of fit. Smoother solutions have less detail or complexity, and an underlying assumption of the minimum-norm approach is that one is seeking the simplest possible solution compatible with the information in the data. If it is known that the data errors correspond to $\mathfrak{f}(\mathbf{x}) = \mathfrak{f}_0$, then, in principle, λ^d should be chosen so that $\mathfrak{f}(\mathbf{x}) = \mathfrak{f}_0$. In practice, one computes solutions $\underline{x}(\lambda^d)$ for a range of values of λ^d and then computes $\vartheta(\lambda^d)$ and $\mathfrak{f}(\mathbf{x}, \lambda^d)$ and studies the trade-off between $\vartheta(\lambda^d)$, which will decrease as λ^d increases, and $\mathfrak{f}(\mathbf{x}, \lambda^d)$, which will increase as λ^d increases. A suitable choice for λ^d is one that gives the most acceptable combination of $\mathfrak{f}$ and ϑ.

The formalism of this section is directly applicable to all of the models described in preceding sections and summarized in Table 5.4. Application simply requires a choice of an appropriate inner product, or $\mathcal{C}$ and f_k.

5.6.2 RIDGE REGRESSION

If $f_k = 1/\mathcal{C}$ [i.e., if the matrix F in equation (5.126) is taken to be I, the identity matrix], then the procedure will look for the shortest solution vector $\underline{x}$ in the usual Euclidean sense (i.e., will minimize $\mathbf{x}^T\mathbf{x}$). This corresponds to *ridge regression*, which has been discussed in more detail by Hoerl and Kennard

(1970a,b) and Marquardt (1970); also see Lawson and Hanson (1974). Hoerl and Kennard (1970a), Leite (1983), Bapat, Singh, and Rajaram (1987), and von Frese et al. (1988) have given criteria for choosing the value for λ^d when ridge regression is used to stabilize equation (5.44), the equivalent source model. For many stabilization applications, this is the norm of choice (Lotter, 1987; Bapat et al., 1987; von Frese et al., 1988).

5.6.3 GEOMAGNETIC NORMS

5.6.3.1 Magnetic Field Norms

Appropriate selection of $\mathcal{C}$ and f_k in (5.128) leads to a norm with particular physical meaning and in some cases results in a norm with an equivalent integral form. For example, suppose $\mathbf{x} = \mathbf{p}$, the vector of spherical harmonic coefficients. Then let

$$\psi(\mathbf{r}) = a \sum_{n=1}^{n^*} \sum_{m=0}^{n} \left(\frac{a}{r}\right)^{n+1} \left[g_n^m \cos(m\phi) + g_{-n}^m \sin(m\phi)\right] P_n^m(\cos\theta), \qquad (5.129)$$

where g_{-n}^m equals the usual h_n^m; note that $h_n^0 = 0$. If ψ_a and ψ_b have spherical harmonic coefficients $\{a_n^m\}$ and $\{b_n^m\}$, respectively, then consider the inner products of the form

$$\langle \psi_a, \psi_b \rangle = a^2 \sum_{n=1}^{\infty} \sum_{m=-n}^{n} f_n a_n^m (b_n^m), \qquad (5.130)$$

where f_n is a function of degree, n, only.

If the factor f_n in (5.130) is appropriately chosen, the resulting norms will describe a property of the magnetic field. For example, consider the mean of the square of B_r over the sphere $r = a$:

$$\frac{1}{4\pi a^2} \int_{r=a} (B_r)^2 \, dS,$$

with $dS = a^2 \sin\theta' \, d\theta' \, d\phi' = a^2 \, d\Omega'$, and $B_r = -\partial\psi/\partial r$. Using (5.129) and taking into account the orthogonality relationships of the circular and associated Legendre functions, it is easy to show that

$$\frac{1}{4\pi a^2} \int_{r=a} [B_r(\mathbf{r}')]^2 \, dS = \sum_{n=1}^{\infty} \sum_{m=-n}^{n} \frac{(n+1)^2}{2n+1} (g_n^m)^2. \qquad (5.131)$$

Comparing (5.131) with (5.127), for the choice $\mathcal{C} = a^2$ and $f_k = f_n = (n+1)^2/[a^2(2n+1)]$, the norm, $\langle \mathbf{x}, \mathbf{x} \rangle = \langle \mathbf{p}, \mathbf{p} \rangle = \langle \psi, \psi \rangle$, is equal to the (truncated) mean value of $(B_r)^2$ over the surface of the earth. Table 5.9 gives a list of some other possible expressions for f_n used with spherical harmonic models and their physical meanings. In practice, only the first and third norms of Table 5.9 have been used in analyses of the field from the lithosphere.

Table 5.9. *Some functions f_n used with spherical harmonic potential models and the physical meanings of the corresponding norms*

f_n	Meaning of norm
1	Euclidean length of truncated model vector $\mathbf{p}^T\mathbf{p}$.
$1/(2n+1)$	Mean value of ψ^2 over earth's surface (ψ norm)
$\dfrac{(n+1)^2}{(2n+1)a^2}$	Mean value of $(B_r)^2$ over earth's surface (B_r norm)
$(n+1)a^2$	Mean value of field magnitude B over earth's surface
$\dfrac{(n+1)}{(2n+1)a^2}$	Total energy in field outside earth's surface
$\dfrac{n(n+1)^3}{(2n+1)a^2}$	Mean value of $\partial B_r/\partial r$ over earth's surface

The results for finite spherical harmonic series generalize immediately to the case where n^* in equation (5.129) is infinity. Then take $\mathcal{H}$ to be the Hilbert space (see Appendix 5.2) of all real-valued, bounded harmonic functions of the form of equation (5.129), with $n^* = \infty$. All of the foregoing results carry over to the infinite-dimensional space.

For spherical harmonic models, as defined by equation (5.102), if $\mathcal{C} = a^2$ and $f_k = f_n = 1/(2n+1)$, then the norm is

$$\langle \underline{\mathbf{h}}, \underline{\mathbf{h}} \rangle = \frac{1}{4\pi a^2} \int_{S'} [A(\theta', \phi')]^2 \, dS'. \tag{5.132}$$

When the approximate calculation of equation (5.110) is used, it is still possible to use minimum-norm stabilization. Considering equations (5.111), (5.126), and (5.128), it is clear that

$$[H^T H + \lambda^d F]_{ij} = \delta_{i,j}\left(\frac{N}{2k+1} + \lambda^d \mathcal{C} f_k\right). \tag{5.133}$$

With the norm chosen such that $\mathcal{C} = 1$ and $f_k = 1/(2k+1)$, the damped form of equation (5.110) is

$$C_k^p = (2k+1)\left(\frac{1}{N} + \frac{1}{\lambda^d}\right)\sum_{i=1}^{N} A(\theta_i, \phi_i)\left[\cos(p\phi_i)P_k^p(\theta_i)\right]. \tag{5.134}$$

In the case of spherical cap harmonic analysis, integral norms do not correspond to simple series, because the two sets of basis functions are not mutually orthogonal. This does not mean that series forms like (5.130) cannot be used in a meaningful way – just that they cannot be associated with an integral norm.

5.6.3.2 A Magnetization Norm

Recall the magnetic potential of the lithosphere in integral form, equation (5.22):

$$\Psi_a(\mathbf{r}) = \frac{\mu_0}{4\pi} \int_V \mathbf{M}(\mathbf{r}') \cdot \nabla' \frac{1}{R} d^3 r',$$

with V the volume of the magnetized region (layer). Following the definitions of equation (5.120), it is convenient (Parker, Shure, and Hildebrand, 1987; Jackson, 1990) to define a norm of $\mathbf{M}(\mathbf{r})$ in the form

$$\langle \mathbf{M}(\mathbf{r}), \mathbf{M}(\mathbf{r}) \rangle = \int_V \mathbf{M}(\mathbf{r}') \cdot \mathbf{M}(\mathbf{r}') \, d^3 r'. \qquad (5.135)$$

In this notation, the kth component of the anomaly field is given by

$$A_k(\mathbf{r}) = \int_V \mathbf{G}_k(\mathbf{r}, \mathbf{r}') \cdot \mathbf{M}(\mathbf{r}') \, d^3 r', \qquad (5.136)$$

where, following Jackson (1990), the kernel function $\mathbf{G}_k(\mathbf{r}, \mathbf{r}')$ is given by

$$\mathbf{G}_k(\mathbf{r}, \mathbf{r}') = -\frac{\mu_0}{4\pi} (\hat{\alpha}_k \cdot \nabla_r) \nabla_r \frac{1}{R}, \qquad (5.137)$$

where $(\hat{\alpha}_k \cdot \nabla_r) = \partial/\partial r, (1/r)(\partial/\partial \theta)$, or $[1/(r \sin \theta)](\partial/\partial \phi)$, and

$$R = |\mathbf{r} - \mathbf{r}'|. \qquad (5.138)$$

As Whaler and Langel (1996) pointed out, minimization of the rms of magnetization is also the basis of equivalent source models.

5.6.4   GENERALIZED INVERSE METHODS
5.6.4.1 Introduction

Equation (5.129), with $n^* = \infty$, and its gradient are the defining equations for models with spherical harmonic coefficients as parameters. The number of parameters is infinite; the problem is underdetermined because of the finite number of data. The traditional method of dealing with the underdetermined problem is to assume a truncation level, n^*, preferably at or near the degree at which noise begins to dominate the solution coefficients. The method described here uses a set of basis functions that are linear combinations of the data, and therefore finite in number. From these functions a Gram matrix is constructed that, when inverted, will give a minimum-norm solution. The basis functions are, in fact, equivalent to infinite-dimensional spherical harmonic expansions. This means that, in principle, the solution could be written in the form of equation (5.129), with infinite number of parameters. This approach gives results formally identical with what Gubbins and Bloxham (1985) called stochastic inversion, provided the norms used are identical.

5.6.4.2 Formalism for Forming the Gram Matrix with Perfect Data

Using the Hilbert space $\mathcal{H}$, defined in Section 5.6.3, let $\mathcal{H}'$ be the space of all linear functionals on $\mathcal{H}$. Calculation of each of the anomaly components A_r, A_θ, and A_ϕ is a linear functional, $\mathcal{H} \to R$, denoted L^r, L^θ, and L^ϕ, respectively. R here is the space of real numbers. The linear functional L^ψ maps ψ into its values in R. Then

$$L^r \psi = -\partial \psi(\mathbf{r})/\partial r = A_r(\mathbf{r}), \tag{5.139a}$$

$$L^\theta \psi = -\frac{1}{r}\frac{\partial \psi(\mathbf{r})}{\partial \theta} = A_\theta(\mathbf{r}), \tag{5.139b}$$

$$L^\phi \psi = -\frac{1}{r \sin \phi}\frac{\partial \psi(\mathbf{r})}{\partial r} = A_\phi(\mathbf{r}), \tag{5.139c}$$

$$L^\psi \psi = \psi(\mathbf{r}). \tag{5.139d}$$

Invoking equation (A5.2.13) from Appendix 5.2, let the elements of $\mathcal{H}$ corresponding to L^r, L^θ, L^ϕ, and L^ψ be l^r, l^θ, l^ϕ, and l^ψ, respectively. Then the functions L^r, L^θ, L^ϕ, and L^ψ can be expressed as inner products over the Hilbert space of functionals as

$$L^r \psi = \langle \psi, l^r \rangle \tag{5.140}$$

for all ψ in $\mathcal{H}$. Similar relationships hold for L^θ, L^ϕ, and L^ψ. The elements l^r, l^θ, l^ϕ, and l^ψ can be displayed explicitly by noting that if l^α, as a function of $\mathbf{r}'$, is in $\mathcal{H}$, where α is r, θ, ϕ, or ψ, then it is of the form

$$l^\alpha(\mathbf{r}, \mathbf{r}') = a \sum_{n=1}^{\infty} \sum_{m=0}^{n} \left(\frac{a}{r'}\right)^{n+1} \left[l_{-n}^{m}(\mathbf{r}) \cos(m\phi') + l_{-n}^{m}(\mathbf{r}) \sin(m\phi')\right] P_n^m(\theta'). \tag{5.141}$$

Note that the coefficients l_n^m are functions of $\mathbf{r}$; when considering inner products, such as (5.140), the l^α are to be considered as functions of $\mathbf{r}'$, so that, for example,

$$\langle \psi, l^r \rangle_{r'} = a^2 \sum_{n=1}^{\infty} \sum_{m=-n}^{n} f_n g_n^m l_n^m = B_r(\mathbf{r})$$

$$= \sum_{n=1}^{\infty} \sum_{m=0}^{n} (n+1)\left(\frac{a}{r}\right)^{n+2} \left[g_n^m \cos(m\phi) + g_{-n}^m \sin(m\phi)\right] P_n^m(\theta). \tag{5.142}$$

Equating the two sides,

$$l_n^m(\mathbf{r}) = \frac{n+1}{a^2 f_n}\left(\frac{a}{r}\right)^{n+2} P_n^m(\theta) \cos(m\phi),$$

$$l_{-n}^m(\mathbf{r}) = \frac{n+1}{a^2 f_n}\left(\frac{a}{r}\right)^{n+2} P_n^m(\theta) \sin(m\phi), \tag{5.143}$$

which gives

$$I^r(\mathbf{r}, \mathbf{r}') = \sum_{n=1}^{\infty} \sum_{m=0}^{n} \frac{n+1}{af_n} \left[\frac{a}{r'}\right]^{n+1} \left[\frac{a}{r}\right]^{n+2}$$

$$\times \, [\cos(m\phi')\cos(m\phi) + \sin(m\phi')\sin(m\phi)] P_n^m(\theta) P_n^m(\theta')$$

$$= \sum_{n=1}^{\infty} \frac{n+1}{rf_n} \left[\frac{aa}{r'r}\right]^{n+1} P_n(\mu), \tag{5.144}$$

where the spherical harmonic addition theorem

$$P_n(\mu) = \sum_{m=0}^{n} [\cos(m\phi)\cos(m\phi') + \sin(m\phi)\sin(m\phi')] P_n^m(\cos\theta) P_n^m(\cos\theta') \tag{5.145}$$

has been used, with μ defined by equation (5.2).

A similar analysis gives

$$I^\psi(\mathbf{r}, \mathbf{r}') = \sum_{n=1}^{\infty} \frac{1}{f_n} \left[\frac{aa}{r'r}\right]^{n+1} P_n(\mu). \tag{5.146}$$

Note that

$$I^r = -\partial I^\psi / \partial r. \tag{5.147}$$

Similarly, it can be shown that

$$I^\theta = -\frac{1}{r} \frac{\partial I^\psi}{\partial \theta}, \qquad I^\phi = -\frac{1}{r\sin\theta} \frac{\partial I^\psi}{\partial \phi}. \tag{5.148}$$

Now suppose ψ_0 is the potential of the actual magnetic field of interest and that a set of N measurements $\{c_i, \, i = 1, \ldots, N\}$ is available. For now, c_i will be assumed noise-free and perfectly accurate. It is desired to derive an estimate $\underline{\psi}_0$ of ψ_0 that will fit the $\{c_i\}$ exactly and will satisfy a minimum-norm condition (i.e., will minimize $\langle \underline{\psi}_0, \underline{\psi}_0 \rangle$). Because ψ_0 is the actual field, and the c_i are noise-free and perfectly accurate,

$$c_i = L_i^\alpha \psi_0, \tag{5.149}$$

where $\alpha = r, \theta,$ or ϕ, according to the component measured, and the subscript i on L_i means that the linear functional is evaluated at $\mathbf{r} = \mathbf{r}_i$, or

$$c_i(\mathbf{r}_i) = L^\alpha \psi_0(\mathbf{r}_i) = L_i^\alpha \psi_0 = \langle \psi_0, l_i(\mathbf{r}_i, \mathbf{r}') \rangle_{r'}. \tag{5.150}$$

For $\mathbf{r}_i \neq \mathbf{r}_j$, l_i and l_j are linearly independent, so it is possible to define $\underline{\psi}_0$ by

$$\underline{\psi}_0(\mathbf{r}') = \sum_{j=1}^{N} t_j l_j(\mathbf{r}_j, \mathbf{r}'), \tag{5.151}$$

Table 5.10. *Expressions for Λ_{ij} matrix elements for the ψ norm*

$$\Lambda_{r_j,r_i} = \Lambda_{r_i,r_j} = \frac{x_{ij}}{r_i r_j}\left(1 + \frac{15(x_{ij})^2 - 1 - 12\mu_{ij}(x_{ij})^3 - 2\mu_{ij}x_{ij}}{(s_{ij})^5} - \frac{5x_{ij}(x_{ij}-\mu_{ij})[5(x_{ij})^2 - 1 - 3\mu_{ij}(x_{ij})^3 - \mu_{ij}x_{ij}]}{(s_{ij})^7}\right)$$

$$\Lambda_{\theta_i,\theta_j} = \Lambda_{\theta_j,\theta_i} = \frac{x}{r_i r_j}\left(\frac{\partial^2 \mu_{ij}}{\partial\theta_i \partial\theta_j} + \frac{\partial\mu_{ij}}{\partial\theta_i}\frac{\partial\mu_{ij}}{\partial\theta_j}\frac{5x_{ij}}{(s_{ij})^2}\right)$$

$$\Lambda_{\phi_i,\phi_j} = \Lambda_{\phi_j,\phi_i} = \frac{x}{r_i r_j \sin\theta_i \sin\theta_j}\left(\frac{\partial^2 \mu_{ij}}{\partial\phi_i \partial\phi_j} + \frac{\partial\mu_{ij}}{\partial\phi_i}\frac{\partial\mu_{ij}}{\partial\phi_j}\frac{5x_{ij}}{(s_{ij})^2}\right)$$

$$\Lambda_{\theta_i,\phi_j} = \frac{x}{r_i r_j \sin\theta_i}\left(\frac{\partial^2 \mu_{ij}}{\partial\theta_i \partial\phi_j} + \frac{\partial\mu_{ij}}{\partial\theta_i}\frac{\partial\mu_{ij}}{\partial\phi_j}\frac{5x_{ij}}{(s_{ij})^2}\right)$$

$\Lambda_{\phi_i,\theta_j}$ can be computed from $\Lambda_{\theta_i,\phi_j}$ by interchanging i and j on $\sin\theta$ and on the θ and ϕ derivatives.

$$\Lambda_{\theta_i,r_j} = \frac{\partial\mu_{ij}}{\partial\theta_j}\frac{x_{ij}}{r_i r_j (s_{ij})^2}\left(\frac{6x_{ij}[1-2(x_{ij})^2]}{(s_{ij})^3} - 5(x_{ij}-\mu_{ij})\mathcal{X}\right)$$

Λ_{r_i,θ_j} can be obtained from Λ_{θ_i,r_j} by interchanging j and i in the θ derivative.

$$\Lambda_{\phi_i,r_j} = \frac{\partial\mu_{ij}}{\partial\phi_j}\frac{x_{ij}}{r_i r_j \sin\theta_i (s_{ij})^2}\left(\frac{6x_{ij}[1-2(x_{ij})^2]}{(s_{ij})^3} - 5(x_{ij}-\mu_{ij})\mathcal{X}\right)$$

Λ_{r_i,ϕ_j} can be obtained by interchanging i and j on $\sin\theta$ and on the ϕ derivative.

$$\mathcal{X} = \frac{3(x_{ij})^2[1-(x_{ij})^2]}{(s_{ij})^5}$$

$$x = \frac{a^2}{r_i r_j}, \qquad (s_{ij})^2 = 1 + (x_{ij})^2 - 2x_{ij}\mu_{ij}$$

For μ_{ij} and its derivatives, see Table 5.1.

where the t_j are constants to be determined. Substituting (5.151) into (5.150) gives

$$c_i(\mathbf{r}_i) = \sum_j t_j \langle l_j(\mathbf{r}_j, \mathbf{r}'), l_i(\mathbf{r}_i, \mathbf{r}')\rangle_{r'} = \sum_j t_j \Lambda_{ji}(\mathbf{r}_j, \mathbf{r}_i), \qquad (5.152a)$$

or, because Λ is symmetric,

$$\mathbf{c} = \Lambda\mathbf{t}, \qquad (5.152b)$$

with solution

$$\mathbf{t} = \Lambda^{-1}\mathbf{c}. \qquad (5.153)$$

Note that the $\langle l_i, l_j\rangle$ are real, so Λ is symmetric. The matrix Λ is called the Gram matrix, and it can be shown that the resulting $\underline{\psi}_0$ is of minimum norm. For particular norms of interest (i.e., particular f_n), the elements $\Lambda_{ij} = \langle l_i, l_j\rangle$ can be expressed in closed form (Shure, 1982; Shure, Parker, and Backus, 1982; Achache, Abtout, and LeMouël, 1987; Whaler, 1994). Such expressions are given in Tables 5.10 and 5.11.

The N linearly independent elements $\{l_i, i = 1, \ldots, N\}$ form a basis for an N-dimensional subspace of $\mathcal{H}$, say $\mathcal{H}_N$. All linear combinations of these l_i are in $\mathcal{H}_N$, which is itself a Hilbert space. The vector $\underline{\psi}_0(\mathbf{r}')$ is the minimum-norm projection of $\psi_0(\mathbf{r}')$ onto $\mathcal{H}_N$.

After obtaining $\underline{\psi}_0$, the result can be used to calculate the corresponding magnetic field at any location (i.e., $\underline{\psi}_0$ is an interpolating or continuation function). The symbol $\mathbf{b}$ will be used for calculated fields at locations other than

Table 5.11. *Expressions for Λ_{ij} matrix elements for the B_r norm*

$$\Lambda_{r_j,r_i} = \Lambda_{r_i,r_j} = \frac{(x_{ij})^2[1-(x_{ij})^2]}{(s_{ij})^3} - (x_{ij})^2$$

$$\Lambda_{\theta_i,r_j} = \frac{\partial\mu_{ij}}{\partial\theta_j}x_{ij}\mathcal{K}(\mathbf{r},\mathbf{r}')$$

Λ_{r_i,θ_j} can be obtained from Λ_{θ_i,r_j} by interchanging j and i in the θ derivative.

$$\Lambda_{\phi_i,r_j} = \frac{1}{\sin\theta_i}\frac{\partial\mu_{ij}}{\partial\phi_j}x_{ij}\mathcal{K}(\mathbf{r},\mathbf{r}')$$

Λ_{r_i,ϕ_j} can be obtained by interchanging i and j on $\sin\theta$ and on the ϕ derivative.

$$\Lambda_{\theta_i,\theta_j} = \frac{1}{r_i r_j}\left[\frac{\partial^2\mu_{ij}}{\partial\theta_i\partial\theta_j}\Lambda'_{\psi_i,\psi_j}(\mu) + \frac{\partial\mu_{ij}}{\partial\theta_i}\frac{\partial\mu_{ij}}{\partial\theta_j}\Lambda''_{\psi_i,\psi_j}(\mu)\right]$$

$$\Lambda_{\phi_i,\phi_j} = \frac{1}{r_i r_j\sin\theta_i\sin\theta_j}\left[\frac{\partial^2\mu_{ij}}{\partial\phi_i\partial\phi_j}\Lambda'_{\psi_i,\psi_j}(\mu) + \frac{\partial\mu_{ij}}{\partial\phi_i}\frac{\partial\mu_{ij}}{\partial\phi_j}\Lambda''_{\psi_i,\psi_j}(\mu)\right]$$

$$\Lambda_{\theta_i,\phi_j} = \frac{1}{r_i r_j\sin\theta_i}\left[\frac{\partial^2\mu_{ij}}{\partial\phi_i\partial\phi_j}\Lambda'_{\psi_i,\psi_j}(\mu) + \frac{\partial\mu_{ij}}{\partial\phi_i}\frac{\partial\mu_{ij}}{\partial\theta_j}\Lambda''_{\psi_i,\psi_j}(\mu)\right]$$

$\Lambda_{\phi_i,\theta_j}$ can be computed from $\Lambda_{\theta_i,\phi_j}$ by interchanging i and j on $\sin\theta$ and on the θ and ϕ derivatives.

$$\Lambda'_{\psi_i,\psi_j}(\mu) = \frac{1}{(\mu_{ij})^2-1}[S_1 + S_2]$$

$$S_1 = \ln\left[\frac{s_{ij}+1-\mu_{ij}x_{ij}}{2}\right] + \frac{2}{s_{ij}} - 2$$

$$S_2 = \mu_{ij}\left\{\ln\left[\frac{s_{ij}+x_{ij}-\mu_{ij}}{1-\mu_{ij}}\right] - \frac{2x_{ij}}{s_{ij}} + x\right\}$$

$$\mathcal{K}(\mathbf{r},\mathbf{r}') = \frac{1}{1-\mu^2}\left[1 - \frac{(3x^2+1-3\mu x-\mu x^3)}{s^3}\right] = \frac{1}{1-\mu} - \frac{\mu}{s(s+x-\mu)} - \frac{3x^2+1-2\mu x}{s^3} = \frac{x^2(1+2s-x^2)}{s^3(1+s-\mu x)}$$

$$x = \frac{a^2}{rr'}, \qquad s^2 = 1 + x^2 - 2x\mu$$

For x_{ij} and s_{ij}, see Table 5.10; for μ_{ij} and its derivatives, see Table 5.1.

measurement locations. For example, to calculate an estimate of $b^\alpha(\mathbf{r})$, that is, component α at position $\mathbf{r}$, take

$$\underline{b}^\alpha(\mathbf{r}) = \langle\underline{\psi}_0, l^\alpha(\mathbf{r})\rangle_{r'} = \sum_j t_j\langle l_j(\mathbf{r}_j,\mathbf{r}'), l^\alpha(\mathbf{r},\mathbf{r}')\rangle_{r'}. \qquad (5.154)$$

Or, if ℓ is the N-dimensional vector with components

$$\ell_j = \langle l_j, l^\alpha\rangle_{r'}, \qquad (5.155)$$

then

$$\underline{b}^\alpha(\mathbf{r}) = \ell^T\mathbf{t} = \ell^T\Lambda^{-1}\mathbf{c}. \qquad (5.156)$$

Suppose it is desired to calculate estimates of the field at P locations, $\{\underline{b}_k(\mathbf{r}_k), k = 1,\ldots,P\}$. If $\underline{\mathbf{b}}$ is the column vector of such estimates, and if $\mathcal{L}$ is the $P \times N$ matrix with rows equal to the corresponding ℓ^T, then

$$\underline{\mathbf{b}} = \mathcal{L}\Lambda^{-1}\mathbf{c}. \qquad (5.157)$$

Computation of the matrix elements Λ_{ij} and $\mathcal{L}_{ij}$ is central to implementing a Hilbert-space solution. In the following discussion, the notation Λ_{ij} will be used to refer to either Λ_{ij} or $\mathcal{L}_{ij}$. Whenever the difference is important, it will be made explicit. For convenience, let

$$x = \frac{a^2}{rr'}, \qquad x_{ij} = \frac{a^2}{r_i r_j} \tag{5.158a}$$

and

$$\bar{\delta} = \frac{a}{r}, \qquad \bar{\delta}_i = \frac{a}{r_i}, \tag{5.158b}$$

so that (5.146) becomes

$$l^\psi(\mathbf{r}, \mathbf{r}') = \sum_{n=1}^{\infty} x^{n+1} \frac{1}{f_n} P_n(\mu), \tag{5.159}$$

and substituting (5.159) into (5.130) and using the addition theorem,

$$\Lambda_{\psi_i, \psi_j} = \sum_{n=1}^{\infty} \frac{1}{f_n} x_{ij}^{n+1} P_n(\mu_{ij}) = l^\psi(\mathbf{r}_i, \mathbf{r}_j). \tag{5.160}$$

For computational purposes, one seeks closed-form expressions for the Λ_{ij}. Finding these can be tedious and difficult and is relegated to Appendix 5.3. In practice, the matrix elements needed are seldom those for ψ. Rather, the elements needed are those corresponding to A_r, A_θ, and A_ϕ and the cross-terms between them. Suppose

$$\frac{\partial}{\partial \alpha_i} = \frac{\partial}{\partial r_i} \quad \text{or} \quad \frac{\partial}{\partial \alpha_i} = \frac{1}{r_i} \frac{\partial}{\partial \theta_i} \quad \text{or} \quad \frac{\partial}{\partial \alpha_i} = \frac{1}{r_i \sin \theta_i} \frac{\partial}{\partial \phi_i}, \tag{5.161}$$

as appropriate. Then we know

$$\Lambda_{\alpha_i, \alpha_j} = \left\langle \frac{\partial}{\partial \alpha_i}[l^\psi(\mathbf{r}_i, \mathbf{r}')], \frac{\partial}{\partial \alpha_j}[l^\psi(\mathbf{r}_j, \mathbf{r}')] \right\rangle \bigg|_{r'}. \tag{5.162a}$$

But because differentiation is a linear operator,

$$\Lambda_{\alpha_i, \alpha_j} = \frac{\partial^2}{\partial \alpha_i \partial \alpha_j} \Lambda_{\psi_i, \psi_j}. \tag{5.162b}$$

Sometimes it is convenient to first calculate the matrix elements corresponding to ψ_i and then differentiate to find those corresponding to the α_i.

5.6.4.3 Imperfect Data

The formalism of Section 5.6.4.2 has two major shortcomings. It assumes perfect data, and the number N of basis functions l_i is the same as the number of data points. In practice, data are never perfect, but rather are subject to random error and, in some cases, systematic error. Also, especially with satellite data, the amount of available data is very large. As formulated, Λ is a data-by-data matrix. This can lead to serious computational problems if all data are to be considered. Both of these shortcomings are easily dealt with following Shure et al. (1982) and Parker and Shure (1982).

The model $\mathbf{t}$ found from (5.153) fits the data exactly. Suppose that each data point has an associated variance σ_i^2, so that the model is required only to approximate the data. A solution $\underline{\psi}_0(\mathbf{r}')$ in $\mathcal{H}$ is sought such that

$$\sum_{i=1}^{N} \left(\frac{c_i - \langle \underline{\psi}_0, l_i \rangle}{\sigma_i} \right)^2 \leq \mathcal{S}^2, \quad (5.163)$$

subject to the constraint that $\underline{\psi}_0(\mathbf{r}')$ has the minimum norm. Substituting equation (5.151) into (5.163) gives

$$(\mathbf{c} - \Lambda \mathbf{t})^T W^T W (\mathbf{c} - \Lambda \mathbf{t}) \leq \mathcal{S}^2, \quad (5.164)$$

where W is the diagonal matrix with elements $1/\sigma_i$ $(i = 1, \ldots, N)$. Here the matrix W is related to V_d by

$$(V_d)^{-1} = W^T W. \quad (5.165)$$

Introducing the Lagrange multiplier λ^d, minimize

$$\mathfrak{K}(\mathbf{t}) = \lambda^d (\mathbf{t}^T \Lambda \mathbf{t}) + [(\mathbf{c} - \Lambda \mathbf{t})^T (V_d)^{-1} (\mathbf{c} - \Lambda \mathbf{t}) - \mathcal{S}^2] \quad (5.166)$$

with respect to $\mathbf{t}$. The solution that minimizes $\mathfrak{K}$ is

$$\mathbf{t} = [\lambda^d \Lambda + \Lambda^T (V_d)^{-1} \Lambda]^{-1} \Lambda^T (V_d)^{-1} \mathbf{c} = [\lambda^d V_d + \Lambda]^{-1} \mathbf{c}. \quad (5.167)$$

Equation (5.157) now becomes

$$\underline{\mathbf{b}} = \mathcal{L} [\lambda^d V_d + \Lambda]^{-1} \mathbf{c}. \quad (5.168)$$

5.6.4.4 Dealing with Large Numbers of Data

In order to deal tractably with very large data sets, replace equation (5.151) with

$$\underline{\psi}_0(\mathbf{r}') = \sum_{j=1}^{P} t_j q_j(\mathbf{r}_j, \mathbf{r}'), \quad (5.169)$$

where the q_j are selected members of $\mathcal{H}$, and $P < N$. It is convenient to choose the q_j to be a subset of the l_j associated with the N actual data points, although this is not required. Define

$$\begin{aligned} Q_{ij} &= \langle q_i, q_j \rangle, & i, j = 1, \ldots, P, \\ K_{ij} &= \langle l_i, q_j \rangle, & i = 1, \ldots, N; \ j = 1, \ldots, P. \end{aligned} \quad (5.170)$$

The functions $\{q_i, i = 1, \ldots, P\}$ form a basis for a P-dimensional subspace. Because $P < N$, this is called a depleted basis. Now equation (5.166) becomes

$$\mathfrak{k}(\mathbf{t}) = \lambda^d(\mathbf{t}^T Q\mathbf{t}) + [(\mathbf{c} - K\mathbf{t})^T(V_d)^{-1}(\mathbf{c} - K\mathbf{t}) - \mathcal{S}^2] \tag{5.171}$$

with solution

$$\mathbf{t} = [\lambda^d Q + K^T V_d K]^{-1} K^T V_d \mathbf{c}. \tag{5.172}$$

Equation (5.157) and (5.168) then become

$$\underline{\mathbf{b}} = \mathcal{L}[\lambda^d Q + K^T(V_d)^{-1}K]^{-1}K^T(V_d)^{-1}\mathbf{c}. \tag{5.173}$$

5.6.4.5 Scalar Data

To this point the generalized inverse formalism is applicable only to the linear field components, A_r, A_θ, and A_ϕ. A straightforward extension is possible for scalar data. To implement this extension, it is necessary to recall that a scalar anomaly is always computed relative to some magnetic field model, as in equation (1.11). For simplicity, assume that all measurements $\{c_i\}$ are scalar. Then the ith measurement can be expressed as

$$c_i(\mathbf{r}_i) = \frac{\mathbf{A}(\mathbf{r}_i) \cdot \mathbf{B}_m(\mathbf{r}_i)}{B_m(\mathbf{r}_i)} = \frac{1}{(B_m)_i}\sum_{\alpha=1}^{3} A_\alpha B_m^\alpha, \tag{5.174}$$

where $\mathbf{B}_m(\mathbf{r}_i)$, with components B_m^α or $(B_m^\alpha)_i$ and magnitude $B_m(\mathbf{r}_i) = (B_m)_i$, is computed from a spherical harmonic model of the main field. As mentioned before, inverse calculations are relatively insensitive to the main-field model used. Using a depleted basis defined at the P positions $\{\mathbf{r}_j, j = 1, \ldots, P\}$, and choosing only those l^α corresponding to the radial field, the approximate solution, following (5.169), is

$$\underline{\psi}_0(\mathbf{r}') = \sum_{j=1}^{P} t_j q_j^\alpha(\mathbf{r}_j, \mathbf{r}'). \tag{5.175}$$

Then, following (5.150),

$$c_i(\mathbf{r}_i) = \frac{1}{(B_m)_i}\sum_{\alpha=1}^{3}(B_m^\alpha)_i\langle\psi_0, l_i^\alpha(\mathbf{r}_i, \mathbf{r}')\rangle, \tag{5.176}$$

$$c_i(\mathbf{r}_i) = \frac{1}{(B_m)_i}\sum_j t_j\sum_{\alpha=1}^{3}(B_m^\alpha)_i\langle q_j^r(\mathbf{r}_j, \mathbf{r}'), l_i^\alpha(\mathbf{r}_i, \mathbf{r}')\rangle_{\mathbf{r}'}, \tag{5.177}$$

$$c_i(\mathbf{r}_i) = \sum_j K_{ij}t_j, \tag{5.178}$$

where, similarly to (5.170),

$$K_{ij} = \frac{1}{(B_m)_i} \sum_{\alpha=1}^{3} \left(B_m^\alpha\right)_i \left\langle l_j^r(\mathbf{r}_j, \mathbf{r}'), l_i^\alpha(\mathbf{r}_i, \mathbf{r}')\right\rangle_{\mathbf{r}'}$$

$$= \frac{1}{(B_m)_i} \sum_{\alpha=1}^{3} \left(B_m^\alpha\right)_i (\Lambda_{ij})^{\alpha,r}, \qquad i = 1, \ldots, N; \; j = 1, \ldots, P. \qquad (5.179)$$

If the matrix Q is defined as

$$Q_{ij} = \langle q_i^r, q_j^r \rangle, \qquad (5.180)$$

then the solution equations are (5.171)–(5.173).

Other choices for a basis, full or depleted, are possible, with similar formalisms.

5.6.4.6 Applications to Satellite Anomaly Data

Once a norm is chosen, the elements of the Λ matrix generally can be found in closed form (see Appendix 5.3 and Tables 5.10 and 5.11).

Achache et al. (1987) and Hamoudi, Achache, and Cohen (1995) downward-continued *Magsat* radial magnetic anomaly data, A_r, to the earth's surface using generalized inverse methods. To discuss their formalism, note first that equation (5.139d) applies to every bounded harmonic function ψ. Obviously, the usual geomagnetic potential function is included. But other functions are also harmonic. An important instance is that $r B_r$ is harmonic. Therefore,

$$r B_r(\mathbf{r}) = L^\psi(r B_r) = \langle r' B_r(\mathbf{r}'), l^\psi(\mathbf{r}, \mathbf{r}')\rangle_{\mathbf{r}'}. \qquad (5.181)$$

Note particularly that the operator in (5.181) is L^ψ, not L^r. If, then, there are N measurements of B_r, the c_j, at positions $\mathbf{r}_j$ $(j = 1, \ldots, N)$, following (5.151),

$$r' \underline{B}_0(\mathbf{r}') = \sum_{j=1}^{N} t_j l_j^\psi(\mathbf{r}_j, \mathbf{r}'), \qquad (5.182)$$

where the t_j are found from (5.153), and where predicted values of $r\underline{B}_r(\mathbf{r})$ are computed using (5.154), (5.156), or (5.157).

Achache et al. (1987) required their estimator to fit the data exactly (i.e., they used the formalism of Section 5.6.4.2, not that of Section 5.6.4.3). As a result, their solutions exhibited instability, which was overcome by use of principal-component analysis.

Whaler (1994) applied the formalism of Sections 5.6.4.3 and 5.6.4.4 to vector data from *Magsat* in order to compute a downward-continued scalar map for Africa. Her input data comprised the corrected $2° \times 2°$ averaged, three-component data of Cohen (1989) and Cohen and Achache (1990) covering latitudes between $\pm 60°$. Basis locations were chosen to be global, rather than local,

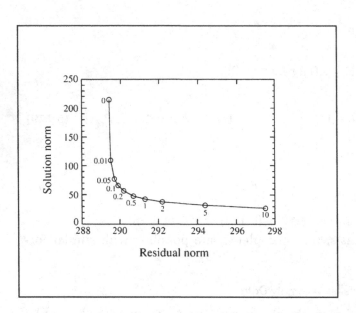

FIGURE 5.11
Trade-off curve for solution norm versus residual norm. Numbers beside points on the curve are values for the damping parameter. The solution norm is in nanoteslas; the residual norm is dimensionless, because the data have been normalized by assumed standard errors to unit variance. (From Whaler, 1994; reprinted with kind permission of Blackwell Science.)

because the norm minimized was global, but the basis locations were highly concentrated over the primary region of interest and highly scattered over most of the rest of the globe. An area of intermediate-density basis locations placed between the high- and low-density regions was found necessary to eliminate edge effects.

The resulting best-fitting model exhibited unrealistic "bull's-eyes" when contoured. Those disappeared when the misfit was relaxed to correspond to 1.6 nT in the X and Z components and 6.4 nT in the Y component, which were reasonable assumptions regarding data accuracy. Figure 5.11 shows the trade-off curve between the goodness of fit, or residual norm, and the smoothness, or solution norm. Eliminating the roughness associated with a higher solution norm does not significantly degrade the goodness of fit. The solution chosen was for a damping parameter of 2, near the knee of the curve.

Langel and Whaler (1996) applied the formalism of Section 5.6.4.4 to scalar data from *Magsat*. A downward-continued map derived for Africa compares well with that derived by Whaler (1994).

5.6.5 MINIMUM-NORM MAGNETIZATION

5.6.5.1 Basic Formalism

Whaler and Langel (1996) have solved for smoothly varying lithospheric magnetization $\mathbf{M}(\mathbf{r})$ in a constant-thickness lithosphere, as follows: If $\mathbf{M}_0(\mathbf{r})$ is the true magnetization of the lithosphere, then, following the notation of (5.135), from (5.136),

$$A_k(\mathbf{r}) = \langle \mathbf{G}_k(\mathbf{r}, \mathbf{r}'), \mathbf{M}_0(\mathbf{r}') \rangle. \qquad (5.183)$$

An estimate, $\underline{M}_0(\mathbf{r})$, of $\mathbf{M}_0(\mathbf{r})$ is sought of the form

$$\underline{M}_0(\mathbf{r}) = \sum_{j=1}^{P} t_j \mathbf{G}_j(\mathbf{r}_j, \mathbf{r}), \qquad (5.184)$$

where the $\mathbf{r}_j$ are basis-function locations. Substituting (5.184) into (5.183) gives

$$A_k(\mathbf{r}) = \sum_{j=1}^{P} t_j \langle \mathbf{G}_k(\mathbf{r}, \mathbf{r}'), \mathbf{G}_j(\mathbf{r}_j, \mathbf{r}') \rangle. \qquad (5.185)$$

The solution formalism is the same as that of Section 5.6.4.4. If $\mathbf{c}$ is the vector of measurements

$$\mathbf{c} = K\mathbf{t}, \qquad (5.186)$$

where

$$K_{ij} = \langle \mathbf{G}_i(\mathbf{r}_i, \mathbf{r}'), \mathbf{G}_j(\mathbf{r}_j, \mathbf{r}') \rangle \qquad (5.187)$$

is the $N \times P$ matrix formed by taking the $\mathbf{r}_i$ ($i = 1, \ldots, N$) to be the data-point locations, and $\mathbf{r}_j$ ($j = 1, \ldots, P$) the basis-point locations. If Q is the Gram matrix with elements

$$Q_{ij} = \langle \mathbf{G}_i(\mathbf{r}_i, \mathbf{r}'), \mathbf{G}_j(\mathbf{r}_j, \mathbf{r}') \rangle, \qquad i, j = 1, \ldots, P, \qquad (5.188)$$

then the estimate $\underline{\mathbf{t}}$ of $\mathbf{t}$ is found, as in (5.172), from

$$\mathbf{t} = [\lambda^d Q + K^T V_d K]^{-1} K^T V_d \mathbf{c}. \qquad (5.189)$$

No assumptions are made about the direction of magnetization (i.e., of induced and remanent magnetizations).

5.6.5.2 Elements of the Three-Dimensional Gram Matrix

In its full formulation, no assumptions about the variation of magnetization with depth are introduced. In this case, from (5.188) and (5.137), following Parker et al. (1987) and Jackson (1990),

$$Q_{ij} = \left(\frac{\mu_0}{4\pi} \right)^2 (\hat{\alpha}_i \cdot \nabla_i)(\hat{\alpha}_j \cdot \nabla_j) \Upsilon(\mathbf{r}_i, \mathbf{r}_j), \qquad (5.190)$$

where

$$\Upsilon(\mathbf{r}_i, \mathbf{r}_j) = \int_V \nabla F_i \cdot \nabla F_j d^3 r', \qquad (5.191)$$

with

$$F_i = \frac{1}{|\mathbf{r}_i - \mathbf{r}'|} = \frac{1}{R_i}, \qquad (5.192)$$

and V is the volume of the magnetized lithosphere. The integral in (5.191) can be evaluated by substituting the gradient of

$$\frac{1}{|\mathbf{r} - \mathbf{r}'|} = \frac{1}{r} \sum_k \left(\frac{r'}{r}\right)^k \sum_m [\cos(m\phi)\cos(m\phi') + \sin(m\phi)\sin(m\phi')] P_n^m(\theta') P_n^m(\theta) \tag{5.193}$$

into (5.191) and using the orthogonality relations for spherical harmonics. An alternative method (Jackson, 1990) is to use Gauss's theorem to transform the volume integral to a surface integral and then make use of the generating function for Legendre polynomials, the addition theorem for spherical harmonics, and the orthogonality relations. If $W_{ij}(\mu, s)$ is defined by

$$W_{ij}(\mu, s) = \sum_{n=0}^{\infty} \frac{n}{2n+1} \left(\frac{s^2}{r_i r_j}\right)^{n+1} P_n(\mu_{ij}), \tag{5.194}$$

and if $b = a - d$, where a is the earth's radius and d the depth of the magnetic layer,

$$\Upsilon(\mathbf{r}_i, \mathbf{r}_j) = 4\pi \left(\frac{W_{ij}(\mu, a)}{a} - \frac{W_{ij}(\mu, b)}{b}\right). \tag{5.195}$$

To find the elements Q_{ij}, expressions are needed for the derivatives $(\hat{\alpha}_i \cdot \nabla_i)$ of $\Upsilon(\mathbf{r}_i, \mathbf{r}_j)$. The general expression to be evaluated is

$$\frac{\partial^2 W}{\partial \alpha_i \partial \alpha_j} = \frac{\partial^2 \mu}{\partial \alpha_i \partial \alpha_j} \frac{\partial Q}{\partial \mu} + \frac{\partial^2 \rho}{\partial \alpha_i \partial \alpha_j} \frac{\partial Q}{\partial \rho} + \frac{\partial \rho}{\partial \alpha_i} \frac{\partial \rho}{\partial \alpha_j} \frac{\partial^2 Q}{\partial \rho^2} + \frac{\partial \mu}{\partial \alpha_i} \frac{\partial \mu}{\partial \alpha_j} \frac{\partial^2 Q}{\partial \mu^2}$$
$$+ \left(\frac{\partial \rho}{\partial \alpha_i} \frac{\partial \mu}{\partial \alpha_j} + \frac{\partial \mu}{\partial \alpha_i} \frac{\partial \rho}{\partial \alpha_j}\right) \frac{\partial^2 Q}{\partial \rho \partial \mu}, \tag{5.196}$$

where

$$\rho = \frac{s^2}{r_i r_j}, \qquad \frac{\partial \rho}{\partial r_i} = -\frac{\rho}{r_i}, \qquad \frac{\partial^2 \rho}{\partial r_i \partial r_j} = \frac{\rho}{r_i r_j}. \tag{5.197}$$

Expressions for the derivatives of Q are given by Jackson (1990), except for $\mu = -1$, which are given by Whaler and Langel (1996).

5.7 AN ESTIMATE OF THE POTENTIAL FUNCTION AND SUSCEPTIBILITY

This section is adapted from Arkani-Hamed and Dyment (1996), which is an extension of Arkani-Hamed and Strangway (1985b). They have devised an innovative method to find a spherical harmonic expression for the magnetic susceptibility corresponding to an anomaly map. The assumptions are that (1) the anomalies are from lateral variations of the induced magnetization in a magnetic layer of constant thickness and (2) the induction is along the direction of the main field. The main field of the earth is decomposed into its dipole ($\mathbf{B}_0$), and nondipole ($\mathbf{B}_1$) parts:

$$\mathbf{B} = \mathbf{B}_0 + \mathbf{B}_1, \tag{5.198}$$

where

$$\mathbf{B}_0(\mathbf{r}_0) = \left(\frac{a}{r_0}\right)^3 g_1^0 [2(\cos\theta_0)\hat{\mathbf{r}} + (\sin\theta_0)\hat{\boldsymbol{\theta}}]. \tag{5.199}$$

The unit direction vector along $\mathbf{B}_0$ is

$$\hat{\mathbf{b}}_0 = -[2(\cos\theta)\hat{\mathbf{r}} + (\sin\theta)\hat{\boldsymbol{\theta}}]/\Gamma, \tag{5.200}$$

where

$$\Gamma = (1 + 3\cos^2\theta)^{1/2}. \tag{5.201}$$

If $\hat{\mathbf{b}}$ and $\hat{\mathbf{b}}_1$ are the unit vectors along $\mathbf{B}$ and $\mathbf{B}_1$,

$$\hat{\mathbf{b}} = \mathbf{B}/B, \qquad \hat{\mathbf{b}}_1 = \mathbf{B}_1/B_1, \tag{5.202}$$

and if

$$\hat{\mathbf{b}} = \hat{\mathbf{b}}_0 + \delta\mathbf{b}, \tag{5.203}$$

then it can be shown that

$$\delta\mathbf{b} = [B_1\hat{\mathbf{b}}_1 - (B - B_0)\hat{\mathbf{b}}_0]/B. \tag{5.204}$$

The anomaly field $\mathbf{A}$ is the usual negative gradient of a potential function, Ψ_a, equation (4.25),

$$\Psi_a = a \sum_{nm} \left(\frac{a}{r}\right)^{n+1} [G_n^m \cos(m\phi) + H_n^m \sin(m\phi)] P_n^m(\theta).$$

5.7.1 THE POTENTIAL

Following Arkani-Hamed and Dyment (1996), let

$$\tau = \Gamma A^s = -\Gamma\hat{\mathbf{b}} \cdot \nabla\Psi_a, \tag{5.205}$$

with the expansion

$$\tau = \sum_{nm} [\tau_{nm}^a \cos(m\phi) + \tau_{nm}^b \sin(m\phi)] P_n^m(\theta). \tag{5.206}$$

Also define

$$B_\delta = \Gamma\delta\mathbf{b} \cdot \nabla\Psi_a, \tag{5.207}$$

with the expansion

$$B_\delta = \sum_{nm} \left[\mathcal{A}_{nm}^a \cos(m\phi) + \mathcal{A}_{nm}^b \sin(m\phi) \right] P_n^m(\theta). \tag{5.208}$$

Substitute (5.203), (5.206), and (5.207) into (5.205) and equate terms in $\cos(m\phi)$ $P_n^m(\theta)$ and in $\sin(m\phi)P_n^m(\theta)$ on both sides, using the identities

$$\sin\theta \frac{dP_n^m(\cos\theta)}{d\theta} = \frac{n[(n+1)^2 - m^2]^{1/2}}{2n+1} P_{n+1}^m(\cos\theta)$$

$$- \frac{(n+1)(n^2 - m^2)^{1/2}}{2n+1} P_{n-1}^m(\cos\theta), \tag{5.209}$$

$$P_k^m(\cos\theta) = \frac{2k-1}{(k^2 - m^2)^{1/2}} (\cos\theta) P_{k-1}^m(\cos\theta)$$

$$- \left(\frac{(k-1)^2 - m^2}{k^2 - m^2} \right)^{1/2} P_{k-2}^m(\cos\theta), \tag{5.210}$$

to find the relationship between the τ_{nm}, the $\mathcal{A}_{nm}$, and the G_n^m and H_n^m:

$$\begin{pmatrix} \tau_{nm}^a \\ \tau_{nm}^b \end{pmatrix} + \begin{pmatrix} \mathcal{A}_{nm}^a \\ \mathcal{A}_{nm}^b \end{pmatrix} = \xi_{nm} \begin{pmatrix} G_{n+1}^m \\ H_{n+1}^m \end{pmatrix} + \zeta_{nm} \begin{pmatrix} G_{n-1}^m \\ H_{n-1}^m \end{pmatrix}, \tag{5.211}$$

where the coefficients are given by

$$\xi_{nm} = -\left(\frac{a}{r} \right)^{n+3} \frac{3(n+2)[(n+1)^2 - m^2]^{1/2}}{2n+3},$$

$$\zeta_{nm} = -\left(\frac{a}{r} \right)^{n+1} \frac{(n+1)(n^2 - m^2)^{1/2}}{2n-1}. \tag{5.212}$$

The equations (5.211), (5.207), and (5.208) are solved iteratively for Ψ_a. Given measurements of the scalar anomaly field at positions $\{\mathbf{r}_i\}$, compute $\tau(\mathbf{r}_i)$, and solve for the τ_{nm} by whatever method is convenient. To initiate the iterative process, set the $\mathcal{A}_{nm}$ in (5.211) equal to zero. Given the τ_{nm}, equation (5.211) can be rearranged into recursion formulas for G_n^m and H_n^m, with $G_{-1}^m = 0$, $G_0^0 = 0$, similar relations for H, and $\zeta_{00} = 0$. If the three components of $\mathbf{A}$ are measured, they are easily converted to scalar anomaly values. The resulting approximation to Ψ_a is then used in (5.207) to obtain an estimate for B_δ and hence for the $\mathcal{A}_{nm}$. Using this estimate for $\mathcal{A}_{nm}$ in (5.211), a second estimate is obtained for Ψ_a. The process is iterated until convergence.

If gridded values of the scalar field are available, then they can easily be converted into gridded values of τ. This may make the approximate methods of deriving spherical harmonic coefficients in Section 5.4.3 attractive for finding the τ_{nm}.

5.7.2 SUSCEPTIBILITY

When deriving a formalism for finding susceptibility, Arkani-Hamed and Dyment (1996) assumed that the inducing main dipole field is constant throughout the magnetic layer [i.e., $r_0 = a$ in equation (5.199)]. In the following, that assumption is not made, and $\mathbf{B}_0(\mathbf{r})$ is given by (5.199). For that reason, the results of this section differ from those of Arkani-Hamed and Dyment (1996), although the basic principles are the same.

With $\mathbf{M}$ and κ as defined in Section 5.1.2, then by the second assumption, equation (5.23) is replaced by

$$\mathbf{M} = \frac{\kappa(\mathbf{r})}{\mu_0}\mathbf{B} = \frac{\kappa(\mathbf{r})}{\mu_0}(\mathbf{B}_0 + \mathbf{B}_1) = \mathbf{M}_0 + \mathbf{M}_1. \qquad (5.213)$$

Recall equation (5.20):

$$\Psi_a(\mathbf{r}) = \frac{\mu_0}{4\pi}\int_V \mathbf{M}(\mathbf{r}') \cdot \nabla' \frac{1}{|\mathbf{r} - \mathbf{r}'|}d^3\mathbf{r}' = \Psi_a^0 + \Psi_a^1, \qquad (5.214)$$

where

$$\Psi_a^0(\mathbf{r}) = \frac{\mu_0}{4\pi}\int_V \mathbf{M}_0(\mathbf{r}') \cdot \nabla' \frac{1}{|\mathbf{r} - \mathbf{r}'|}d^3\mathbf{r}' = \frac{1}{4\pi}\int_V \kappa \mathbf{B}_0(\mathbf{r}') \cdot \nabla' \frac{1}{|\mathbf{r} - \mathbf{r}'|}d^3\mathbf{r}'$$
$$(5.215)$$

and

$$\Psi_a^1(\mathbf{r}) = \frac{\mu_0}{4\pi}\int_V \mathbf{M}_1(\mathbf{r}') \cdot \nabla' \frac{1}{|\mathbf{r} - \mathbf{r}'|}d^3\mathbf{r}' = \frac{1}{4\pi}\int_V \kappa \mathbf{B}_1(\mathbf{r}') \cdot \nabla' \frac{1}{|\mathbf{r} - \mathbf{r}'|}d^3\mathbf{r}'.$$
$$(5.216)$$

Expand κ in terms of spherical harmonics:

$$\kappa = \sum_{nm}\left[\kappa_{nm}^a\cos(m\phi') + \kappa_{nm}^b\sin(m\phi')\right]P_n^m(\theta'), \qquad (5.217)$$

with the κ_{nm} assumed independent of θ' and ϕ'. Also, expand Ψ_a^1 in terms of spherical harmonics:

$$\Psi_a^1(\mathbf{r}) = a\sum_{nm}\left(\frac{a}{r}\right)^{n+1}\left[W_{nm}^a\cos(m\phi) + W_{nm}^b\sin(m\phi)\right]P_n^m(\theta). \qquad (5.218)$$

If (5.217) and the gradient of (5.193) are substituted into (5.215), it can be shown that

$$\begin{pmatrix} G_n^m \\ H_n^m \end{pmatrix} - \begin{pmatrix} W_{nm}^a \\ W_{nm}^b \end{pmatrix} = \frac{g_1^0}{a^{n-1}}\int_{R_1}^{R_2}\left(\gamma_{nm}\begin{pmatrix} \kappa_{n+1,m}^a \\ \kappa_{n+1,m}^b \end{pmatrix} + \delta_{nm}\begin{pmatrix} \kappa_{n-1,m}^a \\ \kappa_{n-1,m}^b \end{pmatrix}\right)(r')^{n-2}dr',$$
$$(5.219)$$

where R_1 and R_2 are the inner and outer radii of the magnetized layer, where the κ_{nm} are allowed to vary with r', and where

$$\gamma_{nm} = \frac{3n[(n+1)^2 - m^2]^{1/2}}{(2n+1)(2n+3)},$$

$$\delta_{nm} = \frac{(n-1)(n^2 - m^2)^{1/2}}{(2n+1)(2n-1)}. \qquad (5.220)$$

It is highly unlikely that the κ_{nm} would have the same variation with r' over the entire globe or that satellite data are able to discern variation of susceptibility with depth. Therefore it is usual to assume that the κ_{nm} are independent of r', so the integral over r' is straightforward:

$$\int_{R_1}^{R_2} (r')^{n-2}\, dr' = \frac{(R_2)^{n-1} - (R_1)^{n-1}}{n-1}, \qquad n > 1. \qquad (5.221)$$

In practice, n will always be greater than unity, because the lithospheric field is not known for $n < 14$. Equation (5.219) now becomes

$$\begin{pmatrix} G_n^m \\ H_n^m \end{pmatrix} - \begin{pmatrix} W_{nm}^a \\ W_{nm}^b \end{pmatrix} = \left(\gamma_{nm} \begin{pmatrix} \kappa_{n+1,m}^a \\ \kappa_{n+1,m}^b \end{pmatrix} + \delta_{nm} \begin{pmatrix} \kappa_{n-1,m}^a \\ \kappa_{n-1,m}^b \end{pmatrix} \right) K_n, \qquad (5.222)$$

where

$$K_n = \frac{g_1^0}{a^{n-1}} \frac{(R_2)^{n-1} - (R_1)^{n-1}}{n-1}. \qquad (5.223)$$

Rearranging (5.222) gives the recursion relations

$$\kappa_{n+1,m}^a = \frac{G_n^m - W_{nm}^a}{K_n \gamma_{nm}} - \frac{\delta_{nm}}{\gamma_{nm}} \kappa_{n-1,m}^a, \qquad (5.224)$$

$$\kappa_{n+1,m}^b = \frac{H_n^m - W_{nm}^b}{K_n \gamma_{nm}} - \frac{\delta_{nm}}{\gamma_{nm}} \kappa_{n-1,m}^b. \qquad (5.225)$$

By definition, $\kappa_{-1} = 0$. Equations (5.224), (5.225), and (5.222) are solved iteratively. On the first iteration, the W_{nm} are set to zero. Then, given the G_n^m and H_n^m, the κ_{nm} can be computed easily.

To carry out the iterative process for κ_{nm} and W_{nm}, represent the product $\kappa \mathbf{B}_1$ in equation (5.216) in spherical harmonics:

$$\kappa \mathbf{B}_1 = \omega(\mathbf{r}) = \sum_{n,m} \left(\begin{bmatrix} \omega_{n,m}^{r,a} \\ \omega_{n,m}^{\theta,a} \\ \omega_{n,m}^{\phi,a} \end{bmatrix} \cos(m\phi) + \begin{bmatrix} \omega_{n,m}^{r,b} \\ \omega_{n,m}^{\theta,b} \\ \omega_{n,m}^{\phi,b} \end{bmatrix} \sin(m\phi) \right) P_n^m(\theta). \qquad (5.226)$$

Expand $\nabla'(1/|\mathbf{r} - \mathbf{r}'|)$ in spherical harmonics, substitute into (5.216) along with (5.226), and then integrate, using the orthogonality of the spherical harmonics to show

$$W_{nm}^a = \frac{R_2^{n+2} - R_1^{n+2}}{4(n+2)a^{n+1}} (\delta_{m,0} + 1) \sum_{k=1}^{N} [nR_{kn}^m \omega_{km}^{r,a} + Q_{kn}^m \omega_{km}^{\theta,a} - S_{kn}^m \omega_{km}^{\phi,b}], \qquad (5.227a)$$

$$W_{nm}^b = \frac{R_2^{n+2} - R_1^{n+2}}{4(n+2)a^{n+1}} \sum_{k=1}^{N} [nR_{kn}^m \omega_{km}^{r,b} + Q_{kn}^m \omega_{km}^{\theta,b} + S_{kn}^m \omega_{km}^{\phi,a}], \qquad m > 0, \qquad (5.227b)$$

where

$$R_{kn}^m = \int_{-1}^{1} P_k^m(\mu') P_n^m(\mu') \, d\mu' = \delta_{k,n} \frac{4}{2n+1}, \qquad k, n \neq 0, \qquad (5.228a)$$

$$Q_{kn}^m = \int_{-1}^{1} [1 - (\mu')^2]^{1/2} P_k^m(\mu') P_n^m(\mu') \, d\mu', \qquad (5.228b)$$

$$S_{kn}^m = \int_{-1}^{1} \frac{P_k^m(\mu') P_n^m(\mu')}{[1 - (\mu')^2]^{1/2}} \, d\mu'. \qquad (5.228c)$$

The solution for the κ_{nm} from the preceding iteration and the known $\mathbf{B}_1(\mathbf{r})$ are used to find the ω_{nm} in (5.226). Then (5.227) is used to find the W_{nm}, from which $[G_n^m - W_{nm}^a]$ and $[H_n^m - W_{nm}^b]$ are found and used in (5.224) and (5.225) to find the next estimate of the κ_{nm}. The process is iterated until convergence. The R_{kn}^m are easily computed; the Q_{kn}^m and S_{kn}^m are computed numerically, once, and stored in a look-up table.

5.8 THE USEFULNESS OF VECTOR DATA

Early satellite magnetic data (e.g., from the POGO satellites) are of scalar magnitude only. Inversion of these data is accomplished using the equivalent source method. Once such a solution is obtained, it can be used to compute the associated vector components. Derivation of models capable of predicting vector fields from scalar data is also possible with other inversion methods (e.g., spherical harmonic potential field analysis and minimum-norm magnetization). This capability of estimating vector anomaly components based only on scalar data raises two questions: (1) How accurate are the estimated vector components? (2) How useful are the measured component data for anomaly studies?

In an attempt to answer the first question, Galliher and Mayhew (1982) compared *Magsat* anomaly data over the United States with vector components synthesized from an equivalent source model. The model was derived using scalar data corresponding to the vector data. They found the following:

1. A satisfactory fit to the scalar data can be achieved regardless of the dipole orientation used in the equivalent source model.

2. The vector components estimated from the model reproduce the measured components to within their measurement accuracy, even when the equivalent source dipole direction is arbitrary.

Those findings imply, at least for the area studied, that the vector anomaly data are uniquely determined by the scalar anomaly data and that neither set of data is sufficient to detect the presence of large-scale remanent magnetization.

Other studies have come to different conclusions. Langel et al. (1982d) compared component anomaly maps derived from averaging measured data and from averaging components at the same locations as the measured data, but estimated from an equivalent source model based on scalar data. While the resulting maps are very similar, they are not identical. For a quantitative comparison, linear regressions between the computed and measured averages are summarized as follows:

Component	Intercept (nT)	Slope	Correlation coefficient
∇B (scalar)	$-0.45\,(\pm 0.04)$	$0.97\,(\pm 0.01)$	0.94
∇X (north)	$-0.52\,(\pm 0.05)$	$0.75\,(\pm 0.02)$	0.84
∇Y (east)	$-0.36\,(\pm 0.06)$	$0.57\,(\pm 0.03)$	0.62
∇Z (down)	$-0.28\,(\pm 0.07)$	$0.91\,(\pm 0.02)$	0.87

The equivalent source model reproduces the measured scalar anomalies very well, and the table shows that the vertical components are also in good agreement. The estimated X component correlates well with the measured X component, but is different in magnitude by about 25%, as shown by the slope factor. Neither the magnitude nor the correlation of the Y component can be considered to indicate good agreement. These results were obtained before the effect of the meridional currents on the dusk data were fully appreciated, and that could have affected the Y-component results.

A study using the Y-component measurements was also carried out by Ravat et al. (1995) in the context of implementing a comprehensive procedure to eliminate the effects of contaminating fields from the *Magsat* data. Their processing was done independently of the Y data (i.e., it uses only X and Z data). As a test of the effectiveness of their procedure, they compared the *Magsat* dawn Y-component data, thought to be nearly free from contamination, with Y-component predictions from models derived from X and Z data before and after application of the procedure. The comparison was made in terms of correlation coefficients calculated in a moving circular window with a great-circle radius of $5°$, at the earth's surface. Figure 5.12 shows the results. Prior to the applied corrections, the correlations are spread over a wide range, with many below 0.5. After the applied corrections, the correlation coefficients are mostly greater than

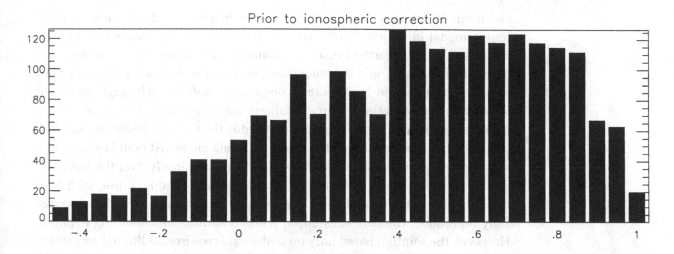

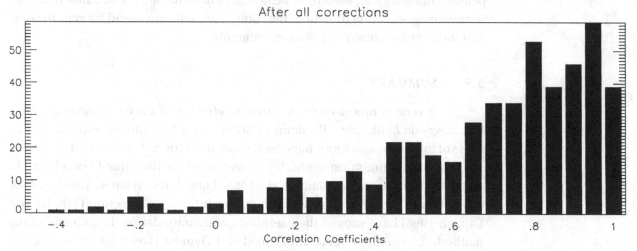

0.75. This indicates that models that assume magnetization directed along the present-day ambient field reproduce most of the measured Y-component data. At the same time, there are some regions where the correlation is 0.6 or less, a possible indication that other directions of magnetization are present.

Indications of an ability to use measured vector data to detect remanent magnetization are also found in the modeling studies of Girdler, Taylor, and Frawley (1992) and Whaler and Langel (1996) for Africa and in those of Ravat, Hinze, and Taylor (1993), Taylor and Ravat (1995), and Pucher and Wonik (1997) for Europe.

Purucker (1990) addressed the question of the usefulness of measured component data in equivalent source modeling. Therein, "data" are generated at satellite altitude from a forward model for two regions, one at low latitude and

FIGURE 5.12
Histograms showing distribution of correlations coefficients between the observed *Magsat* dawn △Y anomaly map and the △Y map computed from an equivalent source solution based on measured △X and △Y. Top: △Y computed from △X and △Y prior to correction for ionospheric fields. Bottom: △Y computed from △X and △Y after all processing has been completed. (From Ravat et al., 1995, with permission.)

one at middle latitude. The resulting data are then used to derive an equivalent source model in which dipoles are 110 km apart and are located within the upper 40 km of the earth's crust. Two solutions are found for each region, one from scalar data only, and the other from combined scalar and vector data, both requiring stabilization by principal-component analysis. Although predicted fields from the equivalent source solutions are very similar to the data, those data are reproduced better at middle latitudes than at low latitudes, and better for scalar plus vector data than for scalar data alone. At both latitudes, the solution using vector data reproduces the data very closely over the entire region. When zero-mean, normally distributed pseudo-random noise, with 1 nT rms, is added to the data, the solution based on scalar and vector data successfully reproduces the data *sans* noise, with no additional smoothing required. However, the solution based only on scalar data requires additional smoothing before satisfactory agreement is achieved. This indicates that, aside from the issue of magnetization direction, the component data are useful for stabilization and increased accuracy of inversion attempts.

5.9   SUMMARY

Inversion processes enter into most efforts to reduce and interpret satellite magnetic field data. Recommendations regarding the appropriate use of some of the various methods have been made in Sections 5.3.4 and 5.4.5. Analyses using minimum-norm methods for downward continuation (Achache et al., 1987; Whaler, 1994; Hamoudi et al., 1995; Langel and Whaler, 1996) and for estimation of magnetization with no constraints on its direction (Whaler and Langel, 1996) have proven the usefulness and computational efficiency of the method. The method of Arkani-Hamed and Dyment (1996) for estimating the potential function and susceptibility also appears sound. Thus a researcher has access to a wide variety of techniques. At the same time, all of these methods involve trade-offs between solution smoothness and goodness of fit. This means that a realistic estimate of the noise level in the data is required. Goodness of fit should not be pressed beyond that level in the quest for extracting increased resolution from the data. In addition to the needed trade-off analysis, the use of local methods requires careful assessment of edge effects. This is particularly important when continuing downward or inverting for magnetization or susceptibility.

Use of vector data together with scalar data lends stability to inversion solutions. Here, care must be taken to weight the data types according to their accuracy, so as not to over-fit the component data. The question whether or not the component data carry any additional information about lithospheric fields beyond that in the scalar data is likely to be debated for some time. Our opinion is that they do carry such information for some regions.

APPENDIX 5.1

CALCULATION OF DERIVATIVES
OF THE TRACE OF MATRIX PRODUCTS

Suppose H and X are matrices. Then, assuming that the matrix multiplications are defined, derivatives of χ, the trace for various matrix products, are as follows:

1. $\chi = \text{trace}(AX)$. Expanding:

$$(AX)_{ii} = \sum_k A_{ik}X_{ki}, \tag{A5.1.1}$$

so

$$\chi = \sum_i \sum_k A_{ik}X_{ki}. \tag{A5.1.2}$$

Then

$$\frac{\partial \chi}{\partial x_{pq}} = A_{qp}, \tag{A5.1.3}$$

so

$$\frac{\partial \chi}{\partial X} = A^T. \tag{A5.1.4}$$

2. $\chi = \text{trace}(XA)$,

$$\frac{\partial \chi}{\partial X} = A^T. \tag{A5.1.5}$$

3. $\chi = \text{trace}(AX^T)$. Expanding:

$$(AX^T)_{ii} = \sum_k A_{ik}X_{ik}, \tag{A5.1.6}$$

so

$$\chi = \sum_i \sum_k A_{ik}X_{ik}. \tag{A5.1.7}$$

Then

$$\frac{\partial \chi}{\partial x_{pq}} = A_{pq}, \tag{A5.1.8}$$

so

$$\frac{\partial \chi}{\partial X} = A. \tag{A5.1.9}$$

4. $\chi = \text{trace}(X^T A)$,

$$\frac{\partial \chi}{\partial X} = A. \tag{A5.1.10}$$

5. $\chi = XAX^T$. Expanding:

$$(XAX^T)_{ijl} = \sum_k \sum_l X_{ik} A_{kl} X_{jl}, \tag{A5.1.11}$$

so

$$\chi = \sum_i \sum_k \sum_l X_{ik} A_{kl} X_{il}. \tag{A5.1.12}$$

Then

$$\frac{\partial \chi}{\partial x_{pq}} = \sum_l A_{ql} X_{pl} + \sum_k A_{kq} X_{pk} = (XA^T)_{pq} + (XA)_{pq}, \tag{A5.1.13}$$

so

$$\frac{\partial \chi}{\partial X} = XA^T + XA, \tag{A5.1.14}$$

$$\frac{\partial \chi}{\partial X} = 2XA \tag{A5.1.15}$$

if A is symmetric.

6. $\chi = X^T A X$,

$$\frac{\partial \chi}{\partial X} = 2AX \tag{A5.1.16}$$

if A is symmetric.

APPENDIX 5.2

VECTOR AND HILBERT SPACES

A5.2.1 VECTOR SPACES

The concepts of vector spaces can be found in any book on linear algebra, or, for example, in Luenberger (1969), with a compete set of proofs. Here the concepts will be summarized without proof. If the set of complex numbers C is denoted by lowercase Greek letters $\{\alpha, \beta, \gamma, \ldots\}$, then a *complex vector space*, say **V**, is a set of elements $\{\mathbf{u}, \mathbf{v}, \mathbf{w}, \ldots\}$ with a well-defined addition operation, $\mathbf{u} + \mathbf{v}$, and a well-defined scalar multiplication, $\alpha\mathbf{u}$. The results of both operations are in $\mathbf{V}$. Further, the elements of $\mathbf{V}$ satisfy

$$\mathbf{u} + \mathbf{v} = \mathbf{v} + \mathbf{u},$$

$$(\mathbf{u} + \mathbf{v}) + \mathbf{w} = \mathbf{u} + (\mathbf{v} + \mathbf{w}),$$

$$\alpha(\mathbf{u} + \mathbf{v}) = \alpha\mathbf{u} + \alpha\mathbf{v}, \qquad (A5.2.1)$$

$$(\alpha + \beta)\mathbf{u} = \alpha\mathbf{u} + \beta\mathbf{u},$$

$$(\alpha\beta)\mathbf{u} = \alpha(\beta\mathbf{u}),$$

and there exists $\mathbf{0}$, a null vector, such that

$$0\mathbf{u} = \mathbf{0}. \qquad (A5.2.2)$$

These relationships hold for all $\mathbf{u}$ and $\mathbf{v}$ in $\mathbf{V}$ and α and β in C.

If for n vectors in $\mathbf{V}$, say $\{\mathbf{u}_i, i = 1, \ldots, n\}$, there are n nonzero scalars $\{\alpha_i, i = 1, \ldots, n\}$ such that

$$\sum_{i=1}^{n} \alpha_i \mathbf{u}_i = \mathbf{0}, \qquad (A5.2.3)$$

then those vectors are said to be *linearly dependent*. A set of vectors that is not linearly dependent is *linearly independent*. Any subset of vectors $\mathbf{W}$ from a vector space $\mathbf{V}$ that themselves form a vector space is called a *vector subspace* of $\mathbf{V}$. Suppose there is a set of linearly independent vectors in $\mathbf{V}$: $\{\mathbf{u}_i, i = 1, \ldots, n\}$. Then the set of all linear combinations of the $\mathbf{u}_i$ is a vector subspace, say $\mathbf{W}$, possibly equal to $\mathbf{V}$ itself. The subspace $\mathbf{W}$ is said to be *spanned* by the $\mathbf{u}_i$. A set of linearly independent vectors that spans a vector space is called a *basis* for that space. If n vectors span a vector space, and if there exist any k linearly independent vectors, then $k \leq n$. All bases of a finite-dimensional vector space have the same number of vectors. Further, any set of n linearly independent vectors in that vector space form a basis.

The set of all linear transformations between two vector spaces is itself a vector space, with addition and multiplication defined as follows: If L and T are linear transformations from $\mathbf{V}$ into $\mathbf{W}$, then the function $(L + T)$ is defined by

$$(L + T)\mathbf{u} = L\mathbf{u} + T\mathbf{u}, \qquad (A5.2.4)$$

and αL is defined by

$$(\alpha L)\mathbf{u} = \alpha(L\mathbf{u}). \qquad (A5.2.5)$$

A *norm* of **V** is a real-valued function, $\|\mathbf{u}\|$, of the elements **u** such that

$$\|\mathbf{u}\| \geq 0,$$
$$\|\mathbf{u}\| = 0 \quad \text{if and only if } \mathbf{u} = 0,$$
$$\|\alpha\mathbf{u}\| = |\alpha|\|\mathbf{u}\|, \tag{A5.2.6}$$
$$\|\mathbf{u} + \mathbf{v}\| \leq \|\mathbf{u}\| + \|\mathbf{v}\|,$$

for all **u** and **v** in **V**. A vector space can have more than one norm.

Suppose that for any pair of vectors **u** and **v** in W there is defined a real-valued function, $\langle \mathbf{u}, \mathbf{v} \rangle$, with the following properties:

$$\langle (\mathbf{u} + \mathbf{v}), \mathbf{w} \rangle = \langle \mathbf{u}, \mathbf{w} \rangle + \langle \mathbf{v}, \mathbf{w} \rangle,$$
$$\langle \alpha\mathbf{u}, \mathbf{w} \rangle = \alpha^*\langle \mathbf{u}, \mathbf{w} \rangle,$$
$$\langle \mathbf{u}, \mathbf{v} \rangle = \langle \mathbf{v}, \mathbf{u} \rangle^*, \tag{A5.2.7}$$
$$\langle \mathbf{u}, \mathbf{u} \rangle \geq 0,$$
$$\langle \mathbf{u}, \mathbf{u} \rangle = 0 \quad \text{if and only if } \mathbf{u} = \mathbf{0},$$

where an asterisk denotes a complex conjugate. Such a function, often denoted $\langle,\rangle$ is called an *inner product*. The function

$$\|\mathbf{u}\| = \langle \mathbf{u}, \mathbf{u} \rangle \tag{A5.2.8}$$

is a norm on **V**. The function $\langle,\rangle$ is also called a *Hermitian bilinear form*.

Two vectors **u** and **v** in **V** are said to be *orthogonal* if $\langle \mathbf{u}, \mathbf{v} \rangle = 0$. A vector **u** is said to be orthogonal to a subset **W** if **u** is orthogonal to **w** for each **w** in **W**.

If **V** and **W** are vector spaces, then a *linear transformation*, L, from **V** into **W** is a function such that

$$L(\alpha\mathbf{u} + \mathbf{v}) = \alpha(L\mathbf{u}) + L\mathbf{v}. \tag{A5.2.9}$$

Let **V** be an n-dimensional vector space with basis $\{\mathbf{u}_i, i = 1, \ldots, n\}$, and **W** a vector space. If $\{\mathbf{w}_i, i = 1, \ldots, n\}$ are any n vectors in W, then there is one and only one linear transformation L from **V** into **W** such that

$$L\mathbf{u}_i = \mathbf{w}_i. \tag{A5.2.10}$$

Further, if L is a linear transformation from **V** into **W**, then the range of L (i.e., the set of all vectors in **W** such that $\mathbf{w} = L\mathbf{u}$ for some **u** in **V**) is a subspace of **W**. The set of vectors **v** in **V** such that $L\mathbf{v} = \mathbf{0}$ is a subspace of **V** called the *null space* of L.

A5.2.2 HILBERT SPACES

A sequence of elements $\{v_1, v_2, v_3, \ldots\}$ in $\mathbf{V}$ is said to be a *Cauchy sequence* if and only if (iff) the limit of $\|v_n - v_m\|$, with $m > n$, is zero as n goes to infinity. In that case, if there is an element $\mathbf{v}$ in $\mathbf{V}$ such that $\|v - v_n\|$ goes to zero as n goes to infinity, then $\mathbf{V}$ is called *complete*. A complete inner-product space, $\mathcal{H}$, is called a *Hilbert space*.

A function F that maps $\mathcal{H}$ to the real numbers R (i.e., $F : \mathcal{H} \to R$) is called a *functional*. $\|v\|$ is a functional. If F satisfies

$$F(\mathbf{v} + \mathbf{w}) = F\mathbf{v} + F\mathbf{w},$$
$$F(\alpha\mathbf{v}) = \alpha(F\mathbf{v}), \tag{A5.2.11}$$

then F is said to be linear. Suppose there exists a real constant $c \geq 0$ such that

$$|F\mathbf{v}| \leq c\|\mathbf{v}\|. \tag{A5.2.12}$$

Then F is said to be bounded, and the smallest value of c that satisfies (A5.2.12) is called the norm of F, $\|F\|$. An important property of functionals is that a linear functional is bounded iff it is continuous.

Suppose $\mathcal{H}$ is a Hilbert space and $\mathcal{H}'$ is the space of all bounded linear functionals on $\mathcal{H}$. $\mathcal{H}'$ is called the *dual space* of $\mathcal{H}$. It can be shown that $\mathcal{H}'$ itself is a Hilbert space.

An important theorem for Hilbert spaces, the projection theorem, states that if $\mathcal{M}$ is a closed subspace of $\mathcal{H}$, and if $\mathbf{v}$ is (any) vector in $\mathcal{H}$, then there exists a unique vector, v_0 in $\mathcal{M}$ such that $\|v - v_0\| \leq \|v - w\|$ for all $\mathbf{w}$ in $\mathcal{M}$, that v_0 is unique, and that if v_0 is in $\mathcal{M}$ it is the unique minimizing vector if $v - v_0$ is orthogonal to $\mathcal{M}$. The projection v_0 is said to be the *minimum-norm* projection of $\mathbf{v}$ onto $\mathcal{M}$.

One of the more important properties in geophysical applications of Hilbert spaces is that if L is a bounded linear functional on $\mathcal{H}$, then there exists an element $\mathbf{l}$ in $\mathcal{H}$ such that

$$L\mathbf{v} = \langle \mathbf{v}, \mathbf{l} \rangle \tag{A5.2.13}$$

for all $\mathbf{v}$ in $\mathcal{H}$.

APPENDIX 5.3

CALCULATION OF ELEMENTS OF THE GRAM MATRIX Λ

The elements of the matrix Λ in the generalized inverse analysis of Section 5.6.4 can be expressed in closed form. Derivation of those forms is given in this appendix for two of the norms of Table 5.9, the ψ and B_r norms.

A5.3.1 MATRIX ELEMENTS FOR THE ψ NORM

If $f_n = 1/(2n+1)$, the norm minimizes the mean value of ψ^2 over the surface of the earth. Consider equation (5.159) for l^ψ. To find a closed form for l^ψ, it is convenient to use the generating function for Legendre polynomials (Copson, 1935; see also Constable, Parker, and Stark, 1993):

$$f = \frac{1}{s} = \frac{1}{(1 + x^2 - 2x\mu)^{1/2}} = \sum_{n=0}^{\infty} x^2 P_n(\mu). \tag{A5.3.1}$$

Then

$$\sum_{n=0}^{\infty}(2n+1)x^{n+1}P_n(\mu) = 2\frac{\partial}{\partial x}(x^2 f) - 3x\, f. \tag{A5.3.2}$$

Taking the appropriate derivatives of $f = 1/s$,

$$l^\psi = \frac{x(1 - x^2)}{s^3} - x. \tag{A5.3.3}$$

From equations (5.160) and (A5.3.3),

$$\Lambda_{\psi_i,\psi_j} = \frac{x_{ij}[1 - (x_{ij})^2]}{(s_{ij})^3} - x_{ij}. \tag{A5.3.4}$$

The elements of Λ corresponding to measurements and/or calculations of A_r, A_θ, and A_ϕ can now be readily computed using (5.162) and are given in Table 5.10. An alternative derivation can be found in Achache et al. (1987).

A5.3.2 MATRIX ELEMENTS FOR THE B_r NORM

If

$$f_n = \frac{(n+1)^2}{a^2(2n+1)},$$

the norm minimizes the mean value of $(B_r)^2$ over the surface of the earth. From equation (5.159),

$$l^\psi = a^2 \sum_{n=1}^{\infty} \frac{2n+1}{(n+1)^2} x^{n+1} P_n(\mu). \tag{A5.3.5}$$

In practice, a closed form for l^ψ is difficult to find, but it is not required. Suppose ψ_a and ψ_b are potential functions in H, with coefficients $\{a_n^m\}$ and $\{b_n^m\}$, respectively. Then the inner product is, by definition,

$$\langle \psi_a, \psi_b \rangle = \sum_{n=1}^{\infty} \sum_{m=-n}^{n} \frac{(n+1)^2}{2n+1} a_n^m b_n^m$$

$$= \frac{1}{4\pi a^2} \int_{r=a} B_r^a(\mathbf{r}')B_r^b(\mathbf{r}')\, dS' = \frac{1}{4\pi a^2} \int_{r=a} \left(-\frac{\partial \psi_a}{\partial r'}\right)\left(-\frac{\partial \psi_b}{\partial r'}\right) dS'.$$

$$\tag{A5.3.6}$$

It is the radial derivatives of l^ψ, and of l^ϕ, l^ϕ, and l^r, that are of importance in forming l^α and the matrix elements Λ_{ij}.

The matrix elements Λ_{ij} are computed from equation (5.160), beginning with

$$\Lambda_{\psi_i,\psi_j} = a^2 \sum_{n=1}^{\infty} \frac{2n+1}{(n+1)^2} x_{ij}^{n+1} P_n(\mu_{ij}), \tag{A5.3.7}$$

from which the other matrix elements can be computed using (5.162b). Thus,

$$\Lambda_{r_i,r_j} = \frac{\partial^2}{\partial r_i \partial r_j} \Lambda_{\psi_i,\psi_j} = \sum_{n=1}^{\infty} (2n+1) x_{ij}^{n+2} P_n(\mu_{ij}), \tag{A5.3.8}$$

$$\Lambda_{r_i,r_j} = \frac{(x_{ij})^2 [1 - (x_{ij})^2]}{(s_{ij})^3} - (x_{ij})^2, \tag{A5.3.9}$$

where the last equality follows from equations (5.159) and (A5.3.3). Similarly,

$$\Lambda_{\theta_i,r_j} = \frac{\partial^2}{r_i \partial \theta_i \partial r_j} \Lambda_{\psi_i,\psi_j} = -\frac{\partial \mu_{ij}}{\partial \theta_i} \sum_{n=1}^{\infty} \frac{2n+1}{n+1} x_{ij}^{n+2} P_n'(\mu_{ij}), \tag{A5.3.10}$$

where

$$P_n'(\mu) = \frac{d}{d\mu} P_n(\mu).$$

But (Copson, 1935)

$$\frac{1}{n+1} P_n'(\mu) = \frac{1}{\mu^2 - 1} [P_{n+1}(\mu) - \mu P_n(\mu)], \tag{A5.3.11}$$

by which (A5.3.10) becomes

$$\Lambda_{\theta_i,r_j} = \frac{1}{1-\mu^2} \frac{\partial \mu_{ij}}{\partial \theta_i} \left[\sum_{n=2}^{\infty} (2n-1) x_{ij}^{n+1} P_n(\mu_{ij}) - \mu \sum_{n=1}^{\infty} (2n+1) x_{ij}^{n+2} P_n(\mu_{ij}) \right]. \tag{A5.3.12}$$

Defining

$$\begin{aligned} \mathcal{K}(\mathbf{r}, \mathbf{r}') &= \frac{1}{1-\mu^2} \frac{1-(3x^2+1-3\mu x - \mu x^3)}{s^3} \\ &= \frac{1}{1-\mu} - \frac{\mu}{s(s+x-\mu)} - \frac{3x^2+1-2\mu x}{s^3} \\ &= -\frac{x^2(1+2s-x^2)}{s^3(1+s-\mu x)}, \end{aligned} \tag{A5.3.13}$$

and using equation (A5.3.1) in a manner similar to that employed in equation (A5.3.2), leads to

$$\Lambda_{\theta_i,r_j} = \frac{\partial \mu_{ij}}{\partial \theta_i} x_{ij} \mathcal{K}(\mathbf{r}_i, \mathbf{r}_j). \tag{A5.3.14}$$

By an almost identical process,

$$\Lambda_{\phi_i, r_j} = \frac{1}{\sin \theta_i} \frac{\partial \mu_{ij}}{\partial \phi_i} x_{ij} \mathcal{K}(\mathbf{r}_i, \mathbf{r}_j). \tag{A5.3.15}$$

To consider the matrix elements $\Lambda_{\theta_i, \theta_j}$, Λ_{ϕ_i, ϕ_j}, $\Lambda_{\theta_i, \phi_j}$, $\Lambda_{\phi_i, \theta_j}$, the following is needed:

$$\Lambda'_{\psi_i, \psi_j} = \frac{\partial}{\partial \mu} \Lambda_{\psi_i, \psi_j} = a^2 \sum_{n=1}^{\infty} \frac{2n+1}{(n+1)^2} x_{ij}^{n+1} P'_n(\mu_{ij}). \tag{A5.3.16}$$

Also needed are the following, using $\int_0^x t^n dt = x^{n+1}/(n+1)$,

$$\sum_{n=1}^{\infty} (2n+1) t^n P_n(\mu) = \frac{2t(\mu - t)}{s^3} + \frac{1}{s} - 1, \tag{A5.3.17}$$

from which

$$\int_0^x \left[\sum_{n=1}^{\infty} (2n+1) t^n P_n(\mu) \right] dt = \int_0^x \left[\frac{2t(\mu - t)}{s^3} + \frac{1}{s} - 1 \right] dt$$

$$= -\left[\ln\left(\frac{s + x - \mu}{1 - \mu} \right) - \frac{2x}{s} + x \right]. \tag{A5.3.18}$$

Then, using (Copson, 1935)

$$\frac{P'_n(\mu)}{n+1} = \frac{a^2}{\mu^2 - 1} [P_{n+1}(\mu) - \mu P_n(\mu)], \tag{A5.3.19}$$

equation (A5.3.16) becomes

$$\Lambda'(\mu)_{\psi_i, \psi_j} = \frac{a^2}{\mu^2 - 1} [S_1 + S_2], \tag{A5.3.20}$$

where, using (A5.3.18),

$$S_2 = \mu \sum_{n=1}^{\infty} \frac{2n+1}{n+1} x_{ij}^{n+1} P_n(\mu_{ij}) = \mu \left[\ln\left(\frac{s + x - \mu}{1 - \mu} \right) - \frac{2x}{s} + x \right], \tag{A5.3.21}$$

$$S_1 = \sum_{n=1}^{\infty} \frac{2n+1}{n+1} x_{ij}^{n+1} P_{n+1}(\mu_{ij}) = \left[\ln\left(\frac{s + 1 - \mu x}{2} \right) + \frac{2}{s} - 2 \right]. \tag{A5.3.22}$$

Derivation of the second equality in (A5.3.22) is achieved by a method similar to the steps from (A5.3.17) to (A5.3.18). Then

$$\frac{1}{r_i} \frac{\partial}{\partial \theta_i} \Lambda_{\psi_i, \psi_j} = \frac{1}{r_i} \frac{\partial \mu_{ij}}{\partial \theta_i} \Lambda'_{\psi_i, \psi_j}(\mu), \tag{A5.3.23}$$

$$\frac{1}{r_i \sin \theta_i} \frac{\partial}{\partial \theta_i} \Lambda_{\psi_i, \psi_j} = \frac{1}{r_i \sin \theta_i} \frac{\partial \mu_{ij}}{\partial \theta} \Lambda'_{\psi_i, \psi_j}(\mu), \tag{A5.3.24}$$

$$\Lambda_{\theta_i,\theta_j} = \frac{1}{r_i r_j} \frac{\partial^2}{\partial\theta_i\partial\theta_j} \Lambda_{\psi_i,\psi_j}(\mu)$$

$$= \frac{1}{r_i r_j}\left[\frac{\partial^2 \mu_{ij}}{\partial\theta_i\partial\theta_j}\Lambda'_{\psi_i,\psi_j}(\mu) + \frac{\partial\mu_{ij}}{\partial\theta_i}\frac{\partial\mu_{ij}}{\partial\theta_j}\Lambda''_{\psi_i,\psi_j}(\mu)\right], \tag{A5.3.25}$$

$$\Lambda_{\phi_i,\phi_j} = \frac{1}{r_i r_j \sin\theta_i \sin\theta_j} \frac{\partial^2}{\partial\phi_i\partial\phi_j} \Lambda_{\psi_i,\psi_j}(\mu)$$

$$= \frac{1}{r_i r_j \sin\theta_i \sin\theta_j}\left[\frac{\partial^2 \mu_{ij}}{\partial\phi_i\partial\phi_j}\Lambda'_{\psi_i,\psi_j}(\mu) + \frac{\partial\mu_{ij}}{\partial\phi_i}\frac{\partial\mu_{ij}}{\partial\phi_j}\Lambda''_{\psi_i,\psi_j}(\mu)\right]. \tag{A5.3.26}$$

The expressions for $\Lambda_{\phi_i,\theta_j}$ and $\Lambda_{\theta_i,\phi_j}$ should now be obvious. Closed-form expressions for all of the B_r-norm matrix elements are collected together in Table 5.11.

CHAPTER 6

ANOMALY MAPS

6.0   INTRODUCTION

Interpretation of geomagnetic anomaly data generally begins with some type of map or with profiles. In satellite magnetic analysis the anomalies generally are derived from sources that must be treated as three-dimensional. Thus, maps are normally used in preference to profiles. Before addressing interpretation, a list of the available maps is presented, together with a summary of their characteristics and the methods used in their derivation.

Besides providing a catalog of maps, this chapter summarizes some of the history of satellite magnetic anomaly maps and provides both qualitative and quantitative (Section 6.5) comparisons of maps derived using different procedures. Aeromagnetic and marine magnetic surveys are available for several large areas for comparison purposes. Although there are difficulties in making comparisons between near-surface data and satellite data, as discussed in

Section 6.6, such comparisons can be meaningful, as described in Section 6.6.2. Understanding the origins of the various maps, the procedures used to derive them, and the agreements/disagreements between them will provide a basis for assessing the accuracy of their estimates of the lithospheric field. Although such an understanding is valuable in using the maps and in assessing the interpretation methods and results described in Chapters 8 and 9, it is not required as background for understanding those chapters. The reader not immediately interested in the topics of this chapter can proceed to the later chapters.

6.1 TABULATION OF PUBLISHED MAPS

Tables 6.1, 6.2, and 6.3 provide a summary of all the published long-wavelength maps derived from satellite data of which we are aware. Included are not only maps based on averaged or gridded data but also maps reduced to common altitude and/or to the pole, as well as magnetization and susceptibility maps. Also included are selected maps derived from aeromagnetic and marine magnetic data. Table 6.1 lists the maps by region, citing the source for each, the satellite on which each map is based, and whether the map is based on scalar (s) data, on some component (x, y, z), or on all vector components (v). Note that for *Magsat*, the scalar data are not derived from a separate scalar magnetometer, but are the square roots of the sums of the squares of the three measured components, as described in Chapter 2. Tables 6.2 and 6.3 supplement Table 6.1. Table 6.2 summarizes some aspects of the data reduction process, namely, how the data were selected for magnetic activity, what main-field model was used as a reference, how the long-wavelength corrections were made, whether or not individual passes were compared with one another, and whether or not a model of external (ionospheric) fields was removed from the data. Table 6.3 contains comments noting special features or aspects of the data reduction for some of the maps.

Table 6.1 is divided into four major sections: (I) maps for low and middle latitudes, generally covering ±50°, although other latitude limits are sometimes used; (II) polar maps, generally poleward of 50° latitude; (III) global maps combining low, middle, and polar latitudes (these may not completely cover the pole, depending on the latitude range of the available data); (IV) regional maps (maps from specific countries or regions). Maps are numbered consecutively within the table. In the rest of this chapter, when a map listed in the tables is being discussed in the text, its number will be given in parentheses.

Historically, data were first acquired from *Cosmos 49*, then from the POGO satellites, and finally from *Magsat*. The first published map (37) was of the United States, using data from *Cosmos 49*. This map was made relative to a degree- and order-9 main-field model and so is a mixture of core and lithospheric fields, with the former larger than the latter. Other maps based on *Cosmos 49*

Table 6.1. *Satellite and long-wavelength surface magnetic anomaly maps: general information*

Map no.	Reference	Coverage and comments	Satellite or aircraft	Map type	Altitude (km)
I. *Maps for low and middle latitudes*					
1	Regan et al. (1975) (Langel, 1993)	±50° lat.	POGO (s)[a]	1° avg.[b]	~540
2	Langel et al. (1982a)	±50° lat.	*Magsat* (s)	2° avg.	~404
3	Langel et al. (1982a)	±50° lat.	POGO (s)	2° avg.	~500
4	Langel et al. (1982c)	±50° lat.	*Magsat* (v)	2° avg.	~430
5	Yanagisawa & Kono (1985)	±50° lat.	*Magsat* (v)	5° avg.	~420
6	Cohen & Achache (1990)	±60° lat.	*Magsat* (v)	2° avg.	~372
7	Langel (1990b)	±60° gm. lat.	*Magsat* (z)	2° avg.	~430
8	Langel (1993)	±50° lat.	*Magsat* (s, z, y)	2° avg.	~430
9	Hamoudi et al. (1995)	±60° lat.	*Magsat* (z)	DC	0
10	Hamoudi et al. (1995)	±60° lat.	*Magsat* (z)	DC, SHA	0
II. *Polar maps*					
11	Langel & Thorning (1982b)	50–87° S lat.	POGO (s)	3° avg.	~500
12	Coles et al. (1982)	40–83° N lat.	*Magsat* (s, v)	2° avg.	~430
13	Ritzwoller & Bentley (1982)	South Pole	*Magsat* (s)	3° avg.	~430
14	Ritzwoller & Bentley (1983)	South Pole	*Magsat* (s)	3° avg.	~430
15	Coles (1985)	40–83° N lat.	*Magsat* (s)	2° avg.	~415
16	Haines (1985b)	40–83° N lat.	*Magsat* (z)	SCHA	335, 535
17	Alsdorf (1991), Alsdorf et al. (1994)	40–90° S lat.	*Magsat* (s)	grid.	375, 485
18	Alsdorf (1991), Alsdorf et al. (1994)	40–90° S lat.	*Magsat* (s)	RTP	375, 485
19	Takenaka et al. (1991)	40–90° S lat.	*Magsat* (s)	3° avg.	~450
20	Alsdorf et al. (1997)	40–83° N lat.	*Magsat* (s)	grid.	400
21	Alsdorf et al. (1997)	40–83° N lat.	*Magsat* (s)	RTP	400
22	von Frese et al. (1997)	40–83° S lat.	*Magsat* (s)	DC, RTP	0
III. *Global maps*					
23	Langel (1990a, 1993)	±90° lat.	POGO (s)	3° avg.	~500
24	Langel (1990a)	±90° lat.	POGO (s)	RTP	500
25	Cain et al. (1984)	±90° lat.	*Magsat* (v)	SHA (14–29)	400
26	Arkani-Hamed & Strangway (1986a)	±78° lat.	*Magsat* (s)	SHA (18–55)	410
27	Arkani-Hamed & Strangway (1985a,b)	±78° lat.	*Magsat* (s)	SHA (18–55)	410
28	Arkani-Hamed & Strangway (1985b)	±78° lat.	*Magsat* (s)	susc.	0
29	Ravat et al. (1995)	±83° lat.	*Magsat* (s, v)	SHA (15–65)	400
30	Arkani-Hamed et al. (1994)	±90° lat.	POGO (s)	SHA (15–60)	400/500
31	Arkani-Hamed et al. (1994)	±90° lat.	POGO/*Magsat* (s)	SHA (15–65)	400
32	This book; based on Cain et al. (1990)	±90° lat.	*Magsat* (s, v)	SHA (14–49)	400
33	Purucker et al. (1996)	±86° lat.	POGO/*Magsat* (s)	magn.	
34	Purucker et al. (1997a)	±86° lat.	POGO/*Magsat* (s)	s-t	
35	Arkani-Hamed & Dyment (1996)	±83° lat.	POGO/*Magsat* (s)	DC, SHA	0
36	Arkani-Hamed & Dyment (1996)	±83° lat.	POGO/*Magsat* (s)	susc.	

(*continues*)

Table 6.1. *(continued)*

Map no.	Reference	Coverage and comments	Satellite or aircraft	Map type	Altitude (km)
IV. *Regional maps*					
A. United States					
37	Zietz et al. (1970)		*Cosmos 49*	3° avg.	~350
38	Langel (1990b)		*Cosmos 49*	5° avg.	?
39	Mayhew (1979)		POGO (s)	ES	450
40	Mayhew (1979)		POGO (s)	magn.	
41	Mayhew (1982a)		POGO (s)	ES	450
42	Mayhew (1982a)		POGO (s)	magn.	
43	von Frese et al. (1982b)		Airmag (s)	UC	450
44	von Frese et al. (1982b)		POGO (s)	ES	450
45	von Frese et al. (1982b)		*Magsat* (s)	2° avg.	347
46	von Frese et al. (1982a)		POGO (s)	ES	450
47	von Frese et al. (1982a)		POGO (s)	RTP	450
48	Won & Son (1982)		*Magsat* (s)	avg.	?
49	Won & Son (1982)		Airmag (s)	avg.	~1
50	Won & Son (1982)		Airmag (s)	UC	300
51	Mayhew & Galliher (1982)		*Magsat* (s)	ES	320
52	Mayhew & Galliher (1982)		*Magsat* (s)	magn.	
53	Mayhew (1984)		*Magsat* (s)	magn.	
54	Carmichael & Black (1986)	Midcontinent	*Magsat* (s)	wt. avg.	400
55	Carmichael & Black (1986)	Midcontinent	*Magsat* (s)	RTP	400
56	Schnetzler & Allenby (1983)	Variable-thickness lower crust	POGO (s)	magn.	
57	Schnetzler (1985)	Variable-thickness lower crust	*Magsat* (s)	magn.	
58	Ruder & Alexander (1986)	Southeast only	Airmag (s)	UC	250, 500
59	Sexton et al. (1982)	Low pass: 200 km	Airmag (s)	hand	
60	Whaler & Langel (1996)		*Magsat* (s)	magn.	
B. South America					
61	Hinze et al. (1982), Longacre (1981)		*Magsat* (s)	RTP	350
62	Yanagisawa & Kono (1984)		*Magsat* (s, v)	5° avg.	~420
63	Ridgway & Hinze (1986)		*Magsat* (s)	2° avg.	~405
64	Ridgway & Hinze (1986)		*Magsat* (s)	ES	350
65	Ridgway & Hinze (1986)		*Magsat* (s)	RTP	450
66	von Frese et al. (1989)		*Magsat* (s)	RTP	300
67	Ravat et al. (1991)		*Magsat* (s)	ES	350
68	Ravat et al. (1991)		*Magsat* (s)	RTP	300
69	Ravat et al. (1991)		*Magsat* (s)	susc.	
C. Africa					
70	Regan et al. (1975)		*Cosmos 49* (s)	5° avg.	~350
71	Kuhn & Zaaiman (1986)	South Africa	*Magsat* (s)	grid. avg.	350

(continues)

Table 6.1. *(continued)*

Map no.	Reference	Coverage and comments	Satellite or aircraft	Map type	Altitude (km)
72	Ravat (1989)	Africa and Europe (50° S–60° N; 35° W–70° E)	*Magsat* (s)	g&c	400
73	Ravat (1989)	Africa and Europe	*Magsat* (s)	RTP	400
74	Ravat (1989)	Africa and Europe	*Magsat* (s)	susc.	
75	Baldwin & Frey (1991)		*Magsat* (s)	2° avg.	~400
76	Baldwin & Frey (1991)		*Magsat* (s)	ES	400
77	Baldwin & Frey (1991)		*Magsat* (s)	RTP	400
78	Girdler et al. (1992)	Central Africa	Magsat (s)	g&c	375
79	Whaler (1994)		*Magsat* (s)	DC	0
80	Langel & Whaler (1996)		POGO/*Magsat* (s)	DC	0
81	Whaler & Langel (1996)		*Magsat* (v)	magn.	
D. Japan					
82	Yanagisawa et al. (1982)		*Magsat* (s)	2° avg.	~450
83	Yanagisawa et al. (1982)		*Magsat* (s)	ES	450
84	Yanagisawa et al. (1982)		*Magsat* (s)	magn.	
85	Yanagisawa & Kono (1984)	Western Pacific	*Magsat* (s, v)	5° avg.	~420
86	Nakatsuka & Ono (1984)		*Magsat* (s, v)	g&c	~430
87	Nakagawa et al. (1985), Nakagawa & Yukutake (1984, 1985)		*Magsat* (s, v)	2° avg.	~430
88	Nakagawa et al. (1985), Nakagawa & Yukutake (1984, 1985)		*Magsat* (s, v)	RHA	300
E. Australia					
89	Mayhew et al. (1980)		POGO (s)	ES	450
90	Mayhew et al. (1980)		POGO (s)	magn.	
91	Mayhew & Johnson (1987)		*Magsat* (s)	magn.	
92	Mayhew & Johnson (1987)		*Magsat* (s)	ES	325
93	Mayhew & Johnson (1987)		*Magsat* (s)	RTP	325
94	Tarlowski et al. (1996)		Airmag (s)	grid.	0.15
F. Canada					
95	Langel et al. (1980a)	Western Canada	POGO (s)	3° avg.	~520
96	Langel et al. (1980a)	Western Canada	POGO (s)	ES	500
97	Langel et al. (1980a)	Western Canada	Airmag (z)	UC	500, 300, 100
98	Arkani-Hamed et al. (1984)	Canada	*Magsat* (s)	DF	390
99	Arkani-Hamed et al. (1985b)	Canada & U.S.A.	*Magsat* (s)	grid., wt. avg.	~390
100	Pilkington & Roest (1996)		Airmag (s)	UC	?
G. India					
101	Mishra (1984)		*Magsat* (z)	avg.	400
102	Mishra & Venkatraydu (1985)		*Magsat* (s)	1° avg.	404

(continues)

Table 6.1. *(continued)*

Map no.	Reference	Coverage and comments	Satellite or aircraft	Map type	Altitude (km)
103	Arur et al. (1985)		*Magsat* (z)	2° avg.	400
104	Agarwal et al. (1986)		*Magsat* (s)	2° avg.	~400
105	Agrawal et al. (1986)		*Magsat* (z)	magn.	
106	Rajaram & Singh (1986)		*Magsat* (s)	4° avg.	~420
107	Negi et al. (1986a)		*Magsat* (s)	magn.	
108	Negi et al. (1986a)		*Magsat* (v)	magn.	
109	Bapat et al. (1987)		*Magsat* (v)	ES	?
110	Singh (1989)		*Magsat* (s, v)	magn.	
111	Singh et al. (1989)		*Magsat* (s)	m-m	
112	Singh et al. (1989)		*Magsat* (z)	m-m	
113	Singh et al. (1989)		*Magsat* (x)	m-m	
114	von Frese et al. (1989)		*Magsat* (s)	RTP	400
115	Singh & Rajaram (1990), Singh et al. (1991, 1992)		*Magsat* (s)	m-m	
116	Singh & Rajaram (1990), Singh et al. (1991, 1992)		*Magsat* (s)	m-m	
117	Rajaram & Langel (1992)		*Magsat* (z)	2° avg.	~400
H. Europe					
118	Arkani-Hamed & Strangway (1986b)	Eastern Europe, Middle East	*Magsat* (s)	DF	~400
119	Arkani-Hamed & Strangway (1986b)	Eastern Europe, Middle East	*Magsat* (s)	susc.	
120	Taylor & Frawley (1987)	Kursk, Russia	*Magsat* (s, v)	g&c	~350
121	Berti & Pinna (1987)	Italy	*Magsat* (s)	2° avg.	~400
122	DeSantis et al. (1989)	Europe	*Magsat* (s, v)	SCHA	400
123	Cain et al. (1989)	Europe	*Magsat* (s, v)	SHA	350, 0
124	DeSantis et al. (1990)	Italy	*Magsat* (v), with observatory, repeat, and survey	SCHA	0
125	Taylor et al. (1992)	Scandinavia	*Magsat* (s)	g&c	400
126	Nolte & Hahn (1992)		*Magsat* (s)	hand	360, 440
127	Nolte & Hahn (1992)		*Magsat* (s)	DIM	
128	Ravat et al. (1993)		*Magsat* (s)	avg.	~400
129	Taylor & Ravat (1995)	Central Europe	*Magsat* (s)	g&c	350
I. China/Southeast Asia					
130	Achache et al. (1987)	Southeast Asia	*Magsat* (z)	1° avg.	~440
131	Achache et al. (1987)	Southeast Asia	*Magsat* (z)	DC	
132	Arkani-Hamed et al. (1988)	China	*Magsat* (s)	DF	~400
133	Arkani-Hamed et al. (1988)	China	*Magsat* (s)	susc.	
134	An et al. (1992)	China	*Magsat* (s, v)	SCHA	400
135	Hamoudi et al. (1995)	0–50° N, 50–150° E	*Magsat* (z)	avg.	372
136	Hamoudi et al. (1995)	0–50° N, 50–150° E	*Magsat* (z)	DC	0

(continues)

Table 6.1. *(continued)*

Map no.	Reference	Coverage and comments	Satellite or aircraft	Map type	Altitude (km)
J. Continental reconstructions					
137	Galdéano (1981, 1983)		*Magsat* (s)	avg.	satellite
138	Frey et al. (1983)		POGO (s)	ES/RTP	400
139	von Frese et al. (1986)		*Magsat* (s)	RTP	400
140	von Frese et al. (1987)		*Magsat* (s)	RTP	400
141	Ravat (1989)		*Magsat* (s)	RTP	400
142	Ravat (1989), Ravat et al. (1992)		*Magsat* (s)	susc.	
K. Miscellaneous continental					
143	Langel & Thorning (1982a)	Nares Strait	POGO (s)	3° avg.	~500
144	Langel & Thorning (1982b)	Greenland	POGO (s)	ES	500
145	Langel & Thorning (1982b)	Greenland	POGO (s)	magn.	
146	von Frese et al. (1987), Hinze et al. (1991)	−40° to 70° lat. −140° to 60° long.	*Magsat* (s)	2° avg.	~400
147	von Frese et al. (1987), Hinze et al. (1991)	−40° to 70° lat. −140° to 60° long.	*Magsat* (s)	RTP	400
148	Counil et al. (1989)	Caribbean	*Magsat* (x, z)	avg.	satellite
149	Counil et al. (1989)	Caribbean	*Magsat* (z)	DC	0
150	Rotanova et al. (1995)	Russia–India (55° S–55° N lat.; 60–100° E long.)	*Magsat* (s, v)	avg.	395
L. Subduction zones/trenches					
151	Clark et al. (1985)	Aleutian arc	*Magsat* (s)	2° avg.	~400
152	Counil & Achache (1987)	Central America	*Magsat* (x, z)	2° avg.	~400, 470
153	Counil & Achache (1987)	Central America	*Magsat* (x, z)	DC	
M. Atlantic Ocean and margins					
154	Hayling & Harrison (1986)	North & Central Atlantic	*Magsat* (s)	magn.	
155	Bradley & Frey (1991)	Labrador Sea	*Magsat* (s)	RTP	400
156	Alsdorf (1991)	South Atlantic	*Magsat* (s)	coll.	375, 485
157	Alsdorf (1991)	South Atlantic	*Magsat* (s)	RTP	375, 485
N. Pacific Ocean					
158	LaBrecque & Cande (1984)	Central Pacific	Marine (s)	avg.	0
159	LaBrecque & Cande (1984)	Central Pacific	Marine (s)	UC	400
160	LaBrecque et al. (1985)	North Pacific	Marine (s)	avg.	0
161	LaBrecque et al. (1985)	North	Marine (s)	UC	400
162	Harrison et al. (1986)	Central Pacific northern Pacific	*Magsat* (s)	magn.	
163	Harrison (1987)	Eastern equatorial Pacific	*Magsat* (s)	magn.	

(continues)

Table 6.1. *(continued)*

Map no.	Reference	Coverage and comments	Satellite or aircraft	Map type	Altitude (km)
O. Miscellaneous ocean					
164	Taylor (1991)	Eastern Indian Ocean	*Magsat* (s)	g&c	~350
165	Bradley & Frey (1988)	Southern Indian Ocean	*Magsat* (s)	RTP	400
166	Antoine & Moyes (1992)	South Africa & surrounding oceans (0–50° E, 0–50° S)	*Magsat* (s)	adaptive filter	400
167	Fullerton et al. (1994)	Southwestern Indian Ocean (0–58° E, 14–70° S)	*Magsat* (s)	RTP	400

[a]Component of map:

s, scalar	y, Y or B_ϕ component
v, vector	z, Z or B_r component
x, X or B_θ component	

[b]Type of map:

avg., average	hand, hand-contoured
wt. avg., weighted average	magn., magnetization
coll., collocation	m-m, magnetic moment
DC, downward-continued	RHA, rectangular harmonic analysis
DF, double Fourier representation	RTP, reduced to pole
DIM, depth-integrated magnetization	SCHA, spherical cap harmonic analysis
ES, equivalent source	SHA, spherical harmonic analysis
grid., gridded	susc., susceptibility
	s-t, susceptibility times thickness
g&c, gridded and contoured	UC, upward-continued

data relative to a degree-9 model have been published by Benkova and Dolginov (1971) and Regan, Davis, and Cain (1973), but are not included in the tables here because they are not good representations of the crustal field, although they are of historical interest.

Langel (1990b) recomputed a U.S. map (38) from *Cosmos 49* data relative to a degree-11 model, as shown in Figure 6.1. This map was derived using a larger bin size because of the high level of noise present in the data. Reliable models for higher degrees are not permitted by the quality and distribution of the data. Comparison of the anomalous field from Figure 6.1 with U.S. maps derived from POGO and *Magsat* data shows substantial agreement, in spite of the fact that the field model subtracted was of only degree 11 and in spite of the

Table 6.2. *Satellite and long-wavelength surface magnetic anomaly maps: summary of data-reduction processes*

Map no.	Quiet-data criteria	Main-field model	Long-wavelength corrections	Pass-by-pass comparisons	External models[a]
1	$Kp \leq 2+$; deleted local time (LT) 9–15 hours	GSFC(12/66)	Potential function	None	None
2	$Kp \leq 2$; 2σ cutoff in 2° block	MGST(4/81)	Potential function; linear trend	None	None
3	$Kp \leq 2$; deleted LT 9–15 hours	POGO(2/72)	Potential function; linear trend	Visual	None
4	$Kp \leq 2$; 2σ cutoff in 2° block	MGST(4/81)	Potential function; linear trend	None	None
5	Not known	MGST(4/81)	Potential function	None	MIF
6	*Am* and *Aj* indices	Unpublished, degree 13	Potential function; linear trend	None	MEA
7	$Kp \leq 2$	GSFC(12/83)	High-pass filter	None	None
8	$Kp \leq 2$	GSFC(12/83)	High-pass filter	None	Z component only
9–10	Derived from map 2				
11	Reject pass if $\lvert \Delta B \rvert \geq 20$ nT anywhere	POGO(2/72)	Linear trend	Visual	None
12	$Kp \leq 1-$ for 9 hours: range at Ft. Churchill Observatory	MGST(4/81)	Quadratic trend	Visual	None
13	$Kp \leq 1-$ for 6 hours; $\lvert \Delta B \rvert \leq 15$ nT	MGST(4/81)	Quadratic trend	?	None
14					
15	$Kp \leq 1-$ for 9 hours: range at Ft. Churchill Observatory	MGST(4/81)	Quadratic trend	Visual	None
16	$Kp \leq 1-$ for 9 hours: range at Ft. Churchill Observatory	MGST(4/81)	Quadratic trend	Visual	None
17–18	Variance limit	GSFC(12/83)	Scaled trend from field model; Fourier bandpass	Fourier	None
19	$Kp \leq 2+$; reject pass if "large" fluctuations in X or Z component	GSFC(12/83)	?	?	MPDF
20–21	Same as 17–18				
22	Based on map 17				
23–24	$Kp \leq 2$; pass rejected if any $\lvert \Delta B \rvert \geq 30$ nT; 9–15 LT deleted ±50°	POGO(2/72)	Potential function; linear trend	Visual	None
25	$Kp < 1$; residual ≤ 100 nT	M051782		None	None
26–28	$Kp \leq 2$; pass rejected if $\lvert \Delta B \rvert \geq 30$ nT anywhere between ±60°; poleward of 60°, $\lvert \Delta B \rvert > 40$ nT rejected	MGST(4/81)	Potential function; quadratic trend	Visual	None

(*continues*)

Table 6.2. *(continued)*

Map no.	Quiet-data criteria	Main-field model	Long-wavelength corrections	Pass-by-pass comparisons	External models[a]
29	$Kp \leq 1+$ for vector: $Kp \leq 2+$, for scalar; $AE \leq 50$ nT; variance ≤ 80 (nT)^2, high latitude	DAWN (6/83-6), GSFC (12/83)	High-pass filter	Visual & Fourier	Yes
30	See map 23				
31	Used data from 18 and 24				
32	$Kp \leq 2$, residual to model <100 nT	M070284	Potential function	None	MIF-like
33–34	Based on map 31, nonstringent criteria				
35–36	Based on map 31, stringent criteria				
37	$|\Delta B| > 60$ nT rejected	Unpublished	None	None	None
38	Outliers rejected	Unpublished, degree 11	None	None	None
39–40	$Kp \leq 2+$; LT 9–15 deleted	?	Potential function; quadratic trend	None	None
41–42	$Kp \leq 2+$; LT 9–15 deleted	POGO (2/72)	Potential function; quadratic trend	None	None
43	"Screeened"	IGS-75	Mean subtracted; low-pass filter		None
44	$Kp \leq 2+$; LT 9–15 deleted	POGO(2/72)	Quadratic trend	None	None
45	$Kp \leq 2-$	MG680982	?	None	None
46–47	?	POGO(2/72)	Quadratic trend	?	None
48	?	?	Linear trend	None	None
49–50	?	GSFC(9/80-2)	3rd-order polynomial		None
51–52	?	?, degree 13	Quadratic trend	?	None
53	Kp selection	?, degree 13	Quadratic trend	None	None
54–55	$Kp \leq 2+$	MGST(4/81)	Wavelength filter	Visual	None
56	Based on map 41				
57	Based on map 52				
58		?	?	?	None
59	$Kp \leq 3$	IGS-75, degree 12			None
60	Based on map 6				
61	?	MG680982	?	?	None
62	?	MGST(4/81)	Potential function	?	MIFC
63–65	$Kp \leq 2+$; low-pass ($\lambda > 4°$) filter	MGST(4/81)	Bandpass filter	Regression analysis	None
66	Based on map 63				
67–69	Based on map 63				
70	?	Unpublished	?	?	None
71	$Kp \leq 3+$	MGST(4/81)	Multiple regression; potential function; linear trend; spatial gradient	?	None

(continues)

Table 6.2. *(continued)*

Map no.	Quiet-data criteria	Main-field model	Long-wavelength corrections	Pass-by-pass comparisons	External models[a]
72–74	$Kp \leq 2+$	GSFC(12/83)	Potential function; bandpass filter; crossover analysis		EE
75–77	$Kp \leq 2+$; *AE* criteria for latitude >60°; outlier test	GSFC(12/83)	Potential function; high-pass filter, 4 km	?	MIFC *Sq* model
78	$Kp \leq 2+$	?	Crossover analysis	Visual	None
79	Used 2° binned data from map 6				
80	Based on map 31				
81	Used 2° binned data from map 6				
82–84	$Kp \leq 2$	MGST(4/81)	Quadratic trend	?	None
85	?	MGST(4/81)	Potential function	?	MIF
86	$Kp \leq 2$	MGST(4/81)	Trigonometric function	?	None
87–88	$Kp \leq 2+$	Core and external: polynomial expansion along track		?	None
89–90	$Kp \leq 2+$; 9–15 LT deleted	POGO(2/72)	Potential function; quadratic trend	None	None
91–93	$Kp \leq 2+$	GSFC(12/83)?	Potential function; quadratic trend	?	None
94	Based on nearby observatories	IGRF 1990	Detrend surface		From variometer
95–96	Rejected pass if $|\Delta B| > 20$ nT	POGO(2/72)	Linear trend	Visual	None
97		POGO(2/72)	Polynomial surface		None
98	$Kp \leq 2$; rejected pass if $|\Delta B| > 30$ nT	GSFC(9/80)	Quadratic trend	Visual	None
99	$Kp \leq 2$; rejected pass if $|\Delta B| > 20$ nT anywhere	GSFC(9/80)	Quadratic trend	Visual	None
100	?	?	?		
101	?	?	?	?	None
102	?	MGST(4/81)	?	Visual	None
103	$Kp \leq 3+$; 3σ from block mean	MGST(4/81)	Potential function; linear trend	?	None
104	$Kp \leq 3$	MGST(4/81)	Potential function; quadratic trend	?	None
105	?	MGST(4/81)	Potential function; linear trend	?	None
106	$Kp \leq 1$; 2σ from block mean	MGST(4/81)	Potential function; quadratic trend	?	None
107	Based on map 2				
108	Based on map 103				
109	Based on map 104				

(continues)

Table 6.2. *(continued)*

Map no.	Quiet-data criteria	Main-field model	Long-wavelength corrections	Pass-by-pass comparisons	External models[a]
110	$Kp \leq 0+$	MGST(4/81)	Potential function; quadratic trend	?	None
111–113	Based on map 104				
114	Based on map 2				
115–116	Based on maps 104				
117	$Kp \leq 1+;?$	GSFC(12/83)	High-pass filter	Visual	None
118–119	$Kp \leq 2$	GSFC(9/80)	Quadratic trend	?	None
120	$Kp \leq 2+$	MGST(6/80)	Crossover analysis; removed mean value	?	None
121	?	?	?	?	None
122	$Kp \leq 2$	GSFC(12/83)	Quadratic trend	?	None
123	?	This paper, degree 14	*Dst* correction	?	Polynomial model
124	Data set same as for map 122	IGRF 1985			
125	?	?	Crossover analysis	Visual	None
126	Hand selection	M102089 (J. C. Cain, personal communication)		Visual	None
127	Based on map 126				
128	Method same as for map 72				
129	Method same as for map 120				
130–131	?	Unpublished, degree 13	Potential function; linear trend	Visual	Polynomial model for EE
132–133	$Kp \leq 2$				
134	$Kp \leq 2$	GSFC(12/83)	Quadratic trend		
135	Subset of map 6				
136	Subset of map 9				
137	Map provenance not specified				
138	Similar to map 3				
139	Based on map 2 for N. and S. America, Euro-Africa, India, and Australia; map 14 for Antarctica				
140	Based on map 14 for Antarctica; elsewhere, "low magnetic activity"	MGST(4/81)	Quadratic trend	?	None
141–142	Method same as for map 72				
143	Same as map 109				
144–145	Rejected pass if $\lvert \Delta B \rvert > 20$ nT anywhere	POGO(2/72)	Linear trend	Visual	None
146–147	?	?	Quadratic trend	?	None

(continues)

Table 6.2. *(continued)*

Map no.	Quiet-data criteria	Main-field model	Long-wavelength corrections	Pass-by-pass comparisons	External models[a]
148–149	?	Degree 13	Linear trend	Visual	None
150	$Kp \leq 2+$; dawn only	MGST(4/81)	Potential function; linear trend	?	None
151	?	?	?	?	None
152–153	?	Unpublished, degree 13	Potential function; quadratic trend	Visual	None
154	Uses map 2				
155	?	?	?	?	None
156–157	Variance limit	GSFC(12/83)	Scaled trend from field model; Fourier bandpass	Fourier	None
158–159	$Ap < 35$; low-pass filter	DGRF 1970, degree 10	*Dst* model		*Sq* model
160–161	$Ap < 35$; Low-pass filter	Unpublished, degree 13	*Dst* model		*Sq* model
162	Uses map 2				
163	?	?	?	?	?
164	$Kp < 2$	MGST(4/81)	Crossover analysis	Visual	?
165	?	?	?	?	?
166	$Kp \leq 3$	MGST(4/81)	Potential function	?	Regression algorithm
167	See maps 75-77				

[a]MIF, mean ionospheric field correction; MEA, mean equatorial anomaly; MPDF, mean polar disturbance field; EE, equatorial electrojet.

high noise level. One other map from *Cosmos 49* has been published, the map of Africa (70) by Regan et al. (1975).

A major breakthrough occurred with publication of the Regan et al. (1975) anomaly map (1) of the world (±50°) (Figure 1.1). It constituted a demonstration that fields whose origins almost certainly were in the crust could be detected and mapped using satellite measurements. Although the map shows clear contamination, in the form of a north–south-trending herringbone pattern, the major anomalies are clear.

Extension to higher latitudes has been difficult, and, indeed, can be said to be still in progress. This is because of the presence of fields from auroral ionospheric currents and from field-aligned currents. The first map at higher latitudes was of western Canada (95), and the first north polar map was (11).

Table 6.1 lists 167 maps, and this includes only maps whose processing has been completed. To avoid confusion, in reading this chapter it is necessary to distinguish two types of spherical harmonic expansions. The first is the usual

Table 6.3. *Satellite and long-wavelength surface magnetic anomaly maps:*
special features and comments

Map no.	Special features and comments
2, 4	Now known to be affected by ionospheric fields at equatorial latitudes.
6	Maps for dawn, dusk, and combined local times.
7	Kaiser high-pass filter, 4-km cutoff. Maps from dawn and dusk. Preliminary map, still contains ionospheric field contamination.
8	Kaiser high-pass filter, 4-km cutoff. Maps for dawn and dusk. Uncorrected B and B_ϕ maps. Z-component map corrected for equatorial electrojet.
9	Minimum-norm downward-continued, following Section 6.5.
10	Downward-continued using SHA; degrees 14–52 for analysis based on data from <425-km altitude; degree 14–42 for analysis based on data from >425-km altitude.
15	Maps for dawn, dusk, and combined local times.
16	Spherical cap analysis of data used in map (13).
17, 18	Maps for dawn, dusk, and combined local times; along-track Fourier filter; double Fourier correlation of dusk and dawn, and maps from different altitude ranges.
20, 21	Along-track Fourier filter; double Fourier correlation of maps from different altitude ranges.
23, 24	Maps for 0, 6, 12, and 18 hours local time, and combined.
25	Spherical harmonic potential analysis, degrees 14–26.
26–28	Maps for dawn, dusk, and combined local times. Spherical harmonic correlation/covariance selection of common features of dawn and dusk maps. Higher SHA resolution in (26) than in (27).
29	Kaiser high-pass filter: 12-km cutoff below 50° latitude, 4-km cutoff poleward of 50°; data corrected for attitude jumps; Fourier correlation selection of features common to adjacent passes; adjustment of level by crossover analysis; SHCA selection of features common to dawn & dusk.
30	Based on scalar data of map (24). SHCA selection of features common to *Magsat* dawn and dusk and POGO.
31	Spherical harmonic potential analysis, degrees 14–49; spherical harmonic correlation/covariance selection of common features of *Magsat* dawn and dusk maps from Ravat et al. (1995) and POGO map (29). Two maps produced using stringent and less stringent correlation criteria.
33	Global equivalent source inversion by conjugate gradient method.
34	Global equivalent source inversion, with a priori, using conjugate gradient method, the SEMM of Section 8.5.
35	Iterative determination of SHA.
36	Iterative determination of susceptibility in 40-km-thick magnetic layer.
37	Relative to degree-9 field model.
38	Relative to degree-11 field model.
56, 57	Magnetization is assumed to reside completely in the lower crust, which is identified using seismic data.
60	Magnetization along field from minimum-norm solution in which magnetization direction is not constrained.
63–65	Dawn only.
67–69	RTP, same as map (63).
72	Gridded by collocation.

(continues)

Table 6.3. *(continued)*

Map no.	Special features and comments
75–77	Maps for dawn, dusk, and combined local times.
78	Method of Taylor and Frawley (1987).
79–80	Minimum-norm downward continuation.
81	Minimum-norm three-dimensional magnetization in 40-km-thick magnetized layer. Three components of magnetization are derived, with no constraint on direction.
86	Gridded by three-dimensional linear regression over 160-km horizontal separation. Maps for dawn, dusk, and combined local times; higher- and lower-altitude maps.
94	Composite aeromagnetic map detrended using long aeromagnetic control traverses corrected for transients by data from continentwide magnetometer array.
95	Spherical-earth upward continuation.
98	Mean of dawn and dusk; bandpass-filtered: 250–1,600 km.
99	Maps for dawn, dusk, and combined local times.
100	Filtered, gridded, and flat-earth upward-continued Canadian Earth Physics Branch high-altitude aeromagnetic data.
103	Data reduced to 400 km by method of Henderson and Cordell (1971)
107, 108	Equivalent source; assumes 40-km-thick magnetic layer.
109	Equivalent source model with ridge regression for stability.
111–113	Ridge regression with annihilator.
115, 116	Ridge regression.
117	Dawn local time only.
118, 119	Frequency-domain analysis; bandpass-filtered: 500–2,300 km; covariant analysis between dawn and dusk.
123	SHA model to degree 63, maps are from degrees 15–50.
124	Combined analysis of satellite and surface data, at the surface.
125	Method of Taylor and Frawley (1987).
126	Uncertainty estimated to be ±5 nT below 64° N latitude, ±10 nT north of 64° N latitude.
127	Two maps, one by trial and error in a forward model, one by linear inversion.
130, 131	Maps for dawn, dusk, and combined local times.
132, 133	Fourier covariant analysis between dawn and dusk; bandpass: 540–2,350 km; correlation analysis.
134	Statistical rejection of spurious points.
148	Separate maps for dawn and dusk; maps of Z at two altitudes.
152, 153	Maps for dawn, dusk, and combined local times; different altitudes.
154	Used 2° × 2° averages from map (2).
156, 157	Same as map (17).
162	Used 2° × 2° averages from map (2).
163	Used 2° × 2° averages from map (2), probably.

potential function representation of equation (4.25), with the magnetic field given by equation (4.26). The second is a spherical harmonic analysis of some component of the anomaly field, or the scalar anomaly field, following equation (4.21). When scalar data are represented, as in equation (4.21), the $(n + 1)$-

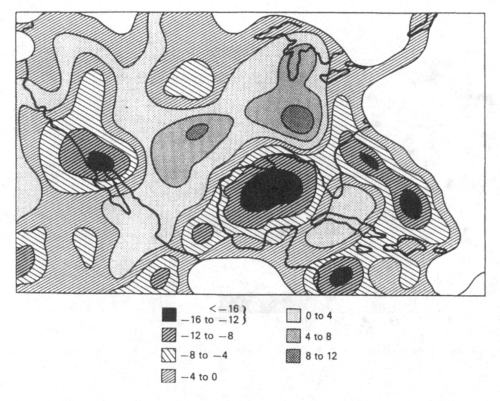

■ < −16 }	□ 0 to 4
▨ −16 to −12 }	▨ 4 to 8
▧ −12 to −8	▨ 8 to 12
◻ −8 to −4	
▨ −4 to 0	

FIGURE 6.1
Scalar anomaly map of the United States derived from *Cosmos 49* data. Selected data are averaged in 5° bins; units are nanoteslas; contour interval 4 nT. (From Langel, 1990b, with permission.)

degree term in the potential function corresponds nearly, but not exactly, to the n-degree term in equation (4.21), as shown in Section 4.5.4. The two types of analysis will be referred to as "potential" [equation (4.25)] and SHA [equation (4.21)].

6.2 SCALAR MAPS AT LOW AND MIDDLE LATITUDES

6.2.1 MAPS BASED ON DATA FROM THE POGO SATELLITES

As already noted, the first scalar anomaly map (1), shown in Figure 1.1, which is restricted, as much as possible, to fields originating in the lithosphere, was derived from POGO data. There are three other maps based on POGO data that have been widely used or are of particular importance: maps (3), (23), and (30). Map (3) and its preliminary forms were widely distributed prior to its publication, and for many years it was the best available map based on POGO data. At low and middle latitudes, map (3) does not differ substantially from map (23), reproduced here in Figure 6.2a. In the process leading to Figure 6.2a, separate maps were derived for four local times centered on 0, 6,

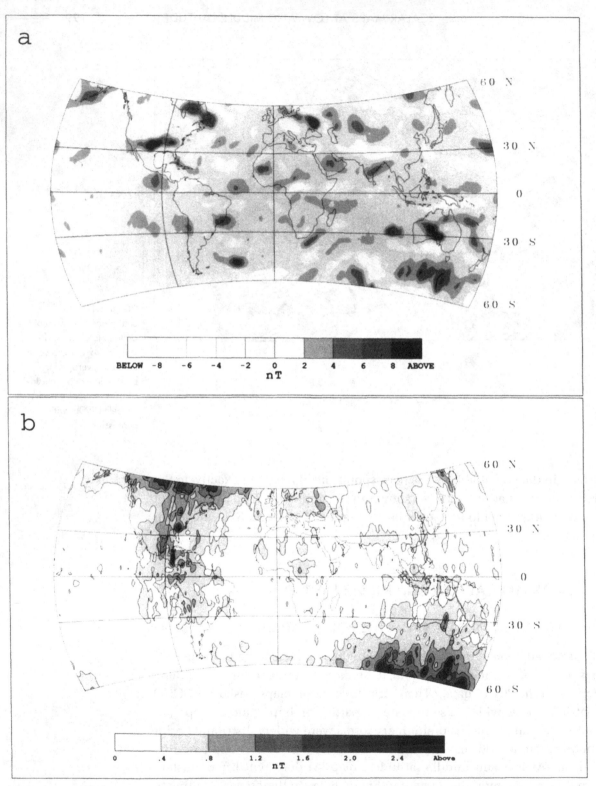

FIGURE 6.2
Average scalar anomaly map (23) from the POGO data. Units are nanoteslas; Van der Grinten projection,
±60°. (a) The average map. Contour interval is 2 nT. On this and other maps in this chapter, the zero
contour is suppressed; one shade is used for the range −2 nT to +2 nT. (b) 95% confidence interval for the
average map; contour interval 0.4 nT. (From Langel, 1990a, with permission.)

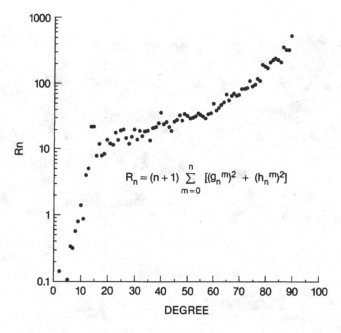

$$R_n = (n+1) \sum_{m=0}^{n} [(g_n^m)^2 + (h_n^m)^2]$$

FIGURE 6.3
Anomaly field spectrum corresponding to the map of Figure 6.2a. R_n is the total mean square contribution to the vector field by all harmonics of degree n. Determined from equation (2.6). (From Arkani-Hamed et al., 1994, with permission.)

12, and 18 hours. For low and middle latitudes, the noon-sector data are heavily influenced by the equatorial electrojet and are discarded. Unlike *Magsat* data, the low- and middle-latitude data from the remaining three local times show no substantial differences, and the final map is the average of these data. The reader should note the scale in Figure 6.2a, because it is a universal scale for all maps in this chapter.

An estimate of the 95% confidence interval for Figure 6.2a is given in Figure 6.2b. The estimate is twice the standard error of the mean, σ_μ, defined by

$$\sigma_\mu = \sigma/\sqrt{N}, \tag{6.1}$$

where σ is the standard deviation of the data (for each bin) about the mean, and N is the number of points in the mean.

The spectrum of R_n, defined and discussed in Section 2.1.1, is a diagnostic tool for anomaly maps. The R_n spectrum corresponding to Figure 6.2 is shown in Figure 6.3. R_n rises steeply for n between 1 and 14, drops a bit between degrees 15 and 16, and then rises slowly until degree 60. Above degree 60, the rise steepens. Two diagnostic features of this spectrum stand out. First is the steepening at higher degrees. This is likely an indicator of the presence of observation and reduction noise in the higher harmonics that is not removed in the process of deriving the map. The second diagnostic feature is at degrees 14 and 15. As expected with removal of the field from a main-field model of degree 13, there is no substantial signal below degree 14. But an unexpected feature is that the power in degrees 14 and 15 appears stronger than is consistent with

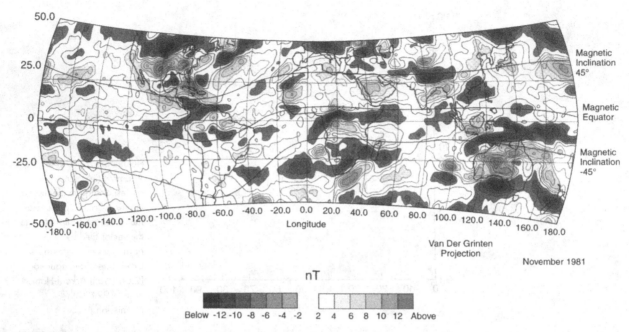

Van Der Grinten
Projection

November 1981

nT

Below -12 -10 -8 -6 -4 -2   2 4 6 8 10 12 Above

FIGURE 6.4

Average *Magsat* scalar anomaly map (2) from combined dawn and dusk data with no correction for fields from the equatorial electrojet. Van der Grinten projection; units are nanoteslas; contour interval 2 nT. (From Langel et al., 1992a, with permission.)

that at higher degrees. It seems probable that the removal of the main field has been imperfect and that the degree-14 and -15 potential terms for the anomaly map still retain substantial contributions from the main field. That remains a matter of some contention that merits further research, and it will be at issue in later sections of this chapter. Based on the foregoing considerations, a map (30) is derived that retains only potential function harmonics of degrees 16 through 60, or harmonics of about degrees 15 through 60 in the non-potential spherical harmonic formulation.

6.2.2   MAPS BASED ON DATA FROM *MAGSAT*

Because of *Magsat's* lower altitude, *Magsat* data were acquired closer to the lithospheric sources than were POGO data, and so have higher resolution. However, they are also closer to the E region of the ionosphere, in which there is a substantial flow of current. Moreover, as altitude is lowered, the E region is approached more rapidly than the lithosphere. This means that the amplitude of fields from ionospheric currents increases proportionately more than that of the lithospheric fields. As a result, the *Magsat* data show differences between dawn and dusk that are not apparent in the POGO data. This was not appreciated immediately, and the initial *Magsat* scalar map (2), Figure 6.4, contains contamination because of the presence of fields from the equatorial electrojet (EE). These are apparent in the figure from trends near the dip equator (zero-magnetic-inclination line) that tend to follow that equator.

Cain et al. (1984) derived a spherical harmonic analysis based on *Magsat* data through degree 29. No attempt was made to treat dawn and dusk data as independent data sets. From terms of degrees 14 through 29 they computed scalar and vector anomaly maps; the scalar map (25) agrees well with Figure 6.4. Cain et al. (1989, 1990) extended the model to potential field degree 63 and degree 49, respectively, using a data set corrected for an external potential function and using an MIF model for fields from ionospheric and other sources. The degree-49 model is preferred (Cain et al., 1990). They did not publish a global anomaly map, but have kindly furnished the spherical harmonic coefficients (J. C. Cain, personal communication, 1993). One advantage of the potential field representation for anomaly fields is that the data apparently need not be filtered along-track individually for each pass, avoiding a source of bias in the anomaly maps.

Arkani-Hamed and Strangway (1985a,b, 1986a) seem to have been the first to treat *Magsat* dawn and dusk data separately. The resulting maps are each fitted with a spherical harmonic series of the form of equation (4.21), together with the corresponding degree power spectrum, P_n, and degree correlation coefficient, ρ_n, between the dawn and dusk spherical harmonic expansions. SHCA (Section 4.5.2) is used to extract common features of the dawn and dusk data. Lower-degree harmonics, $n \leq 13$, have most of the power; high correlations are found between dawn and dusk for $6 \leq n \leq 14$ and for $18 \leq n \leq 41$. The high correlations for the lower degrees were attributed by Arkani-Hamed and Strangway to residual core field in ΔB.

A major difference between the anomaly maps of Arkani-Hamed and Strangway (1986a, 1985a) and those published by earlier workers is the elimination by Arkani-Hamed and Strangway of the longer-wavelength signal, $13 < n < 18$, on the grounds that it should have been removed by subtraction of the model field. Comparison of their map of *discarded* long wavelengths (Arkani-Hamed and Strangway, 1986b, fig. 5a) with, for example, Figure 6.2 shows that the two maps are highly similar. There is a question still to be resolved: How low a degree should be retained in the SHA for the lithospheric fields? When those fields are highly correlated between dawn and dusk data sets, they are unlikely to have originated in the ionosphere or magnetosphere. However, as indicated earlier, some contaminating long-wavelength residuals could be due to inadequate modeling of the main field, though it is unlikely that this is true for degrees greater than 15 in the potential function representation.

The most recent *Magsat* scalar maps, as of this writing, have been derived following the procedures outlined in Section 4.10. There are two resulting scalar maps (29), one derived from processing of the scalar data, and the other from processing of the vector data, followed by computation of the corresponding scalar map.

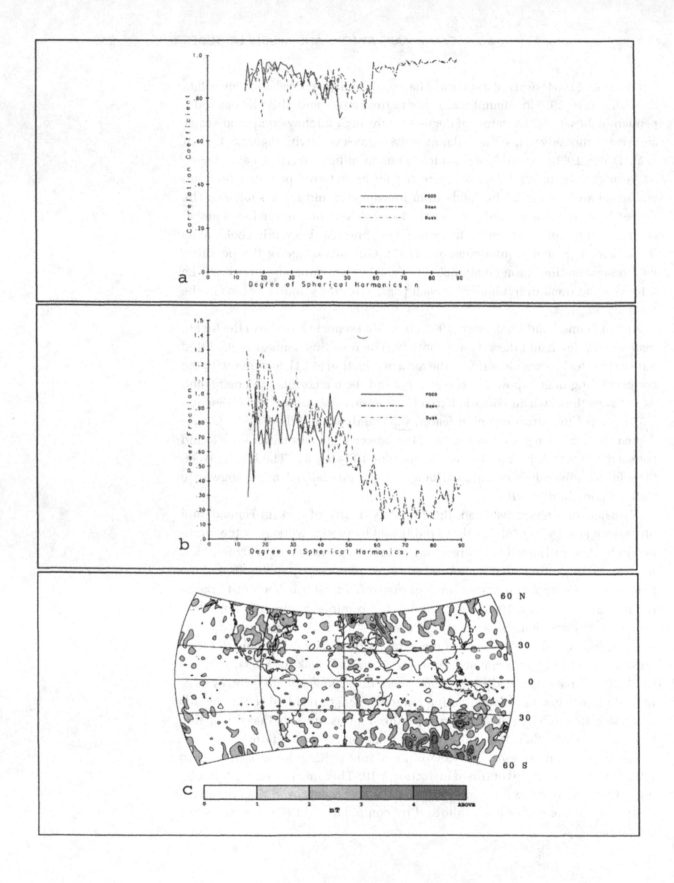

a

b

c

6.2.3 MAPS BASED ON COMBINED *MAGSAT* AND POGO DATA

If SHCA of the independent dawn and dusk *Magsat* data sets is effective in isolating the common anomaly signal, an additional comparison with POGO data should be even more effective. The starting place for such a three-way comparison consists of the separate, corrected, dawn and dusk scalar data sets of Ravat et al. (1995), in SHA form following equation (4.21), designated "Dawn" and "Dusk," and the POGO map (30). Each of these three is considered an independent data set. As for Figure 6.3 for POGO data, the Dawn and Dusk *Magsat* maps showed excess power at degrees 14–15, so harmonics of degree less than 16, or SHA harmonics less than 15, are excluded from the analysis. Above degree 60, only the two *Magsat* data sets are used.

For the three-way correlation of the Dawn, Dusk, and POGO data sets, Arkani-Hamed et al. (1994) adopted two sets of criteria, called stringent and less stringent. Only the latter is considered here. Using a modified version of SHCA, for each degree and order, three degree/order correlation coefficients, ρ_{nm}, and three amplitude ratios, R_{nm}, are computed, the first between POGO and Dawn, the second between POGO and Dusk, and the third between Dawn and Dusk. In each case the comparison is considered acceptable if $\rho_{nm} \geq 0.5$ and the amplitudes are not different by more than a factor of 2. For each degree and order, then, it is possible to have one, two, or three pairs or no pair of coefficients in acceptable agreement. If one pair of coefficients is in agreement, their average is adopted as the final coefficient. If two or three pairs are in agreement, the average for all three is adopted as the final coefficient. If no pair of coefficients is in agreement, the harmonic coefficient is set to zero. Figure 6.5a shows the degree correlations between the selected coefficients from each of the three original maps and the combined coefficients of the final map. Below degree 15, no coefficients are selected. Above degree 60, POGO data are not considered, so its correlations with the final map are zero. The jump in correlation at degree 60 occurs because below degree 60 only two out of the three data sets are required to correlate, whereas above degree 60 two out of two must correlate.

Figure 6.5b shows the ratio of final (combined) power to original (unselected) power versus degree for each of the three original data sets. This fraction can be greater than 1.0 because the contributions of the original sets are averaged for the selection, and the average of the selected coefficients can be greater than one (or even two) of the three contributions to the average. Below degree 50, the output power is almost always at least 50% of each of the three inputs, often 70–90%. Above degree 60 this percentage drops rapidly to between 20% and 30%. This is indicative of the expected higher noise levels at the higher degrees in the original maps. The figure also shows that below about degree 40, POGO anomalies have more energy than *Magsat* anomalies, and the final

FIGURE 6.5 (*opposite page*) Parameters related to map (31), Figure 6.7. (a) Degree correlations between the selected coefficients for each of the contributing data sets and the combined coefficients of the final map. (b) Ratio of final (combined) power to original (unselected) power for each of the contributing data sets. (c) Standard error of the anomaly map; contour interval 1 nT. (From Arkani-Hamed et al., 1994, with permission.)

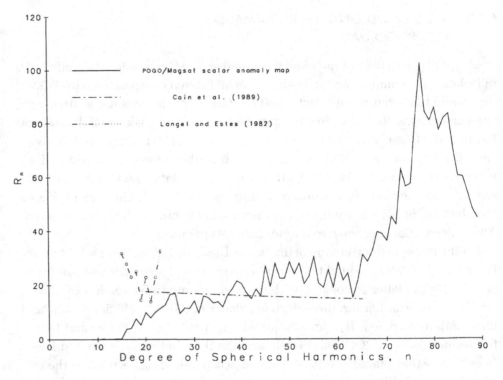

FIGURE 6.6
R_n field spectrum for combined POGO/*Magsat* scalar anomaly map. For comparison, values are included from the determinations of Cain et al. (1989) and Langel and Estes (1982). R_n is the mean square value over the earth's surface of the magnetic field intensity produced by harmonics of the *n*th degree. (From Arkani-Hamed et al., 1994, with permission.)

power has a larger contribution from POGO than from either Dawn or Dusk. Above degree 40, the *Magsat* anomalies have more contribution to the final power than POGO anomalies. These differences in energy contribution probably reflect deletion of lithospheric signal at low degrees from the *Magsat* data during processing and higher resolution of *Magsat* data, that is, power in the higher-degree terms. The resulting coefficients are converted into a potential function, from which R_n is computed, as shown in Figure 6.6, together with values of R_n as found by Langel and Estes (1982) and by Cain et al. (1989). These latter two results were derived from a main-field modeling process using data that were not filtered along track. The analysis of Langel and Estes (1982) extended only to degree 23 and likely was affected by aliasing, whereas that of Cain et al. (1989) was derived by a process of integration that presents some difficulties. Thus, these results should be regarded as indicative rather than definitive.

The most obvious feature of Figure 6.6 is the large increase in amplitude of R_n from the anomaly map above degree 65. The field model R_n, in contrast, is decreasing, albeit slowly, in accord with a source in the lithosphere with a near-white-noise spectrum. It is concluded that the coefficients above degree 65 are dominated by noise. Accordingly, in the final map (31), shown here as Figure 6.7 (in color in Plate 1), only SHA coefficients from degree 15 through degree 65 are retained.

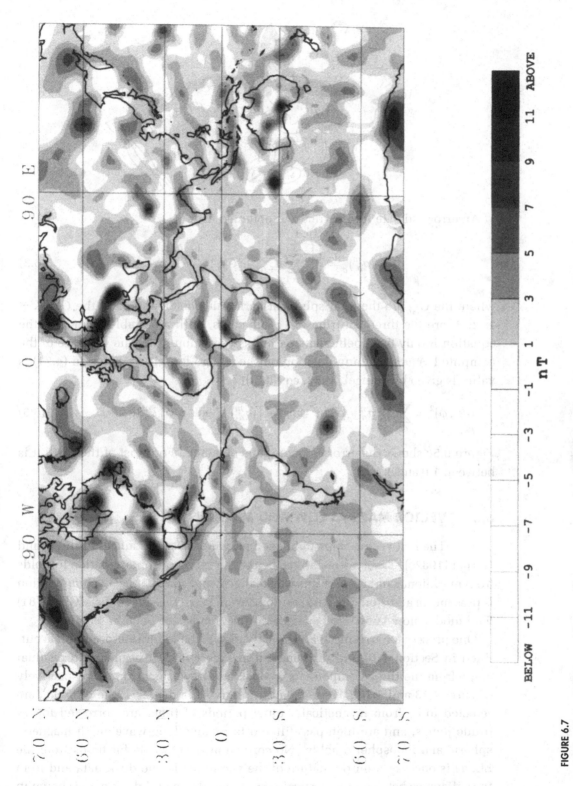

FIGURE 6.7

Combined POGO/Magsat scalar anomaly map (31) at 400 km of altitude; Van der Grinten projection; units are nanoteslas; contour interval 2 nT. See Plate 1 for color version. (From Arkani-Hamed et al., 1994, with permission.)

An error estimate for Figure 6.7 is obtained from

$$(\sigma_n^m)_c = \left\{ \left[\sum_{k=1}^{3} \left(\underline{C}_{nm} - C_{nm}^k \right)^2 \right] \Big/ 3 \right\}^{1/2}, \tag{6.2}$$

where the $\underline{C}_{nm}$ are the final spherical harmonic coefficients, and the C_{nm}^k, $k = 1, 2, 3$, are the three contributing coefficients, with a suitable change in the equation if only two coefficients pass the correlation test. The variance of the computed $A^s(\theta, \phi)$, at any position on the sphere for which equation (4.21) is valid, is given by [Langel, 1987, eq. (173)]

$$\sigma(\theta, \phi)^2 = \sum_{nm} \left\{ \left[(\sigma_n^m)_c \right]^2 \cos^2(m\phi) + \left[(\sigma_n^m)_s \right]^2 \sin^2 m\phi \right\} P_n^m(\cos\theta). \tag{6.3}$$

Figure 6.5c shows the error estimate thus obtained. For most of the earth it is between 1.0 and 1.4 nT.

6.3 VECTOR MAPS AT LOW AND MIDDLE LATITUDES

The first vector maps published (4) were those of Langel, Phillips, and Horner (1982c). These were derived before the extent of contamination by fields from meridional and EE currents was realized, and so equatorial contamination is present. In addition, the data residuals were taken relative to the MGST(4/81) field model, now known to be affected by EE fields (see Section 4.3.4).

One process for arriving at an anomaly map suitable for interpretation is outlined in Section 4.10 and diagrammed in Figure 6.8. It begins with residual maps from the three components at dusk and dawn local times, respectively (Figures 4.13 and 4.14). These data are relative to a degree-13 field model, are selected to be from magnetically quiet periods of time, are corrected for attitude jumps, and are high-pass-filtered to remove long-wavelength magnetospheric and ionospheric fields. No correction is yet made for fields from the EE, as is obvious from the nature of the variations in the dusk data and from the difference between the vertical-component dawn and dusk maps shown in

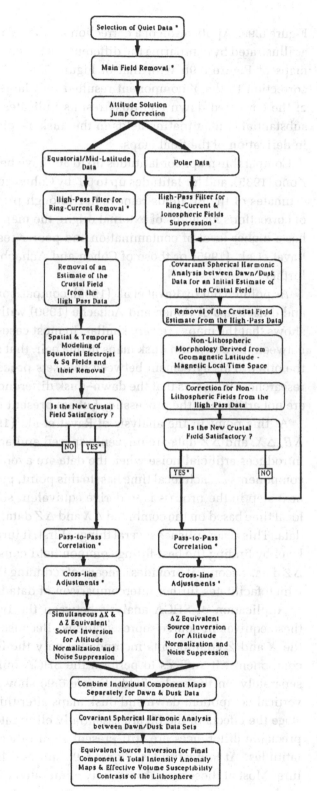

FIGURE 6.8
Schematic diagram of vector
data reduction process.
(From Ravat et al., 1995,
with permission.)

Figure 6.9a. Application of a correction for the EE makes a dramatic difference, as illustrated by comparing the difference between the corrected dawn and dusk maps of Figure 6.9b with that of Figure 6.9a. Although application of this correction to the Y component resulted in a large improvement, comparison of the corrected dawn and dusk results indicates the continuing presence of substantial contaminating fields in the dusk Y-component data. It is not used in derivation of the final maps.

Comparable maps for latitudes up to 50° have been made by Yanagisawa and Kono (1985), and for latitudes up to 60° by Cohen and Achache (1990), also with estimates of the EE fields removed. Although pioneering in the development of corrections for fields of external origin, the map of Yanagisawa and Kono (5) has a higher level of contamination and poorer resolution than either those of Ravat et al. (1995) or those of Cohen and Achache and will not be discussed further.

As pointed out by Ravat et al. (1995), comparison of the fully corrected dawn and dusk maps of Cohen and Achache (1990) with those of Ravat et al. (1995) shows that the maps are very similar. In most cases, where there is a difference between the dawn and dusk maps in one pair, that same difference is evident in the other. This agreement between data sets processed differently by separate researchers indicates that the dawn–dusk differences at this stage of processing are not artifacts of the processing, but are present in the data.

At this point in the analysis of Ravat et al. (1995), the multiply corrected ΔB, ΔX, and ΔZ data are unevenly spaced and are at different altitudes. This introduces artificial noise when the data are averaged. Also, processing of the components at each local time has, to this point, proceeded independently. The next step in the process is to derive equivalent source representations at each local time based on the combined ΔX and ΔZ data, and another based on the ΔB data. This accomplishes several things. First, it further reduces the overall noise level by finding the best-fitting common field consistent with both the ΔX and ΔZ data. Second, it provides a means of reducing the data to a common altitude, which facilitates further intercomparison of data from the two local times.

Application of SHCA analysis between the two local times makes use of these equivalent source representations. Because the representation based on the X and Z components incorporates only the features consistent with both components, it suffices to perform the SHCA only on the Z component and, separately, on the scalar field. Figure 6.9c shows the difference between the vertical-component dawn and dusk maps after this step of the process. At this stage the effects of the EE are essentially eliminated from the data, though appreciable differences are still present from other causes, particularly at high latitudes. At low latitudes, Figure 6.9c shows differences of fine spatial structure. Most of these are subsequently eliminated by application of the crossover analysis.

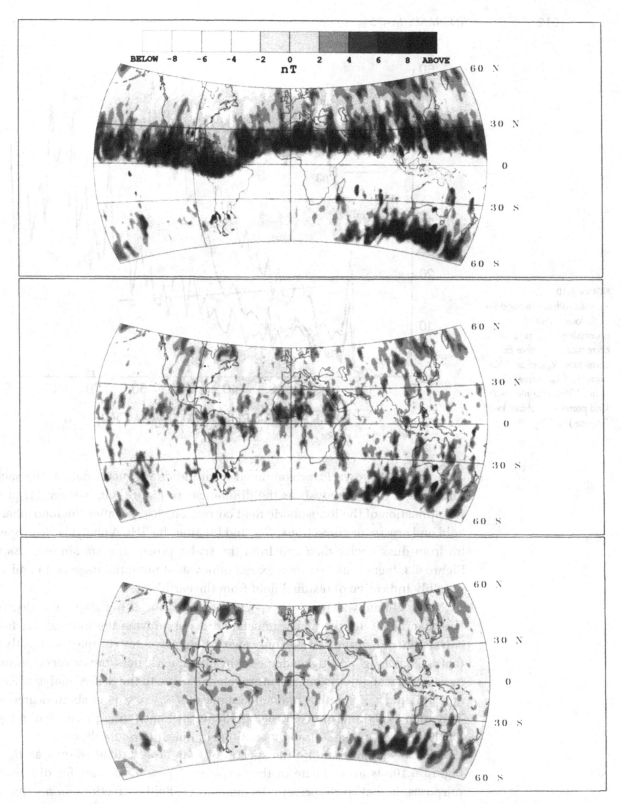

FIGURE 6.9
Differences of dawn and dusk Z-component maps; units are nanoteslas; contour interval 2 nT. Top:
Before ionospheric field corrections. Middle: After ionospheric field corrections. Bottom: After
SHCA correction. (From Ravat et al., 1995, with permission.)

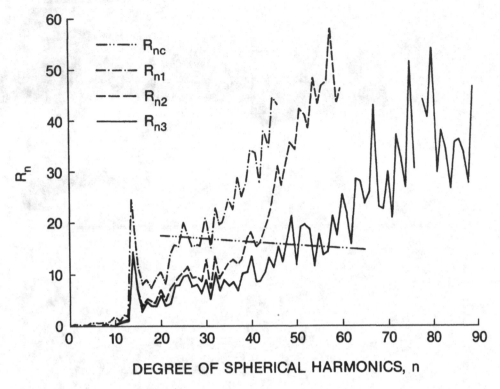

FIGURE 6.10

R_n spectra from vector data at various stages of processing. R_{n1}, prior to EE correction; R_{n2}, after EE correction; R_{n3}, after SHCA analysis. (R_{nc} is from Cain et al., 1989; reprinted with kind permission of Blackwell Science.)

Figure 6.10 shows R_n as computed from the dawn vector data at the same three stages in the analysis as the differences in Figure 6.9, namely, (1) prior to application of the ionospheric field correction, R_{n1}, (2) after the ionospheric field and crossover corrections, R_{n2}, and (3) after the SHCA analysis, R_{n3}. Spectra from dusk vector data and from the scalar processing are similar. As in Figure 6.3, there is an apparent excess of power at potential degrees 14 and 15, possibly indicative of residual field from the earth's core.

Power is removed from the spectra at each processing step. For degrees less than about 40, the major adjustment occurs during the ionospheric field correction, indicating that the ionospheric field correction is operable mostly in that spectral range. At higher degrees, the ionospheric field/crossover correction has some effect, but the largest improvement is due to the SHCA analysis. Even after this process, a rather dramatic increase in R_{n3} occurs at about degree 65. This increase is interpreted as being due to the continuing presence of a high level of noise, probably from fluctuations in ionospheric fields.

It is convenient to adopt the spectra of Cain et al. (1989), shown as R_{nc} in Figure 6.10, as an estimate of the "expected" power, at least for discussion purposes. In that spirit, prior to the ionospheric field correction, for $n > 30$, R_{n1} lies above R_{nc}, indicating dominance of R_{n1} by ionospheric fields. However, for about $15 \leq n \leq 24$, $R_{n1} < R_{nc}$. This is taken to mean that along-track filtering of the data has deleted lithospheric signal in this spectral range. After application of the ionospheric field correction, R_{n2} is significantly less than R_{nc} up to about degree 40. This is interpreted to mean that the ionospheric field correction is deleting lithospheric field in this spectral range.

The final map in the process is obtained by combining the selected spherical harmonics from dawn and dusk. In view of the characteristics of R_n, the selected harmonics are limited to the SHA degree range 15–65. Figure 6.11 (in color in Plate 2) shows the resulting map. The difference map in Figure 6.9b and corresponding maps for other components furnish an approximation of an upper bound for an error estimate for Figure 6.11, and Figure 6.9c furnishes a likely lower bound. Differences in Figure 6.9b are, in general, less than 4 nT, and often less than 2 nT. In Figure 6.9c, the differences are more subdued, mostly less than 2 nT, and randomized, especially at low latitudes.

For comparison, Figure 6.12 shows the combined dawn and dusk maps of Cohen and Achache. When making the comparison, it is important to realize that the maps of Cohen and Achache are averages of dawn and dusk, rather than combinations of common spherical harmonics, and are not restricted to SHA degrees 15–65, but should include all SHA harmonics above degree 12. Cohen and Achache (1990) pointed out the existence, in their maps, of a pattern of anomalies in the ocean basins. Part of that pattern is evident in the maps of Ravat et al. (1995), but part is not. Specifically, in the North Atlantic, some anomalies are elongated in a direction parallel to the ridge, as first pointed out and modeled by LaBrecque and Raymond (1985). These are present in both sets of maps. In the southern Indian Ocean, the vertical-component map of Cohen and Achache (1990) shows both negative and positive anomalies aligned nearly parallel to the ridge axis. These are also evident in Figure 6.9, albeit at weaker magnitudes, not as strongly aligned along the ridge axis as in the map of Cohen and Achache (1990), and broken up into individual anomalies. In the northern Pacific and southern Atlantic, the anomalies in the map of Cohen and Achache (1990) appear in northwest-to-southeast and southwest-to-northeast orientations, respectively. Some of these trends are present in the maps of Ravat et al. (1995) after the ionospheric correction has been applied, but they are mainly eliminated by the crossover analysis. In the final map, Figure 6.11, the trends in the South Atlantic are present, although at lower amplitudes than in the map of Cohen and Achache (1990), but the trends in the Pacific are not seen. This does not necessarily imply that the trends reported by Cohen and Achache (1990), but absent from Figure 6.11, are not real. In the process of trying to derive a map to show anomalies that are almost surely lithospheric in origin, there is always the danger of eliminating some features that should be retained. However, when a trend is eliminated by the crossover analysis, that means it is eliminated by simple leveling adjustments of the passes involved. This would argue against a geologic origin.

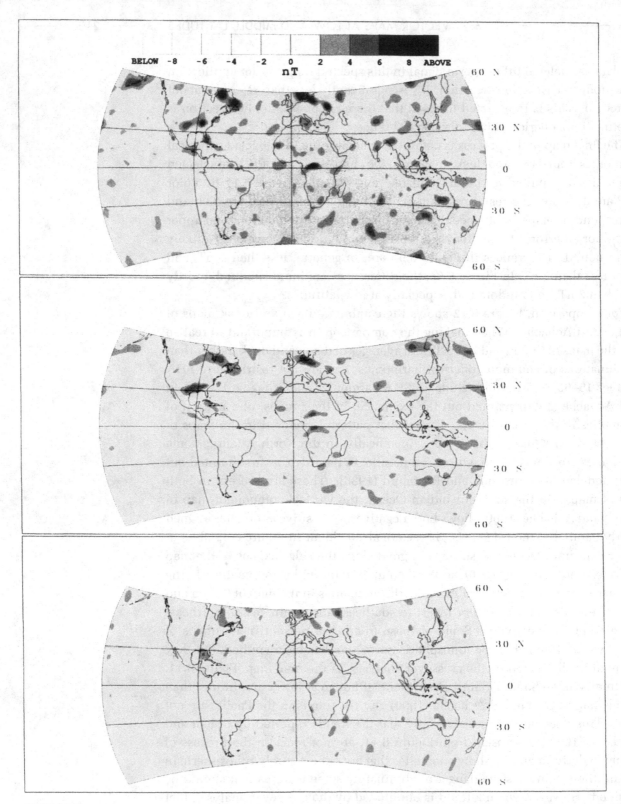

FIGURE 6.11
Combined residual Z (top), X (middle), and Y (bottom) component anomaly maps (29) from *Magsat* dawn and dusk data, based on a SHA analysis retaining degree-15–65 coefficients. Van der Grinten projection; units are nanoteslas. See Plate 2 for color version. (From Ravat et al., 1995, with permission.)

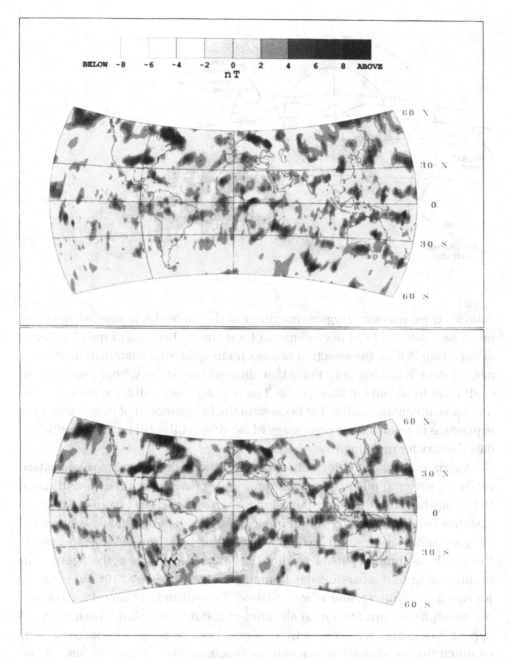

FIGURE 6.12
Combined residual X (bottom) and Z (top) component maps (6) from *Magsat* dawn and dusk data. Units are nanoteslas; contour interval 2 nT. (From Cohen and Achache, 1990, with permission.)

6.4   MAPS OF THE POLAR REGIONS

6.4.1   FACTORS IN HIGH-LATITUDE DATA

Isolation of anomaly fields at high latitudes is much more difficult than at low latitudes because of the ubiquitous presence of fields from ionospheric currents. The situation is exacerbated for *Magsat* data, as compared with POGO

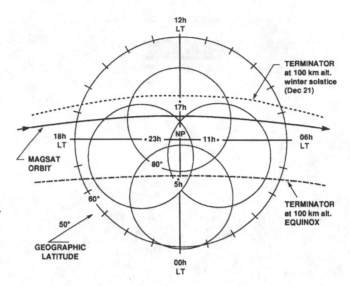

FIGURE 6.13
Schematic representation of relationship among *Magsat* orbit, the E-region terminator, and the auroral oval. Coordinate system is geographic latitude and local time. The poleward edge of the auroral oval is approximated by a circle for each of four representative universal times. (From Coles, 1985, with permission.)

data, for three reasons: *Magsat* was closer to the ionospheric sources, obtained much less data, and did not sample all local times. Because of the abundance of data from POGO, the selection process included visual examination of every pass of data, retaining only those that showed consistency from pass to pass, at all local times, and at all seasons. The selected data still have some level of ionospheric contamination, but because of the large amount of data available, it is possible to reject a large percentage of the data while still retaining sufficient data density for meaningful statistics.

Magsat acquired data for only seven months, which precludes characterization of seasonal effects. Furthermore, ionospheric conditions are different in the northern and southern polar ionospheres. Consider Figure 6.13, which sketches conditions in the Northern Hemisphere. The schematic coordinates are geographic latitude and local time. Recall that the *Magsat* orbit was nearly fixed in local time, in the dawn–dusk meridian, as shown in the figure. The terminator, or line separating day from night, at an altitude of 100 km is shown for two dates, equinox and winter solstice. This altitude is important, because the ionospheric currents flow at altitudes of about 95–120 km. When the ionosphere is in sunlight, its conductivity is enhanced; when it is in darkness, high conductivity occurs only where particle precipitation is high. *Magsat* data acquisition began just after the autumnal equinox and ceased before the summer solstice. This means that for much of its lifetime, the north polar ionosphere was in darkness. On the other hand, the south polar ionosphere was mostly sunlit. As a result, though ionospheric currents are present most of the time in both hemispheres, the currents are stronger and broader in extent in the south than in the north. The isolation problem for lithospheric fields in *Magsat* data is worse in the south than in the north.

Another complicating factor in polar regions is the changing geometry of the auroral currents with respect to the satellite orbit. In low latitudes, at dusk, the fields from the EE organize nicely around the dip equator in a consistent way at all longitudes. Auroral currents generally are concentrated in what is called the auroral oval, which is "centered" near the geomagnetic pole, not the geographic pole, and hence shifts its position relative to the *Magsat* orbit depending on the time of day. In Figure 6.13, the positions of the oval are sketched for 5, 11, 17, and 23 hours UT, showing its shift relative to the orbit of *Magsat*.

Considerable differences exist in the published maps for the polar regions because researchers have taken different approaches to dealing with fields from ionospheric currents. Study of such differences gives an indication of the robustness of anomaly identification and isolation. Hence, in the following sections, a collection of published maps is given for comparison. To facilitate that comparison, Figure 6.14, map (23), serves as a "key" figure for locating 41 and 37 supposed anomalies in the northern and southern polar regions, respectively.

6.4.2 NORTH POLAR SCALAR MAPS

North polar maps are collected together in Figure 6.15. The top two maps are based on POGO data, the rest on *Magsat* data. Each of these maps is from a different researcher with different selection criteria. The criterion for Figure 6.15a, map (23), was visual analysis of POGO data, as described in the preceding section. Figure 6.15b, map (30), was derived by performing a spherical harmonic analysis of Figure 6.15a and, on the basis of the R_n spectrum of Figure 6.3, retaining terms of degrees 15–60. The criterion for Figure 6.15c, map (15), was to strictly limit data to those from magnetically quiet times, "quiet" being defined in terms of the Kp and AE indices and by examination of data from high-latitude magnetic observatories. Data were retained from only 207 passes. Figure 6.15d, map (26), is based on data first selected on the basis of the Kp index, and then reliance on SHCA to select the common features between dawn and dusk. The harmonics retained are between 18 and 55. Figure 6.15e, map (32), is a plot of the field from harmonics 14–49 of the model of Cain et al. (1990). The data used were selected on the basis of Kp and of their nearness to a previous spherical harmonic model. Ionospheric field corrections, the MIF, extended through the polar regions. For Figure 6.15f, map (29), a combination of criteria was used: selection of quiet data based on the AE index and the pass variance, correction with an empirical model of fields of ionospheric origin, harmonic correlation of adjacent passes to select common features, and SHCA between dawn and dusk maps to select the common features. Table 6.4 lists the amplitudes of the 41 anomalies indexed in Figure 6.14 (top), as read from

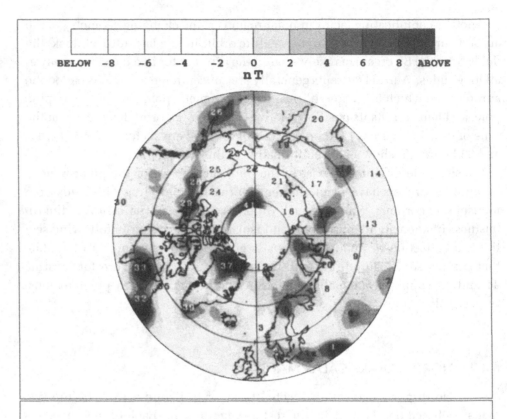

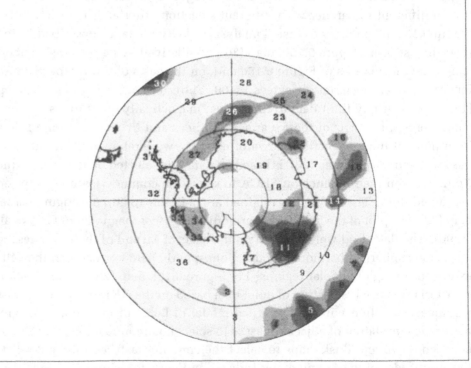

FIGURE 6.14
Key maps. Top, north polar map; bottom, south polar map. Both from map (23), based on POGO data. Anomalies on each map are numbered consecutively and used to reference other maps. For the Northern Hemisphere, zero degrees longitude is at the bottom of the plot, increasing counterclockwise. For the Southern Hemisphere, zero degrees longitude is at the top of the plot, increasing clockwise. Latitude interval is 10°. These are the conventions for all northern and southern polar plots. Equatorward latitude is 50° or −50°. (From Langel, 1990a, with permission.)

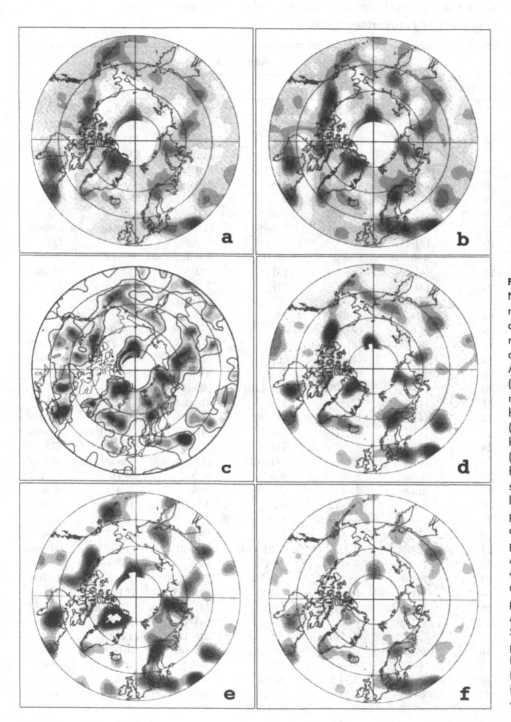

FIGURE 6.15
North polar scalar anomaly maps; data combined from available local times. Top row from POGO; middle and bottom rows from *Magsat.* (a) Map (23). (b) Map (30) derived from map (23) after SHCA, harmonics 15–60 retained. (c) Map (15). (d) Map (26), harmonics 18–55 retained. (e) Map (32). (f) Map (29), from scalar data. For scale, see Figure 6.14. Lowest latitude is 50°, except for part c, where it is 40°. (Part a from Langel, 1993, with permission; part b from Arkani-Hamed et al., 1994, with permission; part c from Coles, 1985, with permission; part d from Arkani-Hamed and Strangway, 1986a, with permission; part e derived from a model kindly furnished by J. C. Cain; part f from Ravat et al., 1995, with permission.)

Table 6.4. *Anomaly amplitudes from selected anomalies in Figure 6.15 ( ) indicates negative*

Anomaly	Amplitude (nT)					
	a	b	c	d	e	f
1	12–14	12–14	>14	12–14	12–14	8–10
2	2–4	8–10	2–4	8–10	8–10	8–10
3	(2)–(4)[a]	(4)–(6)	(4)	(4)–(6)	(8)–(10)	(2)–(4)
4	(8)–(10)	(12)–(14)	(12)–(14)	(12)–(14)	(10)–(12)	(8)–(10)
5	4–6	8–10	10–12	8–10	10–12	6–8
6	2–4	4–6	4–6	8–10	10–12	2–4
7	4	2–4	2	2–4	2–4	2?
8	(4)–(6)	(4)–(6)	(6)	(2)	(4)	(2)
9	(2)–(4)	absent	(4)–(6)	(2)?	(4)–(6)	absent
10	(2)–(4)?	?	(4)	?	?	(2)
11	4–6	6–8	10–12	8–10	10–12	2–4
12	(4)–(6)	(4)–(6)	(6)	(10)S	(8)	absent
13	(2)–(4)	(4)	(2)	?	(4)–(6)	(2)
14	(2)–(4)	(4)–(6)	(6)	(4)–(6)?	(2)–(4)	(2)?
15	2–4	6–8	10	6–8	6–8	2
16	4–6	6–8	8	6–8	8–10	6–8
17	(6)–(8)	(8)	(8)	(4)–(6)	(8)–(10)	(4)–(6)
18	(4)–(6)	(4)–(6)	(6)	(4)	?	absent
19	2–4	6–8	6	4–6	6–8	2–4
20	(4)–(6)	(6)–(8)	(8)–(10)	(6)–(8)	(4)–(6)	(4)–(6)
21	(4)–(6)	(2)	(2)	absent	(4)?	(2)
22	(4)–(6)	(4)–(6)	(4)	(2)	(6)–(8)	absent
23	(6)–(8)	(8)–(10)	(6)	(4)–(6)	(6)–(8)	(4)–(6)
24	(6)–(8)	?	(10)	?	(4)–(6)	(2)–(4)?
25	(6)–(8)	(6)–(8)?	(4)–(6)?	(2)	absent	absent
26	4–6	4–6	8	4–6	8–10	2–4
27	absent	2–4	2	2–4	2–4	4–6
28	6–8	10–12	12	12–14	10–12	4–6
29	6–8	6–8	8	?	6–8	absent
30	absent	2–4	6	6–8	4–6	4–6
31	(2)–(4)	(8)	(12)–(14)	absent	(12)–(14)	absent
32	6–8	8–10	>14	6–8	8–10	6–8
33	8–10	8–10	10–12	10–12	10–12	6–8
34	(4)	(8)	(6)	(2)–(4)	(6)–(8)	(2)
35	(4)–(6)	(8)–(10)	(10)	(8)–(10)	(10)–(12)	(6)–(8)
36	(4)–(6)	(4)	[b]	(4)–(6)	(6)–(8)	(2)
37	10–12	8–10	12	10–12	12–14	4–6
38	(4)–(6)	(6)–(8)	absent	(4)–(6)	(6)–(8)	absent
39	4–6	6–8	8	6–8	4–6	4–6
40	2–4	2–4	8–10	4–6	4–6	2–4
41	8–10	12	>14	12–14	12–14	6–8

[a]Parentheses indicate negative values.
[b]See Labrador Sea.

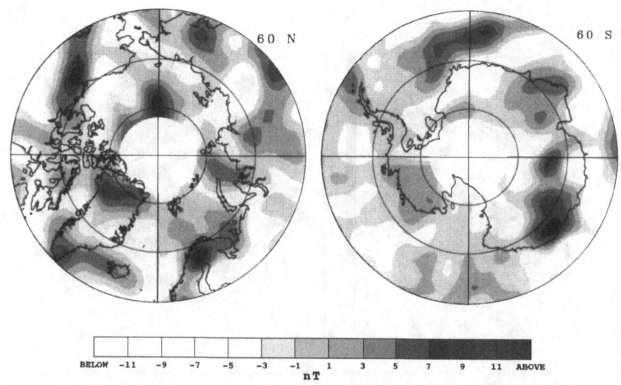

BELOW   −11   −9   −7   −5   −3   −1   1   3   5   7   9   11   ABOVE

nT

FIGURE 6.16
Combined POGO and *Magsat* polar scalar anomaly maps (31). Equatorward latitude is 50° or −50°. See Plate 3 for color version. (From Arkani-Hamed et al., 1994, with permission.)

Figure 6.15. For most anomalies, the amplitudes are all within a factor-of-2 range. Anomalies 9, 18, 21, 22, 25, 27, 29, 30, 31, 34, and 38 are absent from one or more of the maps.

Finally, Figure 6.16 (left) (in color in Plate 3) shows map (31), based on a combined analysis of POGO and *Magsat* data, the polar extension of Figure 6.7. Figure 6.17 shows indicators of the probable errors in the north polar maps. Figure 6.17a shows the standard error for the POGO map of Figure 6.15b (i.e., the northern extension of Figure 6.2b). Figure 6.17b shows an error estimate for the combined POGO/*Magsat* map of Figure 6.16 (left) derived from equation (6.2) (i.e., the northern extension of Figure 6.5c). Figures 6.17c and 6.17d show the differences between the *Magsat* dawn and dusk maps after correcting for ionospheric fields and before and after SHCA (i.e., the northern extension of Figures 6.9b and 6.9c).

6.4.3   SOUTH POLAR SCALAR MAPS

South polar maps are collected together in Figure 6.18. As in Figure 6.15, the top two maps are based on POGO data, the rest on *Magsat* data. The first two rows directly correspond to the same rows in Figure 6.15, with map (14) of Ritzwoller and Bentley (1983) replacing that of Coles (1985). These two maps

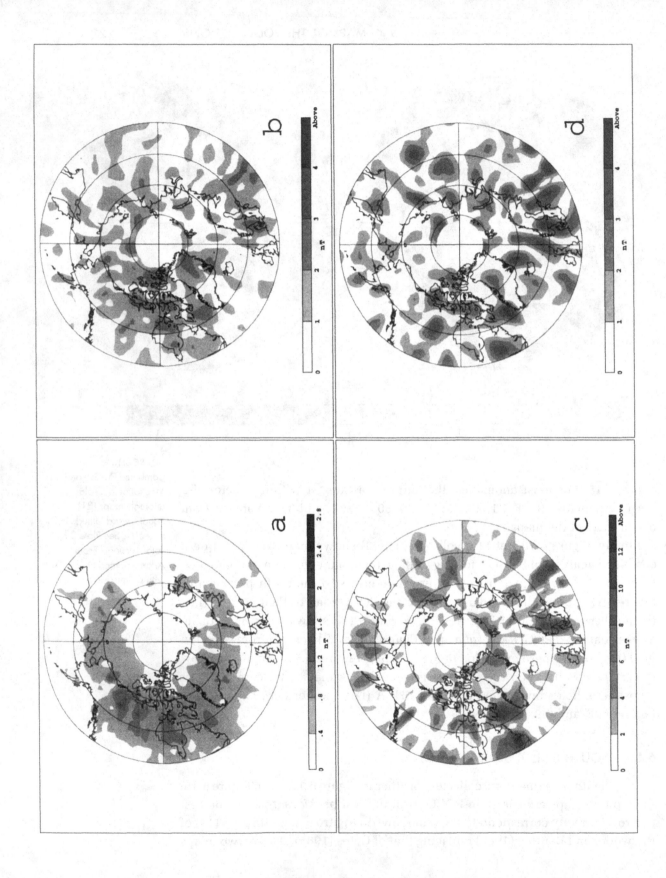

reflect similar selection criteria, so the discussion accompanying Figure 6.15 carries over. Figure 6.18h corresponds with Figure 6.15f. The two additional maps, Figures 6.18f and 6.18g, have no published counterpart for the north polar region. The philosophy behind map (18) of Figure 6.18g is almost the opposite of that used for Figure 6.18c. In deriving Figure 6.18g, as much data as possible were retained at each step. The initial data were selected according to a variance criterion for each pass. Then harmonic comparison was applied to adjacent passes in the two data sets, dawn and dusk, after which the data at each local time were subdivided into four altitude bands, with an anomaly map produced at each. SHCA was then used to select the common portions from dawn and dusk at each altitude band. The resulting four maps are continued to 430 km. SHCA filtering between these four maps resulted in the combined map of Figure 6.18g. For map (19) of Figure 6.18f, the data were selected from passes with low variance in the X and Y components and with $Kp \leq 2$. An empirical model of the disturbance field, the MPDF, was removed from the data. For reference, Table 6.5 shows a list of the amplitudes, from Figure 6.18, of the 37 anomalies indexed in Figure 6.14 (bottom).

Figure 6.16 (right) (in color in Plate 3) is the south polar map (31) based on combined analysis of POGO and *Magsat* data, the extension of Figure 6.7. Figure 6.19 shows the same indicators of the probable errors in the south polar maps as found in Figure 6.17 for the north.

6.4.4 VECTOR MAPS

Vector maps of the vertical component, Z, at polar latitudes are very similar to scalar maps, because the main field is itself mainly radial at these latitudes. The presence of large-amplitude X- and Y-component variations in the auroral belt, due to field-aligned currents (FACs), makes it nearly impossible to derive anomaly maps of these components. Two such attempts have been made, those of Coles et al. (1982) and Cain et al. (1984). Comparison shows many similarities between the two sets of maps, but also some large disagreements. Both seem to be dominated by arc-like bands of positive and negative, suggesting contributions from ionospheric currents or FACs.

6.5 COMPARISONS BETWEEN MAPS

6.5.1 A CORRELATION PROCEDURE

A method is now described for evaluating the correlation between two gridded sets of data for a given field component covering a single geographic region. The two sets of grid points must be common, say at $\{\mathbf{r}_i, i = 1, \ldots\}$.

FIGURE 6.17 (*opposite page*) Error estimates for north polar scalar maps. (a) Standard error for POGO map. (b) Standard error for combined *Magsat* dawn and dusk maps. (c) Difference for *Magsat* dawn and dusk maps after ionospheric field correction. (d) Difference for *Magsat* dawn and dusk maps after SHCA. (Part a from Langel, 1990a, with permission; part b from Ravat et al., 1995, with permission.)

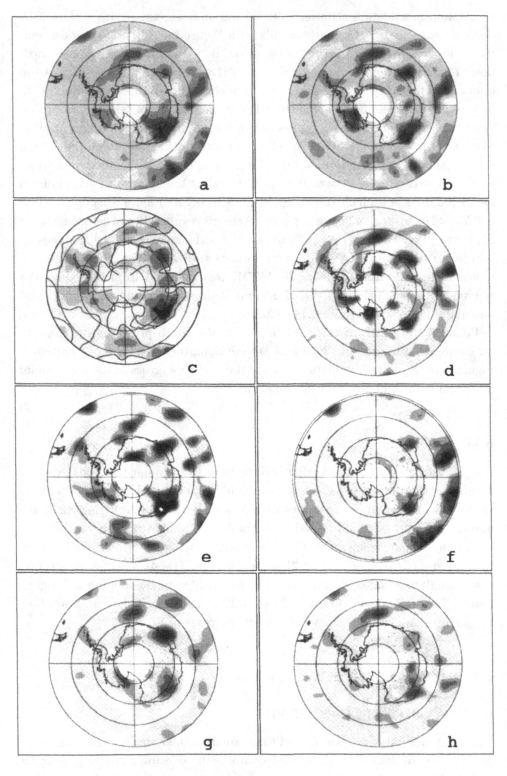

FIGURE 6.18
South polar scalar anomaly maps; data combined from available local times. Top row from POGO; middle and bottom rows from *Magsat*. (a) Map (23). (b) Map (30), derived from map (23) after SHA correlation analysis, harmonics 15–60 retained. (c) Map (14). (d) Map (26), after SHA correlation analysis, harmonics 18–55 retained. (e) Map (32). (f) Map (19). (g) Map (17). (h) Map (29). For scale, see Figure 6.14. Equatorward latitude is −50°. (Part a from Langel, 1990a, with permission; part b from Arkani-Hamed et al., 1994, with permission; part c from Ritzwoller and Bentley, 1983, with permission; part d from Arkani-Hamed and Strangway, 1986a, with permission; part e derived from model kindly furnished by J. C. Cain; part f from Takenaka et al., 1991, reproduced with kind permission of TERRAPUB; part 9 from Alsdorf et al., 1994, with permission; part h from Ravat et al., 1995, with permission.)

Table 6.5. *Anomaly amplitudes from selected anomalies in Figure 6.18*

Anomaly	Amplitude (nT)							
	a	b	c	d	e	f	g	h
1	(2)–(4)[a]	(2)–(4)	(4)	10	(6)	(2)	2	(2)
2	2	4	4	2	4–6	2	absent	2
3	(2)–(4)	(2)–(4)	(4)	(4)–(6)	(4)	absent	(2)–(4)	(2)
4	4	4	—	(2)	6?	absent	2	absent
5	8	8–10	—	2	2	8	2	absent
6	8	2–4	—	2–4	4	6	absent	absent
7	6	2–4	—	absent	6	4	2	absent
8	6	2–4	—	absent	6–8	6–8	8	4
9	(8)–(10)	(8)–(10)	(6)–(8)	(6)–(8)	(12)–(14)	absent	(4)–(6)	(6)–(8)
10	(8)	(8)–(10)	(12)	(8)–(10)	(12)–(14)	(8)	(4)–(6)	absent
11	8–10	10–12	8	10–12	12–14	8	6–8	6–8
12	4	4–6	4	2	absent	(8)	2	6
13	(6)–(8)	(6)	(2)	(10)–(12)	(6)–(8)	4	(4)	absent
14	6	4–6	absent	6–8	absent	absent	absent	4
15	4–6	8–10	2?	10–12	6	12	4	absent
16	3	2	absent	absent	6	4	absent	2
17	(4)–(6)	(6)–(8)	(4)–(6)	(8)–(10)	(6)–(8)	(6)	absent	absent
18	(2)	(4)	(4)–(6)	absent	(8)	(6)	(4)	(4)–(6)
19	(2)	absent	absent	(8)	absent	absent	absent	absent
20	(4)	(6)–(8)	(6)–(8)	(12)–(14)	(10)	(8)	(2)	(6)–(8)
21	absent	absent	(2)	(4)	absent	(8)	(4)–(6)	(8)
22	4–6	8–10	6–8	6–8	10–12	6–8	8–10	4
23	(2)	(4)	(2)	(2)–(4)	(6)–(8)	(4)	(4)	(2)
24	2–4	4	4	4	6–8	4	6	4
25	3	absent	2	2	(4)–(6)	(2)	absent	2
26	4–6	8	4–6	8–10	8	2	6–8	6–8
27	2–4	absent?	2?	4?	absent	absent	absent	2
28	(4)–(6)	(4)–(6)	(4)–(6)	(2)–(4)	absent	(4)–(6)	absent	(4)–(6)
29	(2)–(4)	(2)	(4)–(6)	(4)–(6)	(4)–(6)	(8)	(2)–(4)	(4)–(6)
30	6–8	6–8	—	6	10–12	8–10	4–6	6
31	absent	2–4	4–6	4	6	absent	2–4	2
32	(2)	(4)–(6)	absent	(2)	(4)	(4)–(6)	absent	(2)
33	2	4	4	6	2?	absent	absent	2
34	4	8	2	6	4?	absent	absent	absent
35	2	4–5	2	absent	4	2	absent	2
36	(2)	(4)–(6)	(2)	(2)	(4)–(6)	(4)	(2)	(2)
37	(2)	(2)	(2)	?	(2)?	absent	absent	2

[a]Parentheses indicate negative values.

If $\{A_1(\mathbf{r}_i), i = 1, \ldots\}$ and $\{A_2(\mathbf{r}_i), i = 1, \ldots\}$ are the corresponding anomaly values, then those values are subjected to a nonlinear regression analysis in which their correlation coefficient ρ and the coefficients a (slope) and b (intercept) in

$$A_2(\mathbf{r}_i) = a \cdot A_1(\mathbf{r}_i) + b \qquad (6.4)$$

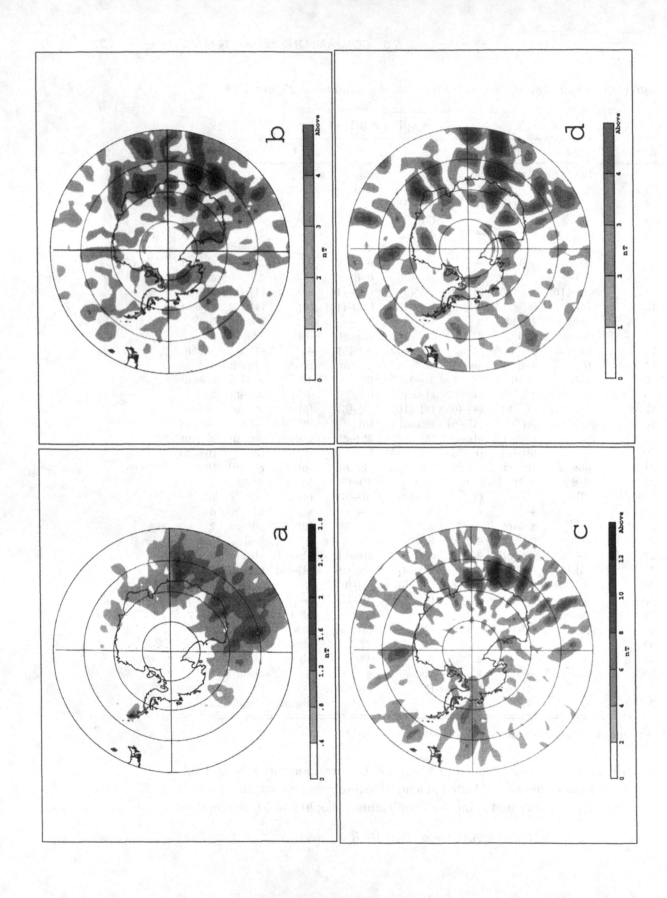

are computed by minimizing

$$\chi^2(a, b) = \sum_{i=1}^{N} \frac{[A_2(\mathbf{r}_i) - b - aA_1(\mathbf{r}_i)]^2}{(\sigma_i)^2(1 + a^2)} \qquad (6.5)$$

with respect to a and b. The value of $(\sigma_i)^2$ is chosen so as to weight the contribution of each position $\mathbf{r}_i$ according to the geographic area that it represents. For a grid evenly spaced in latitude and longitude, $(\sigma_i)^2$ is $1/\sin\theta_i$. A nonlinear, iterative solution (Press and Teukolsky, 1992) is used because both A_1 and A_2 have errors associated with their measurement. Taking into account the accuracy of the anomaly maps, if both points of a data pair have amplitude less than 2 nT, then that pair is excluded from the analysis. On average, if $a = 1.0$, the amplitudes of the two data sets are equal; if $a > 1.0$, the amplitudes in data set A_2 are greater than those in data set A_1; if $a < 1.0$, vice versa. A nonzero intercept indicates that one data set has a level offset from the other. Standard errors for a and b are less than 0.02 nT for all cases cited. Values of b are 0.05–0.6 nT. This is lower than the noise level, so there are no significant level offsets. In each case the probability of obtaining the resulting ρ from an uncorrelated population is less than 10^{-3} (Bevington, 1969). The fit residual to the data is defined as

$$\sigma^2 = \chi^2/N. \qquad (6.6)$$

Maps are always compared at the same altitude, except when a map is of average values, in which case the altitude varies somewhat over the map. Unless not applicable, or otherwise stated, maps are also compared for the spherical harmonic range (SHA) given in the table of correlation values.

Ravat et al. (1995) used this correlation procedure to evaluate the processing that led to their final map and to compare their map with those previously published.

6.5.2 CORRELATIVE INTERCOMPARISON OF FINAL MAPS

6.5.2.1 Low- and Middle-Latitude Scalar Maps

Table 6.6 summarizes the correlation comparison for $\pm 60°$ latitude among five maps based on *Magsat* data and between those maps and a map based on POGO data. Map numbers are taken from Table 6.1. In each case the A_2 and A_1 data sets in equation (6.4) are taken from the map specified in the left column and the map specified in the top row, respectively. Note that the SHA degree range is not the same for all of the maps compared. From Table 6.6, the lowest correlation is 0.66, and 7 of 15 are at 0.8 or higher. In general, having been derived using independent methods and partially independent data

FIGURE 6.19 (*opposite page*) Error estimates for south polar scalar anomaly maps. (a) Standard error of the mean from POGO data. (b) Standard error of POGO/*Magsat* combined map. (c) Dawn–dusk difference after ionospheric field correction and pass-by-pass correlation. (d) Dawn–dusk difference after SHCA (Ravat et al., 1995). (Part a from Langel, 1990a, with permission; part b from Arkani-Hamed et al., 1994, with permission; parts c and d from Ravat et al., 1995, with permission.)

Table 6.6. *Correlations among final scalar maps, ±60° latitude*

Map no. SHA degree	(31)[a] 15–65			(29a)[b] 15–65			(29b)[c] 15–65			(32)[d] 15–49			(26/27)[e] 18–55			(23)[f] 15–60		
Map no. SHA degree	ρ	a	σ	ρ	a	σ	ρ	a	σ	ρ	a	σ	ρ	a	σ	ρ	a	σ
(31) 15–65	×	×	×	0.95	1.15	0.42	0.87	1.49	0.80	0.73	0.61	2.06	0.77	0.86	1.84	0.94	0.90	0.56
(29a) 15–65				×	×	×	0.85	1.29	0.92	0.66	0.48	2.11	0.72	0.70	1.95	0.84	0.75	1.17
(29b) 15–65							×	×	×	0.70	0.39	1.40	0.73	0.53	1.49	0.86	0.60	0.86
(32) 15–49										×	×	×	0.67	1.44	2.85	0.76	1.43	2.12
(26/27) 18–55													×	×	×	0.80	1.02	1.76

[a]Map (31): POGO/*Magsat* combined (Arkani-Hamed et al., 1994).
[b]Map (29a): *Magsat*, from scalar data (Ravat et al., 1995).
[c]Map (29b): *Magsat* from vector data (Ravat et al., 1995).
[d]Map (32): *Magsat*, all data (Cain et al., 1990).
[e]Map (26/27): *Magsat* from scalar data (Arkani-Hamed and Strangway, 1985a,b, 1986a).
[f]Map (23): POGO (Langel, 1990a).

sets, the basic agreement among these six maps is good. Some specifics of comparison are as follows:

1. The combined POGO/*Magsat* map (31) correlates most highly with, and has the smallest residual σ from, the *Magsat* map derived from scalar data (29a) and the POGO map (23), the data sets from which the POGO/*Magsat* map (31) was derived. The worst correlation for the POGO/*Magsat* map is with map (32). Large differences in amplitude are noted between the *Magsat* map derived from vector data (29b) and map (32).
2. Maps (29a) and (29b), derived from *Magsat* scalar and vector data, respectively, correlate well with each other and correlate better with map (26/27) than with map (32).
3. Map (32) generally has the lowest correlations and largest values of σ. This may reflect the fact that the least correction has been applied to this map for external fields.
4. Considering the POGO map (23) as an independent data source, which is not true for map (31), its correlation with the other maps decreases, and its σ increases, in the following order: (31), (29b), (29a), (26/27), (32). The first three of these all have smaller amplitudes than the POGO map; the amplitude of map (26/27) is about equal to that of the POGO map (23), in spite of the difference in degree range, and the amplitude of map (32) is much higher than that of the POGO map (23).

6.5.2.2 Low- and Middle-Latitude ΔZ Maps

Table 6.7 summarizes the correlation comparisons among three vertical-component anomaly maps for latitudes between $\pm 60°$. As in the preceding section, the A_2 and A_1 data sets in equation (6.4) are taken from the map specified in the left column and the map specified in the top row, respectively. The rightmost column will be discussed later. Intercomparing maps (29), (6), and (32) pairwise, maps (29) and (32) are in closest agreement, and maps (6) and (32) in greatest disagreement. Of these, map (32) has the greatest amplitude, and map (29) the least amplitude.

Under certain assumptions it is possible to compare component maps to an independent data set. In the present case, the equivalent source representation of Langel (1990a) is used to derive a set of gridded ΔZ points that, after SHA analysis, can be used for comparisons. The resulting comparisons are shown in the rightmost column of Table 6.7.

Map (29) correlates most highly with the POGO map, $\rho = 0.80$, with the lowest σ, 1.28 nT, followed by map (32). Of the three maps, (6) has the worst correlation, $\rho = 0.63$, and largest σ, 3.73 nT. From the slope values, map (29)

Table 6.7. *Correlations among final ΔZ maps, $\pm 60°$ latitude*

Map no. SHA degree		$(29)^a$ 15–65			$(6)^b$ 15–65			$(32)^c$ 15–65			$(23)^d$ 15–60		
Map no.	SHA degree	ρ	a	σ	ρ	a	σ	ρ	a	σ	ρ	a	σ
(29)	15–65	×			0.67			0.74			0.80		
			×			0.45			0.43			0.56	
				×			2.02			1.52			1.28
(6)	15–65				×			0.58			0.63		
						×			0.86			1.14	
							×			4.51			3.73
(32)	15–65							×			0.73		
									×			1.27	
										×			2.83

[a]Map (29): *Magsat* (Ravat et al., 1995).
[b]Map (6): *Magsat* (Cohen and Achache, 1990).
[c]Map (32): *Magsat* (Cain et al., 1990).
[d]Map (23): POGO (Langel, 1990a).

has considerably lower amplitude than the POGO map, whereas map (32) has greater amplitude.

These results are strong evidence that the procedures followed by Ravat et al. (1995) are thus far the most successful for eliminating contamination by ionospheric fields. The price is significantly reduced amplitude or power in the resulting map.

6.5.2.3 Polar Scalar Maps

Table 6.8 summarizes the correlation comparisons among scalar polar maps. The upper triangular portion of the table pertains to the north pole, and the lower triangular portion to the south pole.

For the north, the table shows correlations among four scalar maps based on *Magsat* data [maps (29a), (29b), (32), and (26/27)], the combined POGO/*Magsat* map (31), and a map (23) based only on POGO data. Data sets for A_2 and A_1 in equation (6.4) are from the maps indicated in the left column and top row, respectively.

All of the northern polar maps correlate among themselves with $\rho \geq 0.69$, indicating good agreement in the overall anomaly patterns. Variations in slope, a, indicate considerable variability in amplitude. The map derived from both POGO and *Magsat* data (31) is not independent of the POGO data and, as expected, shows good correlation with map (23), $\rho = 0.93$, although of somewhat lower amplitude. It also shows good correlation with all of the other maps, $\rho \geq 0.79$, and has exceptionally low σ values with respect to

Table 6.8. *Correlations among high-latitude scalar maps: upper triangle, north polar maps; lower triangle, south polar maps*

Map no. SHA degree	(31)[a] 15–65			(29a)[b] 15–65			(29b)[c] 15–65			(32)[d] 15–49			(26/27)[e] 18–55			(18)[f] 15–60			(19)[g]			(23)[h] 15–60		
	ρ	a	σ	ρ	a	σ	ρ	a	σ	ρ	a	σ	ρ	a	σ	ρ	a	σ	ρ	a	σ	ρ	a	σ
(31) 15–65	×			0.95	1.21	0.24	0.88	1.32	0.53	0.80	0.79	0.53	0.72	0.72	1.24							0.93	0.79	0.41
(29a) 15–65	0.91	1.06	0.34	×			0.90	1.07	0.53	0.72	0.39	1.23	0.74	0.55	1.22							0.83	0.61	0.79
(29b) 15–65	0.74	1.24	0.69	0.77	1.16	0.72	×			0.73	0.37	1.29	0.70	0.49	1.28							0.81	0.57	0.79
(32) 15–49	0.64	0.43	1.41	0.55	0.36	1.60	0.72	0.72	1.26	×			0.69	1.12	1.45							0.85	0.85	0.82
(26/27) 18–55	0.60	0.52	1.65	0.48	0.41	0.52	0.58	0.58	1.60	0.40	0.32	1.26	×									0.80	1.43	1.25
(18) 15–60	0.71	1.02	1.02	0.61	0.93	0.77	0.59	0.77	1.20	0.61	2.44	1.94	0.61	1.90	3.18	×								
(19)	0.52	0.59	1.88	0.39	0.45	1.30	0.44	0.40	2.17	0.53	1.70	2.74	0.42	1.34	1.58	0.39	1.49	1.67	×					
(23) 15–60	0.82	0.72	0.75	0.60	0.58	0.58	0.64	0.51	0.49	0.68	1.56	3.20	0.61	1.24	2.25	0.60	2.20	1.89	1.01	1.45	2.10	×		

[a] Map (31): POGO/*Magsat* combined (Arkani-Hamed et al., 1994).
[b] Map (29a): *Magsat*, from scalar data (Ravat et al., 1995).
[c] Map (29b): *Magsat*, from vector data (Ravat et al., 1995).
[d] Map (32): *Magsat*, all data (Cain et al., 1990).
[e] Map (26/27): *Magsat*, from scalar data (Arkani-Hamed and Strangway, 1985a,b, 1986a).
[f] Map (18): *Magsat*, from scalar data (Alsdorf et al., 1994); south only.
[g] Map (19): *Magsat*, from scalar data (Takenaka et al., 1991); south only.
[h] Map (23): POGO (Langel, 1990a).

the POGO map (23) and *Magsat* maps (29a) and (29b). Map (31) has appreciably greater amplitude than maps (29a) and (29b), but lower amplitude than maps (32) and (26/27). In general, the combined map (31) seems to retain the greatest degree of commonality among all maps without undue sacrifice of amplitude.

Of the maps derived from *Magsat* data alone, (29a) has the highest set of correlations, and (26/27) the lowest. Map (32) has the highest amplitude, and map (29a) the lowest. Considering the map derived from POGO data (23) as an independent data source, its correlation with the *Magsat* maps decreases in the following order: (32), (29a), (29b), (26/27). The order for increasing σ is different: (29a), (29b), (32), (26/27). As for their low-latitude counterparts, maps (29a) and (29b) are of considerably lower amplitude than the POGO map (23), map (26/27) is about the same amplitude as the POGO map, and map (32) is of considerably greater magnitude than the POGO map.

For southern polar latitudes, the lower left triangular portion of Table 6.8 shows correlations among six scalar maps based on *Magsat* data [maps (29a), (29b), (32), (26/27), (18), and (19)], the combined POGO/*Magsat* map (31), and a map (23) based only on POGO data. Data sets for A_2 and A_1 in equation (6.4) are from the maps indicated in the top row and left column, respectively (i.e., reversed from the convention used for the north pole).

The situation for the south is much more confused than that for the north. Overall, correlation coefficients range from 0.39 to 0.91. Moreover, most maps have similarly wide ranges of correlations with the other maps. As in the north, there is wide variability in map amplitudes. With the exception of map (29b), in the north, each individual map based on *Magsat* data correlates more highly with the combined POGO/*Magsat* map (31) than with other *Magsat*-based maps. That is true only for three of the southern maps [maps (29a), (32), and (18)], although map (31) does correlate moderately well with most maps: ρ values from 0.52 to 0.91.

Making comparisons among only the maps derived from *Magsat* data, note the following:

1. Map (19) has the lowest correlation with each of the other maps. Its range of correlation is 0.39–0.53.
2. Map (18) has the most consistent correlation with the other maps. If the correlation with (19), $\rho = 0.39$, is not counted, its range of correlations is 0.59–0.61.
3. If the high correlation between (29a) and (29b), $\rho = 0.77$, is discounted, because of the commonality of data selection and methods used, and if (19) is excluded, the total range of correlations is 0.48–0.61.

4. Comparison with the POGO map (23) as an independent data source shows map (32) with the highest correlation, $\rho = 0.68$, map (29b) next, with $\rho = 0.64$, and all other maps with ρ either 0.60 or 0.61.

There is no clear-cut best map or sets of maps for the south polar region. Map (18) of Alsdorf et al. (1994) is the most consistent in its correlations with other maps; maps (32) and (29b) are most consistent with the POGO data. Application of SHCA to a combination of maps (32), (18), (29b), and (23) might result in as optimal a map as is currently possible.

6.6 COMPARISONS OF SATELLITE AND SURFACE SURVEYS

Magnetic surveys have been conducted from ships and aircraft and on land, here collectively called "surface" surveys. In principle, if a surface survey encompasses sufficient area, it will contain the identical long-wavelength information as a satellite survey over the same area. Because they are taken nearer to the sources, surface data are inherently of higher resolution. Comparison of the two is of interest to verify the integrity and compatibility of the data sets and methods used to extrapolate, or continue, them to other altitudes, to better understand the relationships among the geologic features giving rise to long- and short-wavelength magnetic anomalies, and, in some cases, to provide corrections to the long-wavelength portions of the surface surveys. The process of comparison is itself difficult, so before consideration of published results, possible problems related to preservation of the long-wavelength components are discussed.

6.6.1 PROBLEMS WITH INTERCOMPARING SURFACE AND SATELLITE SURVEYS

If a surface survey and satellite survey contain identical long-wavelength information, it should be possible to upward-continue the surface measurements to satellite altitude and make a direct comparison between the two data sets. In practice, this process is difficult to implement with precision, because (1) typically the extents to which the main field is removed in the two cases are not equivalent, (2) spurious long-wavelength information is easily introduced into a surface survey, and (3) when smaller surface surveys are stitched together to form a larger survey, long-wavelength information is often lost or distorted. These problems are discussed in more detail in the following sections. Useful discussions of these topics can also be found in Schnetzler et al. (1985), Arkani-Hamed and Hinze (1990), Grauch (1993),

Pilkington and Roest (1996), and Tarlowski et al. (1996). Despite these problems, the studies of Langel et al. (1980a), Sexton et al. (1982), von Frese et al. (1982b), Won and Son (1982), Schnetzler et al. (1985), and Pilkington and Roest (1996) have shown the validity of lithospheric anomalies in satellite elevation data.

6.6.1.1 Differences in Main-Field and Regional-Field Removal

By definition, the anomaly field $\mathbf{A}$ is the residual field after the main field $\mathbf{B}_m$ and any fields of external origin have been removed from the measured field $\mathbf{B}$. For the present, assume that there are no fields of external origin. For satellite data, $\mathbf{B}_m$ is represented by a spherical harmonic analysis, and $\mathbf{A}$ is taken to be composed of all fields of degree greater than, say, n^*, which typically is taken to be 13. Later in the analysis, components of $\mathbf{A}$, or A^s in the case of scalar data, may be individually spherically harmonically analyzed, with additional terms of specific degree and/or order deleted. Based on the POGO/*Magsat* map described in Section 6.2.3, satellite magnetic anomaly maps encompass spherical harmonic degrees 15 to about 65. Using $360°/n$, these limits correspond to a wavelength range of 5.5° to 24°, or approximately 600 to 2,640 km along a great circle. However, as pointed out in Section 4.3.5, all values of m from zero to n are present in spherical harmonic degree n, so that wavelengths longer than 24° may be present in the maps, especially in the east–west direction, if the low-order terms are significant. Spherical harmonic analysis and double Fourier analysis are not equivalent! The difference can result in incompatibility when comparing surface and satellite surveys.

In many cases it is not practical to reduce a surface survey in the same way as a satellite survey, because (1) an accurate main-field model of sufficient degree, n^*, may not be available at the time the survey is conducted, and (2) it is not possible to represent regional surface surveys in terms of a global spherical harmonic analysis. The field model of choice for reduction of surface surveys is generally the IGRF of the appropriate epoch, that is, a model of degree 10. This leaves main-field degrees 11–13, and perhaps 14, remaining in the residual, or anomaly, data. Subsequent filtering, sometimes called "removing the regional," say with a bandpass of 600–2,640 km, does *not* result in a data set of exactly the same wavelength range as the satellite analysis, although it may be comparable. Upward continuation of such a data set must not automatically be assumed to give a data set valid for comparison with satellite anomaly maps, although it may be the best approximation available.

Figure 6.20 shows the results of a simulation involving four anomaly maps with long-wavelength differences due to differences in field model and/or filtering. Figure 6.20a shows the "actual" anomaly field in the vertical component at a 400-km altitude. This map contains SHA degrees 15–65, corresponding

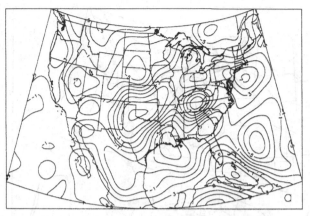
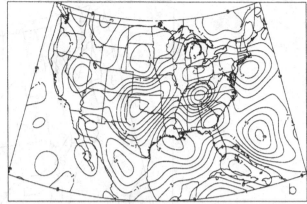
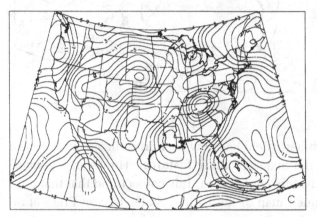
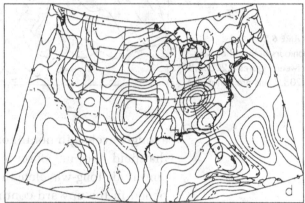

FIGURE 6.20

U.S. anomaly maps with different bandpass characteristics: (a) computed from ES; (b) bandpass version of the map in part a filtered to pass 6° to 24°; (c) main-field degrees 11–13 added to the map in part a; (d) bandpass version of the map in part c filtered to pass 6° to 24°. Units are nanoteslas.

to the great-circle wavelength range 5.5°–24°. Figure 6.20b shows the map resulting from applying a two-dimensional Fourier filter with bandpass 6°–24° to the map in Figure 6.20a. Major features are relatively unchanged, but there are noticeable differences. For example, in Figure 6.20b the positive anomaly over Nevada is not as clearly defined; the negative over Mexico is stronger; the pattern over the northern plains states is subtly shifted; and the negative over Georgia and to the southeast in the Atlantic is stronger. Figure 6.20c has the field from main-field potential degrees 11–13 added to the field of Figure 6.20a. The "anomalies" are now somewhat overlapping; some new features have appeared off the West Coast, and the patterns off the East Coast are changed. Figure 6.20d shows the result of applying the bandpass filter to Figure 6.20c. Supposedly, Figure 6.20d has the same wavelength content as Figure 6.20b, and, in general, the major anomaly patterns of Figures 6.20a and 6.20b are reproduced in Figure 6.20d. But, as shown in the difference map of Figure 6.21,

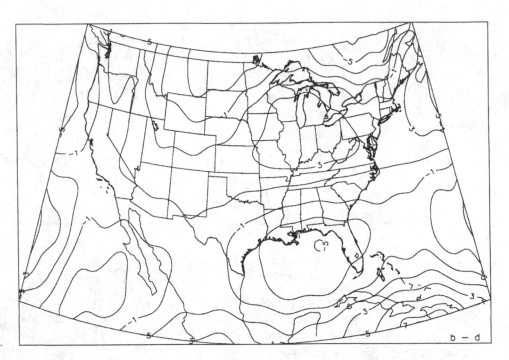

FIGURE 6.21
Contours of the difference between Figures 6.20b and 6.20d. Units are nanoteslas.

a long-wavelength nonlithospheric field has now been introduced. In this case that field does not obscure the major anomaly features. However, a greater-amplitude long-wavelength feature could do so. It is conjectured that that happened in the upward-continued map of the United States by von Frese et al. (1982b), discussed in Section 6.6.2.2.

If either Figure 6.20b or Figure 6.20d were the result of upward-continuing an aeromagnetic survey, one would say that the agreement with the satellite map, Figure 6.20a, is very good, although not perfect. This example illustrates that different reduction procedures give diverse results, but if those procedures are done carefully, to try to ensure equivalent wavelength content, then the main anomaly features are comparable.

6.6.1.2 Missing and Spurious Long Wavelengths

When removing an estimate of the main field based on, for example, an IGRF model, the main-field estimate must be accurate at the epoch of the survey. Because most surveys are not conducted precisely at the epoch of a main-field model, a secular variation model is required. Errors in the secular variation model translate directly to errors in the main-field estimate. Unfortunately, definitive (e.g., DGRF) models are not generally available until typically 5 years or more after an epoch. It is not practical to wait several years for a "best" model in order to reduce the measurements from a survey, so one uses the best model available at the time. This is usually a model based on a data set that has no data at, or even within 2–3 years of, the epoch of the survey. Such models are predictive and are notorious for their inaccuracy. Thus, long-wavelength errors are introduced into surface surveys.

In order for such surveys to be useful in studying the longer wavelengths, and for comparison with satellite data, the original data must be preserved until a proper main-field model can be made available. At that time the data can be reanalyzed to obtain more realistic long-wavelength information.

Another source of long-wavelength errors in surface surveys is the presence of time-varying fields of external origin. Such fields pose problems for satellite data also. However, the problem these fields bring to surface surveys is different from that for satellite surveys. In the case of satellite surveys, each region is overflown many times by the satellite, and the problem is to select the common signal. In the case of surface surveys, multiple data from a single location or nearby locations are generally not available. Rather, the data are acquired along specific tracks. If the field from external sources changes as the data are acquired along the tracks, a spurious long-wavelength field is introduced. Many surveys make an attempt to avoid collecting data during times of magnetic disturbance, which reduces the problem but still leaves the substantial fields due to Sq, unless the data are acquired in the hours near local midnight, when Sq is small.

The effects of external fields can be compensated in two ways. First, if fixed observatory or even variometer data are available within a few tens of kilometers from the position of the data tracks, then those data can be used to make an estimate of the time-varying portion of the field, and a correction can be applied to the data. Such corrections are invariably imperfect because of spatial variability in Sq. Second, data tracks may occasionally overlap, particularly in marine data and when tie lines are flown in an aeromagnetic survey. In these cases, a polynomial can be removed from each track so as to minimize the crossover differences. This tends to reduce long-wavelength contamination. It may also remove long-wavelength information from the data.

Long-wavelength information can be both lost and spuriously introduced in the process of combining adjacent local surveys into a larger, unified data set. A typical "large-scale" aeromagnetic survey corresponds roughly to a single state of the United States. For example, Kansas (Yarger, 1985) has an area roughly equivalent in size to a rectangle $8°$ in longitude by $3°$ in latitude. In principle it is possible to "stitch" together adjacent surveys into a larger whole. However, in doing so one encounters all of the problems discussed earlier, plus others. Two problems are common. The first comes about when the survey is initially reduced. The best possible main-field model is removed, but its degree is less than 13, and the secular variation is not well known, and so long-wavelength

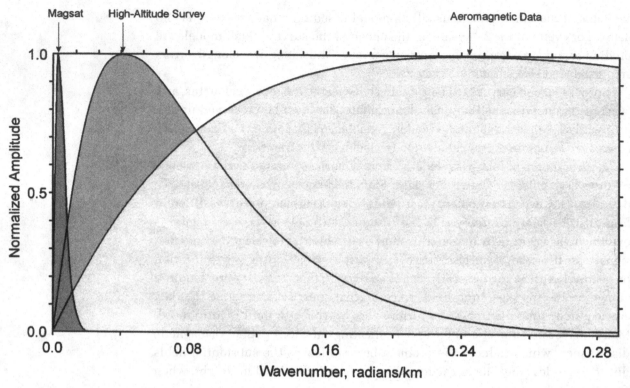

FIGURE 6.22
Amplitude spectra from magnetic surveys flown at three altitudes. Each spectrum has been normalized to a maximum of 1. The dark-shaded area is the spectrum that is expected from a survey flown at 450 km. The light-shaded area is the spectrum at 1 km, the altitude of a typical aeromagnetic survey. The medium-shaded area is the spectrum that is expected from an aeromagnetic survey at 20 km. The magnetic layer is assumed to be 10 km thick. The shapes of the spectra depend only on the depth to the top of the layer and on the thickness of the layer. Thus, they are applicable for both total-field and vertical-field surveys. (*continued*)

components remain in the resulting map. Often an additional polynomial surface, a regional field, will be removed. The resulting map will then be stripped of information of longer wavelength than that corresponding to the size of the area surveyed. Unless the original data are preserved, that longer-wavelength information is lost. Stitching that survey together with an adjacent survey cannot add the information deleted. If one subsequently flies tie lines, the content of the long-wavelength information in the stitched survey is limited to the long-wavelength information in the tie lines.

The second problem comes from trying to stitch together adjacent surveys acquired at widely separated epochs. Suppose the original data are retained. How does one remove the main field in a consistent way? If a degree-13 main-field model derived from satellite data is available for both epochs, there is little or no problem. Suppose DGRF models are available. Both surveys can then be reduced relative to a degree 10-model at epoch, leaving degrees 11–13 (14?) their attendant and unknown secular variations as long-wavelength contaminants. If polynomial surfaces are removed from each survey to account for those long wavelengths, there is no way known to us to constrain the adjacent polynomials to be equivalent in the long wavelengths they represent.

The report by Whaler (1994) regarding a stitched-together set of surveys for part of Africa is illuminating. The individual surveys were very diverse, many with unknown background removal, flown over a period of 40 years. Whaler (1994) reported that because the long-wavelength content of the stitched map was regarded as unreliable, it was deleted, and in its place a downward-continued map from *Magsat* was substituted. Comparison of the power spectra of the downward-continued *Magsat* data and of the stitched surveys shows that the *Magsat* data have essentially no power at higher wavenumbers, and the stitched survey has essentially no power at low wavenumbers, with minimal overlap.

A gap in the power spectra from satellite and surface surveys is expected on theoretical grounds. Blakely (Hildenbrand et al., 1996) calculated normalized amplitude spectra for three altitudes, as shown in Figure 6.22. Although computed in a flat-earth approximation, the results should apply to the spherical-earth case. They note that the information in any magnetic survey is contained in a spectral band defined by the altitude and areal extent of the survey. The figure shows spectra for satellite altitudes (*Magsat*, 450 km), for typical aeromagnetic-data altitudes (1 km), and for a proposed high-altitude aeromagnetic survey of the United States (20 km). Note the data gap between the spectra from *Magsat* and aeromagnetic data.

6.6.2 PUBLISHED COMPARISONS

Several direct comparisons of satellite and surface surveys have been published, some of which are described in this section.

6.6.2.1 Canada

Comparison of the upward-continued Earth Physics Branch (EPB) aeromagnetic survey data for all of Canada, map (100), with a satellite magnetic anomaly map derived by taking a simple average of the POGO and dawn and dusk *Magsat* data of Arkani-Hamed et al. (1994) has been reported by Pilkington and Roest (1996). Upward continuation of the aeromagnetic data used a fast Fourier, flat-earth algorithm, and the data were bandpass-filtered for consistency with the satellite data. Anomalies were taken with respect to the appropriate DGRF main-field model. Figure 6.23 shows the upward-continued EPB data and the satellite anomaly map. As they noted, both anomaly amplitude and position are generally in agreement (e.g., the positive anomalies over the Alpha Ridge and east of Hudson Bay), although from Lake Superior to the west, along the 50° latitude line, the maps in Figure 6.23 are very different. M. Pilkington (personal communication, 1995) suggests that this is not likely an edge effect. The reason for the discrepancy is not known at present, but the agreement for most of the region indicates basic compatibility between the two data sets.

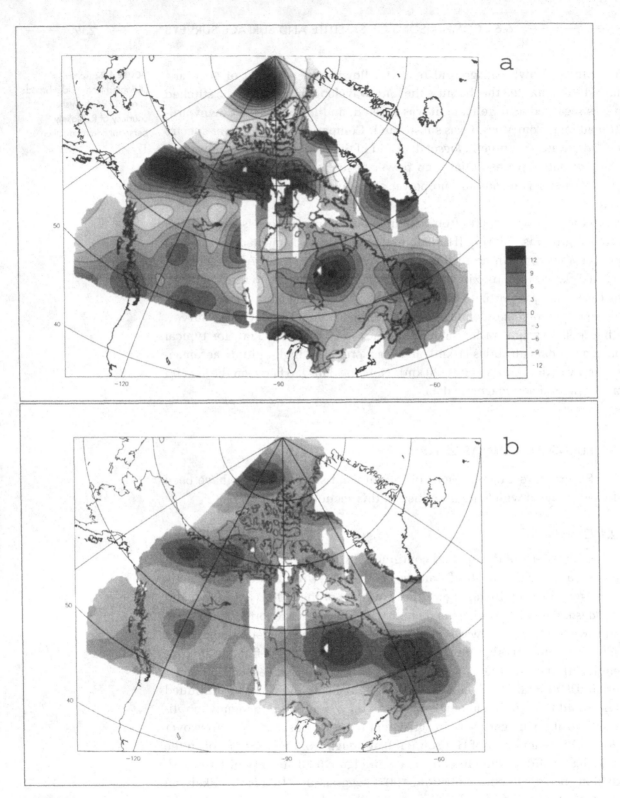

FIGURE 6.23
(a) EPB data upward-continued to 400-km altitude (100), using a flat- earth approximation. (b) Satellite
magnetic data at ~400-km altitude. (From Pilkington and Roest, 1996; reprinted with kind permission of
NRC Research Press, Ottawa, Canada.)

6.6.2.2 United States

Project Magnet of the U.S. Naval Oceanographic Office (NOO) conducted an aeromagnetic survey of the United States between September 1976 and September 1977. Those data were reduced independently by Sexton et al. (1982) and Won and Son (1982) and were compared to satellite data by von Frese et al. (1982b) and Won and Son (1982). Of the two upward-continued U.S. maps, that of Won and Son (1982) is in better agreement with the satellite data. Figure 6.24 reproduces their map (50), together with an equivalent source representation of the scalar field (51) from Mayhew and Galliher (1982); see also Schnetzler et al. (1985). Both maps clearly show the high over Kentucky and

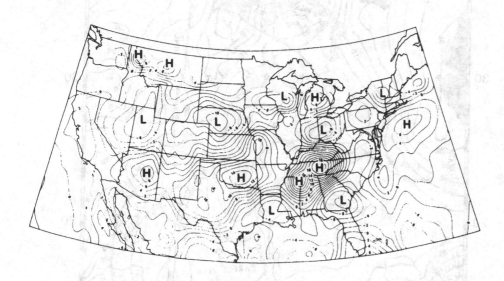

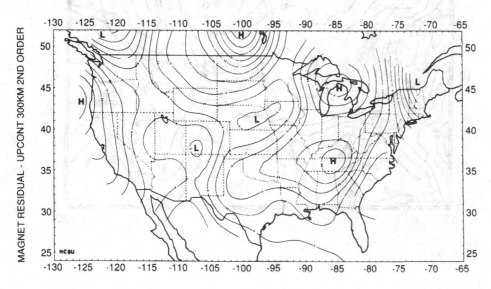

FIGURE 6.24
Comparison of (top) equivalent source representation of scalar anomaly field at 320 km derived from *Magsat* data (52) with (bottom) upward-continued aeromagnetic survey data at a 300-km altitude (50). (Top map from Mayhew and Galliher, 1982; bottom map from Won and Son, 1982; reproduced with permission.)

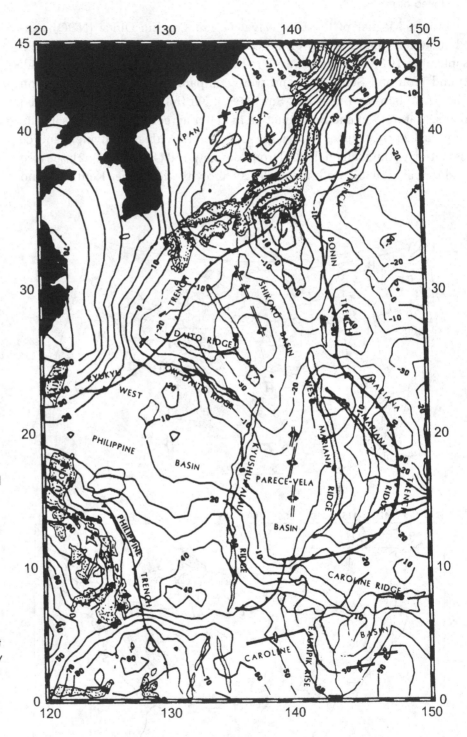

FIGURE 6.25
Filtered sea, surface anomaly field over the western Pacific; 45 nT have been added to the observed anomalies. Note the strong correlation between the region's ridge systems and the magnetic field. A negative follows the abandoned back-arc systems of the Japan Basin and the Parece Vela Basin. In general, a positive anomaly is observed over the abandoned and present island arc systems. The only exception is the Marian arc system. (From LaBrecque et al., 1985, with permission.)

Tennessee, the low trending SW-to-NE through Nebraska and Iowa, the high over Michigan, the low over Vermont and New Hampshire, and a relative high in California. At the same time, the high over Oklahoma and low over Utah are much more well defined in the *Magsat* data. Both maps show a relative low in the area of the Rio Grande Rift.

6.6.2.3 Marine Regions

Well over 10 million miles of scalar ship-track data have been collected by various marine geophysical organizations, particularly during the 1960s and 1970s. LaBrecque and Raymond (1985), Hayling and Harrison (1986), Raymond and LaBrecque (1987), Arkani-Hamed (1988, 1989, 1991), and Yañez and LaBrecque (1997) found that some observed satellite anomalies are essentially due to those seafloor spreading anomalies acquired during several especially long periods when the earth's main field did not reverse (e.g., the Cretaceous Quiet Zone, KQZ).

LaBrecque and Cande (1984) and LaBrecque, Cande, and Jarrard (1985) derived a long-wavelength anomaly map for the North Pacific. No really adequate main-field model was available, and diurnal variations were accounted for by use of the model of Malin (1973), adequate for the average Sq, but not for its considerable day-to-day variations. Inadequacies in the available secular variation model were corrected for by taking linear regressions with respect to time of Sq corrected residuals collected in $6° \times 6°$ bins over the 1960–80 time span. We regard that as a heroic effort, with significant results, but believe that the longer-wavelength features of the resulting map (160), part of which is shown here as Figure 6.25, should be treated with caution, particularly in areas of sparse data. LaBrecque et al. (1985) noted positive anomalies in the marine and satellite maps along the Japan and Bonin trenches and the Mariana Ridge (near 35° N, 130° E) and in the Indonesian islands (5° N, 130° E), and a negative anomaly (relative negative in Figure 6.25) near the Caroline Basin. In the central Pacific, near (30° N, 170° E), LaBrecque and Cande (1984) noted that a positive anomaly in the sea-surface map near the intersection of the Hawaiian and Emperor seamount chains corresponds to the largest-amplitude anomaly observed in both the POGO and *Magsat* anomaly maps.

6.7   GLOBAL INTERPRETIVE MAPS

To this point, the satellite-data-based maps discussed have shown variations in the scalar, including reduced-to-pole, or vector component magnetic anomaly variations at satellite altitude. For interpretive purposes, such maps are often reduced to the earth's surface or transformed into magnetization and/or susceptibility maps using the methods described in Chapter 5. Figures 6.26

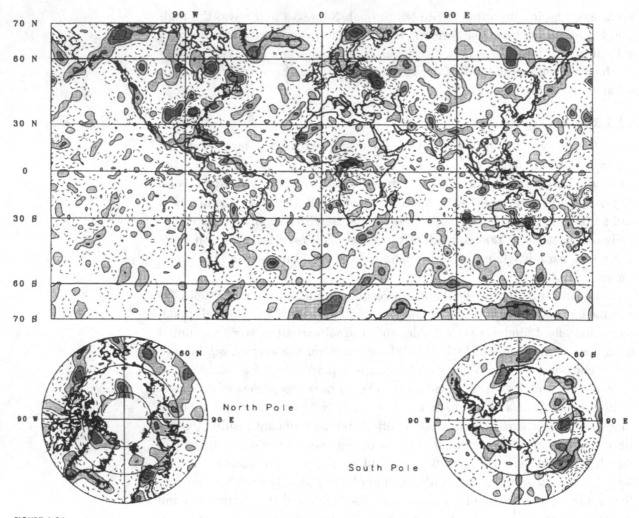

FIGURE 6.26

Magnetization contrast map (33) derived from the combined POGO/*Magsat* map. Continuous lines with a grey scale overlay are used for the positive contours, and dashed lines without any overlay are used for negative contours. Contour levels are at ±0.1, 0.3, 0.5, 0.7, and 0.9 A/m. These magnetizations are calculated on the assumption of a 40-km-thick magnetic layer. (From Purucker et al., 1996, with permission.)

and 6.27, maps (33) and (36), show a relative magnetization map and a susceptibility map, respectively, each assuming a 40-km-thick magnetic layer. These are based on the less stringent and more stringent spherical harmonic maps (31) of Arkani-Hamed et al. (1994), respectively. The variations in the two maps are close, but not exactly the same.

6.8 REGIONAL MAPS

6.8.1 GENERAL COMMENTS

Maps (37) through (167) of Tables 6.1, 6.2, and 6.3 are regional rather than global. They are of varying quality and were derived for various purposes.

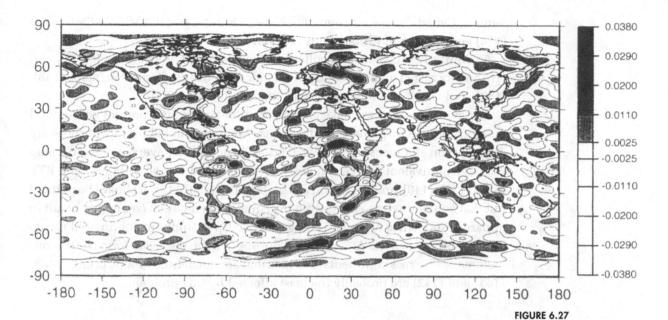

FIGURE 6.27
Magnetic susceptibility
contrast of the earth's
lithosphere (36) in SI units
assuming a 40-km-thick
magnetic layer. (From
Arkani-Hamed and Dyment,
1996, with permission.)

Many were developed for geologic/tectonic study of a particular region or feature, generally identified in the "Coverage and comments" column in Table 6.1, and some of these are discussed in Chapter 9. Such maps are often reduced to the pole, or are transformed to maps of magnetization or of susceptibility. Some summary comments are in order:

1. Mayhew and colleagues pioneered in the application of equivalent source methods to satellite magnetic field data. In a series of analyses of the United States, maps (39), (40), (41), (42), (51), (52), and (53), and Australia, maps (89)–(94), the ES method was developed and refined using both POGO and *Magsat* data. The resulting series of magnetization maps increased in resolution as the analyses developed and as *Magsat* data became available. These are among the most reliable and detailed magnetization maps available from satellite data. The references cited include discussions of correlations of magnetization variations with geologic and tectonic features, a topic that is discussed in Chapter 9.

2. A series of increasingly sophisticated studies of South America has been published by Hinze and co-workers, maps (61) and (63)–(69).

3. Frey and colleagues have conducted a series of studies of particular tectonic or geologic features and/or localized areas, some of which have included derivation of localized maps, such as maps (75)–(77) and maps (151), (155), (165), and (167).

4. Arkani-Hamed and colleagues not only have contributed to derivation of global maps (26), (27), (28), (30), (31), (35), and (36) but also have derived more localized maps while conducting geologic/tectonic analyses of several regions,

namely, Canada, maps (98) and (99), Eastern Europe and the Middle East, maps (118) and (119), and China, maps (132) and (133).

5. Similarly, Achache and colleagues have derived global maps (6), (9), and (10) and localized maps for study of specific regions: the Caribbean, maps (148), (152), and (153); and Southeast Asia, maps (130) and (131).

6. Several researchers have examined anomalies across present continental margins in tectonic plate reconstructions. Maps assembled for this purpose are all based on the measured scalar field and include an average anomaly map (137), an equivalent source map only partly reduced-to-pole (138), several RTP maps (139), (140), and (141), and a susceptibility map (142). It should be pointed out that contamination effects from the equatorial electrojet commonly result in apparent anomalies elongated along lines of dip latitude. In continental reconstructions between the Americas and Europe/Africa, these can form continuous "anomalies" across continental margins that are not of lithospheric origin. Maps (141) and (142) are probably the least affected by this problem.

6.8.2 MAPS OF HIGHER RESOLUTION

A few of the regional maps have been derived using special methods designed to increase the resolution and reliability of the map for that region beyond what is possible with a global approach. Two approaches have been used, both discussed in previous chapters, namely, rectangular harmonic analysis (RHA) (Section 5.3.2) and crossover analysis (Section 4.4.4) after special selection of the data.

The region of the Japanese Islands was the subject of RHA by Nakagawa et al. (1985). In that study, increased resolution was achieved, map (88), by restricting the analysis to a relatively small area. This, in fact, is equivalent to applying a bandpass filter to the data.

The region surrounding Kursk, Russia, was the subject of analysis by Taylor and Frawley (1987). In that method, data tracks through the area of study were limited to those with altitudes in a restricted range, as low as is possible while still retaining enough data for the analysis (e.g., 340–360 km). The low altitude assures selection of passes as close as possible to the source, and the limited altitude range minimizes variations due to altitude differences. Differences between dawn and dusk passes are resolved via crossover analysis.

It is instructive to compare results obtained in such an analysis with a map obtained by global analysis. Figure 6.28 shows the Kursk map (120) of Taylor and Frawley (1987) together with the equivalent map from the combined POGO/*Magsat* map (31) of Arkani-Hamed et al. (1994). The main features of the two maps are very similar. However, the map of Taylor and Frawley (1987) has higher amplitudes and shows more detail. Some of the finer details probably are spurious, but even if their map were smoothed to remove some of the

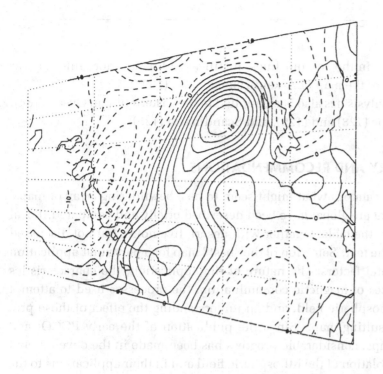

FIGURE 6.28

Comparison of scalar anomaly maps for the Kursk, Russia, region: left map (120); right map (31); both maps at a 350-km altitude. (Part a reprinted from P. T. Taylor and J. J. Frawley, *Magsat anomaly data over the Kursk region, U.S.S.R. Physics of the Earth and Planetary Interiors, 45,* 255–65, 1987, with kind permission of Elsevier Science NL, Sara Burgerhartstraat 25, 1055 KV Amsterdam, The Netherlands.)

finer-scale "kinks" in the contour lines, it still would retain more information than the segment of the map of Arkani-Hamed et al. (1994).

This type of analysis has also been applied to the eastern Indian Ocean (164), Africa (78), Europe (128) and (129), and Scandinavia (125).

6.9   SUMMARY AND RECOMMENDATIONS

Table 6.1 catalogs what might seem to be a bewildering array of maps. Many of those that are global have been described briefly and compared. Local maps are listed in the table, together with descriptive notes, but not discussed at any length in the text. Selection of the "best" map for a particular application depends on several factors: the nature of the application, the characteristics of possible sources of magnetic contamination, the processes used to attempt to isolate the lithospheric field, and an understanding the effect of those processes on the resulting map. Since the publication of the early POGO- and *Magsat*-based maps, considerable progress has been made in the development of methods for isolation of the lithospheric field and in their applications to the data. Later maps are likely to be better approximations of the actual lithospheric field.

For a comprehensive view of the magnetic anomaly field, one needs a global map. However, maps of this type based on bin averages tend to subdue and broaden the anomalies and reduce resolution. Also, because the average altitude of the data varies from bin to bin, they contain artificial variations due to differing altitudes over the map. These problems are largely eliminated in maps computed at satellite altitude from inverse methods (e.g., the equivalent source method), provided they are based on measured or gridded (e.g., by collocation) data points and not on averages. Such maps best preserve the integrity of the measurements and can readily be reduced to pole. Downward continuation and inversion to magnetization or susceptibility can enhance shorter-wavelength variations, but can also amplify data and reduction noise, and so must incorporate smoothing or regularization. These maps are useful and often are easier to compare with geologic or tectonic maps, provided that intepretation of subtle features is firmly based on satellite altitude data such as profiles. It should be noted that downward continuation together with reduction to pole is equivalent to inversion to magnetization along the direction of the ambient main field in a constant-thickness magnetic layer.

For study of local regions and/or individual anomalies, resolution is a fundamental issue. The inherent resolution of satellite data is addressed in Chapter 8. But that resolution is inevitably degraded by some types of processing. Maps derived by averaging over an area, even an area smaller than the basic data resolution, tend to broaden and subdue the anomalies and obscure subtle features that are apparent in profile data. As mentioned earlier, maps based on inverse methods are better. For the highest resolution, there is no substitute for study of individual profiles. This should be done keeping in mind the possibility of external-field contamination and in conjunction with a map to preserve the two-dimensional context. With respect to maps, local analyses such as RHA, SCHA, and crossover analysis will generally preserve detail lost in global maps. We prefer the method of crossover analysis described in Section 4.4.4.

CHAPTER 7

MAGNETIC ANOMALIES AND THEIR SOURCES

7.0 INTRODUCTION

The complex pattern of the earth's regional magnetic anomalies covering a broad spectrum of wavelengths is caused by the wide variety of geologic features in the lithosphere resulting from the long and varied geologic history of the earth. These anomalies are caused by lateral contrasts in the magnetic polarization, or magnetization, **M**, of the lithosphere that are derived primarily from variations in the type, history, and distribution of ferrimagnetic minerals, generally magnetite. This iron oxide and other minerals that are the sources of magnetic anomalies occur, with few exceptions, only in trace amounts (but highly variable quantities) in the crystalline igneous/metamorphic rocks of the lithosphere. For the most part, these rocks are hidden from direct human view by the overlying, generally nonmagnetic sediments and sedimentary strata of both the continents and oceans.

This chapter provides a review of the sources of magnetization contrasts within the lithosphere. This includes consideration of induced and remanent magnetizations and their origins and the relationships of magnetizations to crustal rocks (petrology) and their constituent minerals. A discussion of lithospheric magnetizations in continental and oceanic realms follows. The chapter is intended to introduce both physicists and geologists to the connection between magnetic fields and the nature and processes of the lithosphere. The literature on this topic is extensive, and the reader who seeks a more comprehensive view of this topic is directed to the numerous references cited herein.

7.1   SOURCES OF MAGNETIZATION CONTRASTS

This section provides a brief discussion of the sources of magnetic anomalies (the magnetic minerals), the chemical, physical, and geologic processes affecting the magnetism of these minerals, and the magnetic characteristics of the various components of the lithosphere. Although considerable effort has been put into investigating the magnetic properties of minerals, such studies have focused largely on minerals of direct or indirect importance in magnetic surveys exploring for natural resources and on minerals and processes of importance for the measurements and interpretations of paleomagnetic studies. Much less is known about the synergism of geologic processes and the magnetic properties of rocks. Even more important is our lack of knowledge of the magnetic character of the lower crust and the upper mantle. Even with the increased attention resulting from satellite altitude magnetic observations and compilations of near-surface magnetic anomaly data on a continental scale, our knowledge of the magnetic properties of the deeper portions of the lithosphere is rather rudimentary, largely because of limited access to samples from those depths and concerns with the representativeness of the available samples, as well as the potential modifications of such samples during the processes of bringing them to the surface.

7.1.1   FUNDAMENTAL TYPES OF MAGNETISM

Magnetism is derived from electrical current associated with the movement of an electrical charge. Thus an atom, which consists of a positively charged nucleus surrounded by "shells" of orbiting and spinning electrons, will produce a magnetic moment if the shells are incompletely filled with an uneven number of electrons. A magnetic moment also may be present if an incompletely filled shell has an even number of electrons, because spin vacancies in one direction are preferentially filled first. Imposition of a magnetic field upon an atom causes precession of an electron orbit about the original orbital direction, resulting in an additional angular momentum in the direction of the applied field and an **M** directed opposite to the field **B**. This type of magnetization is called *diamagnetism*. Diamagnetism is universal to all atoms, because the orbits of all electrons experience the precessional effect. However, the diamagnetic effect is counteracted by other magnetic moments induced by the field, except in the case of atoms in which the shells are completely filled by electrons.

An additional magnetic field is produced when many atoms are subjected to an external magnetic field. This field is caused by the spin of an unpaired electron in an atom. Spinning electrons act as magnetic dipoles. Thus, in the presence of an inducing field, the magnetic moment of the dipole aligns with the imposed field. Where there is an uneven number of electrons in an atom, a net magnetic moment is produced. This is termed *paramagnetism*. If there is no imposed magnetic field, thermal effects randomize the orientations of the atoms, resulting in no net magnetic moment. However, in the presence of an imposed **B**, the moments favor an orientation parallel to **B**. The magnetic susceptibility (see Section 5.1.2) of paramagnetic substances is temperature-dependent, with susceptibility decreasing with increasing temperature. Many common rock-forming minerals, such as biotite, amphibole, pyroxene, garnet, and olivine, contain Fe_2, Fe_3, and Mn_2, which are strong carriers of paramagnetism.

Although paramagnetism and, to a lesser extent, diamagnetism contribute in a minor way to the magnetism of rocks, the principal rock magnetism is derived from a third general class of materials, called ferromagnetic. Ferromagnetism and its subclasses are found in only a few naturally occurring minerals. Nonetheless, these few minerals are the dominant carriers of magnetism in rocks. They have a strong magnetic susceptibility, in comparison with diamagnetic and paramagnetic minerals, and they exhibit permanent magnetism, even in the absence of an external field. In ferromagnetic minerals, quantum-mechanical exchange forces come into play, so that the magnetic moments of nearby atoms tend to line up in parallel. Such regions are called *domains*. The sizes and polarities of domains tend toward the configuration that will minimize the total system energy, which is a combination of a variety of atomic-level energies. The minimum-energy configuration is highly dependent upon the crystal structure of the particular mineral and, to a lesser extent, also depends on grain shape and deformation. These minerals are classified into three categories:

1. *ferromagnetic* minerals, in which all the magnetic moments in the crystal are in parallel
2. *antiferromagnetic* minerals, in which there are two sublattices with equal antiparallel magnetic moments
3. *ferrimagnetic* minerals, in which there are two sublattices with antiparallel magnetic moments that are unequal

The latter are important as sources of magnetic anomalies.

Geophysically important magnetization is temperature-dependent because the exchange energy must work against the randomizing effects of thermal fluctuations. Beyond a critical temperature called the *Curie temperature* (T_C) or *Neel temperature*, a mineral becomes paramagnetic.

Ferrimagnetism is named after ferrites, which exhibit this property. Ferrites have a general chemical composition of $XO \cdot Fe_2O_3$, where X is a divalent metallic ion such as Fe, Co, or Mg. The magnetic susceptibilities of ferrimagnetic minerals are much greater than those of paramagnetic minerals because of the interactions between adjacent atoms.

Ferrimagnetic properties change with the shapes of the magnetic grains, because of internal demagnetization, and with their sizes. Grains the size of single magnetic domains, which are of the order of 0.1 μm in magnetite, have been used by Neel (1949, 1955) to explain the ferrimagnetic properties of materials. Single-domain grains have intense and stable remanent magnetization, as compared with coarser-grained, multidomain materials. Multidomain grains have characteristics similar to (but muted) those of single-domain elements, prompting theories to explain this behavior based on pseudo-single-domain models (Verhoogen, 1959; Stacey, 1962, 1963; Stacey and Banerjee, 1974). Those theories and more recent studies attempting to bring theory and experimental data together have been reviewed by Dunlop (1995). Grain sizes smaller than single domains exhibit no hysteresis and behave as paramagnetic materials. However, these grains take on a *superparamagnetism*, where thermal agitations maintain the magnetization in a free state, and thus the grain is much more sensitive to the external field.

Remanent magnetism contributes to the sources of magnetic anomalies observed over the continents and particularly over the oceans. Over 20 kinds of remanent magnetization are recognized (Merrill and McElhinny, 1983). The remanent magnetization of a rock in place is *natural remanent magnetization* (NRM), which consists of primary remanent magnetizations acquired during the rock's formative stage and any number of secondary magnetizations subsequently imposed upon the rock. The ratio of remanent magnetization to induced magnetization, Q in equation (1.19), is highly variable among rock types. Rocks containing fine-grained magnetic minerals that acquire intense, stable remanent magnetization commonly have Q values greater than 10, whereas coarse-grained rocks have values that seldom reach 1.

Although there are numerous physical and chemical processes that lead to a wide variety of remanent magnetizations in the earth's materials, only a few are significant contributors to magnetic anomaly sources. *Thermoremanent magnetization, chemical remanent magnetization,* and *viscous remanent magnetization* are widely recognized as major sources of geomagnetic anomalies, although other types, such as *detrital remanent magnetization* (due to alignment of magnetic grains in sedimentary rocks) and *piezoremanent magnetization* (which is stress-induced), may locally contribute to near-surface magnetic anomalies. It should be understood that in the natural setting of the earth these processes overlap among each other, producing complex relationships. More comprehensive discussions of the sources of remanent magnetization are available (e.g., Nagata, 1961; Stacey and Banerjee, 1974; Tarling, 1983; O'Reilly, 1984; Dunlop, 1995).

In discussing the sources of remanent magnetization, it is useful to introduce two concepts regarding the acquisition and loss of magnetization in magnetic grains. These are the concepts of blocking temperature (and volume) and relaxation time, as discussed by Neel (1949, 1955). In the presence of a large magnetic field, magnetic domains will readily overcome internal energy barriers and align in the direction of the field. However, at lower fields, such as the geomagnetic field, the magnetic energy imposed by the field will be insufficient to cause immediate reorientation of the domains into the direction of the field. That is the case until thermal vibrations allow reorientation of the domains. At any specific temperature there is a range of thermal vibrations and thus atoms of greater thermal energy that will overcome internal energy barriers and align with the ambient field. At the same temperature, over time, all atoms will eventually have sufficient energy to achieve this condition of alignment. The time required for each individual domain to reach alignment is called the *relaxation time*. The relaxation time is measured as the ratio between the internal energy barrier (which is dependent on the volume of the domain) and the

FIGURE 7.1
Relationships among temperature, volume, and relaxation time. This graph is based on titanomagnetite, although the general shape and relationships are valid for all magnetic minerals, but with different values for different minerals. The behavior as the grain size changes from single-domain to multidomain is very strongly controlled by the presence of imperfections in the crystal lattice, making actual physical measurements difficult. The estimated values are shown by dashed lines. The onset of such multidomain behavior can take place at smaller grain sizes than indicated here. (Reprinted with kind permission of Chapman and Hall from Figure 2.6, page 26, of D. H. Tarling, *Paleomagnetism*, Chapman and Hall, 1983.)

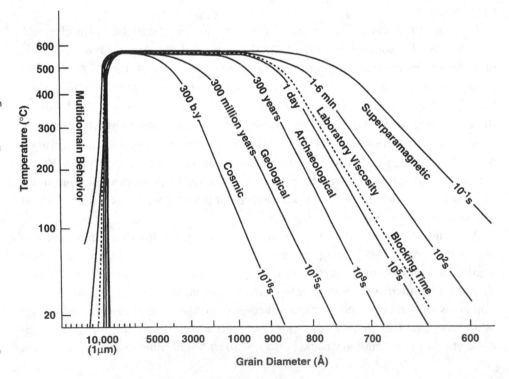

thermal vibrations (energy) necessary to cause reorientation. Neel, considering single-domain grains, showed that the relaxation time t is

$$t = t_0 e^{VM_s(T)H_c(T)/2kT},\tag{7.1}$$

where t_0 is an atomic reorganization time (ca. 10^{-10} s), V is the grain volume, T is the absolute temperature, $H_c(T)$ is the coercivity at temperature T, $M_s(T)$ is the saturation magnetization at temperature T, and k is Boltzmann's constant. Thus, the log of the relaxation time is proportional to the ratio between the volume of the grain and its temperature. The temperature at which a domain exceeds the energy barrier preventing alignment with the ambient field is the *blocking temperature*, or the Curie temperature. In the case of a changing grain size, this effect is related to the *blocking volume*. Factors controlling the blocking temperature include the size and composition of the domain and the intensity and duration of the applied field. The foregoing discussion applies equally well to the reverse situation of the loss of remanence.

Thermoremanent magnetization (TRM) leads to intense, stable magnetization within magnetic minerals that reliably records the direction of the ambient field during its acquisition. These characteristics are especially profound in fine-grained rocks such as volcanic rocks, and less so in coarse-grained, multidomain magnetic minerals that occur in plutonic rocks. TRM is acquired by magnetic minerals as they cool through their Curie temperatures and the associated range of blocking temperatures controlling their relaxation times. The converse situation also occurs; that is, TRM is lost as a rock is heated to above its Curie temperature. Magnetic minerals in rocks generally exhibit a broad range of grain sizes and thus a wide range of blocking temperatures, so that only minerals of specific grain sizes will acquire TRM at a particular temperature (Figure 7.1). Remanence acquired over a specific temperature interval is referred to as *partial thermoremanent magnetization* (PTRM). The TRM is the summation of the individual PTRMs vectorially added, because each PTRM reflects the intensity and direction of the ambient field over the specified cooling interval.

It should be noted that self-reversing magnetization occurs in nature, but there is no evidence that it occurs on anything but a local basis, and thus it is unlikely to be a major source of magnetic anomalies.

Chemical remanent magnetization (CRM) is analogous to TRM, but in this case the temperature remains constant and the grain size changes. Initially, precipitating mineral grains are sufficiently small that they are superparamagnetic, but as they grow, the relaxation time increases exponentially, until the blocking volume is reached, a condition analogous to the blocking temperature. As the grains grow, the relaxation time increases until the dimensions reach multidomain size. At grain sizes above the blocking volume, changes in the applied field have little impact upon the magnetization. The CRM behavior of multidomain grains as they grow is complex, but the magnetic moment gradually decreases as the grains continue to grow.

Another form of remanent magnetization of potential importance as a source of magnetic anomalies is a time-dependent isothermal remanent magnetization identified as *viscous remanent magnetization* (VRM), sometimes called thermal VRM (TVRM). It is known from laboratory studies that under ambient conditions rock samples can acquire a magnetization that is parallel to the inducing field. The rate of acquisition of VRM increases with temperature and is enhanced by the presence of coarse grains of magnetic minerals. These controls correspond to their effects on the length of the relaxation time and depend on composition, temperature, grain size, and previous magnetic history. With the acquisition of a field direction parallel to the currently existing ambient field there is a decay in the remanence preserved from previous field directions. The relaxation times for VRM in magnetic minerals are of the magnitude of geologic periods, measured in millions of years.

7.1.2 MAGNETIC MINERALS

Most important rock-forming minerals, such as quartz and feldspars, are diamagnetic; thus the majority of the content of rocks is essentially nonmagnetic. Paramagnetic minerals also have low magnetic susceptibilities and no remanent magnetization. Nonetheless, as Clark and Emerson (1991) have pointed out, the paramagnetic minerals cause systematic differences in the magnetic susceptibilities of felsic and mafic rocks. Those authors have shown that the volume susceptibilities for a granite with 2% FeO by weight and a gabbro with 12% FeO, assuming no other sources of magnetic susceptibility, are 12×10^{-5} SI and 80×10^{-5} SI, respectively. Such differences will have only minor effects on most magnetic anomalies, except for sources that have large volumes and whose adjacent rocks have very low magnetite contents.

The principal magnetic minerals in rocks are elements of the ternary system FeO-Fe_2O_3-TiO_2, with additional minerals of the Fe-Ni-S system and the metal alloy system Fe-Ni-Co (e.g., Creer, Hedley, and O'Reilly, 1976; Haggerty, 1979; O'Reilly, 1984). Quantitatively the Fe-Ti-O ternary system is by far the most important. However, not even all the components of this system are magnetic. The mineralogic and magnetic data for the Fe-Ti-O system are shown in Figure 7.2. The most significant solid-solution series is shown by the ulvöspinel (Fe_2TiO_4)/magnetite (Fe_3O_4) join, which is referred to as the titanomagnetite series. The open section of the join indicates the solubility gap in the series that occurs below a temperature of 600°C. This miscibility gap grows with decreasing temperature to the indicated gap at room temperature. Under most geologic conditions, cooling from the high-temperature solid solution causes exsolution to essentially pure magnetite and either ilmenite ($FeTiO_3$) or ulvöspinel at room temperature. As this process proceeds, ulvöspinel tends to oxidize to ilmenite and magnetite if dissociated water is present. Ulvöspinel has no significance as a source of magnetic anomalies because it has a Neel temperature

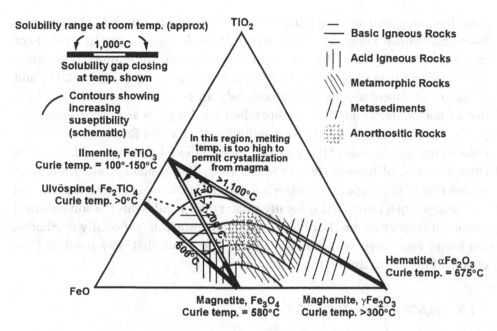

FIGURE 7.2
Review of mineralogic and magnetic data pertaining to the Fe-Ti-O system. (Reprinted from F. S. Grant, Aeromagnetics, geology, and ore environments. I. Magnetite in igneous, sedimentary and metamorphic rocks: an overview. *Geoexploration*, 23, 303–33, 1984/1985a, with kind permission of Elsevier Science NL, Sara Burgerhartstraat 25, 1055 KV Amsterdam, The Netherlands.)

less than room temperature. Magnetite, on the contrary, is the principal source of magnetic anomalies. It has a Curie temperature of 580°C and thus is ferrimagnetic throughout much or all of the normal crust of the earth. The saturation magnetization is strong, and it has an intense magnetic susceptibility ($\sim$1 cgs, 4π SI). Although magnetite is magnetically anisotropic, that property is overshadowed by grain-shape anisotropy, which reduces its apparent magnetic susceptibility.

The ilmenohematite series [$X\,\mathrm{FeTiO_3} \cdot (1 - X)\mathrm{Fe_2O_3}$] has as its end members ilmenite ($X > 0.9$), which is antiferromagnetic, with a Neel temperature below room temperature. Thus, under geologic conditions it behaves paramagnetically. Hematite ($X < 0.4$) is also antiferromagnetic, but exhibits weak (parasitic) ferromagnetism on one plane of its crystalline structure. The products of the solid-solution series are ferrimagnetic for $0.9 > X > 0.45$. However, the normal geologic situation (Tarling, 1983) leads to exsolution to its end members, ilmenite and hematite. Low-temperature (<200°C) oxidation of magnetite can produce maghemite ($\mathrm{Fe_2O_3}$); maghemite, which has the same chemical composition as hematite, but different crystal structure, is ferrimagnetic. It forms a complete solid-solution series with magnetite and has overall properties similar to those of magnetite. Other components of the $\mathrm{FeO\text{-}TiO_2\text{-}Fe_2O_3}$ ternary system have only paramagnetic properties and thus are not sources of magnetic anomalies.

Other naturally occurring minerals that may play roles in contributing to magnetic anomalies are native elements and iron sulfides. Native Fe is seen only rarely in nature because of its propensity to oxidize in the earth's environment. Nevertheless, this strongly ferromagnetic mineral is occasionally observed in rocks [e.g., Haggerty and Toft (1985) found native iron in xenoliths (foreign rock fragments) from a kimberlite pipe]. Iron sulfides (FeS_{1+X}) occur in a variety of forms and compositions. Pyrrhotites ($0 < X < 1.0$) are antiferromagnetic when $X > 0.1$. The Curie temperature for magnetic pyrrhotites is roughly 320°C, and its magnetic susceptibility is approximately an order of magnitude less than that of magnetite, although it is dependent on the grain size (coarser-grained pyrrhotite has higher magnetic susceptibility) (Clark and Emerson, 1991). It is noteworthy that studies of the magnetization in the 9-km-deep drill hole in the Oberpfalz area of Bavaria (Bosum et al., 1997; Berckhemer et al., 1997) have shown that in that region pyrrhotite is a dominant carrier of magnetization, and its Koenigsberger ratio (Q) is generally greater than 1 and is due to a soft chemical remanent magnetization. Further, the magnetic minerals, especially pyrrhotite, are found associated with fracture zones, suggesting that they resulted from chemical processes during fluid circulation.

7.1.3 MAGNETIC PETROLOGY

Magnetic petrology, as defined by Wasilewski and Warner (1994), "is an extension of petrology, integrating magnetic property studies with conventional petrology for the purpose of understanding the development and modification of the magnetizations in rocks." In this section we provide an overview of this topic, focusing on the petrologic controls on magnetic properties. The interested reader will find detailed discussions in the following publications and their reference lists: Haggerty (1979); Grant (1984/1985a,b); Wasilewski (1987); Shive, Frost, and Peretti (1988); Reynolds et al. (1990); Henkel (1991). The acquisition of rock magnetization sufficient to produce observable magnetic anomalies either through induction or remanent magnetization is complex and depends on several coupled factors, including the bulk chemistry, the initial temperature and the rate of change of temperature, and the availability of volatiles. As a result, the magnetic properties of rocks (igneous, metamorphic, and sedimentary) cover a broad range, commonly several orders of magnitude. This is well illustrated in compilations of magnetic susceptibility data for rocks and their Koenigsberger ratios (e.g., Figures 7.3 and 7.4). As a result, it is difficult to make any but the broadest generalizations regarding the magnetic properties of specific rock types. In the simplest situations, mafic rocks with normal magnetite content and rocks with intense normal remanent magnetization will produce significant magnetic anomalies.

Igneous rocks. Rocks that have solidified from the molten state either within or on the surface of the earth are referred to as igneous rocks. They make up a prominent part of the continental crust and the vast majority of the oceanic crust, as well as the upper mantle. Thus, they are important sources of magnetic

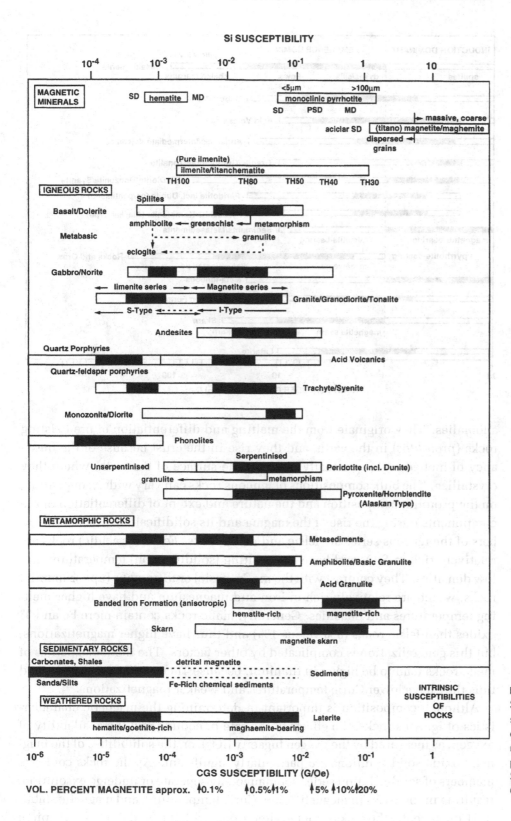

FIGURE 7.3
Summary of rock susceptibilities. (From Clark and Emerson, 1991; reprinted with kind permission of the Australian Society of Exploration Geophysicists.)

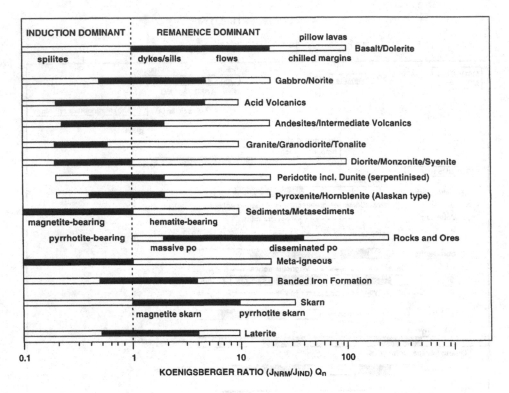

FIGURE 7.4
Summary of Koenigsberger ratios. (From Clark and Emerson, 1991; reprinted with kind permission of the Australian Society of Exploration Geophysicists.)

anomalies. They originate from the melting and differentiation of pre-existing rocks (*protoliths*) in the earth, and they rise in the crust because of the buoyancy of melted rock, eventually reaching the surface of the earth, where they crystallize. The bulk compositions of igneous rocks can vary widely, depending on the protolith composition and the nature and extent of differentiation of the components during the rise of the magma and its solidification. The end members of the igneous series are felsic and mafic rocks. *Felsic* (or acidic) rocks are relatively rich in SiO_2 and have low melting (solidification) temperatures and low densities. They contrast with the *mafic* (basic) or *ultramafic* (upper-mantle) rocks, which are relatively rich in iron and magnesium and have higher melting temperatures and densities. Generally, mafic rocks contain more Fe and Ti oxides than felsic rocks (5% versus 1%) and thus have higher magnetizations, but this generalization is complicated by other factors. The titanomagnetites of mafic rocks tend to be higher in titanium, because Ti crystallizes out early, and thus they have lower Curie temperatures and weaker magnetizations.

Although composition is important in determining the magnetic characteristics of igneous rocks, the effects of the rate of cooling and the availability of oxygen, as measured by the oxygen fugacity (fO_2), on the solubilities of the magnetic oxide solid solutions are particularly significant. As the rocks cool, the members of solid-solution series spontaneously separate or undergo exsolution, resulting in increases in magnetization, Curie temperature, and magnetic stability. Concurrently the rocks undergo alterations, with the oxidation of ulvöspinel toward ilmenite and the production of magnetite. The result is a more magnetic rock. However, if the high-temperature oxidation continues to an advanced oxidation stage, as it does in subaerial mafic rocks (i.e., basalts), the magnetization may be decreased by the formation of non-ferrimagnetic components (Reynolds et al., 1990). Finally, low-temperature oxidation leads to the destruction of magnetite and to the formation of low-magnetization hematite.

The rate of cooling also is an important factor in controlling the magnetic properties of igneous rocks. In basaltic rocks, where the rate of cooling is high and most of the volatiles, including oxygen, escape, the iron and titanium oxides remain in solid solution. The result is that the magnetic susceptibility is low, but the small-grain iron oxides are magnetically stable, thus retaining their imposed remanent magnetization. The Koenigsberger ratios for volcanic rocks are high ($Q > 1$), whereas plutonic rocks that cool slowly within the crust have Q values that seldom reach 1 (Figure 7.4).

As illustrated in Figure 7.3, granitic-type rocks are bimodal in their magnetic susceptibility. The more intensely magnetic granitic rocks belonging to the magnetite series are of the I type, reflecting a dominantly igneous or metaigneous protolith, whereas the lower-magnetization granitic rocks, the S type, are derived from sedimentary or metasedimentary rocks and experience strong interactions with crustal materials (Chappell and White, 1974). The principal distinction between the types is that the magnetite-bearing I type is more highly oxidized than the ilmenite-bearing S type. This higher oxidation state is believed to be primarily a result of the oxidation level of the source rock. In any event, the felsic rocks generally are more highly oxidized than mafic rocks, resulting in nonmagnetic oxides and thus lower magnetizations.

Metamorphic rocks. Rocks whose mineralogy, texture, and structure have been modified by temperature and pressure are described as metamorphic rocks. These effects may be local, as around a specific igneous pluton, or regional, where broad, low-gradient temperatures and pressures have altered the original rocks, the protoliths. The former can produce intense magnetic effects, but only over limited distances measured in a few tens of kilometers at most. In contrast, regional metamorphic effects can cover extensive areas measured in hundreds of kilometers. Metamorphism due to burial within the earth, with the attendant increases in temperature and pressure (especially in high-stress regions affected by the compression associated with converging crustal plates), is termed *prograde metamorphism.* Low-temperature metamorphism, which leads to breakdowns in mineralogy and texture, rather than to the constructive effects of prograde metamorphism, is called *retrograde metamorphism.* Both types of metamorphism can be extensive and can lead to profound changes in the magnetic oxides and the rock magnetization. Gradients in regional metamorphic effects change gradually, both horizontally and vertically, in comparison with the metamorphic effects due to local temperature and pressure perturbations.

Magnetic characteristics of metamorphic rocks depend on the composition of the protolith (especially the iron content of the source rocks), the temperature regime, the chemical effects brought about by the temperature/pressure conditions, and especially the oxidation state, which controls the amounts and types of iron oxides that can form and the partitioning of iron between the silicates and oxides. It is important to emphasize that rocks of the same chemical composition can have quite different magnetizations as a result of metamorphic processes. Thus the magnetizations of rocks arising from a single protolith will vary both vertically and horizontally as a result of these processes.

Metamorphism of igneous rocks usually decreases the magnetization because of the destruction of magnetite (Haggerty, 1979). However, there are numerous notable exceptions to this generalization. Grant (1984/1985a) described the production of magnetite from iron-bearing silicates over a wide range of metamorphic conditions. Secondary magnetite is formed by reactions that are accelerated by increasing temperatures, causing the breakdown of hydrous (Fe, Mg) silicates. However, under extreme metamorphism, such as occurs with granitization and migmatization, iron and titanium oxides recombine to form magnetite-ilmenite solid solutions, causing a decrease in overall magnetization (Grant, 1984/1985a). Magnetite is produced during serpentinization of mafic and ultramafic rocks, leading to some of the very highest rock magnetizations. There is also a tendency for an increase in magnetic susceptibility with increasing metamorphic grade, as a result of the formation of coarser magnetite, which permits easier domain wall movement. Remanent magnetization may increase with temperature up to the Curie point, with Q values on the order of 1, and the direction of remanent magnetization similar to the earth's ambient magnetic field. Thermal enhancement of magnetic susceptibility near the Curie point as a result of thermal fluctuations, the Hopkinson effect, has been suggested as a source of increased magnetization deep within the crust, but it remains unproven. For example, the studies of Kelso, Banerjee, and Teyssier (1993) on the magnetic properties of deep crustal rocks from central Australia showed no Hopkinson effect. Shive et al. (1992) reported that the effect was not found in presumed samples of lower crustal rocks.

Sedimentary rocks are unimportant as sources of long-wavelength magnetic anomalies, but with prograde metamorphism, the reactions can produce secondary magnetite. In iron-oxide-bearing sedimentary rocks, constructive metamorphism will cause reduction of the oxides to magnetite. The result is a metamorphosed iron formation with both intense magnetic susceptibility and remanent magnetization. However, under extreme metamorphism, the iron oxides may be modified to silicates with minimal magnetic properties.

The foregoing discussion provides some generalities regarding the effects of metamorphism upon the magnetic properties of rocks. However, it is clear that the generalizations have numerous exceptions, depending upon the coupling of chemistry and environment. Nonetheless, generally we can anticipate an increase in magnetization with depth until the Curie point is reached, as a result of increases in iron oxides, magnetic susceptibility, and viscous remanent magnetization.

7.2 LITHOSPHERIC MAGNETIZATION

Harrison et al. (1986) noted that the magnetizations required to produce many of the satellite-measured anomalies are greater than expected. Shive et al. (1992), expanding on Mayhew, Johnson, and Wasilewski (1985b), summarized estimates of magnetization and magnetic layer thickness from studies not only of satellite data but also of aeromagnetic data. The range of magnetizations they cited was 2–10 A/m, with 4 A/m a typical value. Magnetic layer thicknesses corresponding to those estimates varied from 10 to 50 km for continental areas, with most in the 25–40-km range. They noted that those magnetizations were several times higher than expected on the basis of rock measurements. This has occasioned considerable debate as to the location of the "missing" magnetization, a problem that Shive et al. (1992) considered to be unresolved.

7.2.1 CONTINENTAL LITHOSPHERE

It is convenient to consider continental lithosphere and oceanic lithosphere separately. Dealing with continental lithosphere, Wasilewski, Thomas, and Mayhew (1979) and Wasilewski and Mayhew (1992) argued that the mantle is nonmagnetic relative to the crust. The basis for their argument was that in the crust the dominant magnetic mineral is titanomagnetite, whereas the magnetic mineralogy of the upper mantle consists of nonmagnetic chromium spinels and magnesium ilmenites. In contrast, Haggerty (1979) has argued that serpentinized zones in the upper mantle could be sources of long-wavelength magnetic anomalies, because the serpentinization process can produce magnetic oxides as well as metallic phases with high Curie temperatures. Mayhew et al. (1985b) and Wasilewski (1987) pointed out that the magnetic phases described by Haggerty have not been observed in mantle xenoliths, which they have studied in detail, and that those phases would be expected to be unstable mineralogically and mechanically.

A major source of magnetization is found in lower crustal mafic granulites, as sampled by xenoliths, such as at converging plate margins, rift valleys, and continental intraplate region settings (Wasilewski and Mayhew, 1982; Wasilewski, 1987), and as found in exposed crustal cross sections, such as the Ivrea zone (Wasilewski and Fountain, 1982). Wasilewski (1987) noted that magnetism found in lower crustal xenoliths is most commonly due to magnetite, with ilmenite-hematite-based magnetism found in rift environments and convergent plate regions. Metamorphic processes are very important in creating and destroying magnetic mineralogies. Mayhew, Wasilewski, and Johnson (1991) illustrate the effect of such processes

for a basic rock carried up through the metamorphic grades. The transition to greenschist grade destroys original magnetic materials, iron being tied up on non-magnetic silicate minerals or weakly magnetic spinels. Amphibolite-grade rocks are rather unpredictable, containing highly variable amounts of magnetite, depending on the specific petrogenetic conditions under which they are formed. Magnetizations of these rocks vary over orders of magnitude; commonly, the iron in these rocks is likewise tied up in minerals other than magnetite, so that they are much more weakly magnetic than the same rocks in the granulite facies. As the granulite grade is reached, iron is released from non-magnetic minerals to form grains of nearly pure magnetite, producing strongly magnetic rocks, especially those having more mafic compositions. Magnetizations of several A/m are common, and the Curie points are close to that for magnetite. Thus, the granulite-grade levels of the deep crust are very likely the most important regionally extensive zones of enhanced magnetization. At eclogite grade, iron again enters silicate minerals such as garnet, and magnetization falls to low levels.

Shive et al. (1992) reviewed possible sources of lower crustal magnetization. In agreement with Wasilewski and co-workers, they concluded that magnetite is the dominant magnetic mineral. They calculated that in order to produce magnetizations of the order of 5 A/m by induction, 5% magnetite by volume would be required. The presence of strong, coherent remanent magnetization would remove that requirement, but in their view the lower crustal rocks are hot and coarse-grained and hence lacking strong or coherent remanence.

Wasilewski and Mayhew (1992) and Wasilewski and Warner (1994) reported on results from studies of about 400 xenoliths. Continental samples were available from Japan, Antarctica, South Africa, the western United States, the Aleutians, and Australia. Magnetization in some of those samples reaches 100 A/m; initial susceptibilities are as high as 0.14 SI; and 5–10-A/m levels of induced magnetization are common. They found that the predominant magnetic domain structure is pseudo-single-domain (PSD), which can acquire significantly higher viscous magnetization (Worm, 1989) than can the coarse-grained structure assumed by Shive et al. (1992). Xenolith samples from ultramafic rocks thought to originate in the upper mantle were found to be nonmagnetic.

Kelso et al. (1993) studied an exposed cross section of continental crust in the Arunta Block of Australia. They also found that the dominant magnetic mineral is magnetite. In that case, NRM dominates over induced magnetization. Much of the remanence is found in single-domain grains, small pseudo-single-domain grains, or relatively stable regions within multidomain grains, and regardless of the bulk rock composition, granulite-grade rocks are much more magnetic than amphibolite-grade rocks. The median NRM is 4.1 A/m (the mean would be higher), with Q of 7.2. Calling attention to the VRM experiments of Kelso and Banerjee (1991), which indicate that granulites could acquire magnetizations of about 4 A/m over the Brunhes epoch, they concluded that the present-day lower crustal magnetization could be predominantly due to VRM acquired along the present ambient main-field direction. Such magnetization cannot be distinguished from induced magnetization in analyses of magnetic anomaly data.

In view of the foregoing discussion, the nature of the transition between crust and upper mantle may have a large effect on magnetization variations. The early study of Wasilewski et al. (1979) assumed a sharp boundary at the Mohorovicic discontinuity (Moho). Recent seismic reflection results indicate the presence of highly reflective structures in the lower crust in some regions. In such cases the Moho may not be causing the strongest reflection, but rather may be highly variable and elusive, perhaps best explained by Mooney and Meissner (1992) as a 3–5-km-wide laminated transition zone at the bottom of a laminated lower crust. They noted that the Moho may be modified by igneous intrusions and ductile deformation and may be the youngest part of the continental crust. Arndt and Goldstein (1989) proposed mechanisms by which material might be transported across the crust–mantle boundary.

A model for the composition of the lower continental crust and the upper mantle was proposed by O'Reilly and Griffin (1985) and Griffin and O'Reilly (1987a). It is based mainly on studies of xenoliths, particularly from the young basaltic fields of eastern Australia. In granulite- and eclogite-facies xenoliths thought to be derived from the lower continental crust, Griffin and O'Reilly (1987a,b) found that the rocks are overwhelmingly mafic. In their model, an ultramafic (spinel lherzolite) mantle is assumed to be overlain by a crust of felsic to intermediate composition. They distinguish the "petrologic" crust–mantle boundary, taken to be the depth where spinel lherzolite becomes dominant, from the Moho. Part of the basis for this model is a xenolith-derived geotherm (O'Reilly and Griffin, 1985) for eastern Australia, shown in Figure 7.5c, which gives a higher temperature at any pressure than conventional continental or oceanic geotherms. Based on experimental measurements of seismic compressional and shear wave velocities, respectively V_p and V_s, in granulites, eclogites, and lherzolites, O'Reilly and Griffin (1985) proposed a crust–mantle rock distribution with depth as shown in Figure 7.6. The figure shows a measured V_p profile and the corresponding profile calculated from their model. In this model, the petrologic crust–mantle boundary is at about 25 km, whereas the seismic Moho is between 55 and 60 km. The gradual increase in V_p from about 6 km/s near 20 km to about 7.5 km/s near 45–50 km is thought to be due to the intermixing of mafic rocks (as discussed later), the lower V_p in the lherzolites at 25–50 km because of the abundance of pyroxene and/or amphibole, and the high temperature. At 55–60-km depths, the transition from spinel to garnet lherzolite results in the seismic Moho. Toft, Hills, and Haggerty (1989), studying xenoliths from kimberlites in the West African craton, also found a transition zone between a granulite lower crust and eclogite upper mantle at depths of about 40–60 km.

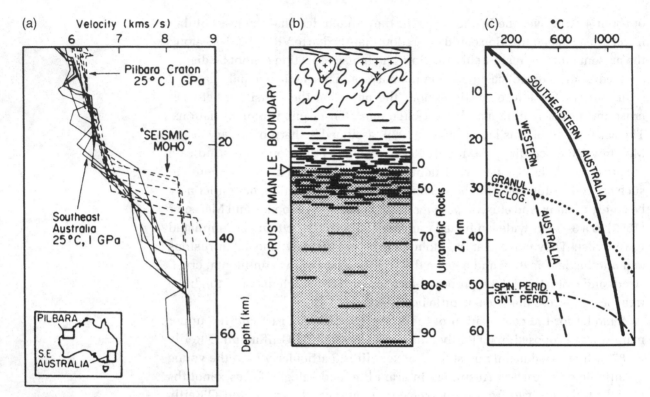

FIGURE 7.5

(a) V_p–depth models for Western Australia (Pilbara) craton and for southeastern Australia. (Adapted from Drummond, 1982.) (b) Model for crust–mantle boundary in noncratonic regions. (c) Geotherms for southeastern Australia. (Heat-flow data from Sass and Lachenbruch, 1979; reproduced with kind permission of Academic Press.) The granulite–eclogite boundary is a conservative (high) estimate based on the spinel-pyroxenite-to-garnet-pyroxenite transition (Griffin et al., 1984). (From W. L. Griffin and S. Y. O'Reilly, Is the Moho the crust–mantle boundary? *Geology, 15,* 241–4, 1987; reproduced with permission (*continued*)

As shown in Figure 7.5b, O'Reilly and Griffin (1985) visualized layers of mafic rocks, the black horizontal lines, with highest concentration near the petrologic crust–mantle boundary, consistent with the presence of multiple horizontal layering found in some seismic reflection data. As shown in Figure 7.6, the model gives a good fit to the measured V_p profile in eastern Australia, also shown in Figure 7.5a. In cratonic areas, such as western Australia, V_p profiles commonly show a shallower, more sharply defined Moho, as in Figure 7.5a. They noted that the main difference between cratonic and other regions is the nature of the geotherm, as illustrated in Figure 7.5c. Temperature was considered the key factor in determining whether granulite or eclogite facies would predominate, as shown in Figure 7.7, which then would determine V_p. According to Furlong and Fountain (1986), additions of 10 km or more to the base of the crust could have been possible by an underplating process. Griffin and O'Reilly (1987a) pointed out that such a process would entail an elevated, strongly curved geotherm, as they found for eastern Australia.

If it is assumed that the magmatic underplating and the accompanying thermal input eventually ceased, then the geotherm would have moved toward one determined by conduction (e.g., as in western Australia), with a time constant of about 10 million years (Griffin and O'Reilly, 1987a, based on Sass and Lachenbruch, 1979). Griffin and O'Reilly (1987a) envision that as cooling takes place, mafic rocks will progress from granulite to eclogite at increasingly shallower depths (Figures 7.7 and 7.8), resulting in an increase of 0.5–1.0 km/s in V_p. Simultaneously, cooling will result in increased V_p for other rock types present: "The Moho moves upward and becomes much more pronounced as a result of cooling." This is shown by the model V_p profiles of Figure 7.9. Under their model assumptions, in cratonic regions the petrologic crust–mantle boundary and Moho will nearly coincide. However, they point out that the xenolith data imply a more dominantly mafic lower crust than that assumed in the model. In areas where that is the case, the result is that V_p will be greater than 8 km/s in what is regarded petrologically as crust, but which will then be defined seismically as mantle (i.e., the seismic Moho will be above the petrologic Moho).

In the model of Griffin and O'Reilly (1987a) there are several important implications for crustal magnetization (Mayhew et al., 1991). First, the model predicts that the rocks in the transition zone can be either granulite or eclogite and, because the granulites are expected to contain magnetite the bottom of the magnetic layer, depends on the position of the Curie isotherm relative to the granulite–eclogite phase transition. Second, in the geotherm determined by O'Reilly and Griffin (1985), the Curie isotherm in eastern Australia is at a depth of about 12 km, well above either the petrologic crust–mantle boundary or the Moho. This may be indicative of the situation in other regions of high heat flow. Third, the model indicates that whereas in cool, cratonic areas the

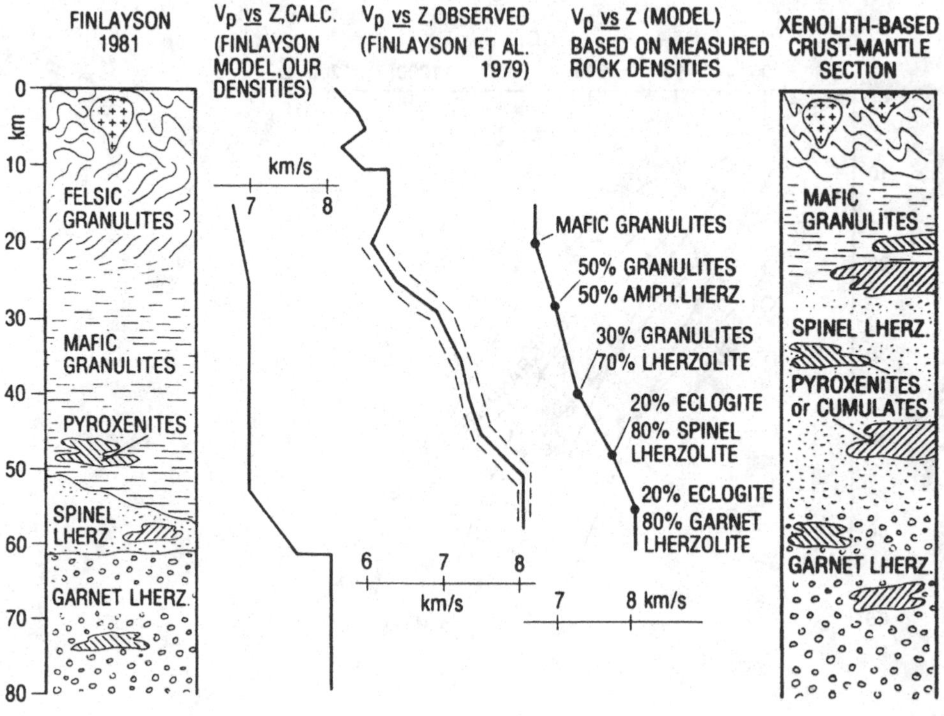

FIGURE 7.5 (continued) of the publisher, The Geological Society of America, Boulder, Colorado, USA. Copyright © 1987, The Geological Society of America, Inc.)

FIGURE 7.6
Interpretation of seismic refraction data (Finlayson et al., 1979) using data on distribution, geochemistry, and density of xenoliths. Left column shows interpretation based on early model of Ferguson et al. (1979); this does not actually fit the observed seismic velocity profile. Right-hand columns show the interpretation of O'Reilly and Griffin (1985) and one possible mixture of rock types that could explain the observed seismic velocities. (Reprinted from S. Y. O'Reilly and W. L. Griffin, A xenolith-derived geotherm for southeastern Australia and its geophysical implications. *Tectonophysics, 111,* 41–63, 1985, with kind permission of Elsevier Science NL, Sara Burgerhartstraat 25, 1055 KV Amsterdam, The Netherlands.)

petrologic and seismic crust–mantle boundaries may be nearly coincident, it is also possible that a highly mafic lower petrologic crust exists below the seismic Moho with a temperature below the Curie isotherm.

Also of importance are the results of Durrheim and Mooney (1991), who found that Proterozoic crust is generally thicker than Archean crust. They estimate that Archean crust is typically about 35 km thick, away from collisional boundaries, whereas Proterozoic crust is typically about 45 km thick. The additional thickness seems to be mostly due to a thicker high-velocity (>7.0 km/s) layer at the base of the crust. The universality of this result is unknown, but it implies a likelihood that, other things being equal, Proterozoic crust has a higher integrated magnetization than Archean crust.

7.2.2 OCEANIC LITHOSPHERE

The oceanic crust is divided into three layers (Keary and Vine, 1990), though the boundaries may be gradational rather than sharp. Layer 1, 0–1 km thick, is sediment. Layer 2, the basement layer, is between 0.7 and 2.0 km thick and is made up of an upper portion of extruded basalt, layer 2A, and a lower portion of basalt massively intruded by dikes, layer 2B. Layer 3 is gabbroic, with an average thickness of about 5 km. Layer 3 is the plutonic foundation for the oceanic crust.

FIGURE 7.7
Position of eclogite–granulite transition relative to geotherms for southeastern Australia (SEA) and Western Australia (WA); the 10-Ma and 30-Ma curves represent decay of the SEA thermal anomaly, if a time constant of 10 million years is assumed. The transition field is based on experimental determinations of the disappearance of plagioclase in a wide range of mafic-to-intermediate compositions (Green and Ringwood, 1972), extrapolated to lower T, assuming a pressure change of 15 bar/°C. (From W. L. Griffin and S. Y. O'Reilly, Is the Moho the crust–mantle boundary? *Geology*, 15, 241–4, 1987; reproduced with permission of the publisher, The Geological Society of America, Boulder, Colorado, USA. Copyright © 1987, The Geological Society of America, Inc.)

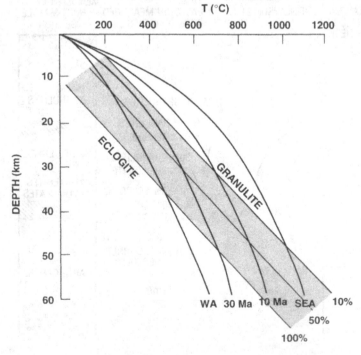

Remanent magnetization dominates in the oceanic crust, giving rise to the seafloor spreading anomalies. Harrison (1987) reviewed the evidence concerning the distribution of magnetization within oceanic lithosphere. The early view was that magnetization is confined to layer 2, with models of the magnetized layer thickness varying from about 0.5 to 2.0 km. For crust older than 4 million years, in which the magnetization has decayed and is representative of most ocean basins, an average magnetization of about 8 A/m in a 0.5-km layer is needed to model the spreading anomalies.

Seafloor spreading anomalies exhibit what is called skewness, which refers to shape asymmetry relative to a square-wave anomaly pattern. Skewness is expected because of deviations of the relative directions of the inclination and declination of the ambient field relative to the remanent magnetization and relative to the strike of the magnetic lineation. An anomaly reduced to the pole should show no skewness; anomalous skewness is that remaining after reduction to the pole. To account for the observed skewness, models have been proposed with sloping reversal boundaries, with tectonic rotation, and with magnetization in multiple layers, including layer 3 (e.g., Harrison, 1987; Arkani-Hamed, 1988; Tivey, 1996).

The vertical magnetic structure of a steep submarine escarpment in crust about 2 million years old was studied by Tivey (1996). He found that the extrusive layer is highly magnetized, about 10 A/m in an 0.8-km-thick layer, and that forward models with dipping polarity boundaries reproduce the amplitude and shape of the measured sea-surface anomalies.

Oceanic serpentinite has been proposed as a contributor to long-wavelength magnetic anomalies (Arkani-Hamed, 1988; see also Nazarova, 1994). Furthermore, on the basis of a model of the subducted oceanic lithosphere along the

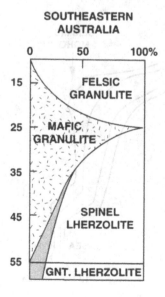
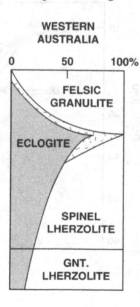

FIGURE 7.8
Distribution of rock types in the model crust and mantle of Figure 7.5b for two end-member thermal regimes illustrated in Figure 7.7. (From W. L. Griffin and S. Y. O'Reilly, Is the Moho the crust–mantle boundary? *Geology, 15,* 241–4, 1987; reproduced with permission of the publisher, The Geological Society of America, Boulder, Colorado, USA. Copyright © 1987, The Geological Society of America, Inc.)

Kuril–Kamchatka trench based on *Magsat* data, Arkani-Hamed and Strangway (1986c) concluded that approximately the upper 35 km of oceanic upper mantle has an effective magnetic susceptibility of 0.0085 SI. Thomas (1987) has proposed a model for the magnetization of the oceanic crust in which viscous and induced magnetizations are the main sources of the long-wavelength anomalies seen in satellite data, except for the Cretaceous quiet zone.

Several studies of satellite magnetic anomaly data have examined the characteristics of Cretaceous quiet zone (KQZ) anomalies in the satellite data, combined with data on the seafloor spreading anomalies measured by marine magnetic surveys, in an effort to determine the distribution of magnetization in the oceanic lithosphere. The resulting models address questions concerning the decay of magnetization with distance from the spreading center, as documented by Raymond and LaBrecque (1987) and the references therein, and observations of anomaly skewness. LaBrecque and Raymond (1985) and Raymond and LaBrecque (1987) have examined *Magsat* anomaly data along four profiles in the North Atlantic. In the initial study, LaBrecque and Raymond (1985) reproduced the main features of the data with a source model using a 500-m-thick magnetized layer, with remanent magnetization intensity of 8–15 A/m, with depth varying as the square root of crustal age, and with addition of a shift in skewness. Raymond and LaBrecque (1987) were able to replace the ad hoc addition of anomalous skewness with a more sophisticated model of magnetization acquisition. They proposed that the original TRM acquired at spreading

FIGURE 7.9
$V_p–Z$ depth profiles for the model crust and mantle of Figure 7.5b, under thermal regimes shown in Figure 7.7. V_p was calculated using lithologic columns constructed as in Figure 7.8 and the average densities and rock compositions given for Australian xenoliths by O'Reilly and Griffin (1985). CMB is the crust–mantle boundary. The dashed part of the WA curve shows the effect of moving the 100% eclogite contour in Figure 7.7 upward by only 3 km. (From W. L. Griffin and S. Y. O'Reilly, Is the Moho the crust–mantle boundary? *Geology, 15*, 241–4, 1987; reproduced with permission of the publisher, The Geological Society of America, Boulder, Colorado, USA. Copyright © 1987, The Geological Society of America, Inc.)

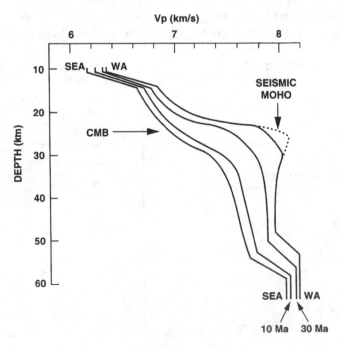

centers decays with time, whereas CRM is acquired in a manner determined by the reversal history during the 20 million years after emplacement.

The models of LaBrecque and Raymond (1985) and Raymond and LaBrecque (1987) retain the structure of the standard lineated anomaly model, with magnetization in oceanic layer 2A. Their modifications consist in lateral variations within that layer. Their models require magnetizations higher than the magnetizations measured in rocks from locations away from spreading centers, though see Tivey (1996). In contrast, Arkani-Hamed (1988, 1989, 1991) proposed models in which TRM is acquired not only in layer 2A but also in the lower crust and upper mantle. In those models, magnetization is acquired by the crust as it cools below its blocking temperature. Because the cooling time is longer at depth, sloping boundaries of magnetization contrast are predicted. Central to the models of Arkani-Hamed (1988, 1989, 1991) is the two-dimensional model for thermal evolution of the lithosphere of Arkani-Hamed and Strangway (1986c, 1987) and Arkani-Hamed (1991). In those models, the amount of skewness is determined by the ratio of magnetization in layer 3 (lower crust) or layer 4 (upper mantle) to that in layer 1 (oceanic layer 2A), using oceanic lower crust for marine anomalies, and upper mantle for satellite altitude anomalies.

All of the studies just described utilized two-dimensional models. Yañez and LaBrecque (1997) compared three-dimensional models of the three types for both the North Atlantic Ocean and northeastern Pacific Ocean. They referred to the three types as (1) TRM, the original model of LaBrecque and Raymond (1985), (2) CRM, the model of Raymond and LaBrecque (1987), and (3) TVRM (for thermoviscous remanent magnetization), the model of Arkani-Hamed (1988, 1989, 1991). In their models, layer 2A magnetizations are taken to be 3.833 A/m and 5 A/m in 1.5-km-thick layers in the Atlantic and Pacific, respectively. Magnetization is uniform below layer 2A and is taken to be 1.6 A/m in the North Atlantic and 2.4 A/m in the Pacific. The magnetic source region is subdivided into a two-dimensional grid of elements about 1° square, each assigned an age based on Cande et al. (1989), with ages in the Cretaceous Quiet Zone estimated by linear interpolation. Each block is rotated to its estimated position at the time of formation at a spreading center and is then moved stepwise to its present location. At each position, any acquired magnetization is taken along the direction of an axial dipole field. In the Pacific, the Hawaiian and mid-Pacific seamounts are assumed to have acquired TRM of 6 A/m at the time of their formation in a volume determined from bathymetric contours.

From the resulting three models, Yañez and LaBrecque (1997) computed maps for the North Atlantic and northeastern Pacific. Plate 4 shows the results for the northeastern Pacific Ocean. This region is much more complicated than the Atlantic because of the considerable northward displacement of the Pacific lithosphere since formation, so that the directions of the acquired magnetizations have changed significantly over time. In this case, the TRM model is inadequate, whereas both the CRM and TVRM models are credible – particularly considering the model complexities and uncertainties. One difference between these two models, as noted by Yañez and LaBrecque (1997), is the location of the positive anomaly at 205–215° E longitude near the Murray Fracture Zone. The CRM model correctly reproduces its shape and location, whereas in the TVRM model the anomaly is located about 5° south of the measured anomaly. East of 195° E longitude, neither the CRM model nor the TVRM model is satisfactory. This is attributed either to errors in estimated ages or to the effects of anomalies associated with the Hawaiian Islands and mid-Pacific mountains.

7.3   SUMMARY

Contrasts in lateral magnetizations, and thus in magnetic anomaly sources, occur throughout the crust as a result of a broad range of geologic processes that have operated upon the earth. Modifications in the structure of the crust caused by a wide variety of factors – vertical and horizontal forces, intrusions and extrusions of magma from the mantle or deeper in the crust, metamorphism by thermal and pressure variations, and alteration of the magnetic minerals to nonmagnetic forms by chemical processes, primarily oxidation – have led to spatial variations in the magnetization of crustal rocks. Magnetization measurements and petrologic studies of rock samples brought to the surface from the mantle by structural and igneous processes indicate that ferrimagnetic minerals, which are the sources of most magnetic anomalies, generally are not present in the mantle (Wasilewski et al., 1979; Wasilewski and Mayhew, 1992), at least in continental regions. However, the existence of sub-crustal lithospheric sources of magnetic anomalies cannot be ruled out, because of the possibility of localized zones in which mantle processes lead to the production of ferrimagnetic minerals and the possible presence of ferrimagnetic elemental iron in the upper mantle (Haggerty, 1978). A further important source of lateral magnetization contrasts consists in the vertical variations in the depth of the Curie isotherm. This isotherm is the depth at which temperatures in the earth reach approximately 580°C, where magnetite loses its ferrimagnetic properties and thus its intense magnetization. The Curie isotherm generally is located within the upper mantle under both the continents and oceans. However, increases in heat transfer from the mantle to the crust in tectonically active regions, leading to an upward warping of the Curie isotherm, will produce a thinning of the magnetic crust and a magnetic anomaly.

Magnetization of lithospheric rocks can result either from induction in the earth's present-day magnetic field or from permanent or remanent magnetization reflecting ancient geomagnetic fields. The majority of anomalies mapped in near-surface magnetic surveying is caused by induction, because the character of the anomalies is consistent with the direction of earth's magnetic field (Hinze and Zietz, 1985). Nonetheless, sources in which remanent magnetization predominates are evident in near-surface survey maps. In fact, this magnetization predominates in the ocean basin rocks. As a result, anomalies derived from remanent magnetization are prominent over the oceans and may be important, at least locally, in the continents. The latter situation has been especially attractive to some investigators, because the lower crust, which is a primary source of regional anomalies, has highly variable regional magnetizations, though on the average they are an order of magnitude greater than those in the upper crust (e.g., Hall, 1974).

CHAPTER 8

METHODS IN INTERPRETATION

8.0   INTRODUCTION

The purpose of gathering and analyzing magnetic anomaly data is to gain an understanding of the earth's lithosphere. In contrast to the situation for many aeromagnetic and ground surveys that have specific local geologic objectives, the goal for satellite surveys is to study the large-area or long-wavelength characteristics of the crust and upper mantle. This is accomplished by deriving models that are consistent with the observations and are believed to be plausible representations of some of the characteristics of the lithosphere. There are many kinds of models; their common purpose is to generalize about the observations and lead to better understanding of the present state of the lithosphere and of the processes that have led to that state. However, all models are limited by the amount of information contained in the available data. Because they are taken at high altitude, satellite magnetic data can be used to detect only anomalies of large spatial scale and large magnetization contrast. Furthermore, the high altitudes of the satellite surveys and the generally limited range of depths of

the magnetic sources tend to minimize the importance of the interpretations of source depths that are so prominent in the analysis of near-surface surveys.

When considering interpretational methods, the simplest are very qualitative, often consisting of visual comparison of an anomaly map with geologic or tectonic maps. Many of the earliest interpretive studies of satellite anomaly maps were of that type. Often, more quantitative interpretations begin with such a qualitative comparison as a guide. Direct comparisons (e.g., by cluster analysis or other methods of correlation analysis) between geophysical quantities such as gravity, heat flow, and depth to the Moho can lead to insights into geophysical processes operative in particular regions. Such methods will be called semi-quantitative. Finally, the methods of inverse and forward modeling result in hypotheses about specific distributions of magnetization. These are the quantitative models. Through synthesis of all the relevant types of available data, quantitative models of the lithosphere can be constructed in terms of structural and compositional variations and the movements of material and energy. Inferences sometimes are drawn about lithospheric evolution from these models.

In the following sections, the more widely used interpretive methods are described, proceeding from the qualitative to the quantitative. Extended treatment is given to spherical-earth forward modeling because it is widely used and is considered an important analysis tool. Most of the sections are relatively self-contained and independent, so the reader interested in a particular method can turn directly to the pertinent section. Section 8.5 introduces a new methodology that has not been used extensively and is expected to stir debate: A priori long-wavelength information is built into a global magnetization model. This simulates the anticipated anomalies resulting from a marked contrast between continental and oceanic magnetizations.

8.1 QUALITATIVE/SEMI-QUANTITATIVE METHODS

8.1.1 MAP COMPARISONS

Commonly, interpretation begins with qualitative comparisons of different data types in the form of maps or profiles. An example using *Magsat* data is shown in Figure 8.1, from Hastings (1982), in which a preliminary version of the scalar *Magsat* anomaly map overlies a broad tectonic map of Africa, with major features indexed by numbers and letters. His discussion assumes that the majority of the anomalies are induced. Then, because the anomaly map is not a reduced-to-pole (RTP) map, and much of Africa is at low inclinations of the main magnetic field, except in the southern part, negative anomalies correspond to high positive magnetizations, and except near the equator the anomalies are offset from their sources. Thus, considerable care must be exercised in

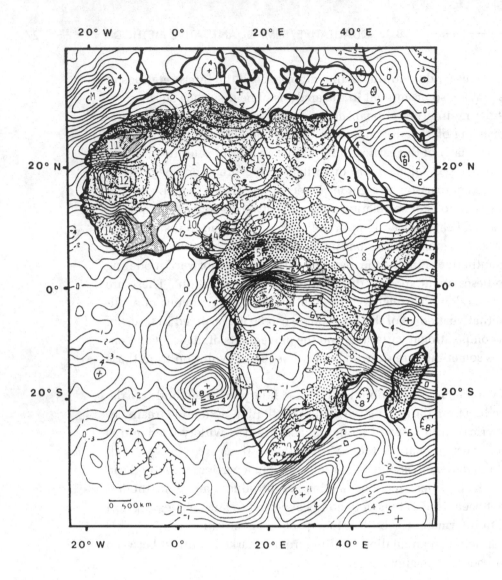

CONTINENTAL FEATURES
    1. Ahaggar plateau
    2. Arabian basin
    3. Atlas Mountains
    4. Benue trough
    5. Central African uplift
    6. Chad basin
    7. Congo basin
    8. East African rift
    9. Karroo basin
  10. Pan-African shield (of Nigeria)
  11. Reguibat shield
  12. Taoudeni basin
  13. Tibesti plateau
  14. West African craton (nucleus)

OCEANIC FEATURES
    A. Agulhas plateau
    M. Madeira-Torre rise
    W. Walvis ridge

KEY

UPLIFTED AREAS

Archaean shields (over 2400 million years old)

Middle Precambrian shields (2400-1000 m.y.)

Uplifted areas less than 1000 m.y. in age

DEPRESSED AREAS

Moderate depressions

Deep portions of basins

MAGSAT contours are 1 nanotesla (nT).

evaluating these qualitative correlations. In the following discussion, numbers in parentheses correspond to the index numbers in Figure 8.1. Hastings (1982) noted strong negative anomalies associated with the Reguibat shield (11), the West African craton (14), and the Central African uplift (5), known as the Bangui anomaly. A positive anomaly is associated with the Taoudeni basin (12), with amplitude roughly proportional to sediment thickness. Weak anomalies are noted in the region of the Pan-African shield (10) and a saddle in the enclosure of the Ahaggar (1) and Tibesti (13) plateaus. Anomalies are not obviously associated with the Atlas Mountains (3), the Benue trough (4), the Chad basin (6), or the East African rift (8). On the basis of such comparisons, he noted the tendency of older Precambrian shields to cause intense negative anomalies, which would be positive if reduced to the pole, whereas progressively younger uplifted areas cause progressively weaker anomalies. He also noted positive anomalies (negative magnetizations at equatorial latitudes) in proportion to sediment thickness associated with depressions.

On a global basis, Frey (1982a) compared scalar anomalies from *Magsat* data with known geologic and tectonic structures. He demonstrated that on a global scale there is a tendency for anomalies detected at satellite altitude to be associated with large features such as shields, platforms, and subduction zones (all with mainly negative anomalies) and basins and abyssal plains with mainly negative anomalies and to be bounded by "linear" features such as sutures, rifts, folded mountains, and age province boundaries. At equatorial latitudes these signs are reversed, unless the maps are reduced to the pole. These anomaly/geologic feature associations are illustrated for Asia by Frey (1982b), for South America by Hinze et al. (1982), and for India by Achache et al. (1987).

The geologic basis for the generalized magnetic anomaly signatures for a variety of major geologic features has been discussed in numerous reports (e.g., Frey, 1982a; Schnetzler, 1989). Examples include positive anomalies (1) over old continental shield, which may occur because these are areas of thick, cold crust, (2) over subduction zones, because "cold" magnetic ocean crust is penetrating into a relatively nonmagnetic mantle, (3) over ancient rifts, because mafic material from the mantle has intruded into the base of the crust in the extensional regime of the rift, and (4) over submarine plateaus, because the oceanic crust has been thickened. Relative negative anomalies over basins and abyssal plains may be due to crustal thinning, and over young rifts because the Curie isotherm is elevated. Suture regions where crustal blocks with different characteristics have come together might be expected to (and often do) have magnetic signatures.

Though useful as a first look at the general tendencies of anomaly occurrences, closer examination reveals that the associations discussed above are oversimplifications to which there are numerous exceptions. The more detailed studies summarized in Chapter 9 will be used to expand on these simple associations

FIGURE 8.1 (*opposite page*) Preliminary total-field *Magsat* anomaly map for Africa, showing significant tectonic features. Anomaly contours are 1 nT. Continental areas without shading are covered with relatively shallow sedimentary sequences. Tectonic features modified from United Nations Educational, Scientific and Cultural Organization (1971, map in pocket). (From Hastings, 1982, with permission.)

and examine more closely the complicated relationships of anomalies to local tectonics and geology.

8.1.2 APPLICATION OF POISSON'S THEOREM

Global magnetic anomaly data sets derived from satellite measurements are useful as complements to the increasingly available global data sets of gravity, topography, heat flow, crustal thickness, seismic velocities, and so forth. Although the integrated use of these data sets commonly is restricted to visual spatial correlation, they can be combined in a much more quantitative manner. For example, Poisson's theorem can be used in the analysis of potential-field data. In the idealized situation where there is a single source body with uniform distribution of density, magnetization, and direction of magnetization, the magnetic and gravity potentials from that source are related by Poisson's theorem (Poisson, 1826; Chandler et al., 1977, 1981; Lugovenko and Pronin, 1982; Lugovenko et al., 1986; Blakely, 1995):

$$\Psi = \frac{M}{\sigma G} \frac{\partial \Psi_g}{\partial i}, \tag{8.1}$$

where Ψ is the magnetic potential, Ψ_g the gravitational potential, i the direction of source magnetization, M the source magnetization, σ the source density, and G the universal gravitational constant. If the magnetic field is reduced to the pole, it is meaningful to take the radial derivative of (8.1) to give

$$A_r = \frac{M}{\sigma G} \frac{\partial g_r}{\partial r}, \tag{8.2}$$

where g_r is the radial component of the gravity anomaly. Chandler et al. (1977, 1981) and von Frese, Hinze, and Braile (1982a) suggested that when dealing with real anomaly data, (8.2) be put in the form

$$A_r = \mathcal{A}_0 + \frac{\Delta M}{\Delta \sigma G} \frac{\partial g_r}{\partial r}, \tag{8.3}$$

where ΔM and $\Delta \sigma$ are magnetization and density contrasts, and $\mathcal{A}_0$ is an intercept term to account for anomaly base-level variations. It has been shown (von Frese et al., 1981a, 1982a) that (8.2) can be extended to higher derivatives relating the nth magnetic derivative to the $n+1$ gravity derivative. Study of higher derivatives emphasizes shorter wavelength features of the fields under study. Blakely (1995) discussed some of the conditions for validity of Poisson's theorem and some of its implications and gave examples for some simplified geometries.

Crustal gravity and magnetic anomalies do not generally meet the assumptions leading to (8.3), and the foregoing equations should not be applied to measurements at a single location. A regional approach has been described by

Chandler et al. (1977, 1981) and von Frese et al. (1982a), using gridded data at constant altitude, either along a profile or in map form. The basic equation is (8.3), or a higher derivative, which is applied to subgroups of points in a series of moving windows along the profile or over the map grid. Within each window, a least-squares linear regression to the magnetic and gravity anomaly data (or their derivatives) is used to calculate the intercept (A_0), slope ($\Delta M/\Delta \sigma G$), and correlation coefficient between the two quantities. The slope multiplied by G gives the $\Delta M/\Delta \sigma$ ratio. Correlations between gravity and magnetic anomalies are complicated, because magnetic sources reside only in the lithosphere (once isolated from the main and external fields), whereas density variations occur throughout the mantle, and identification of those gravity anomalies due to only lithospheric sources is not always possible. Nevertheless, the correlation between gravity and magnetic anomaly data often is useful in study of the lithosphere (e.g., Chandler and Carlson Malek, 1991). The actual $\Delta M/\Delta \sigma$ ratio depends upon properties of both the anomaly source and the surrounding rock and, even when known, is not unique to any particular type of rock or structure. However, it can furnish a useful constraint on the nature of the lithosphere. In a series of simulations, Chandler et al. (1977, 1981) investigated conditions under which it may (or may not) be possible to find a meaningful estimate of $\Delta M/\Delta \sigma$. The best results are obtained over the high-gradient edges of spatially coincident gravity and magnetic anomalies and when the correlation coefficient is near either -1.0 or 1.0, when the $\Delta M/\Delta \sigma$ ratio is relatively constant, and when the intercept coefficient is zero, or at least constant – all over a region of meaningful size. Chandler et al. (1977, 1981) have shown that in the case of superposed sources, the estimated $\Delta M/\Delta \sigma$ will be a linear combination of the source ratios. They also indicated that spikes in a plot of the slope are indicators of interference from multiple anomalies and that the choice of window size is important.

8.2 QUANTITATIVE METHODS, FORWARD MODELING

8.2.1 INTRODUCTION

Inverse methods, as described in Chapter 5, offer one approach to estimating the distribution of magnetization in the lithosphere. However, it is not always straightforward to account for geophysical information from other than magnetic field data during the inversion process. Nor are the effects of varying the model parameters easily investigated. Forward modeling is generally most useful when studying a particular anomaly or group of anomalies in a limited geographic region. The model parameters typically include specification of the boundaries of candidate source regions or bodies. Those boundaries can be based on the known or assumed geology and tectonics, on the results of seismic and/or gravity surveys, and on other geophysical models of, say, heat

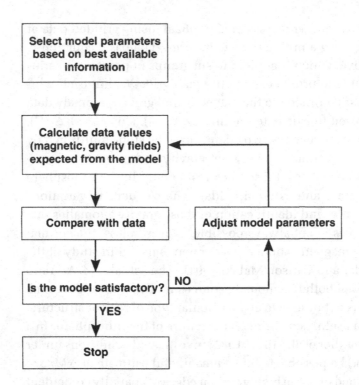

FIGURE 8.2
Idealized flow diagram of the forward modeling process.

flow or of the magnetization derived from an inverse model. Each body region is then assigned a magnetization or susceptibility. This assignment can be based on consideration of typical magnetizations (e.g., Sections 7.1 and 7.2) or on estimates of magnetization from inverse models. Often, gravity and magnetic data are modeled together. In that case, the parameters will also include mass density. Among the more important uses of forward modeling are testing of hypotheses (i.e., whether or not some assumed magnetization contrast will result in a particular anomaly) and sensitivity studies (i.e., testing the effects of changing the parameters in a model).

The forward modeling procedure is outlined in Figure 8.2. To begin, an initial set of parameters is specified based on the best available information. The expected magnetic field is calculated. Typically, this may be in map form for comparison with a satellite magnetic anomaly map. If the model field reproduces the measured field, then the chosen parameters furnish an acceptable model. Such models are inherently nonunique. To explore the suite of models capable of explaining the data, the parameter space is varied. If no information aside from a magnetic survey is available, the possible parameter space is essentially infinite, hence the importance of additional information for constraint of the available parameter space. Formulation of the magnetic forward problem typically begins with equation (5.20), with a corresponding equation for gravity.

8.2.2 FLAT-EARTH MODELS

There are many approaches to forward modeling (e.g., Bhattacharyya, 1978). Most have been developed for interpretation of ground, aircraft, or ship surveys of the magnetic field and/or gravity field. In comparison with satellite data, such surveys are very local, and generally it suffices to assume that the earth is flat and that the ambient magnetic field is constant over the region being modeled. Blakely (1995) contains a good fundamental description of such models, and his references point to additional discussion.

In general, forward models assume that the source responsible for the anomaly can be modeled by a collection of smaller bodies of regular known geometry. Density and magnetization are generally assumed constant throughout each of these bodies. Body geometry is selected for tractability of analysis. Typical geometries are dipoles (for the magnetic field), rectangular prisms, stacks of laminae, and polyhedrons. The resulting formulae and the software for computation can be found in Blakely (1995) and the references cited therein.

If the length of an anomaly source is much greater than its width, as is often the case with geologic features, and the altitude at which the anomaly is measured is small compared with the source length, then a two-dimensional model may be sufficiently accurate. Such models are then preferred because of their simplicity and computational efficiency. Such a method was devised by Talwani and Heirtzler (1964) and has been described by Blakely (1995). The method has been generalized by Shuey and Pasquale (1973) to what is called a "$2\frac{1}{2}$-dimensional" analysis. In this case the body is no longer assumed to be infinite in the y direction. Other widely used two-dimensional methods are found in Talwani (1965) and Plouff (1976).

Using satellite magnetic anomaly data, Hastings (1982) derived a two-dimensional model of the Bangui magnetic anomaly in Africa. In the model, shown in Figure 8.3, the Chad (13-km-thick) and Congo (10-km-thick) sedimentary basins, with zero susceptibility, lie on either side of a highly magnetic, 1.5-relative-susceptibility, 40-km-thick Central African uplift, all surrounded by 40-km-thick normal crust with a 1.0 relative susceptibility. As shown in the figure, this model successfully reproduces the anomaly pattern from a single traverse of the POGO satellite. A full three-dimensional model for this region of Africa was derived by Regan and Marsh (1982), as described in Section 1.2. Their model is made up of a large number of prisms, and in order to take into account the variations in direction and strength of the ambient main field, the field inclination and strength are computed separately for each prism. They considered that the area analyzed was small enough that the curvature of the earth could be neglected.

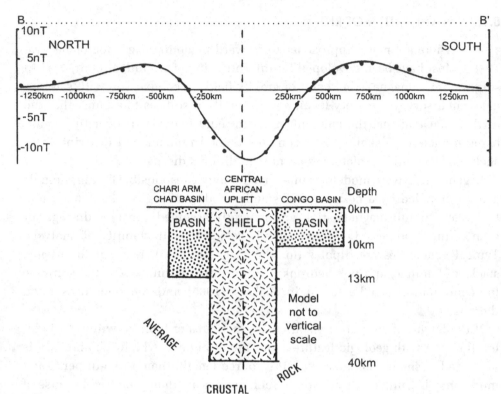

FIGURE 8.3

Two-dimensional model (solid line) and observed POGO data (dots) over the Bangui anomaly at a 550-km altitude. Normalized susceptibilities are near zero for basin sediments (dot pattern), 1.0 for average crustal rock (no shading), and 1.5 for shield rocks (random hatchured pattern). (From Hastings, 1982, with permission.)

8.2.3 A SPHERICAL-EARTH MODEL

Although some analyses of satellite magnetic anomaly data claim success using flat-earth techniques (e.g., the rectangular harmonic analyses described in Chapter 5; Mayhew, Estes, and Myers, 1985a; Ruder and Alexander, 1986), the underlying geometry remains spherical. A useful and very general method that uses spherical-earth geometry and accounts for changes in the ambient **B** over the region of analysis was developed by von Frese et al. (1980, 1981b), on which the following discussion is based.

Consider equation (5.20). The vector magnetic anomaly field $\mathbf{A}(\mathbf{r}_i)$ at $\mathbf{r}_i$ is found by taking the negative gradient of the magnetic potential, and the scalar anomaly field is found by taking $\hat{\mathbf{b}}(\mathbf{r}_i) \cdot \mathbf{A}(\mathbf{r}_i)$, where $\hat{\mathbf{b}}(\mathbf{r}_i)$ is computed from a suitable model of the main field. Assume that $\mathbf{M}(\mathbf{r}_j)$, the magnetization or dipole moment per unit volume, is given by

$$\mathbf{M}(\mathbf{r}_j) = M(\mathbf{r}_j)\hat{\mathbf{m}}(\mathbf{r}_j), \tag{8.4}$$

where $M(\mathbf{r}_j)$ is the scalar magnitude, and $\hat{\mathbf{m}}(\mathbf{r}_j)$ is the unit direction of $\mathbf{M}(\mathbf{r}_j)$. It is then easily shown that the anomalous magnetic field can be expressed in the form

$$A^k(\mathbf{r}_i) = \int_V M(\mathbf{r}_j) Q^k(\mathbf{r}_{ij}) d^3 r_j \tag{8.5}$$

for suitable $Q^k(\mathbf{r}_{ij})$, where $k = r, \theta, \phi$, and s for A_r, A_θ, A_ϕ, and A^s, respectively. Expressions for Q^k are

$$Q^r(\mathbf{r}_{ij}) = -\hat{\mathbf{r}}_i \cdot \nabla_i \left[\hat{\mathbf{m}}(\mathbf{r}_j) \cdot \nabla_j \frac{1}{r_{ij}} \right], \tag{8.6a}$$

$$Q^\theta(\mathbf{r}_{ij}) = -\hat{\boldsymbol{\theta}}_i \cdot \nabla_i \left[\hat{\mathbf{m}}(\mathbf{r}_j) \cdot \nabla_j \frac{1}{r_{ij}} \right], \tag{8.6b}$$

$$Q^\phi(\mathbf{r}_{ij}) = -\hat{\boldsymbol{\phi}}_i \cdot \nabla_i \left[\hat{\mathbf{m}}(\mathbf{r}_j) \cdot \nabla_j \frac{1}{r_{ij}} \right], \tag{8.6c}$$

$$Q^s(\mathbf{r}_{ij}) = -\hat{\mathbf{b}}_i \cdot \nabla_i \left[\hat{\mathbf{m}}(\mathbf{r}_j) \cdot \nabla_j \frac{1}{r_{ij}} \right]. \tag{8.6d}$$

It has been shown (von Frese et al., 1981b) that the gravity field $g(\mathbf{r}_i)$ can be expressed in the same form as (8.4), with $\rho_m(\mathbf{r}_j)$, replacing $M(\mathbf{r}_j)$, and Q^k becoming

$$Q^g(\mathbf{r}_{ij}) = -G \frac{\partial}{\partial r_i} \left(\frac{1}{r_{ij}} \right). \tag{8.6e}$$

The expressions Q^k are source functions that depend only on the problem geometry. Given specified directions for $\hat{\mathbf{m}}(\mathbf{r}_j)$, and using (5.8) and the results of Section 5.1.1.2, the Q^k are easily computed.

Suppose that the geometry of the body responsible for the measured anomaly can be specified by

ϕ^l, ϕ^u = lower and upper longitude limits of the source volume,
θ^l, θ^u = lower and upper co-latitude limits of the source volume,
r^l, r^u = lower and upper radial limits of the source volume,

so that (8.5) becomes

$$A^k(\mathbf{r}_i) = \int_{\phi^l}^{\phi^u} \int_{\theta^l}^{\theta^u} \int_{r^l}^{r^u} M(\mathbf{r}_i) Q^k(\mathbf{r}_{ij}) \, dr_j \, d\theta_j \, d\phi_j. \tag{8.7}$$

A scheme is sought for efficient numerical evaluation of the integral. Consider the innermost integral, over r_j. The interval $r^l \leq r_j \leq r^u$ can be transformed into the interval $-1 \leq r_j' \leq 1$ by taking

$$r_j = [r_j'(r^u - r^l) + (r^l + r^u)]/2. \tag{8.8}$$

A change of variables then gives

$$\int_{r^l}^{r^u} M(\mathbf{r}_j) Q^k(\mathbf{r}_j) \, dr_j = \frac{r^u - r^l}{2} \int_{-1}^{1} M(\mathbf{r}'_j) Q^k(\mathbf{r}'_j) \, dr'_j. \tag{8.9}$$

The integral over $[-1, 1]$ can then be evaluated by Gauss-Legendre quadrature, as described in Section 5.4.3.2. In particular, following equation (5.109),

$$\int_{-1}^{1} M(\mathbf{r}'_j) Q^k(\mathbf{r}'_j) \, dr'_j \doteq \sum_{j=1}^{N} \mathcal{A}_j M(\mathbf{r}'_j) Q(r'_j), \tag{8.10}$$

where the N values of $\mathcal{A}_j$ are weights assigned to the N values of $M(r'_j) Q(r'_j)$ at the interpolation coordinates, or node points, r'_j. The r'_j are the roots, or zeros, of the Legendre polynomial of degree N. Tables of r'_j and $\mathcal{A}_j$ may be found in Abramowitz and Stegun (1964); see also Carnahan, Luther, and Wilkes (1969) and Stroud and Secrest (1966). Values of $\mathcal{A}_j$ can also be computed from

$$\mathcal{A}_j = \frac{2\left[1 - \left(r'_j\right)^2\right]}{N^2 \left[P_{N-1}\left(r'_j\right)\right]^2}. \tag{8.11}$$

The approximate integral is then given by

$$\int_{r^l}^{r^u} M(\mathbf{r}_j) Q^k(\mathbf{r}_j) \, dr_j = \frac{r^u - r^l}{2} \sum_{j=1}^{N} \mathcal{A}_j M(\mathbf{r}'_j) Q^k(\mathbf{r}'_j). \tag{8.12}$$

Extension to evaluation of (8.7) gives

$$\int_{\phi^l}^{\phi^u} \int_{\theta^l}^{\theta^u} \int_{r^l}^{r^u} M(\mathbf{r}_i) Q^k(\mathbf{r}_{ij}) \, dr_j \, d\theta_j \, d\phi_j$$

$$= \frac{(\phi^u - \phi^l)}{2} \sum_{n=1}^{N_n} \left\{ \frac{(\theta_n^u - \phi_n^l)}{2} \sum_{m=1}^{N_m} \left[\frac{(r_{nm}^u - r_{nm}^l)}{2} \right. \right.$$

$$\left. \left. \times \sum_{j=1}^{N_j} M(r'_j, \theta'_m, \phi'_n) Q^k(r'_j, \theta'_m, \phi'_n) \mathcal{A}_j \right] \mathcal{A}_m \right\} \mathcal{A}_n, \tag{8.13}$$

where (see Figure 8.4)

$\mathcal{A}_j, \mathcal{A}_n, \mathcal{A}_m$	are the Gauss-Legendre coefficients,
ϕ^u, ϕ^l	are the upper and lower longitude limits of the body,
θ_n^u, θ_n^l	are the upper and lower latitude limits of the body for the nth longitude
r_{nm}^u, r_{nm}^l	are the upper and lower radial limits of the body for the nth longitude and mth latitude.

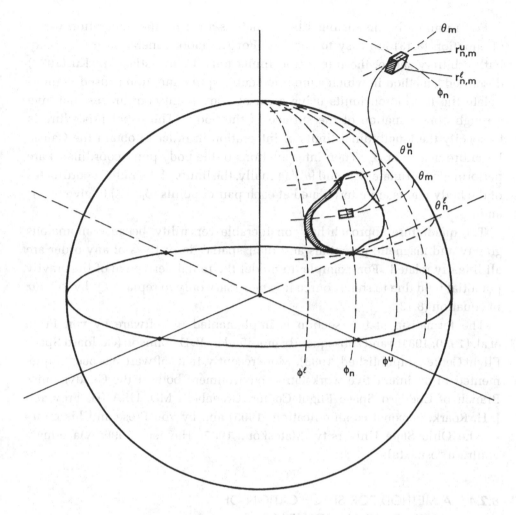

ϕ_n : n^{th} Longitude Step

θ_m : m^{th} Latitude Step

FIGURE 8.4
Diagram of geometry used
in spherical-earth analysis.

Equation (8.13) computes gravity or magnetic anomalies accurately by summing, at each observation point, the effect of $N_n \cdot N_m \cdot N_j$ sources, appropriately weighted. Inclusion of $\hat{\mathbf{m}}(\mathbf{r}_j)$ in (8.6) permits specification of any magnetization direction. If only induced magnetization is considered, then $\hat{\mathbf{m}}(\mathbf{r}_j)$ is simply the field direction computed from a suitable model of the main field. If both induced and remanent magnetizations are to be modeled, $\hat{\mathbf{m}}(\mathbf{r}_j)$ is taken in the direction of the vector sum of the induced and remanent magnetizations. Once $\hat{\mathbf{m}}(\mathbf{r}_j)$, $M(\mathbf{r}_j)$, and the geometry of the source body are specified, calculation of the Q^k and evaluation of (8.13) are straightforward. The process can be speeded by judicial use of tabulated quantities, particularly the geometric factors that might be identical for different versions of a model.

For a uniformly dimensioned body such as a prism, the integration limits of equation (8.13) are easy to specify. For the more general case of a body with arbitrary shape, the integration limits must be investigated. Ku (1977) described a method in which a modified cubic spline function is used to interpolate the integration limits using a set of body point coordinates that give a rough approximation of the surface of the body. The usual procedure is to specify the longitudinal limits of integration in order to obtain the Gauss-Legendre nodes for ϕ'_n. Then interpolations of the body point coordinates are performed to estimate $(\theta'_n)^u$ and $(\theta'_n)^l$. Finally, the limits of the radial coordinates of the body points are interpolated at each pair of points (ϕ_n, θ_m) to give $(r'_{n,m})^u$ and $(r'_{n,m})^l$.

The quadrature approach has considerable versatility, because anomalous gravity and magnetic potentials and their spatial derivatives of any order are all linearly related. For example, to model the radial derivative of the gravity potential field due to some source it is necessary only to replace Q^r by $\partial Q^r/\partial r$ in equation (8.13).

The formalism of this section is implemented in software by von Frese et al. (1980, 1981b) and subsequently modified by B. D. Johnson (Goddard Space Flight Center, unpublished, 1983). More recently, that software has been implemented in an interactive workstation environment both at the Geodynamics Branch of Goddard Space Flight Center, Greenbelt, MD, USA (H. Frey and J. H. Roark, personal communication, 1995) and by von Frese and his group at The Ohio State University (Mateskon, 1985; Hayden, 1996; via e-mail: vonfrese@osu.edu).

8.2.4 A METHOD FOR SPECIFICATION OF MAGNETIZATION CONTRASTS

Many analyses are conducted in terms of anomaly contrast with respect to some surrounding average or level, from which magnetization or susceptibility contrasts are deduced. Further, it is generally considered that satellite data are unable to resolve magnetization variations, or layering with depth. Consequently, what is typically modeled is the variation, or contrast, of depth-integrated magnetization as a function of location at the earth's surface. The assumption is made that summing the products of the magnetization contrast and the thicknesses of the layers of the earth results in negligible error because of the low ratio of layer thickness to observation altitude.

Computationally, it is easier to use a single body or prism than many. Frey (1985) proposed a method for reducing magnetization contrast variations with depth to a single body by summing the susceptibility times thickness contrast contributions from expected lithospheric layers. This gives a column-integrated magnetization or susceptibility contrast. For simplicity, suppose it is desired to

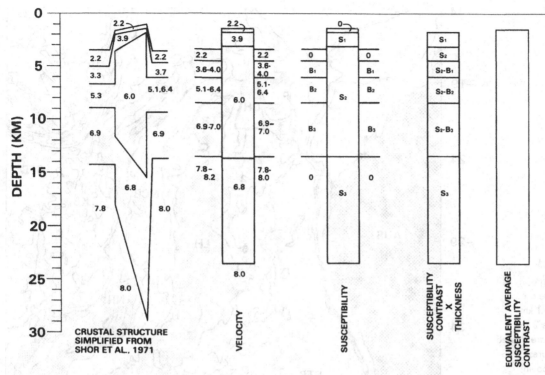

CRUSTAL STRUCTURE
SIMPLIFIED FROM
SHOR ET AL., 1971

VELOCITY

SUSCEPTIBILITY

SUSCEPTIBILITY
CONTRAST
X
THICKNESS

EQUIVALENT AVERAGE
SUSCEPTIBILITY
CONTRAST

model a particular anomaly that is spatially isolated from other anomalies and is caused by a simple body. If seismic data are available, layers in the lithosphere can be defined, both within the body and within the surrounding lithosphere, as, for example, in Figure 8.5, corresponding to the Lord Howe rise and the adjacent lithosphere. Five cross sections are sketched. The leftmost or first column corresponds to the known seismic crustal structure. Next is shown an approximate structure with horizontal boundaries. To specify the column magnetization, one then assigns susceptibility values to each layer in the body and in the surrounding lithosphere, as in the middle column of the figure, where the S_i are the susceptibilities of the body layers and the B_i (for background) are the susceptibilities of the surrounding lithosphere. The assigned layer susceptibilities are converted to susceptibility contrast by subtracting the susceptibility of the surrounding lithosphere from that of the magnetized body, as in the fourth column of Figure 8.5. Finally, the susceptibility contrasts are multiplied by the thickness of the layer, and the results for the column are totaled to give the equivalent magnetization contrast of the anomalous body, as in the right hand column of the figure.

An example of this method is furnished by the modeling study of the Lord Howe rise by Frey (1985). Figure 8.6 shows the area studied: 146° to 176° E longitude and −14° to −44° latitude. Figure 8.7 shows a *Magsat* scalar anomaly map and reduced-to-pole (RTP) scalar map based on POGO data. Note the

FIGURE 8.5
Block modeling procedure for calculating an equivalent average susceptibility contrast. The illustration is based on the Lord Howe rise submarine plateau and surrounding lithosphere. The crustal structure, from Shor et al. (1971), is simplified into average-thickness blocks. Susceptibilities are assigned to each layer in the source body (S) and background (B). Anomalous blocks are determined by subtracting the susceptibilities of the background from those of the source body. These susceptibility contrasts are multiplied by the thickness of the anomalous layer, the products are totaled, and the sum is divided by the total thickness of the source body. (From Frey, 1985, with permission.)

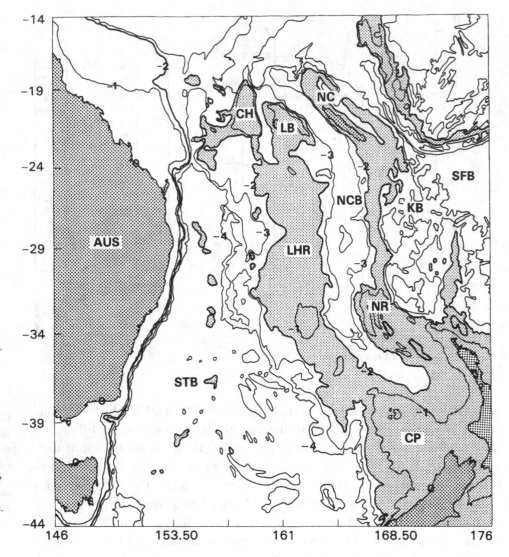

FIGURE 8.6
Submarine topography of the Lord How rise area. Contour interval 1 km. Elevations above 2 km below sea level are shaded. Designated features: AUS, Australia; STB, South Tasman basin; CH, Chesterfield group; LB, Landsdowne bank; LHR, Lord Howe rise; CP, Challenger plateau; NR, Norfolk ridge; NCB, New Caledonia basin; NC, New Caledonia; KB, Kingston basin; SFB, South Fiji basin. (From Frey, 1985, with permission.)

positive RTP anomaly over the Lord Howe rise. Following the preceding discussion, Figure 8.8 shows the model adopted for the lithosphere surrounding the Lord Howe rise (i.e., the B_i of Figure 8.5). Susceptibilities in the background model are assumed to be zero in the sediment layer ($V_p = 2.2$ km/s) and the mantle ($V_p = 7.8$–8.2 km/s), following Wasilewski et al. (1979) and Thomas (1984). A susceptibility value of 0.006 cgs (0.075 SI) is adopted for the upper two crustal layers, and the lowest layer is assigned a susceptibility typical of gabbros, namely, 0.002 cgs (0.025 SI).

Figure 8.5 shows the seismic crustal layer for the Lord Howe rise as defined by Shor, Kirk, and Menard (1971), the simplified horizontal layer structure

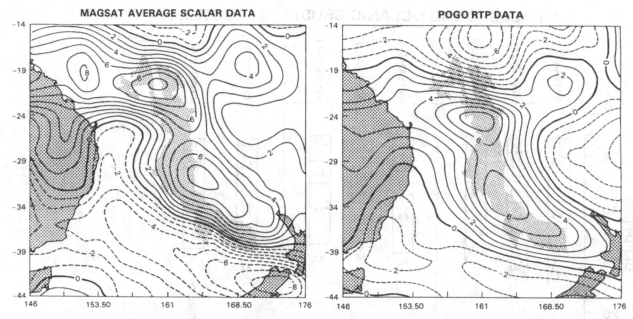

MAGSAT AVERAGE SCALAR DATA

POGO RTP DATA

FIGURE 8.7
Magsat and POGO
magnetic anomalies over the
Lord Howe rise area. The
Lord Howe rise is shown
shaded, as outlined by the
2,000-m isobath. Left:
Magsat average scalar data
at a mean altitude of
approximately 400 km.
Right: POGO
reduced-to-pole (RTP) data,
scaled to a constant
50,000-nT field throughout
the map. (From Frey, 1985,
with permission.)

adopted by Frey (1985), and a schematic of, from left to right, the assigned layer susceptibilities for both the Lord Howe rise and the surrounding lithosphere, the resulting layer susceptibility contrast times thickness, and the equivalent average susceptibility contrast. The background or surrounding lithosphere susceptibilities, B_1, B_2, and B_3, are taken as 0.006, 0.006, and 0.002 cgs (0.075, 0.075, and 0.025 SI), respectively, as described in the preceding paragraph. Susceptibilities for the Lord Howe rise, S_1, S_2, and S_3, are assigned in one of four ways corresponding to four candidate models for the rise. In three of these cases the susceptibilities assigned are derived from those expected of certain types of continental crust, whereas in the fourth the susceptibilities are those expected of oceanic crust. Each of these is illustrated in Figure 8.9.

For the oceanic model, Figure 8.9A, the values for S_1, S_2, and S_3 are the same as for the oceanic background layers, B_1, B_2, and B_3. The resulting susceptibility contrast times thickness is 0.0565 cgs (0.710 SI), as indicated in Figure 8.9A. For a Lord Howe rise thickness of 22 km, the equivalent average susceptibility of the Lord Howe rise with respect to the surrounding lithosphere is 0.0026 cgs (0.033 SI). A positive susceptibility contrast produces a positive relative magnetic anomaly. From this, and an earlier study (Frey, 1983), it is concluded that thickened oceanic crust will always produce positive magnetic anomaly contrasts, if the anomaly is induced.

Three "continental"-type models are considered in Figure 8.9B–D. These are designated "ordinary," "upper volcanics," and "altered crust." In the ordinary continental model the susceptibilities S_1, S_2, and S_3, are taken to be 0.0005,

BACKGROUND OCEANIC CRUSTAL STRUCTURE

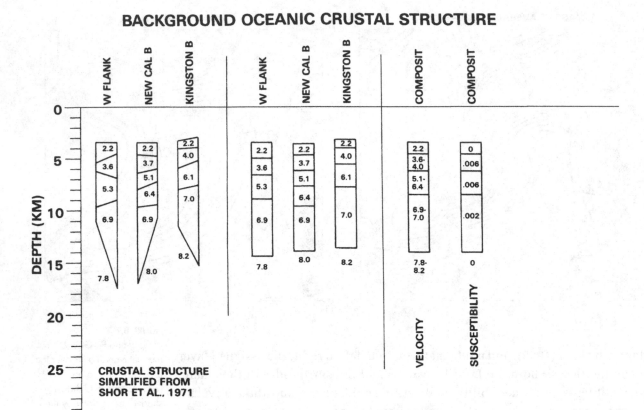

FIGURE 8.8
Composite background crustal model for the Lord Howe rise source body. Structure was derived from averaging the crustal sections of the West Flank (of the Lord Howe rise), the New Caledonia basin, and the Kingston basin. Original crustal sections are from Shor et al. (1971). (From Frey, 1985, with permission.)

0.0005, and 0.001 cgs (0.006, 0.006, and 0.013 SI), respectively. For the upper two layers these are representative of granitic rocks, and for the lowest layer of more intermediate rocks. The resulting sum of susceptibility contrast times thickness for the four layers is −0.01675 cgs (−0.21 SI). For a 22-km total thickness, the equivalent average susceptibility contrast is −0.00076 cgs (−0.0096 SI). In the second (upper volcanic) crustal model, Figure 8.9C, the upper crustal susceptibility is changed from 0.0005 cgs (0.006 SI) to 0.008 cgs (0.100 SI), a value that would be suitable if the upper layer were made up of basaltic volcanics. The sum of the susceptibility contrast times thickness is then −0.00738 cgs (−0.093 SI), or an average susceptibility contrast for the Lord Howe rise of −0.00034 cgs (−0.004 SI). Both the ordinary and upper-volcanic models result in negative anomalies, contrary to Figure 8.7.

Because under the assumption of this model that the lower crustal layer of the Lord Howe rise is surrounded by nonmagnetic material, the easiest way to achieve a net positive susceptibility contrast is to increase the susceptibility of the lowest layer. If such a layer were of basaltic composition, a reasonable susceptibility might be 0.008 cgs (0.10 SI). The altered-crust continental

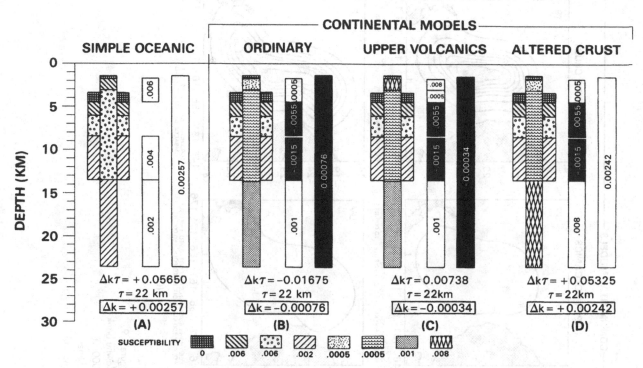

FIGURE 8.9
Crustal sections, anomalous blocks (source body minus background), and equivalent average source bodies for four different Lord Howe rise crustal models. Negative susceptibility contrasts are shown as black with white lettering. Resulting equivalent average susceptibility contrasts shown at the bottom (Δk). (From Frey, 1985, with permission.)

model of Figure 8.9D is the same as the ordinary continental model except that the susceptibility of the lowest layer is changed from 0.002 cgs (0.025 SI) to 0.008 cgs (0.10 SI), giving a positive equivalent average susceptibility contrast of 0.00242 cgs (0.030 SI), almost equivalent to that for the simple oceanic model.

Figure 8.10 shows the resulting magnetic anomaly contrasts at 500 km, as computed using the formalism of Section 8.2.3. The upper four panels of the figure show results from models incorporating a model anomaly corresponding only to the Lord Howe rise, as indicated by the shaded areas of the figure. In the lower four panels, the model also incorporates those portions of the Challenger plateau (CP in Figure 8.6) and the Chesterfield Group–Lansdowne bank (CH and LB in Figure 8.6) outlined by the 2,000-m isobath. The oceanic and altered-crust model anomalies correspond well with the RTP anomaly from the POGO data, assuming that the adjacent negative anomalies over the Tasman, South Fiji, and Solomon basins and the negative overlying New Zealand are due to unmodeled sources.

Clearly the model is simplified, in that it assumes strictly blocklike structure, neglects Moho topography, and assumes typical and constant susceptibilities in the various layers and zero susceptibility below the Moho. Yet it seems sufficient to allow the conclusion that the anomaly source may indeed correspond to a crustal structure similar either to the oceanic model or to the altered-crust model. Independent evidence (Carlson, Christensen, and Moore, 1980; Nur

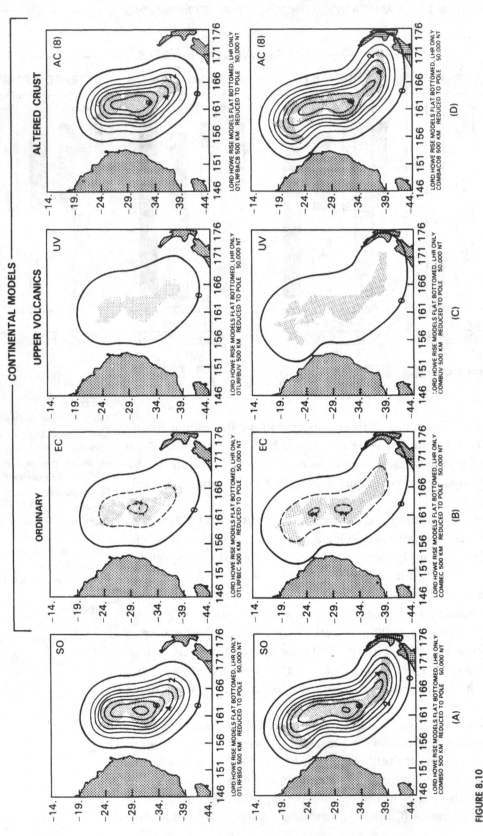

FIGURE 8.10

Calculated magnetic anomalies for the Lord Howe rise, based on the crustal models shown in Figure 8.9. Lower panels show extended Lord Howe rise structure, including the Chesterfield Group–Landsdowne bank and Challenger plateau elevations. Anomalies calculated at a 500-km altitude in a vertical 50,000-nT inducing field (see Figure 8.7, right). Contour interval is 1 nT. Negative values shown by dashed lines. Block model used to represent structure is shown by light shading. (From Frey, 1985, with permission.)

and Ben-Avraham, 1982) indicates that the Lord Howe rise is composed of sub-merged continental crust. Frey (1985) concluded that if that is true, then the continental crust is not ordinary, but must have been altered in some way, and that the anomaly is then explained in a satisfactory way by assuming that the lowest crustal layer has an unusually high, but by no means unreasonable, susceptibility. The Lord Howe rise model is presented as an example of a particular approach to forward modeling. Other examples of this approach can be found in Bradley and Frey (1988), Clark, Frey, and Thomas (1985), Fullerton et al. (1989, 1994), and Vasicek, Frey, and Thomas (1988).

8.3 TYPICAL MAGNETIC ANOMALIES

It is instructive to consider suites of magnetic anomalies derived from idealized sources as a guide to illustrating the critical attributes controlling the signatures of observed satellite magnetic anomalies. In this section we describe only a few suites that were calculated by procedures outlined in previous discussions. An extensive collection of calculations of anomalies from prismatic bodies with various combinations of induced and remanent magnetizations has been published by Andreasen and Zietz (1969). Here, the presentations include two-dimensional (strike-infinite) flat-earth total magnetic intensity profiles calculated by a modification of the method of Talwani and Heirtzler (1964), Section 8.2.2, and three-dimensional, spherical-earth magnetic anomalies derived from the method of von Frese et al. (1980, 1981b), Section 8.2.3.

Figures 8.11–8.13 show the total magnetic intensity anomaly profiles for three suites of sources that are illustrated in the lower portion of each figure. The sources are idealized, having a rectilinear shape and a homogeneous magnetic susceptibility contrast of 0.01 SI. The amplitudes can be adjusted linearly for any deviation from the assumed susceptibility contrast. To illustrate the effect of altitude above the earth, profiles are shown for 10, 100, and 500 km. All profiles are based on an earth field of 60,000 nT.

Figure 8.11 shows the effect of varying source widths from 10 to 5,000 km as a function of observation altitude for a 90° field inclination (i.e., reduced-to-pole). The 10-km-width source, despite its vertical extent of 40 km, produces essentially no observable anomaly at 500 km for any geologically reasonable magnetization, whereas the other sources cause anomalies of amplitudes comparable to those observed in the *Magsat* mission. The gradients of the anomalies decrease, as do the amplitudes of the anomalies, with altitude (note the scale change). The inflection points of the gradients for the positive anomaly roughly correspond to the vertical edges of the sources. All the anomalies have marginal negative components due to the dipolar nature of the magnetic sources. These minima will distort the amplitude and gradients of anomalies derived from adjacent sources. The anomaly amplitude increases at a decreasing rate with

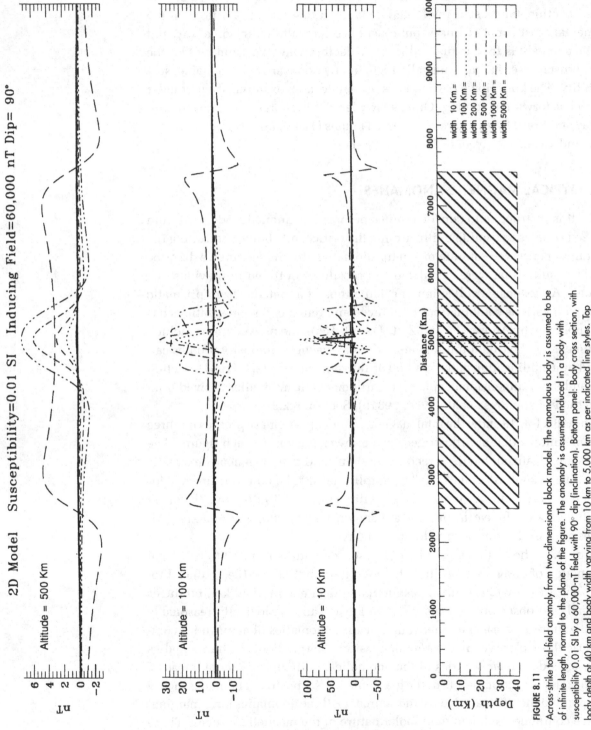

FIGURE 8.11

Across-strike total-field anomaly from two-dimensional block model. The anomalous body is assumed to be of infinite length, normal to the plane of the figure. The anomaly is assumed induced in a body with susceptibility 0.01 SI by a 60,000-nT field with 90° dip (inclination). Bottom panel: Body cross section, with body depth of 40 km and body width varying from 10 km to 5,000 km as per indicated line styles. Top three panels: Scalar anomaly at 10 km, 100 km, and a 500-km altitude. Anomaly plotted with same line style as used for width in bottom panel.

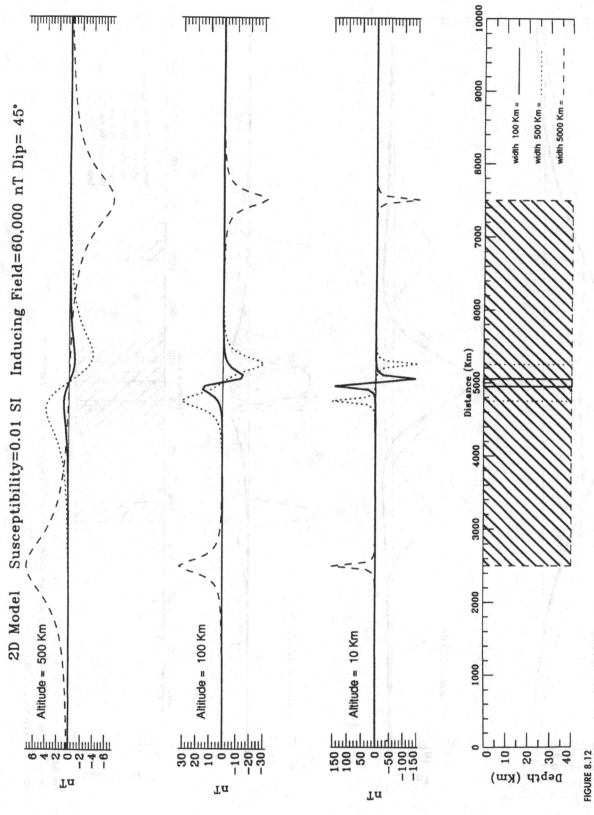

FIGURE 8.12
Same as Figure 8.11, except that the inducing field has a dip (inclination) of 45°, and body widths are 100 km, 500 km, and 5,000 km.

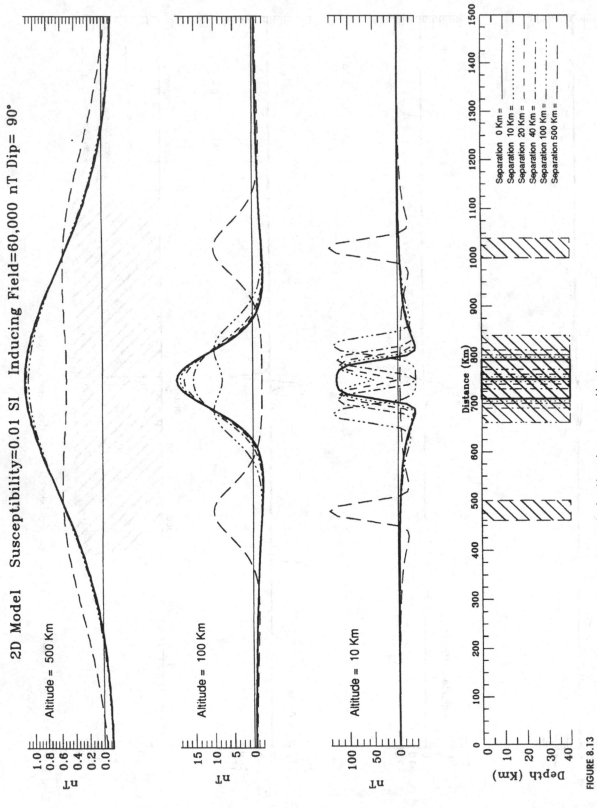

FIGURE 8.13

Same as Figure 8.11, except for two anomalous bodies, each of width 40 km, separated by distances varying from zero to 500 km.

increasing source width until the anomaly turns into a double-peaked anomaly with a peak located adjacent to each vertical edge of the source. This effect, which is observed on all the profiles, increases with source width, so that for the 100-km-altitude profiles, the anomaly due to the 5,000-km-wide source approximates the zero level over the central portion of the source. This of course makes it difficult to identify anomalous lithospheric sources whose widths are measured in thousands of kilometers in satellite magnetic data.

As the inclination of the inducing field changes, the anomaly pattern also will change. To illustrate this effect, Figure 8.12 repeats the results of Figure 8.11 for bodies of widths 100, 500, and 5,000 km, except that the inducing field has an inclination of 45°. For this northern magnetic hemisphere representation, north is to the right of the figure, so the negative component of the anomaly is to the right. The positions of the maximum and minimum of the anomaly approximate the edges of the source, but these increase in their distances from the edges of the source as the anomaly broadens with increasing observation altitude. Note that for the widest source, the continental-size 5000-km-wide source, the anomaly is restricted to isolated positive and negative signatures over the south and north margins of the source, respectively. The complex relations between source and anomaly as functions of geomagnetic inclination pose a deterrent to interpretation. Thus, mathematical reduction to the pole (Section 5.1.4) is used in many analyses to convert the observed anomaly field to a normalized vertical magnetization. However, this procedure can lead to erroneous results because of instability effects in the mathematical conversion near the geomagnetic equator and where remanent magnetization is prominent in sources and its direction is not coincident with the assumed geomagnetic field.

The effect of observational altitude on the horizontal resolution is clearly illustrated in Figure 8.13. Total magnetic intensity anomalies from identical 30-km two-dimensional sources separated at varying distances are plotted for altitudes of 10, 100, and 500 km. The inducing magnetic field is 90° to avoid the complication of the nonvertical magnetization. We can generalize from this figure that the horizontal separation between identical sources must exceed the altitude in order to identify separate sources from the characteristic dual maxima.

The strike-infinite assumption in two-dimensional analysis is usually violated in satellite data. Hence, the profiles and contour maps of Figures 8.14 and 8.15 show the total magnetic intensity (B) and the three orthogonal components of the anomalous magnetic field at an altitude of 400 km for a three-dimensional (vertical cylinder) source based on a spherical earth and the magnetization parameters specified in the figure. In Figure 8.14 the cylinder (circular plug) is located off the coast of west Africa at a geomagnetic inclination of $-23°$. As a result, the negative component of the B field is located to the south of the

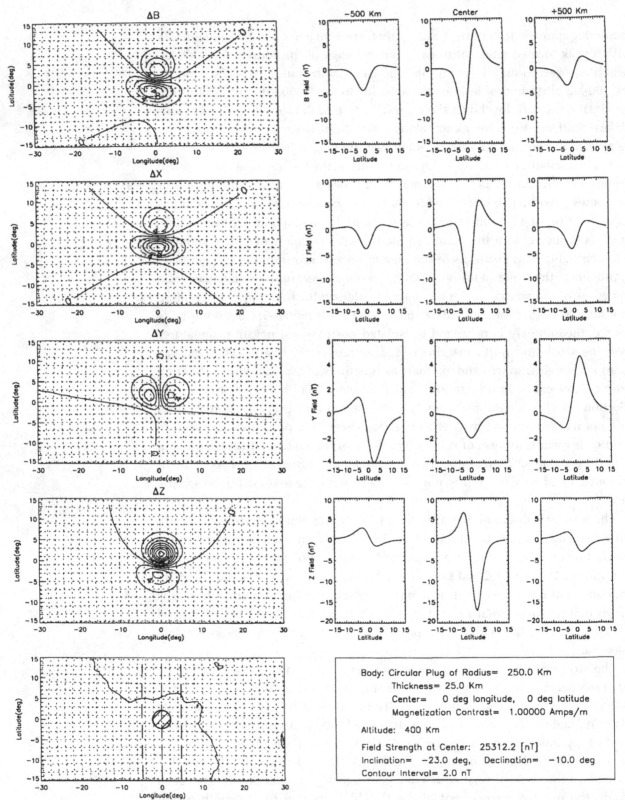

Body: Circular Plug of Radius= 250.0 Km
        Thickness= 25.0 Km
        Center=   0 deg longitude, 0 deg latitude
        Magnetization Contrast= 1.00000 Amps/m

Altitude: 400 Km

Field Strength at Center: 25312.2 [nT]
Inclination= −23.0 deg,   Declination= −10.0 deg
Contour Interval= 2.0 nT

positive and is greater in amplitude. The other components (X, Y, and Z) also are shown for comparison. At this latitude, the ΔX and ΔB anomalies are nearly identical; ΔZ is strongly negative to the north of the body, and ΔY shows anomalies to the east (positive) and west (negative), offset to the north. Such patterns cannot be discerned from individual profiles, as simulated in the right panels, but must be synthesized from many passes. Note that the profiles to the side of the body show an anomaly signature even though they do not pass over the body.

The magnetic anomalies of an equivalent source located between the British Isles and Norway are shown in Figure 8.15. In this case the geomagnetic inclination is $+72.4°$, and the magnetization contrast of the source is half that of the source used in Figure 8.14. Note the intense positive magnetic anomalies in B and Z and the negligible negatives to the north of the source. Lines connecting the minima and maxima of the anomalies in both B and Z in both figures are in the azimuth of the magnetic meridian, parallel to the declination of the inducing field. These maps and profiles illustrate the profound effects that the directional characteristics of the inducing field have on the magnetic anomalies and the care that is needed in comparing anomalies of different regions unless they are reduced to the pole.

8.4 HORIZONTAL RESOLUTION

When undertaking interpretive studies, it is important to have an estimate of the resolving capability of the observations. The resolution for satellite magnetic field data has been studied by Schnetzler, Taylor, and Langel (1984) and, for *Magsat* in particular, by Sailor et al. (1982).

8.4.1 RESOLUTION OF PAIRS OF MAGNETIC BLOCKS IN SATELLITE SCALAR MAGNETIC FIELD DATA

Schnetzler et al. (1984) studied the capacity of satellite scalar magnetic field data [i.e., the A^s of equation (1.11)] to distinguish two bodies with identical shapes and of equal induced magnetization contrasts. Modeling was carried out using the formalism described in Section 8.2.3. The model is shown in Figure 8.16a. Two identical prismatic bodies, 200 km by 200 km in area, and 40 km in thickness, have equal magnetic susceptibilities that contrast with the susceptibility of the surrounding material by 0.0005 cgs (0.0063 SI). Contours of A^s at various altitudes from the two blocks shown in Figure 8.16a are shown in Figure 8.16b.

At sufficiently low altitude, A^s exhibits two peaks in amplitude, say A_1 and A_2, as seen in Figure 8.16b. Suppose A_m is the minimum amplitude between the two bodies, measured on the profile connecting the centers of the blocks

FIGURE 8.14 (*opposite page*) Anomaly computed from three-dimensional spherical-earth model assuming induction in an anomalous source body, a circular plug, located at 0° longitude and 0° latitude. The lower right caption panel gives details about the anomalous source body, the altitude at which the anomalous field is computed and displayed, the magnetization contrast of the body with respect to its surroundings, the inducing field at the center of the body, and the contour interval of the anomaly contour plots. The lower left reference panel gives the outline of the top of the body, nearby continental outlines, and hypothetical satellite tracks. The upper four panels in the left column show contours of the anomaly in (from the top) ΔB, ΔX, ΔY, and ΔZ. The sets of three panels in the right column show anomaly profiles of the indicated components along the hypothetical satellite tracks shown in the reference panel.

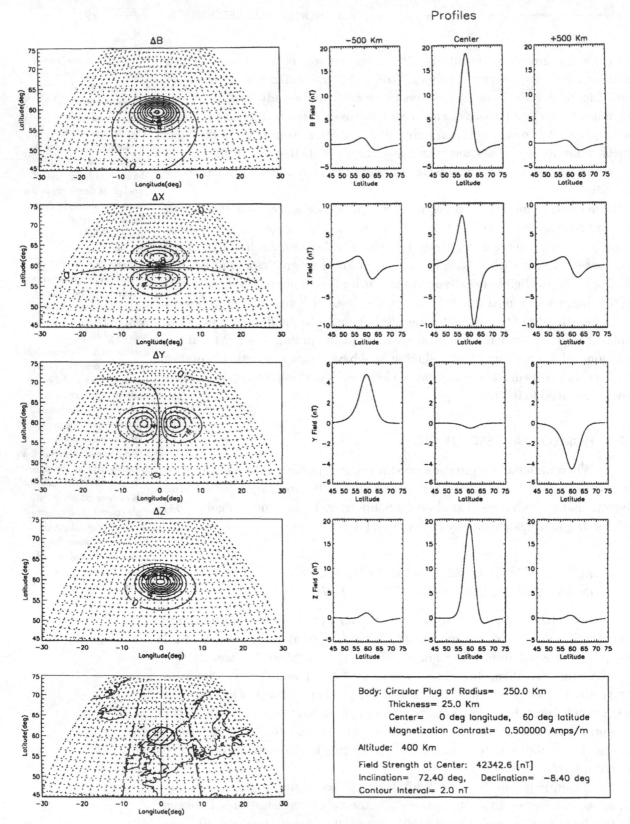

FIGURE 8.15
Same as Figure 8.14, except that the body is located at 0° longitude and 60° latitude.

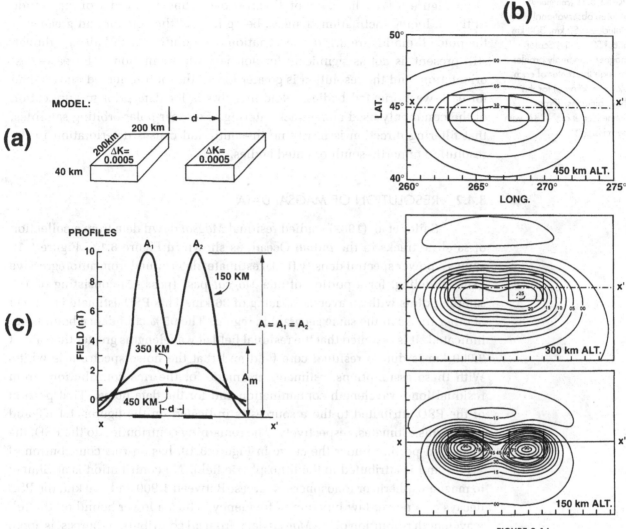

FIGURE 8.16
(a) Model of geologic blocks with anomalous susceptibility. The two blocks have a magnetic susceptibility contrast with their surroundings of 0.0005 cgs (0.0063 SI), and a variable edge-to-edge separation of d. (b) The scalar anomaly field, A_s, generated by modeled blocks separated by 150 km. The anomaly is induced by a main field with $B = 57,000$ nT and $I = 75°$. A_s is shown (continued)

(Figure 8.16c). Then the resolution, R, is defined as

$$R = \frac{(A_1 + A_2)}{2} - A_m.$$ (8.14)

From Figures 8.16b and 8.16c it is clear that R is a strong function of altitude.

Figure 8.17 shows the variation of R as a function of altitude for various separation distances d, as a function of d for various altitudes, as a function of d for several inducing inclinations I, and as a function of body orientation. In parts a and b, $I = 75°$; in parts c and d, the altitude is 150 km; in parts a–c, the two bodies are separated in an east–west direction. At any altitude, resolution remains relatively constant over a range of d, provided that $d >$ altitude. When d decreases to a distance less than the altitude, R falls off rapidly.

FIGURE 8.16 (*continued*)
at three observational
altitudes: 450 km, 300 km,
and 150 km. (c) Scalar
magnetic anomaly profiles
through the center of each
block (*x–x'* on b) for the
three altitudes. (From
Schnetzler et al., 1984, with
permission.)

From Figure 8.17c, the value of R varies over almost an order of magnitude as the inducing inclination changes, being least at the equator and greatest at the pole. From Figure 8.17d, the variation of resolution with latitude, though still present, is not as significant for north–south orientation as for east–west orientation, and the resolution is greater for north–south-oriented bodies than for east–west-oriented bodies. Note that this is for data prior to application of the commonly used along-track filtering. For near-polar-orbiting satellites, that filtering direction is nearly north–south and causes deterioration in the resolution of north–south-oriented bodies.

8.4.2 RESOLUTION OF *MAGSAT* DATA

Sailor et al. (1982) studied residual *Magsat* dawn data from a collection of satellite tracks in the Indian Ocean, as shown in Figure 8.18. Figure 8.19 shows a power spectral density (PSD) estimate, determined from autoregressive (AR) modeling, for a portion of one *Magsat* pass (pass 45) consisting of 180 sample points with an average spacing of 36 km. This PSD estimate is typical of other tracks in the same geographic region. The 95% confidence bounds are indicated. It is assumed that the residual field at wavelengths greater than about 3,000 km is due to residual core field and that the noise spectrum is white. With these assumptions, estimates are made for the rms contributions from residual long-wavelength contamination and for the data noise. That portion of the PSD attributed to these sources is indicated on the figure: 1.1-nT and 0.3-nT rms estimates, respectively. The remaining contribution to the PSD, the unshaded portion under the curve in Figure 8.19, has an rms contribution of 2.4 nT and is attributed to the lithospheric field. No contribution is attributed to magnetospheric or ionospheric sources. Between 1,900 and 250 km, the PSD decays as a power-law function of frequency. Thus, a lower bound on the full wavelength resolution of the *Magsat* data, from all contributing sources, is about 250 km.

8.5 GLOBAL MAGNETIZATION WITH A PRIORI INFORMATION

Surface magnetic anomalies illustrate the intense magnetization of the global crust and variations in these magnetizations. Thus, a marked magnetic anomaly might be anticipated in the relatively abrupt, roughly threefold thickening of the crust from the oceans to the continents. In fact, the highly filtered maps of Arkani-Hamed and Strangway (1986a) show distinct differences between oceanic and continental areas. In statistical studies, Hinze, von Frese, and Ravat (1991), building upon an earlier qualitative evaluation (Hinze et al., 1982), compiled the mean magnetic anomaly amplitude for crustal anomalies

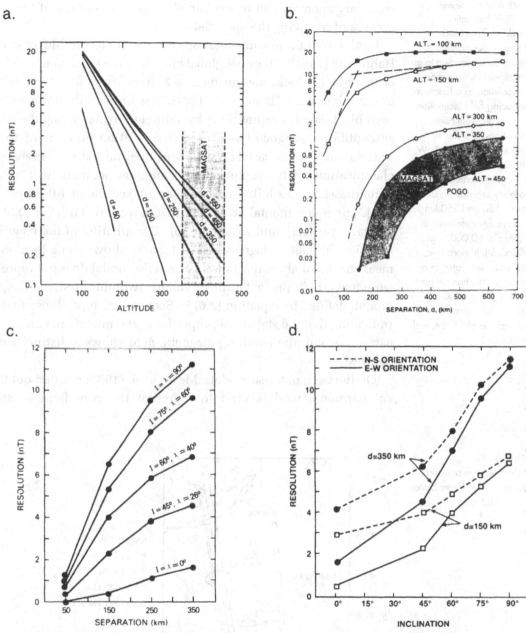

FIGURE 8.17

(a) Resolution as a function of altitude for bodies of varying separation d. (b) Resolution as a function of block separation, d, for varying observational altitude. (c) Resolution as a function of body separation, d, for varying (*continued*)

and for oceanic anomalies off the coast of South America and found a statistically significant increase in continental anomaly amplitudes. Similar studies have found equivalent results. However, a ubiquitous continent–ocean contrast is absent from magnetic anomaly maps prepared from satellite altitude data. This lack may be attributable to the unlikely occurrence of nearly equivalent total lithospheric magnetizations because of enhanced magnetization of the oceanic crust or suboceanic mantle. More likely, the lack of a continental

FIGURE 8.17 (*continued*)
main-field inclination *I*.
Altitude of observation is
150 km. Inducing-field
strength corresponding to an
axial dipole at inclination *I*.
(d) Resolution as a function
of inducing-field inclination
I for two different body
orientations: parallel to
geomagnetic latitude, E–W,
and perpendicular to
geomagnetic latitude, N–S.
Two body separations are
shown: 150 and 350 km.
Susceptibility difference in
all cases is +0.0005 cgs
(0.0063 SI). In parts a–c,
bodies are separated in an
east–west direction. In parts
a and b, B = 57,000 nT,
and *I* = 75°. (From
Schnetzler et al., 1984, with
permission.)

boundary anomaly can be explained by the elimination of the anomaly in the process of removing the main field.

Pertinent to these considerations are the models of Meyer et al. (1983) and Hahn et al. (1984). These are global models of the crust composed of 16,200 two- or three-layer blocks measuring $2° \times 2°$ (two layers for Meyer et al., 1983; two- or three layers for Hahn et al., 1984). Assuming only induced magnetization, each block-layer is represented by a dipole directed along the main field, with susceptibility assigned based on a classification into one of 16 crustal types. The classification is based on surface geology and seismic investigations. Layer classifications, with magnetization in amperes per meter in a 50,000-nT field in parentheses, are as follows: sea water (0), sediments (0), upper oceanic layer (1.0), upper continental layer (1.0–2.0), lower crustal layer (0.5), intermediate crustal layer (1.2), and sub-Moho (0). Examination of the magnetic field from spherical harmonic degrees greater than 13 shows some local agreement with measured anomaly maps, but in general the model does not reproduce the measurements. In its present form, its value is twofold. First, the R_n spectrum of this model, defined by equation (2.6) in Section 2.1, reproduces that of Figure 2.3, indicating that the statistical properties of the model are similar to those of the earth. Second, the resulting magnetic field shows a distinct continent–ocean contrast.

On the basis of those models, Meyer et al. (1985) pointed out that the spherical harmonic models used to represent the core field must also include

FIGURE 8.18
Study area with selected
Magsat tracks. (From Sailor
et al., 1982, with
permission.)

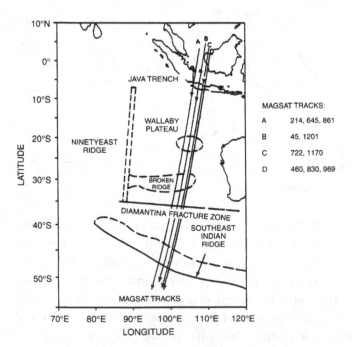

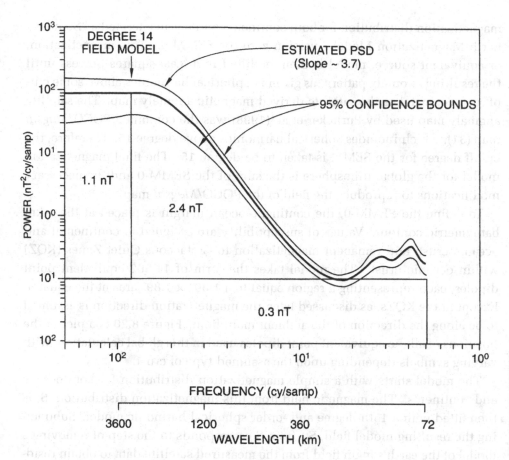

FIGURE 8.19
Double-sided power spectral
density for scalar residual
field from pass 45. Noise
floor is 0.3 nT, rms. (From
Sailor et al., 1982, with
permission).

representation of the longer-wavelength fields of lithospheric origin. They made
a convincing case that most of the expected continent–ocean difference is contained in these long wavelengths. Because of the dominance of the main field,
the long-wavelength lithospheric fields are simply not observable. This conclusion is verified in the studies of Cohen (1989) and Counil, Cohen, and Achache
(1991), who modeled the continent–ocean boundary as a step function in susceptibility. They showed that the field variations represented by the higher-
degree harmonics of the field from this simple continent–ocean model have
amplitude and wavelength characteristics similar to those observed on typical
anomaly maps derived from satellite data. An obvious conclusion is that some
of the observed anomalies may not be due to localized variations in magnetization, but may reflect elements of a continent–ocean boundary anomaly.

In an effort to explore the possible effects of changes in integrated magnetization at the continent–ocean boundary and to account for such effects in modeling the lithosphere's magnetization or susceptibility, Purucker et al. (1998)
derived a magnetization model that includes a built-in first-order continent-
ocean boundary effect. This is accomplished by assuming a simple a priori

magnetization distribution for both continents and oceans, termed a "Standard Earth Magnetization Model," version zero, or SEMM-0. This magnetization, or equivalent source, model is then modified in a least-squares process until the resulting anomaly pattern, as given in spherical harmonics above some cutoff degree, matches the satellite-derived magnetic anomaly map. The satellite anomaly map used by Purucker et al. (1998) was the combined POGO/*Magsat* map (31), which includes spherical harmonics above degree 15. Therefore, the cutoff degree for the SEMM is taken to be degree 15. The final magnetization model for the global lithosphere is the sum of the SEMM-0 and the necessary modifications to reproduce the field of the POGO/*Magsat* map.

To define the SEMM-0, the continent–ocean margin is placed at the 1-km bathymetric contour. Values of susceptibility are assigned to continental and oceanic crust, and remanent magnetization to Cretaceous Quiet Zones (KQZ) within oceanic crust. The model takes the form of 11,562 equivalent point dipoles, each representing a region equal to a $1.89° \times 1.89°$ area at the equator. Except in the KQZs, as discussed later, the magnetization direction is assumed to be along the direction of the ambient main field. Figure 8.20 is a plot of the positions of all the equivalent point dipoles used in this global calculation, with varying symbols depending upon the assumed type of crust.

The model starts with a simple magnetization distribution for both oceans and continents. The magnetic field from this magnetization distribution, $\mathbf{S}$, is then fitted with a 15th-degree and -order spherical harmonic model. Subtracting the resulting model field, $\mathbf{S}_c$, from $\mathbf{S}$ corresponds to the step of removing a model of the earth's main field from the measured satellite data to obtain residuals corresponding to the lithospheric field [i.e., as in equation (1.6)]. In the usual data reduction procedure for isolating the lithospheric signal, the satellite residual field is filtered along the track of the orbit (Section 4.4.3), very nearly a north–south direction. Correspondingly, the residual, $\mathbf{A}_s' = \mathbf{S} - \mathbf{S}_c$, is filtered in the north–south direction. The resulting filtered difference field, $\mathbf{A}_s$, corresponds to the "measured" anomaly field, $\mathbf{A}$ and is called the anomaly field of the SEMM-0:

$$\mathbf{A}_s = \text{filtered}(\mathbf{S} - \mathbf{S}_c). \qquad (8.15)$$

Then the magnetization distribution specified in the SEMM-0 is modified until the resulting anomaly pattern, with the spherical harmonic model field removed, matches the satellite-derived magnetic anomaly map $\mathbf{A}$. To accomplish this task, a modified data set is created by subtracting $\mathbf{A}_s$ from the anomaly field $\mathbf{A}$, as isolated by methods described in Chapter 4:

$$\Delta\mathbf{A} = \mathbf{A} - \mathbf{A}_s. \qquad (8.16)$$

Then a set of dipole modifications is derived that will reproduce $\Delta\mathbf{A}$ in a least-squares sense. The new model, called SEMM-1, is the sum of these

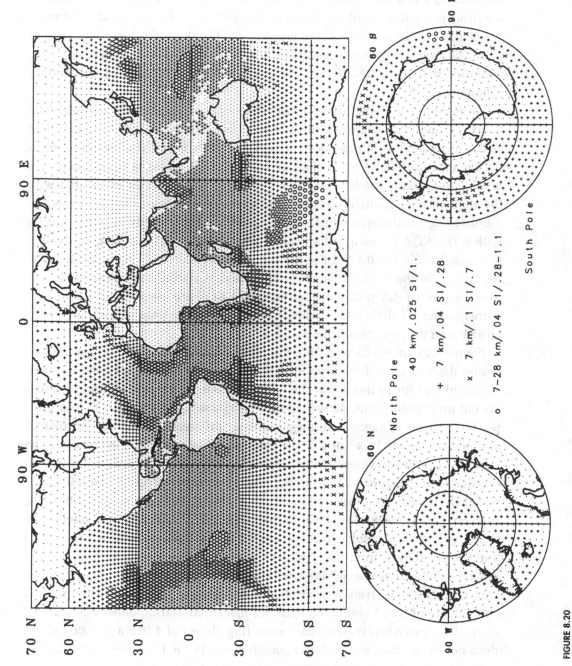

FIGURE 8.20

Continental margins, with the positions of the equivalent point dipoles used in this global analysis. Dots indicate dipole locations in continental crust; plus symbols, in oceanic crust; ○ symbols, in oceanic plateaus; × symbols, in KQZs. Following the symbols are the assumed attributes of that crust: thickness (in km), magnetic susceptibility (in SI), and ζ, the product of susceptibility times thickness in (10 · SI · km). Mercator and polar stereographic projections. (From Purucker et al., 1998, with permission.)

North Pole

   · 40 km/.025 SI/1.

   + 7 km/.04 SI/.28

   × 7 km/.1 SI/.7

   ○ 7–28 km/.04 SI/.28–1.1

South Pole

modifications and the dipoles of SEMM-0. It is a global magnetization model of the lithosphere that, unlike earlier models, excludes edge effects at continent–ocean and KQZ boundaries. Purucker et al. (1998) adopted values of 0.025 SI in a continental crust of thickness 40 km, and 0.040 in a 7-km-thick oceanic crust, assuming that layers 2A, 2B, and 3A contribute to the magnetic susceptibility (Thomas, 1987).

To account for their thicker crust, the oceanic plateaus, as identified by Sandwell and MacKenzie (1989) and Marks and Sandwell (1991), are included as distinct features. In most oceanic regions, fields from the alternating remanent magnetizations of the oceanic seafloor spreading magnetic anomalies are nearly cancelled at satellite altitude. However, oceanic crust associated with the KQZ is wide enough to result in measurable anomalous fields at satellite altitude (e.g., LaBrecque and Raymond, 1985). Accordingly, dipoles that lie within the KQZ are assigned an initial magnetization of 3.3 A/m, equivalent to a susceptibility of 0.1 SI in an ambient field of 42,000 nT. Also, because the KQZ crust in the western Pacific was formed at significantly lower latitudes than its present-day position, dipoles in the KQZ are assigned an inclination corresponding to their estimated paleolatitude of formation, rather than the inclination of the present-day field.

Figure 8.21 shows the ΔZ from the SEMM-0 after removal of spherical harmonic degrees 15 and less and after the north–south filter has been applied. It should be kept in mind that the SEMM-0 is an oversimplification of actual crustal properties, particularly for continental crust. Its purpose is to introduce possible effects of large-scale magnetization differences between continental and oceanic crust. It is not a priori clear that this contrast is everywhere the same, or even everywhere a sharp discontinuity, as portrayed in the model. Implications of Figure 8.21 for interpretation are discussed in Section 9.1.

SEMM models are presented in terms of the product of susceptibility times thickness of the magnetized layer, called ζ, because the data taken at satellite altitude are unable to resolve the depth distribution of magnetization. Figure 8.22 and Plate 5 show ζ from the SEMM-1. To measure the degree of fit of the ΔZ anomaly field from the SEMM-1 to that of the POGO/*Magsat* map, a regression/correlation analysis was run between the two. This yielded correlation coefficients of 0.993 and 0.963 between measured and modeled ΔX and ΔZ anomalies, respectively, with corresponding slopes of 1.016 and 1.065. This does not mean that the short-wavelength result is the true magnetization, but only one plausible contender among many possibilities.

Purucker et al. (1998) tested the sensitivity of the SEMM-1 results to the crustal parameters chosen for the SEMM-0 and found that, for end-member SEMM-0 assumptions that are physically unlikely, degrees 17–65 of spherical harmonic analyses of ζ from the resulting SEMM-1 models are almost identical. These wavelengths of the SEMM-1 are robust.

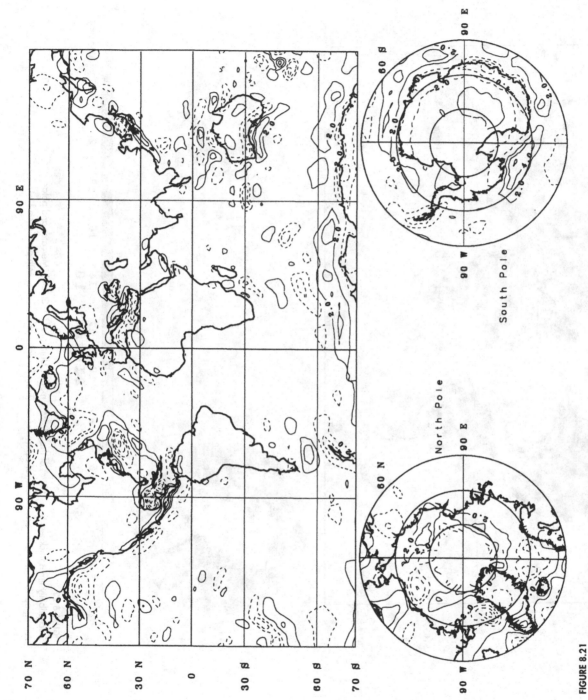

FIGURE 8.21
Vertical field (of degree > 15), $(A_s)_z$, at 400 km, from the SEMM-0 model after north–south filtering. Units are nanoteslas. The contour interval is 2 nT; negative contours dashed, and zero contour suppressed. Mercator and polar stereographic projections. (From Purucker et al., 1998, with permission.)

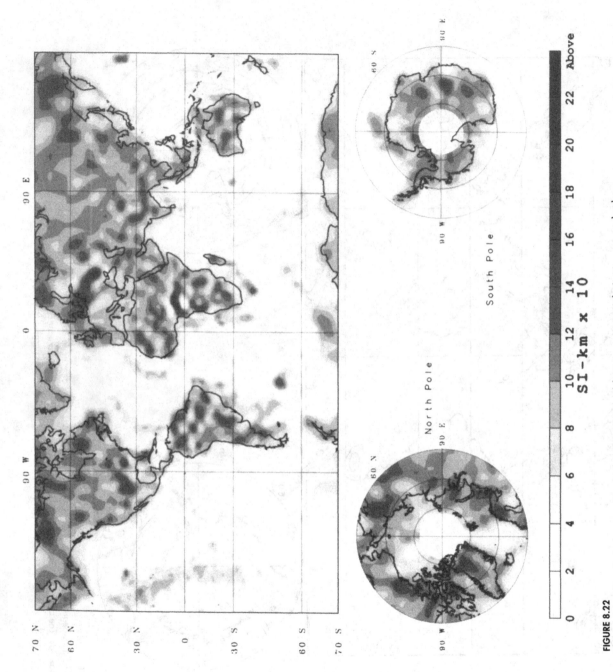

FIGURE 8.22
Susceptibility times thickness times 10, ζ, of the SEMM-1 model. Units are (10 · SI · km). Mercator and polar stereographic projections. See Plate 5 for color version. (From Purucker et al., 1998, with permission.)

$$SI-km \times 10$$

0 2 4 6 8 10 12 14 16 18 20 22 Above

In Chapter 9, maps of ζ are used to give an overall picture of the variations of susceptibility for the regions discussed. It is then appropriate to ask how those variations differ from the variations in a typical equivalent source model without an a priori model or from those represented in an RTP map. Because the continental crustal thickness used in the SEMM is everywhere the same, variations in ζ within continental regions are essentially the same as variations in other representations. Some exceptions occur near the edge of the continent.

8.6   DISCUSSION

The scope of the methods discussed in this chapter is broad. All are useful in some way within their inherent limitations. Qualitative comparison (Section 8.1.1) can be a useful reconnaissance tool for gaining an initial evaluation of possible anomaly sources. Particularly useful comparisons often can be made not only with known geology but also with crustal thickness, heat flow, and other geophysical parameters. To the extent that such comparisons can be made quantitative (e.g., Section 8.1.2), additional constraints are available for forward modeling. Three-dimensional forward models are preferred for satellite altitude data, and the methodology of Section 8.2.3 is an effective, efficient tool. Although many forward models specifically include magnetizations for several crustal layers (e.g., models for Africa in Chapter 9), it is generally agreed that because satellite data are acquired at altitudes much greater than the thickness of the magnetized layer, variations of magnetization with depth cannot be resolved. Thus it is often possible to simplify model construction by dealing with the contrast in depth-integrated magnetization. A useful procedure for specifying such magnetization is given in Section 8.2.5.

Sections 8.3 and 8.4 serve to illustrate the types of anomaly signatures resulting from idealized, isolated anomaly sources and the capability of noise-free satellite data to resolve individual sources. It is important not to extend interpretation beyond the limits of what can reasonably be resolved in the data.

Finally, with regard to the SEMM, as noted in the introduction to this chapter, debate as to its meaning and applicability is anticipated. However, away from continent–ocean and KQZ boundaries, it shows the same variations as the usual equivalent source magnetization with no a priori model or as seen in an RTP map. As a tool for analysis, the SEMM should be used in conjunction with RTP and the usual magnetization maps. Its importance lies in calling attention to possible edge-effect anomalies that could otherwise be interpreted in terms of local variation in magnetization, as will be discussed further in Section 9.1. The SEMM also is intended to provide a tool for conducting local analyses while taking into account the probable effects of nearby anomalies from sources not being studied. Examples of this use are found in Chapter 9, though its general usefulness as such a tool remains to be proved.

GLOBAL SATELLITE MAGNETIC ANOMALY INTERPRETATION

Langel, Benson, and Orem (1991) documented more than 400 publications based on *Magsat* data, including contributions to 9 books, 20 theses, and over 250 peer-reviewed journal articles. That does not include studies based on data from other satellite missions (e.g., *Cosmos 49* or POGO). Also, additional papers have been published since the compilation by Langel et al. (1991). The portion of that literature addressing study of the lithosphere is diverse in quality and is neither comprehensive nor unified. In this chapter, selected results from that literature are discussed, chosen because of their contributions to our understanding of the lithosphere and because they illustrate interpretive methods. It is beyond the scope of this book to consider all regions or all published results. Four regions are considered: North America, Australia, Russia/Europe, and selected topics concerning oceanic areas. For each region, the distribution of either ζ (susceptibility times thickness), from the SEMM-1 model (Figure 8.22), or the equivalent source magnetization is discussed as it relates to the known geology and tectonics, followed by a review of published interpretations of the satellite data for the region. A simplified version of the geologic time scale is shown in Table 9.1 for reference. For cross-reference purposes, numbers in curly brackets, {}, refer to the maps listed in Table 6.1. Lists of publications by region are given in Appendix 9.1.

9.1 IMPLICATIONS OF THE SEMM FOR INTERPRETATION

Figure 8.21 shows a map of vertical-component magnetic anomalies resulting from the hypothetical continent–ocean magnetization contrast of the SEMM-0 described in Section 8.5. The possible existence of such anomalies has important implications for interpretation of satellite magnetic anomaly data. The figure shows some small localized anomalies and some elongated anomalies, mainly in coastal regions. Many of these model anomalies approximate features found in the satellite anomaly maps of Chapter 6 (e.g., the positives

along the Aleutian arc, off the south coast of Australia, and at the southern tip of Greenland, and the negative in the Labrador Sea). This confirms the contention of Meyer et al. (1985) and the results of Cohen (1989) and Counil et al. (1991) that most of the continent–ocean contrast is contained in the low-degree field and is removed along with the main field when residuals are computed relative to a model and when along-track filtering is applied to the data.

Over oceanic plateaus, the SEMM-0 anomalies correspond to local magnetization variations. Otherwise, they do not, but rather reflect the edge contrast at continent–ocean and KQZ boundaries. To the degree that the regional contrasts in the SEMM-0 ζ correspond to real geophysical contrasts, the resulting anomalies should be found in the maps of Chapter 6. Correlation analysis and visual comparison indicate positive correlations between numerous SEMM-0 anomalies and measured anomalies, particularly near coastal boundaries. The amplitudes of highly correlated anomalies often agree within a factor of 0.7.

Table 9.1. *Geologic time scale*

Age (Ma)	Eon	Era	Period	Epoch
0.01	Phanerozoic	Cenozoic	Quaternary	Holocene
2				Pleistocene
5			Tertiary	Pliocene
24				Miocene
37				Oligocene
57				Eocene
65				Paleocene
144		Mesozoic	Cretaceous	
208			Jurassic	
245			Triassic	
286		Paleozoic	Permian	
360			Carboniferous	
408			Devonian	
438			Silurian	
505			Ordovician	
570			Cambrian	
	Precambrian			
2500	Proterozoic			
4600	Archean			

Not all coasts show such anomalies, nor do all SEMM-0 anomalies correlate highly with the measured anomalies. When the two do not agree, it is taken as an indication of differences between the true and SEMM-0 magnetizations.

The SEMM-0 model a priori is fictitious in the sense that it is inferred, not measured, so one should be cautious when drawing conclusions about any specific anomaly, or lack thereof, in Figure 8.21. Yet the possible implications need to be considered. Some of these are as follows:

1. Anomalies at coastlines may reflect the continent–ocean contrast rather than local variations in magnetization. For example, the measured low and high anomaly pair at the south coast of Australia may indicate typical onshore and offshore magnetizations. Similarly, the positive at the southern tip of Greenland may not indicate higher magnetization in that portion of the Greenland crust. Similar considerations apply to portions of the Antarctic margins and to the north coast of Alaska.
2. Anomalies associated with some oceanic plateaus can be accounted for with the SEMM assumptions of thickened crust and, in some cases, enhanced KQZ magnetization, as, for example, Broken ridge and Iceland. However, this is not universally true. Measured anomalies associated with the Walvis ridge and Agulhas plateau, for example, are not reproduced in the SEMM-0.
3. Some anomalies studied in explicit lithospheric models may be due, in part or in whole, to the continent–ocean edge effect and the way the data

are processed (i.e., the "SEMM effect"). Possible examples include the anomalies associated with the Aleutian and Mid-American trenches. In these cases, the possible SEMM effect should be examined as to its impact on models such as those by Clark et al. (1985) for the Aleutian arc and those by Counil and Achache (1987) and Vasicek et al. (1988) for the Mid-American trench.

4. However, caution is needed in considering such implications from the prediction of SEMM-0 anomalies, such as the elongated positive near Japan with no corresponding measured anomaly. This indicates that the SEMM a priori does not always correspond to the actual continent–ocean contrast.

5. In some cases, the SEMM effect serves to confirm interpretations. For example, the conclusion of Bradley and Frey (1991) that the negative anomaly in the Labrador Sea is due to the changes in susceptibility and thickness at the continent–oceanic transition is directly confirmed.

As noted in Chapter 7, estimates of magnetization from long-wavelength satellite and aeromagnetic anomaly studies are in the range of 2–10 A/m, with 4 A/m a typical value (Shive et al., 1992). Continental and oceanic magnetizations from the SEMM-1 are in the range 0–2.2 A/m and 0–7.5 A/m, respectively.

9.2 NORTH AMERICA

9.2.1 OVERVIEW

At the heart of the present-day North American craton are several former microcontinents welded together by what are thought to be early Proterozoic collisional orogens. Figure 9.1 illustrates the main tectonic elements of the continent, and Table 9.2 summarizes some of the microcontinent characteristics. Strong compressional deformation, plutonism, and metamorphism occurred between 2.8 and 2.6 Ga in all of the Archean provinces. Table 9.3 summarizes the early Proterozoic orogens that welded the microcontinents together.

Although satellite magnetic anomaly data for North America have provided new insights into the nature of the lithosphere of this continent (e.g., Kentucky anomaly, Mississippi embayment signature, Rio Grande rift minimum), one of the more important aspects of the North American data lies in its utility in providing calibration for the interpretation of satellite data. The North American continent consists of a wide range of geologic features of variable ages and has been studied extensively by both surveys of surface geology and geophysical investigations of the lithosphere. There are numerous gaps in these studies, especially in the geophysical surveys; nonetheless, the existing knowledge base provides the opportunity to calibrate global satellite magnetic signatures with actual geologic elements. This is useful in developing hypotheses for the origins of magnetic signatures in less well known or exposed regions of the earth. In

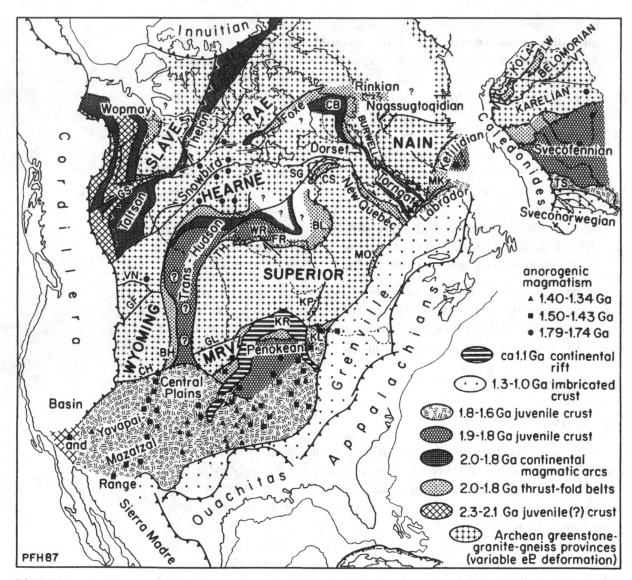

FIGURE 9.1
(see caption on p. 317.)

addition, the North American data can be used to extrapolate from exposed features to less well exposed and poorly studied regions of the continent, a method widely used in aeromagnetic interpretation.

Figure 9.2 shows contours for susceptibility times thickness, ζ, from the SEMM-1, with anomalies numbered for easy reference. Numbers in parentheses in the text refer to these numbered anomalies. Contours of ζ below the continental average of 1.0 are dashed. With some exceptions, the boundaries of the Precambrian craton have lower ζ values than the adjacent craton [e.g., (17), (18), (35), (31)]. Exceptions occur over the northwestern United States (29), Labrador (15), and offshore of the northeastern United States (24). The Appalachian–Ouachita front, probably the southern rifted continental margin

Table 9.2. *Archean microcontinents of North America*

Province	Age (Ga)	Comments
Superior	2.8–2.7	Small areas predating 3.5 Ga. Subparallel E–NE-trending belts with incremental ages across strike. Possibly Archean equivalent of island arcs, assembled north to south. Province then collided to south against 3.5-Ga Minnesota foreland.
Nain	≥2.9	Relics from 3.8–3.6 Ga in SW Greenland. Possibly a collage of terranes with independent histories, aggregated about 2.7 Ga.
Slave	2.7–2.5	Small areas predating 3.5 Ga. Plutonic-metamorphic terrane with gneissic inliers; tightly folded metasediments.
Wyoming	≥3.1	Assembled about 2.7 Ga. Not well exposed. Retrogressed granulite-grade orthogneisses and paragneisses; shelf-type metasediments. Intrusions of granite-granodiorite batholiths. May be transected by latitudinal late Archean suture zone southward of the Wind River range.
Hearne	2.9–2.8 (?)	Juvenile late Archean crust with infolded remnants of platform cover and foreland basins. Variation in metamorphic grade increasing outward from central core of greenschist grade. Underlain by submarine volcanic and volcanic-derived turbiditic rocks penetrated by gabbro, diorite, tonalite, and granitic plutons.
Rae	ca. 2.8	Felsic gneisses dominant; poorly known age and origin; sample ages about 2.8 Ga.
Burwell	ca. 2.8	Mainly granulite-grade granitoid gneiss, mafic gneiss, and paragneiss. Possible extension into central western Greenland.

of the craton, is outlined [(17), (18)] (Carmichael and Black, 1986). Within the craton, the southern Superior province, (2) and (3), the Slave province (7), the northern Wyoming province, and much of the Hearne province (4) show low values for ζ. The low over the Nain province and Burwell province is at least in part due to the SEMM edge anomaly. Thus, with the major exceptions of the northeastern Superior province, (1), and the Rae province, (5) and (6), most of the Archean terrane shows low values for ζ. The northern part of the positive, (5), lies over the Thelon magmatic arc and Queen Maud uplift, both of mainly granulite grade, whereas to its south the positive, (5), lies over the southern Rae province. Although ζ for (5) and (6) is particularly high, all of the Rae province except its southern tip shows above-average ζ values. The negative, (8), lies over the junction between the Taltson magmatic arc and the Rae province. Unlike the Thelon arc to its north, extensive granulites are absent from the Taltson arc.

To the west of the Archean craton, over a series of magmatic arcs and accreted terranes, are anomalies (12), (13), and (14): (12) apparently over the west-turning

FIGURE 9.1 (*opposite page*) Precambrian tectonic elements of the North American craton (platform cover removed) and Baltic shield. Names spelled with all-capital letters are Archean provinces; names spelled with an initial capital and then lowercase letters are Proterozoic and Phanerozoic orogens. BH, Black Hills inlier; BL, Belcher foldbelt; CB, Cumberland batholith; CH, Cheyenne belt; CS, Cape Smith belt; FR, Fox River belt; GF, Great Falls tectonic zone; GL, Great Lakes tectonic zone; GS, Great Slave Lake shear zone; KL, Killarney magmatic zone; KP, Kapuskasing uplift; KR, Keweenawan rift; LW, Lapland–White Sea tectonic zone; MK, Makkovik orogen; MO, Mistassini and Otish basins; MRV, Minnesota foreland; SG, Sugluk terrane; TH, Thompson belt; TS, Trans-Scandinavian magmatic zone; VN, Vulcan tectonic zone; VT, Vetrenny tectonic zone; WR, Winisk River fault. (From P. F. Hoffman, Precambrian geology and tectonic history. In: *The Geology of North America: An Overview*, ed. A. W. Bally and A. R. Palmer. Reproduced with permission of the publisher, The Geological Society of America, Boulder, Colorado, USA. Copyright 1989, The Geological Society of America, Inc.)

Table 9.3. *Comparison of early Proterozoic orogens in Laurentia*

Orogen	Foreland	Hinterland	Magmatism (U-Pb age, Ga)
Thelon	Slave	SW–Rae	2.02–1.91
Wopmay	Slave	arcs	1.95–1.84
Snowbird	Hearne (?)	SW–Rae (?)	post-1.92/pre-1.85
S. Alberta	Hearne	Wyoming	pre-1.85
Trans-Hudson	Superior	Hearne (+)	1.91–1.81
Cape Smith	Superior	Hearne (?)	1.96–1.84
New Quebec	Superior	SE–Rae	2.14–1.81
Torngat	Nain	SE–Rae	post-2.3/pre-1.65
Foxe	NE–Rae	SE–Rae	1.90–1.82
Nagssugtoqidian	Nain	NE–Rae	(?)
Makkovik	Nain	arcs(?)	1.86–1.80
Labrador	Nain (+)	arcs (?)	1.71–1.63
Penokean	Superior (+)	arcs	1.89–1.82
Yavapai	Wyoming (+)	arcs (+)	1.71–1.62

Source: Based on Hoffman (1988).

Table 9.4. *Metamorphic grades and ζ anomalies in Canada*

Anomaly	Positive or negative[a]	Metamorphic grade
(1)	+	Amphibolite/amphibolite overprinted on granulite
(2)	–	Amphibolite and felsic plutons mixed with greenschist to the north; upper amphibolite with some granulite to the south
(3)	–	Amphibolite and felsic plutons
(4)	–	Amphibolite and felsic plutons in the north, changing to greenschist in the south
(5)	+	Mainly granulite
(7)	–	Upper amphibolite mixed with felsic rocks and greenschist
(8)	–	Granulites absent
(9)	+ (weak)	Amphibolite
(10)	– (weak)	Upper amphibolite, some granulite
(15)	+	Amphibolite with an area of granulite

[a]Positive/negative defined relative to the "10" contour (i.e., $\zeta = 1.0$, the continental average) in Figure 9.2.

arm of the Fort Simpson magmatic arc, (13) over part of the Hottah arc, and (14) over the Ksituan arc, Wabamun region, and part of the Buffalo Head region. One of the most extensive Proterozoic terranes in Canada, the Trans-Hudson orogen, has no coherent magnetic signature. It is overlain by the positive, (9), in the north, next to Hudson Bay, it is overlain by the negative, (10), where the orogenic belt bends southward; and is relatively featureless to the south.

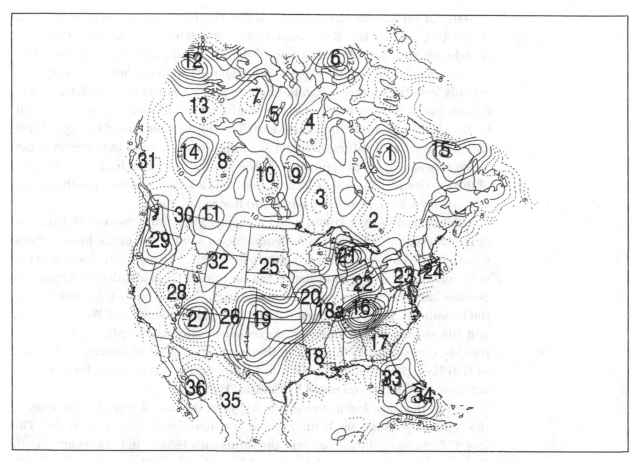

FIGURE 9.2
Contours of susceptibility
times thickness, ζ, for North
America. Units are (10 · SI ·
km).

These changes perhaps indicate crustal variations along the length of the
province.

According to Hoffman (1989), the region overlain by the high, (15), is mainly
mafic granulite-grade paragneiss and orthogneiss, with U-Pb ages predating the
Grenvillian orogeny. Extensive mafic granulites seem to be absent from the
Canadian Grenville province south of (15).

Because metamorphic grade is one of the important factors in determining
the magnetic properties of rocks, comparison is made between the metamorphic map of the Canadian shield (Geological Survey of Canada, 1978) and the
anomalies of Figure 9.2. Some care must be exercised in using these associations because metamorphism is mapped at the surface and thus may not reflect
conditions deeper in the crust. Table 9.4 summarizes the results, together with
some metamorphic grade information from Hoffman (1989). Examination of the
table reveals no generalization, but there is a tendency for association between
high ζ and granulite and low ζ and greenschist/lower-amphibolite grades (e.g.,
none of the positive anomalies are related to felsic rocks, but four of the six
negative anomalies involve felsic rocks, in part).

Many of the Proterozoic terranes in the United States are marked by high values of ζ – (27), (19), (20) – interrupted by a relative low, (26), at the Rio Grande rift and the substantial low of the Mississippi embayment, (18), and its northeastward extension, (18a), (22). The high with the largest extent, (19), extends from the southern Granite-Rhyolite province in Texas and Oklahoma to the northeast, (20), cutting across the southern Central Plains orogen. A small high, (21), lies over the Michigan basin. The northeastern trend of high ζ, (20) and (21), and its curvature through Lake Superior are similar to the trend of the Keweenawan rift, KR in Figure 9.1. To the southwest, the high, (27), to the west of the Rio Grande rift, (26), is located over the Colorado plateau, and the low to its northwest, (28), is over the Basin and Range province.

Mooney and Weaver (1989) noted a change in tectonic framework from the strike-slip regime south of Cape Mendocino to subduction of the Juan de Fuca plate beneath Oregon and Washington to the north. The high, (29), located in the subduction regime and truncated to the south near Cape Mendocino appears to be associated with the volcanic arc region (e.g., Cascade Range). Its peak is near the location of the thickest crust (46 km) shown by Mooney and Weaver (1989), and the westward and eastward decreases in anomaly amplitude in Oregon roughly correspond to thinning of the crust. The trend of lower ζ to the east of (29) [i.e., (30)] slants NW–SE into Idaho, similar to the trend for Cenozoic active-margin tectonism noted by Condie (1989).

One of the most intense magnetic highs in the world map, (16), occurs at the Kentucky–Tennessee boundary and is called the Kentucky anomaly. The low, (17), to its southeast lies over the Brunswick terrane in Georgia and South Carolina, and a prominent high is found over Florida, (33), and extending south over the Bahamas platform, (34).

Apparent correlations of magnetic anomalies with suture zones of orogens and sedimentary basins were noted by Arkani-Hamed et al. (1984). Calling attention to the magnetic lows over the Rocky Mountains, Appalachian Mountains, and the supposed Grenville suture zone, they hypothesized that thick sediments "may have depressed the magnetized layer into a high-temperature region and partially demagnetized it."

They noted that certain relative magnetic highs are located over basins associated with failed arms of old rifts: The Michigan and Lake Superior basins, anomaly (21) and its NW trend, are associated with the midcontinent rift. The Thelon basin is under part of (5). The Quebec basin, Pennsylvania embayment, and Baltimore canyon, together overlain by anomaly (23), are all associated with late Precambrian rifting of North America. The Sydney, Magdalen, and Acadian basins, overlain by the unnumbered low located along the eastern Canadian coast south of Newfoundland, probably are related to rifting along the Atlantic margin. Magnetic highs are associated with what they termed craton-margin basins [e.g., Eagle plain and Peel plain basins lie under part of anomaly (12), and the Peace River formation lies under part of anomaly (14)]. Magnetic lows were found in association with some craton-center basins, namely, Hudson Bay, (4), and the Williston basin, north part of (25).

Arkani-Hamed et al. (1984) set forth a hypothetical model that accounts for the noted associations: A basin formed by subsidence resulting from sedimentary loading might be depressed below the Curie isotherm and become demagnetized, resulting in a magnetic low. Basins associated with rift arms might be formed by thinning of the lithosphere and penetration of asthenospheric material into the lithosphere and crust, accompanied by uplift. Later the material would cool and subside. The hot mantle material would initially heat the lithosphere, and that part heated above the Curie isotherm would be demagnetized. On the other hand, some of the intruding asthenospheric material might emplace magnetic material and/or metamorphose existing lower crustal rocks and enhance their magnetization. In this scenario, the magnetization might be expected to be low early in the history of the basin, but high after cooling.

9.2.2 ANOMALY STUDIES

9.2.2.1 Ungava Anomaly

The northeastern part of the Superior province, the Minto and Bienville subprovinces in Figure 9.3, is located on the Ungava peninsula of northern Quebec, Canada, and is a region of positive satellite magnetic anomaly, (1). Noble (1983), as summarized by Hall, Noble, and Millar (1985), modeled this anomaly using a group of spherical prisms 36 km thick, in accord with the mean crustal thickness, located as shown in Figure 9.4. Prism magnetizations of 5.2 A/m provide a model matching the observed anomaly, with prisms of -1.0 A/m added to model the nearby anomaly lows. Noble (1983) interpreted the source of the positive anomaly as associated with the tectonics of the formation of the Richmond Gulf aulacogen (Chandler and Schwarz, 1980; Chandler, 1982; Hoffman, 1989, Fig. 9.16).

An alternative model of the positive anomaly can be given in the context of the SEMM. In Figure 9.3, to the south and west of the Superior province are large areas underlain by granite-greenstone belts, whereas to the northeast, the Minto and Bienville regions are mainly high-grade granite-gneiss of plutonic and volcano-plutonic affinity. From Figure 9.2, low values of ζ, (2) and (3), characterize the southern and western regions of the province, whereas a high value, (1), is located to the northeast in the Minto and Bienville regions.

In order to incorporate localized information into the SEMM-1, a forward modeling approach is taken in which additional source dipoles are added to the Minto-Bienville region. Each of these more localized dipoles represents a region equal to about 0.25° by 0.25° at the equator. A local susceptibility distribution is selected for the additional dipoles, based on Figure 9.3 over the land and

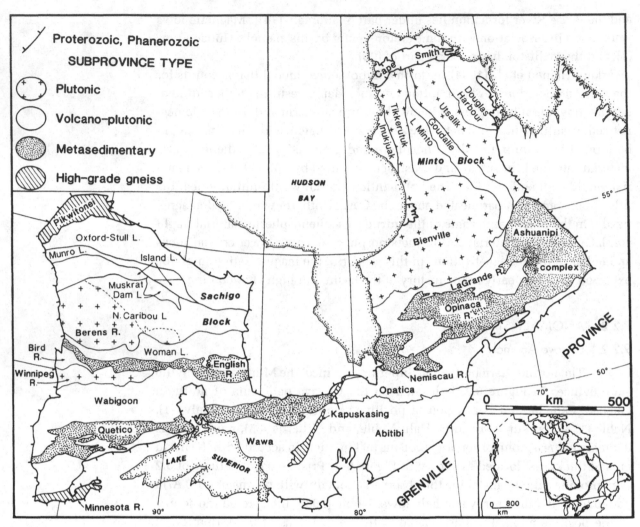

FIGURE 9.3
Generalized geologic map of the Superior Province. (Adapted from Percival et al., 1992; reprinted with kind permission of NRC Research Press, Ottawa, Canada.)

the aeromagnetic map of Pilkington and Roest (1996) over Hudson Bay. Plutonic terranes are assigned the highest susceptibility, volcano-plutonic terranes are assigned near-average susceptibility, and metasedimentary terranes are assigned the lowest susceptibility. This selection is constrained by the SEMM in that the net ζ for the more localized sources is required to be a near approximation to the SEMM-determined value on the larger scale. Over Hudson Bay and the surrounding lowlands the susceptibility distribution is based on the susceptibility-times-thickness contrast that would produce a given aeromagnetic anomaly. The final susceptibility values are adjusted by trial and error. Figure 9.5 (middle) shows the resulting more localized ζ distribution, the vertical field calculated from that distribution, and (top) the vertical field from the combined POGO/*Magsat* anomaly map {31}.

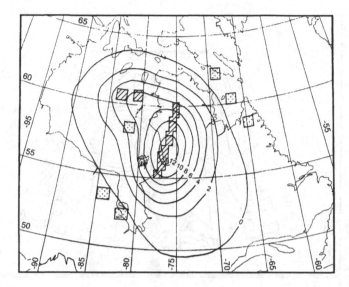

FIGURE 9.4
Prismatic model and calculated Ungava anomaly field; spherical prisms are 36 km in thickness and have magnetizations of 5.2 (crosshatched) and -1.0 (dot pattern) A/m; contour interval 2 nT. (From Hall et al., 1985, after Noble, 1983, with permission.)

Comparison of the aeromagnetic map of Pilkington and Roest (1996) and Figure 9.5 over the land indicates some discrepancy for the Tikkerutuk volcano-plutonic terrane (Figure 9.3) between the aeromagnetic data and the susceptibility assigned only on the basis of the geologic information. This may indicate the presence of unidentified sources in the unexposed middle and lower crust. Figure 9.5 (bottom) shows a revised model that is in harmony with the aeromagnetic result and still reproduces the satellite anomaly. Satellite data alone are insufficient to distinguish between these two models.

9.2.2.2 Crustal Thickness/Free-Air Gravity and Magnetization in the United States

Anomaly sources are complex combinations of lithospheric characteristics. Wasilewski et al. (1979), Mayhew et al. (1985b), and Wasilewski and Mayhew (1992) have argued that, at least for continental crust, the Moho is the lower boundary of the magnetized crust in areas where the Curie isotherm lies below the Moho. It is also argued (e.g., Hall, 1974; Wasilewski and Mayhew, 1992) that the lower crust is the seat of the long-wavelength anomalies. To verify these hypotheses or to make use of them in interpretation requires estimates for the thickness of the crust and its composition. Using seismic refraction data, Braile et al. (1989) compiled such information for the United States. Figure 9.6 shows contours for inferred depth below sea level to the Moho, and Figure 9.7 shows contours for the average seismic velocity in the crust, which has been shown to be an indicator of mean crustal composition (Smithson, Johnson, and Wong, 1981). These maps are based on roughly 250 refraction profiles poorly

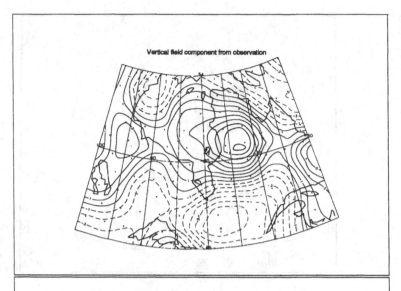

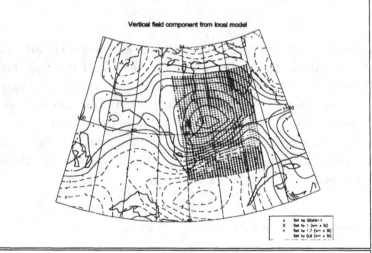

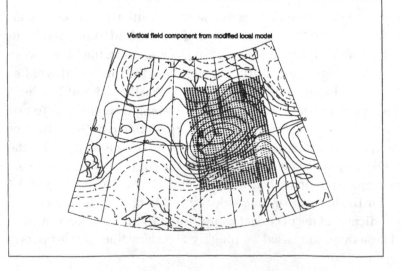

FIGURE 9.5
Comparison of the observed vertical anomaly field with the vertical field produced by local versions of the SEMM over the northeastern Superior province. Top: Observed field. Middle: Magnetization distribution and field from the model based on the geologic map in Figure 9.3. Bottom: Magnetization distribution and field from the model modified to be in agreement with aeromagnetic data. The contour interval is 1 nT, with solid lines indicating positive values of the magnetic field, and dashed lines indicating negative values. The symbols (×, ○, +, −) on the maps locate dipoles. Dipoles set by the SEMM-1 (indicated by × symbol) are separated by the equivalent of 1.89° at the equator. Dipoles set by the local model (indicated by ○, +, or − symbols) are separated by 0.25°. Azimuthal equidistant projection.

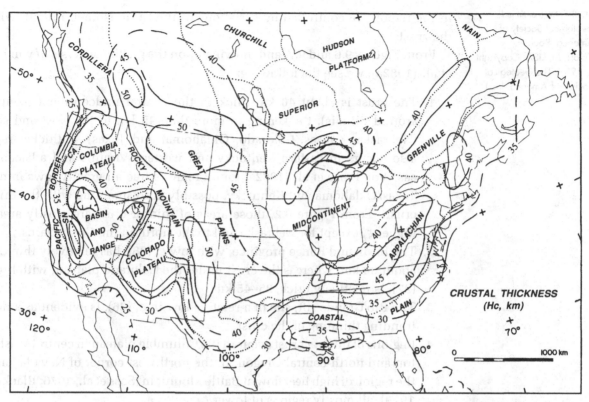

FIGURE 9.6

Contour map of crustal thickness for a portion of the North American continent. Numbers show crustal thickness in kilometers measured from the surface to the inferred Moho discontinuity; contour interval 5 km. (From L. W. Braile, W. J. Hinze, R. R. B. von Frese, and G. R. Keller, Seismic properties of the crust and uppermost mantle of the conterminous United States and adjacent Canada. In: *Geophysical Framework of the Continental United States*, ed. L. C. Pakiser and W. D. Mooney pp. 655–80, 1989. Geological Society of America Memoir 172. Reproduced with permission of the publisher, The (*continued*)

distributed across the area. The Great Plains and the central Great Lakes regions, in particular, are poorly covered.

Variations in crustal thickness may be reflected in variations of free-air gravity. For example, greater crustal thickness can correspond to lower gravity, because crustal rock is of lower density than mantle rock (e.g., Carmichael and Black, 1986). It has been noted (von Frese et al., 1982a) that higher crustal temperatures are often associated with geodynamic processes that result in crustal thinning, intrusions from the mantle, and isostatically uncompensated topography. This leads to the expectation of an inverse relationship between free-air gravity anomalies and magnetic anomalies. To investigate relationships between gravity and magnetic anomalies, von Frese et al. (1982a) described application of "internal correspondence analysis" (Section 8.1.2) to the second radial derivative of the free-air gravity data and the first radial derivative of the reduced-to-pole (RTP) magnetic anomaly data. Negative correlations are particularly high in much of the midcontinent, (19), (20), in the western Gulf of Mexico, (18), and along the East Coast, (17) and northward. Positive correlations are evident over the Dakotas and West Coast. The belt of positive correlation extending along the West Coast from Washington through Oregon, California, and much of Utah is contrary to what is expected if the magnetization variations

FIGURE 9.6 (*continued*)
Geological Society of
America, Boulder,
Colorado, USA. Copyright
© 1989, The Geological
Society of America, Inc.)

in the region are controlled by variations in the Curie isotherm depth within the crust.

From Figures 9.6 and 9.7, and drawing upon the gravity analysis of von Frese et al. (1982a), we see the following:

1. The crust is about 40 km thick in the central midcontinent (centered around Illinois), thickening to more than 45 km to the west and south (Nebraska, Kansas, Colorado, Oklahoma, and Texas). A thicker region also occurs in eastern Kentucky and western Virginia and adjacent areas. Von Frese et al. (1982a) found relative free-air gravity lows in north Texas/Oklahoma and Kentucky, possibly corresponding with the thicker crust. From Figure 9.2, those areas of thicker crust are mostly areas of higher ζ, except for Nebraska and to the north in the Great Plains.
2. The Basin and Range province, with low ζ, (28), has relatively thin crust, about 30 km, whereas the crust under the Colorado plateau, with high ζ, (27), is relatively thick, 35–45 km.
3. The Rio Grande rift, site of a relative low in ζ, (26), is evident as a north-trending thinning of the crust.
4. Regions of thinner crust occur in the Columbia plateau in central Washington and north-central Oregon, in the northwest corner of Nevada, and in the region of high heat flow at Battle Mountain (Sass et al., 1976; Blackwell, 1979), all mostly regions of lower ζ.

With regard to the nature of the crust as portrayed in the average compressional (P) wave velocity of the crust (Figure 9.7), abnormally high velocities indicative of more mafic composition are observed in northern New York, associated with the Adirondack massif, in the south-central United States, where the crust is thicker than normal, in the northern Great Plains, and in the Midcontinent rift region. Lower velocities are recorded in the Basin and Range province and northward across eastern Oregon and Washington and adjacent Idaho, as well as in the Rio Grande rift. Crustal velocities, as reflected in the smoothed contouring of Figure 9.7, also decrease markedly from continental crust to oceanic crust. Comparison of Figures 9.2 and 9.7 shows direct correlations between mean crustal velocities and ζ values. This follows directly from the increased ferrimagnetism of more mafic rocks, as anticipated on the basis of laboratory measurements and empirical correlations. Though very much a generalization and an oversimplification in detail, nonetheless the observed correlation is striking.

Schnetzler and Allenby (1983) developed a method for combining dipole moment maps with crustal thickness maps to estimate the average magnetization in the crust. Using that method, with magnetic moment estimations, {52}, based on *Magsat* data by Mayhew and Galliher (1982) and lower crustal

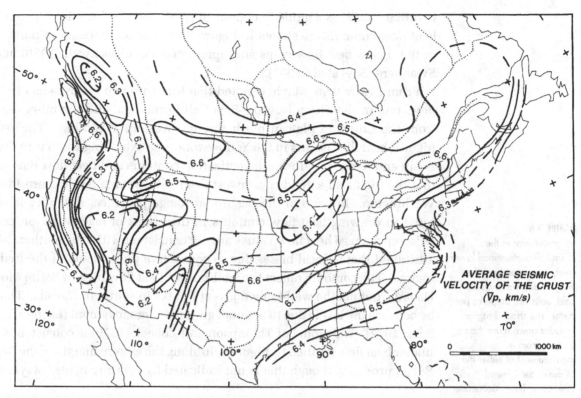

FIGURE 9.7
Contours for average seismic velocities in the crust for a portion of the North American continent. Contours give values velocity in kilometer per second; contour interval 0.1 km/s. (From L. W. Braile, W. J. Hinze, R. R. B. von Frese, and G. R. Keller, Seismic properties of the crust and uppermost mantle of the conterminous United States and adjacent Canada. In: *Geophysical Framework of the Continental United States*, ed. L. C. Pakiser and W. D. Mooney, pp. 655–80, 1989. Geological Society of America Memoir 172. Reproduced with permission of the publisher, The Geological Society of America, Boulder, Colorado, USA. Copyright © 1989, The Geological Society of America, Inc.)

thickness estimates from Allenby and Schnetzler (1983), Schnetzler (1985) derived the map of magnetization {57} shown in Figure 9.8. Although there are reasons to doubt the assumption that all magnetization causing satellite anomalies resides in the lower crust, and measured thicknesses of the lower crust are subject to considerable error, yet important insights are gained from these models (Schnetzler and Allenby, 1983; Schnetzler, 1985). It seems clear that variations in both the magnetization and the thickness of the magnetized layer are important in producing long-wavelength magnetic anomalies. For example, the magnetic high extending across the south-central United States from Kentucky to New Mexico, a region of generally greater crustal thickness (Figure 9.6), is also a region of overall increased magnetization of the lower crust (Figure 9.8).

9.2.2.3 Correlation of Magnetization with Heat Flow

Mayhew (1984) (see also Mayhew, 1982b, 1985) noted that in the younger lithosphere of the western United States, variations in magnetization are often inversely correlated to variations in measured heat flow. To make a quantitative comparison, he used a then-current heat flow data set to derive a smoothed contour map of heat flow, using the magnetization contours of

Figure 9.9a {53} as a guide in regions of sparse data on heat flow. The resulting heat flow contours are shown in Figure 9.9b. For comparison, in parts c and d of the figure, heat flow maps are reproduced from Blackwell (1979) and from Swanberg (Sass et al., 1981).

From Figure 9.9b, Mayhew noted that there are two major belts of low heat flow, one in the Sierra Nevada–Baja California area, and the other extending from the Colorado plateau north through western Wyoming. The region of highest heat flow occurs in the Yellowstone area extending south into the entire Basin and Range province. Magnetization contours and heat flow data indicate that the high also extends northwest from Yellowstone into eastern Idaho and Washington. In parts of the Basin and Range province, difficulty is encountered in drawing heat flow contours in directions of magnetization contours; adjacent average heat flow values are not consistent with one another. Mayhew attributed that to local biases due to convective phenomena in the hydrologic regime. The contours reflect the well-known heat flow high of Battle Mountain, Nevada, but not the well-known low at Eureka in southern Nevada. That could be because the low is due to regional groundwater movement (Lachenbruch and Sass, 1977). The model of Thompson and Zoback (1979) is consistent with the higher heat flow implied by Figure 9.9b along the eastern margin of the Basin and Range province, though that is not indicated by measurements. Mayhew (also,

FIGURE 9.8
Magnetization in the seismically determined lower crust, calculated from the *Magsat*-derived anomaly field; units are amperes per meter. The shaded region includes areas more than 225 km from a measurement of either the Moho or the Conrad boundary. (From Schnetzler, 1985, with permission.)

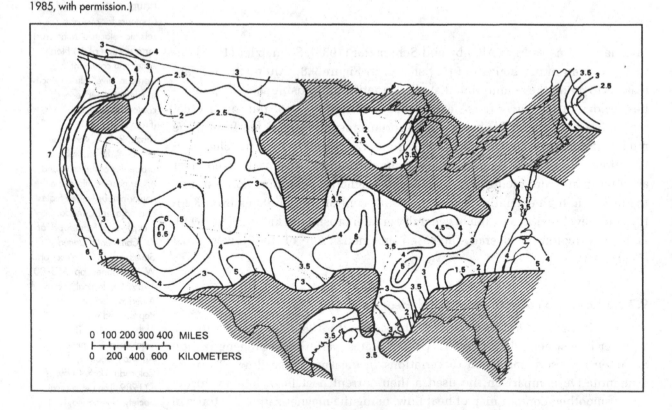

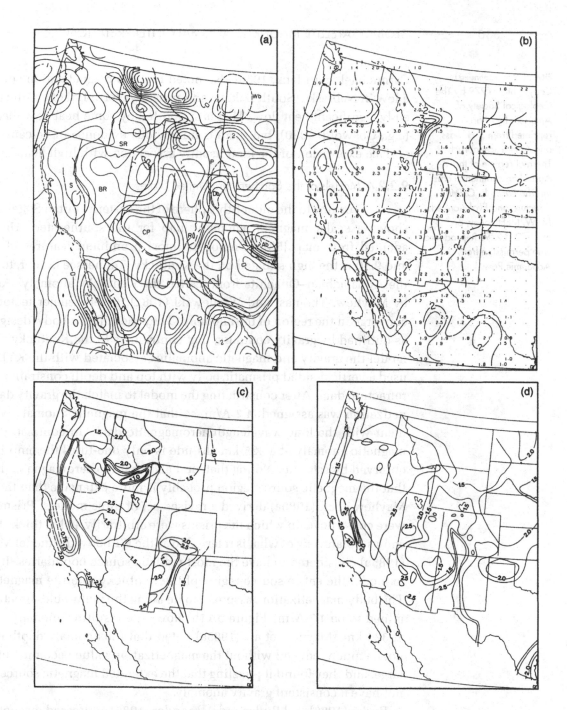

FIGURE 9.9
(a) Magnetization contours for western United States. (b) Contours for averaged heat flow, guided by the contours of part a. (c) Heat flow from Blackwell (1979), to be compared with part b. (d) Heat flow from Swanberg (Sass et al., 1981), to be compared with part b. Contour interval for parts b–d is 0.5 HFU. (Reprinted from M. A. Mayhew, *Magsat* anomaly field inversion for the U.S., *Earth and Planetary Science Letters, 71,* 290–6, 1984, with kind permission of Elsevier Science NL, Sara Burgerhartstraat 25, 1055 KV Amsterdam, The Netherlands. Part c reprinted from Blackwell, D. D., Heat flow and energy loss in the western United States. In: *Cenozoic Tectonics and Regional Geophysics of the Western Cordillera,* ed. R. B. Smith, and G. P. Eaton, pp. 175–208, 1979. Geological Society of America Memoir 152. (Reproduced with permission of the publisher, The Geological Society of America, Boulder, Colorado, USA. (*continued*)

Carmichael and Black, 1986) also noted that the magnetization contours show a low in Nebraska, South Dakota, and possibly North Dakota, indicative of possibly significant heat flow. That is supported by some heat flow measurements (e.g., Goswold, 1980), but there is no obvious evidence of volcanic activity or surface expression of the kind usually associated with high heat flow.

9.2.2.4 Models of the Southeastern United States

Turning to the older lithosphere of the eastern United States, Figure 9.10 shows a scalar magnetic anomaly map for the southeastern United States, adapted from map {52}. One of the most prominent features of Figures 9.2 and 9.10 is the high situated on the Kentucky–Tennessee border, to the west of the Appalachian–Ouachita front, called the Kentucky anomaly. As discussed by Mayhew, Thomas, and Wasilewski (1982), several major tectonic features converge in the region of the anomaly. They identified a body, designated KYB, as outlined by gravity contours, that contributes to the Kentucky anomaly. To model the gravity and magnetic anomalies associated with the KYB body, they used a vertical-sided prismatic body with top and depth constrained by seismic refraction data. After constructing the model to match the gravity data, the magnetization was assigned, 4.2 A/m, so that the magnetic anomaly was in agreement with the long-wavelength aeromagnetic data. In contrast, the resulting magnetic anomaly at a 325-km altitude is only one-third the amplitude of that observed by satellite. Noting that, and also that the aeromagnetic data indicate that the magnetic source region probably is more extensive than the KYB body, Mayhew et al. (1985a) derived a revised set of three models. Prismatic models were constructed in which the prisms were arbitrarily 40 km thick. Figure 9.11a shows a plan view of what is referred to as the KT body, the model Mayhew et al. (1985a) considered to have the most realistic source boundaries. It is intended to model the entire source region of the Kentucky satellite magnetic anomaly. The body magnetization, assumed to lie along the main-field direction, is determined to be 4.2 A/m. Figure 9.11b shows the resulting anomaly in total field, at 325 km. Mayhew et al. (1985a) noted that the anomaly depth extent could not be much reduced without the magnetization value becoming unreasonably large, and they found it puzzling that the extended magnetic source region does not have a consistent gravity anomaly.

Ruder (1986) and Ruder and Alexander (1986) addressed magnetic modeling of the southeastern United States and, in particular, the negative in Georgia, (17), of amplitude about −10 nT (Figure 9.10). Their regional model was constructed to be consistent with the known tectonics, following Hatcher (1978), Cook et al. (1979), Cook and Oliver (1981), and Williams and Hatcher (1982), and incorporates the paleosuture suggested by Arnow et al. (1985) from COCORP data, located due north of the Brunswick terrane. As shown in Figure 9.12, the

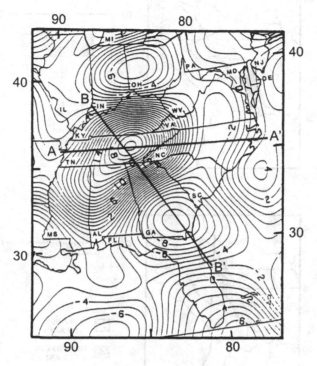

FIGURE 9.10
Equivalent source representation of *Magsat* magnetic anomaly field at a 325-km altitude; contour interval 1 nT. Albers equal-area projection. (Reprinted from M. A. Mayhew, R. H. Estes, and D. M. Myers, Magnetization models for the source of the "Kentucky anomaly" observed by *Magsat. Earth and Planetary Science Letters, 74*, 117–29, 1985, with kind permission of Elsevier Science NL, Sara Burgerhartstraat 25, 1055 KV Amsterdam, The Netherlands.)

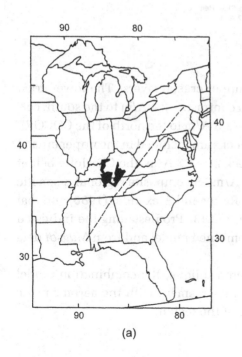

(a)

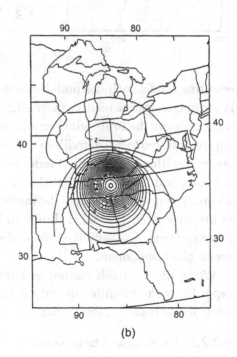

(b)

FIGURE 9.11
(a) Plan view of the source model for the Kentucky magnetic anomaly, the K T body. (b) Magnetic anomaly from the KT body at a 325-km altitude. (Reprinted from M. A. Mayhew, R. H. Estes, and D. M. Myers, Magnetization models for the source of the "Kentucky anomaly" observed by *Magsat. Earth and Planetary Science Letters, 74*, 117–29, 1985, with kind permission of Elsevier Science NL, Sara Burgerhartstraat 25, 1055 KV Amsterdam, The Netherlands.)

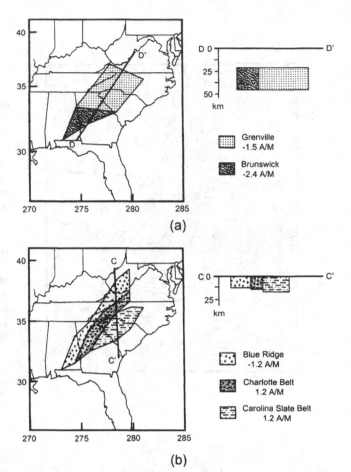

FIGURE 9.12
(a) Upper crustal, three-body model for the southern Appalachian magnetic crust. (b) Lower crustal, two-body block model for southern Appalachian magnetic crust. The anomalous region is bounded on the west by the New York–Alabama lineament (King and Zeitz, 1978) and on the east by the fall line. (Reprinted from M. E. Ruder and S. S. Alexander, *Magsat* equivalent source anomalies over the southeastern United States: implications for crustal magnetization. *Earth and Planetary Sciences Letters*, 78, 33–43, 1986, with kind permission of Elsevier Science NL, Sara Burgerhartstraat 25, 1055 KV Amsterdam, The Netherlands.)

resulting model includes both lower and upper crustal layers. The lower crust is assigned negative relative magnetization contrast: −2.4 A/m to the south, corresponding to the Brunswick terrane, and −1.5 A/m to the north of the COCORP suture, inferred to be Grenvillian. Three blocks are included in the upper crustal layer: the Blue Ridge–Inner Piedmont block at −1.2 A/m, the Charlotte belt at 1.2 A/m, and the Carolina Slate belt at 1.2 A/m. A composite regional magnetic anomaly map including both the model of Ruder and Alexander (1986) and that of Mayhew et al. (1985a) is shown in Figure 9.13. Profiles along the indicated cross sections, A–A′ and B–B′, from the combined model and from *Magsat* data are in good agreement.

While these models cannot be considered unique, the combination model reproduces the satellite anomaly data, is in agreement with the aeromagnetic data, and is based on the known tectonics of the region.

9.2.2.5 The Mississippi Embayment

At the time of the breakup of Laurentia, 0.8–0.6 Ga, several aulacogens developed along the rifted margin, one of which was the Mississippi embayment, a broad reentrant of Mesozoic and Cenozoic sedimentary rocks. Its axis

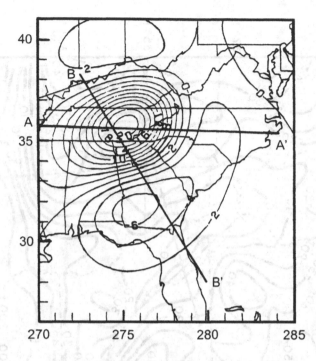

FIGURE 9.13
Composite magnetic anomaly at 325 km for southern Appalachian and Kentucky–Tennessee crustal models; contour interval 2 nT. (Reprinted from M. E. Ruder and S. S. Alexander, *Magsat* equivalent source anomalies over the southeastern United States: implications for crustal magnetization. *Earth and Planetary Sciences Letters,* *78,* 33–43, 1986, with kind permission of Elsevier Science NL, Sara Burgerhartstraat 25, 1055 KV Amsterdam, The Netherlands.)

parallels the Mississippi River, and it tapers northeastward into the New Madrid seismic zone (von Frese et al., 1981b). Studies in this region have been summarized by Keller et al. (1983). Burke and Dewey (1973) suggested the embayment as a possible failed arm rift, and Ervin and McGinnis (1975) modeled the underlying Reelfoot rift as a late Precambrian–early Paleozoic aulacogen, using gravity, seismic, stratigraphic, and petrologic data. Cordell (1977) refined the Bouguer gravity map for the area by removing the effects of the sedimentary cover. Hildenbrand, Kane, and Stauder (1977) mapped an apparent deep-seated graben in the gravity and magnetic data in the upper embayment region, correlating with the New Madrid seismic zone.

Examination of Figure 9.2 shows a strip of variable but low magnetization, (18) and (18a), extending from southern Louisiana northeastward along the Mississippi River to southern Illinois, where it turns more eastward, (22), and extends to Lake Erie. The southern portion of this strip corresponds to the Mississippi embayment. In the Bouguer gravity map it is evident as a high extending to the south of Illinois.

Figure 9.14 summarizes the model of Ervin and McGinnis (1975), a revised model of Austin and Keller (1979), the corrected Bouguer gravity map of Cordell (1977), and a model derived by von Frese et al. (1981b) using the method described in Section 8.2.3. The model cross sections in parts A and B of the figure are for the profile marked G in part D of the figure. Based on the model of Austin and Keller (1979), von Frese et al. (1981b) used the four-body model of part C of the figure. To extend the model in three dimensions, the cross section was projected to the south and north within the region outlined in D, with a

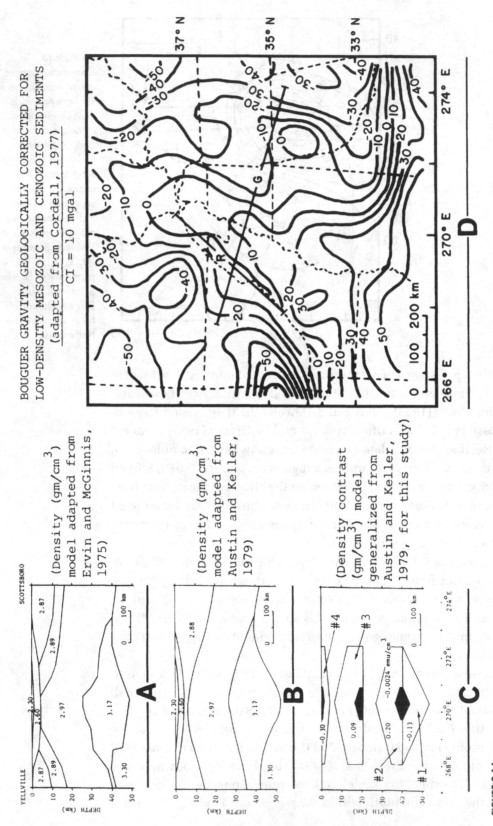

BOUGUER GRAVITY GEOLOGICALLY CORRECTED FOR
LOW-DENSITY MESOZOIC AND CENOZOIC SEDIMENTS
(adapted from Cordell, 1977)
CI = 10 mgal

(Density (gm/cm³)
model adapted from
Ervin and McGinnis,
1975)

(Density (gm/cm³)
model adapted from
Austin and Keller,
1979)

(Density contrast
(gm/cm³) model
generalized from
Austin and Keller,
1979, for this study)

FIGURE 9.14
(see caption on opposite page.)

northward taper based on the gravity map. The blacked-in regions on part C of the figure show the cross sections of the model bodies at their north ends. Using the density values shown in C, the resulting gravity anomaly was computed and found to be in reasonable agreement with the measurements.

Under the assumptions that the long-wavelength sources are mainly in the lower crust and that the Curie isotherm depth is 40 km or more, von Frese et al. (1981b) looked for the source of the negative magnetic anomaly in body 2. Possible reasons for a negative magnetization contrast were that the body is depleted of magnetic minerals or, less likely, that its temperature exceeds the Curie temperature. Accordingly, a relative magnetization of -2.4 A/m was assigned to body 2, giving a calculated anomaly at a 450-km altitude with amplitude consistent with the satellite data, but spatial characteristics that differed, so that a more refined analysis is needed. Von Frese et al. (1982a) noted that the $(\Delta M/\Delta\sigma)$ ratio of the model compared well with the ratio found from internal correspondence analysis of the POGO satellite magnetic anomalies and upward-continued free-air gravity measurements. An alternative distribution of magnetization in the various layers was proposed by Thomas (1984).

9.2.2.6 Florida and the Bahamas Platform

It is now considered that in the late Paleozoic, prior to the closing of the Iapetus Ocean, the Florida block was part of Africa (Opdyke et al., 1987; Venkatakrishnan and Culver, 1988), adjacent to the Bove basin near the Leo uplift. Susceptibility times thickness from the SEMM shows a prominent positive over the Leo uplift (discussed in Section 9.3.2) and another over the Florida peninsula and Bahamas platform. However, the nature of the Florida and Bahamas crust is debated (e.g., Meyerhoff and Hatten, 1974; Mullins and Lynts, 1977; Klitgord, Popenoe, and Schouten, 1984; Ladd and Sheridan, 1987; Sheridan et al., 1988; and references therein). Following Uchupi et al. (1971), Mullins and Lynts (1977) recognized northwestern and southeastern sectors of the platform, with a dividing line as indicated in Figure 9.15. The availability of seismic reflection data (Sheridan et al., 1981; Ladd and Sheridan, 1987) led to the interpretation that thinned, rifted continental or transitional crust extends southeast from Florida to about the location of the extension of the Blake Spur magnetic anomaly, coincident with the "N. E. Prov." channel in Figure 9.15 (Ladd and Sheridan, 1987; Sheridan et al., 1988). In contrast, the southeastern part of the platform is built on oceanic crust. Ladd and Sheridan (1987) suggested that this southeastern crust may have originated as a shallow oceanic basaltic ridge, similar to the Iceland–Faeroe ridge.

Ridgway (1984) and Ridgway and Hinze (1986) noted that the satellite magnetic high in this region indicates that the magnetic crust is either thicker than or more magnetic than the surrounding crust and that the Dietz, Holden, and Sproll (1970) model of a thin oceanic crust overlain by 10–15 km of clastics

FIGURE 9.14 (*opposite page*) Development of Mississippi embayment density and magnetization models. (D) Index map of the embayment (shaded contour), where G is the gravity profile studied by Ervin and McGinnis (1975). R is the seismic refraction line studied by McCamy and Meyer (1966). In the model of Austin and Keller (1979), the following seismic velocities and densities were found for each layer: I: $V_p = 8.0$–8.1 km/s; $V_s = 4.6$ km/s; $\rho = 3.30$ g/cm^3. II: $V_p = 7.3$–7.4 km/s; $V_s = 4.0$ km/s; $\rho = 3.17$ g/cm^3. III: $V_p = 6.5$–6.6 km/s; $V_s = 3.8$ km/s; $\rho = 2.97$ g/cm^3. IV: $V_p = 6.1$–6.3 km/s; $V_s = 3.6$ km/s; $\rho = 2.88$ g/cm^3. V: $V_p = 4.9$–5.0 km/s; $V_s = 2.9$ km/s; $\rho = 2.60$ g/cm^3. VI: $V_p = 4.2$–4.3 km/s; $V_s = 2.3$ km/s; $\rho = 2.30$ g/cm^3. (Adapted from fig. 6 of von Frese et al. (1981b); reprinted with kind permission of Springer-Verlag; © 1981. Part A reprinted from C. P. Ervin, and L. D. McGinnis, Reelfoot rift: reactivated precursor to the Mississippi embayment. *Geological Society of America Bulletin*, 86, 1287–95, 1975; reproduced with permission of the publisher, The Geological Society of America, Boulder, Colorado, USA. Copyright © 1975, The Geological Society of America, Inc.)

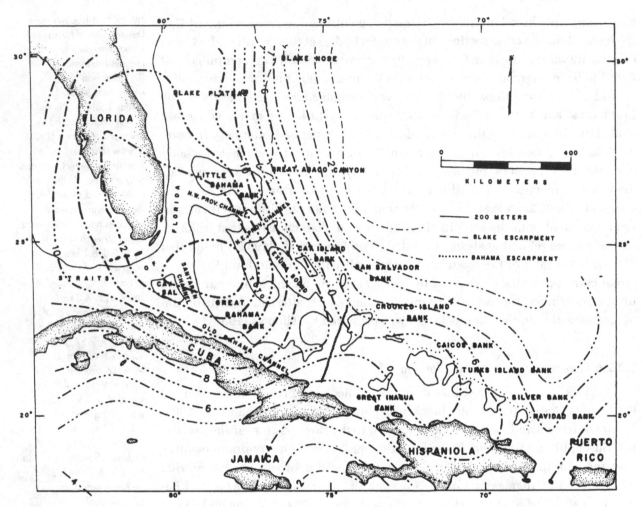

FIGURE 9.15
(*see caption on opposite page.*)

and carbonates seems unlikely. Following Pindell (1985), Counil, Achache, and Galdéano (1989) considered the region of positive anomaly to be an allochthonous block of pre-Mesozoic continental origin, possibly similar in nature to the Yucatan and Chortis blocks in Central America.

Figure 9.15 shows that ζ is high over all of peninsular Florida and extending southward toward Hispaniola. There is a pinching-out, or saddle, of susceptibility times thickness at the line on the figure dividing the northwestern and southeastern platform sections. To the southeast of the saddle, susceptibility times thickness increases again slightly. The positive anomaly is a robust feature of both POGO and *Magsat* data, and continental-like magnetizations are found whenever these data are inverted. If a modified SEMM-0 a priori is assumed, with low magnetization, then upon solving for a new SEMM-1 the entire Bahamas platform retains elevated values of ζ; that is, the high values of ζ are required by the satellite magnetic data. Although such values are commonly associated with continental lithosphere, they are also found associated with some oceanic plateaus and ridges. Some of these (e.g., Agulhas plateau, Broken ridge, Crozet plateau, Kerguelen plateau) are considered to be composed of thickened oceanic crust, whereas others (e.g., Campbell plateau, Rockall plateau) are considered to be continental crust. Note particularly that ζ for the Iceland–Faeroe ridge is comparable to that for the Bahamas platform.

Continuity of high ζ between the Leo uplift and Florida lends support to an African affinity for Florida and would be consistent with such affinity for at least the northwestern part of the Bahamas platform. The suggestion of Ladd and Sheridan (1987) that the Bahamas platform crust may be similar to that of the Iceland–Faeroe ridge is also plausible. Hayling and Harrison (1986) pointed out that the elongation of the Bahamas anomaly follows the path of the San Fernando hotspot. In their model, the Bahamas passed over this mantle plume at 155–170 Ma, thickening the crust during a period of largely normal polarity.

9.2.3 DISCUSSION

Whereas the broad associations between satellite magnetic anomalies and geologic and tectonic features drawn in Section 8.1.1 may allow useful generalizations, it is clear that there are many exceptions and that real understanding of the source of a particular anomaly will require detailed analysis. This is particularly clear from the observations of relatively low magnetizations over many Archean shields and relatively high magnetizations over several basins. On the other hand, it is encouraging that in many regions fairly clear associations are found between the satellite anomaly patterns and the known tectonics and geology. These associations, though certainly more detailed and perhaps more robust than earlier studies, should still be regarded with caution. Their meanings can be established only by detailed modeling that includes constraints from other data types.

The association of high-grade metamorphic regions with magnetic highs, and vice versa, while not definitive, is at least indicative. Amphibolite-grade regions show both magnetic highs and lows, though most of the lows seem to occur in regions with an admixture of felsic rocks. It is also apparent that crustal thickness and Curie isotherm depth, when above the Moho, are both important factors in determining the lateral variation of depth-integrated magnetization. While it seems probable that much of the magnetization lies in the lower crust, this cannot be true universally, particularly in regions of an elevated Curie isotherm. Thus, any individual anomaly is due to a complex combination of many factors. This means that it could be misleading to transfer a model or analysis from one region to another without first verifying that the prevailing conditions and geologic history are at least similar.

As is usual with interpretation of potential field data, the models described are nonunique. Plausibility is increased when data from known geology, tectonics, and other geophysical measurements are included. For example, the

FIGURE 9.15 (*opposite page*) Overlay of ζ in $(10 \cdot SI \cdot km)$ on a map of the Bahamas platform. The heavy dark line divides the platform into northwestern and southeastern sectors. Susceptibility times thickness times 10 is plotted on the overlay, with a contour interval of 1. Mercator projection. (The map of the Bahamas platform is from H. T. Mullins and G. W. Lynts, Origin of the northwestern Bahama Platform: review and reinterpretation. *Geological Society of America Bulletin*, 88, 1447–61, 1977; reproduced with permission of the publisher, The Geological Society of America, Boulder, Colorado, USA. Copyright © 1977, The Geological Society of America, Inc.)

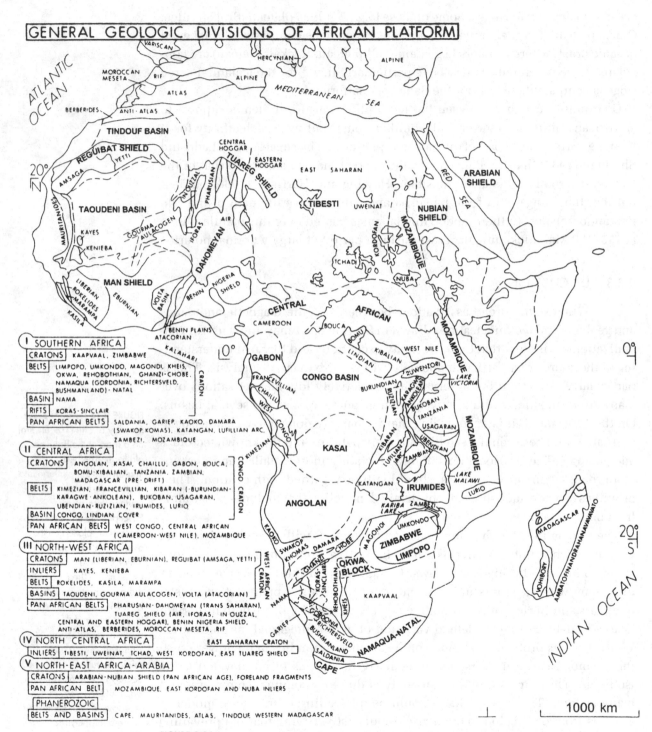

FIGURE 9.16

General geologic divisions of the African platform. Locations referred to in the text are in bold print.
(Adapted from Goodwin, 1991; reprinted by permission of Harcourt Brace and Company, Ltd.)

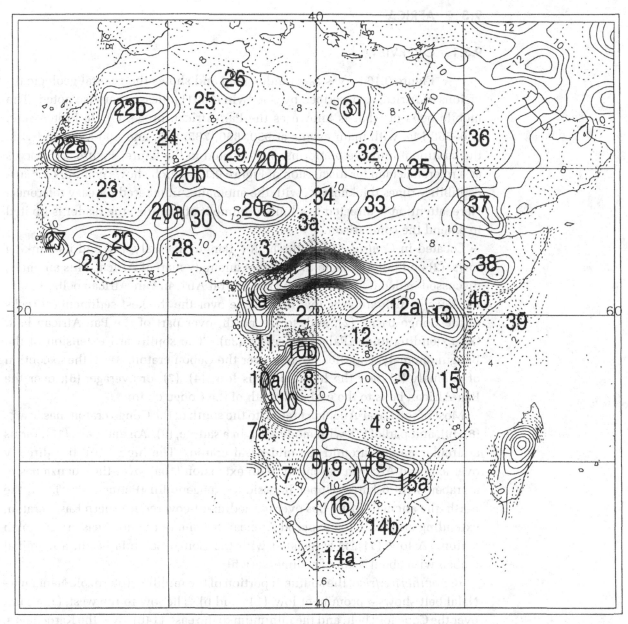

FIGURE 9.17
Contours of ζ for Africa; units are (10 · SI · km).

availability of extensive heat flow data for the western United States makes possible the inferences of Figure 9.9 regarding regional heat flow distribution. And those inferences are bolstered by the complementary information from gravity–magnetic correlations. As another example, studies of the southeastern United States furnish a good example of the incorporation of seismic refraction and reflection results, as well as geologic and tectonic constraints, into the modeling process.

9.3 AFRICA

9.3.1 OVERVIEW

Figure 9.16 serves as an index map and shows the general geologic divisions of Africa, as described by Goodwin (1991) unless otherwise noted. The mobile belts of what is known as the Pan-African orogeny (late Proterozoic, 950–550 Ma, with main events at 950, 860, 785, 685, and 600 Ma) subdivide the continent into five platforms, as indicated in the figure. Figure 9.16 serves as a reference map for discussion of Figure 9.17, which shows contours for susceptibility times thickness, ζ, with the anomalies numbered for easy reference. Here, attention is directed to a limited number of features; the reader is invited to extend the comparison.

Located in *Central Africa*, the high-ζ Bangui anomaly, (1), and its flanking lows, (2) and (3), dominate the satellite anomaly map of Africa. This anomaly, discussed in detail later, lies along the central African Pan-African belts. Flanking lows, which are very low indeed, lie over the thickest sedimentary rocks of the Congo basin, (2), and, to the north, over part of the Pan-African belt, (3), extending under Paleozoic cover, (3a). The southward extension of the Bangui anomaly, (1a), lies directly over the Gabon craton. With the exception of anomaly (6) over the Irumides, ζ is low, (4), (7), or average, (5), over the Pan-African belts to the east and south of the Congo craton.

Most of the Kasai-Angolan craton, to the south of the Congo craton, has low ζ, (8), (9), and part of (7), with moderate ζ in a saddle, (9). An anomaly, (10), forms a circular pattern around the mid-Kasai craton. The high, (10), lies directly over exposed Archean basement, with extension (10a), over the Cuanza basin, a graben with 6 km of Upper Jurassic to Neogene fill (Petters, 1991). To the north, the anomaly (10b) lies over the sediment-covered northern Kasai craton, extending around to the exposed Archean terrane in the northeastern Angolan craton. A low, (11), is associated with the Congo-Cabinda basin, a marginal graben with about 3 km of sedimentary fill.

In *Southern Africa*, the southern portion of the middle Proterozoic Namaqua–Natal belt shows a prominent low, (14a and b). The low to the west, (14a), lies over the Cape fold belt, and the minimum to the east, (14b), over the Karoo lavas. The eastern boundary of the Kalahari craton, and indeed the entire Mozambique belt south of 5° S latitude, shows low to average values of ζ, (15).

Over the entire Kaapvaal craton, ζ is above average, with a distinct high, (16), over the western craton and a lower (saddle) region, (17), in the area of the Ventersdorp-Transvaal-Waterberg epicratonic basins and associated greenstone belts. The Archean Limpopo belt shows high ζ, (18). Over the Zimbabwe craton, ζ is mostly low to the north, part of (4), and moderately high to the south, the northern edge of (18). A low, (19), is located directly over the mid-Proterozoic Okwa block, mostly obscured by sedimentary cover.

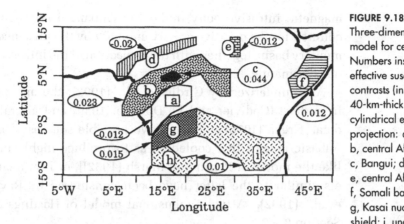

FIGURE 9.18
Three-dimensional magnetic model for central Africa. Numbers inside the map are effective susceptibility contrasts (in SI units) for 40-km-thick polygons; cylindrical equidistant projection: a, Congo basin; b, central African shields; c, Bangui; d, Benue rift; e, central African rift (?); f, Somali basin (shield); g, Kasai nucleus; h, Angola shield; i, undefined on the tectonic map, but the source is required to model the magnetic anomalies. (From Ravat, 1989, with permission)

In *Northwest Africa*, a band of high ζ, (20), extends over the Man shield, (20), (20a), over the Dahomeyan belt, (20b), part of the Tuareg shield, and across the Tibesti inlier, (20d). Highs, (20) and (20a), are clearly associated with the early Proterozoic portion of the Man, or Leo shield, with a relative low, (21), over the Archean terranes to the west. Over the Reguibat shield, ζ is consistently high, (22). Two peaks are evident, one over the Archean terrane to the west, (22a), and one to the east, (22b), over Proterozoic terrane and extending over the Tindouf basin to the north. The saddle between (22a) and (22b) lies over a region of low metamorphic grade. Extension of anomalies (22a) and (22b) over the adjacent basins may mark extension of the exposed terranes into the basement of the basins.

The northern Taoudeni basin is overlain by high ζ extending southward from the Reguibat shield. However, to the south, at the location of thick sedimentary rocks in the Hodh area, ζ shows a substantial low, (23). Extending northeastward from the Taoudeni basin is a series of other basins; values of ζ are low or near average all along this trend, (23), (24), (25), and (26), with subdued relief to the northwest.

In *Northeast Africa–Arabia*, a moderate high, (35), is associated with the Nubian shield, and a moderate low, (36), with the Arabian shield, a puzzling combination, because they are considered to have a common origin and geologic history.

9.3.2 ANOMALY STUDIES

9.3.2.1 Central Africa

Historically, the large anomaly in central Africa, (1) in Figure 9.17, was the first to be detected in satellite data. Several models have been proposed to explain the origin of the anomaly, beginning with that of Regan and Marsh (1982) described in Section 1.2, in which they suggested a highly magnetic intrusive body under the Oubangui sedimentary basin. They accounted for the negative gravity anomaly by the presence of the overlying sedimentary basin, together with the projection of the intrusive body into the higher-density mantle.

As summarized by Girdler et al. (1992), the anomaly was first observed in surface (Godivier and Le Donche, 1956) and aircraft (Green, 1973, 1976) data. Green (1976) suggested two possible sources: a major upwelling of ultrabasic magma that cooled so as to produce highly magnetic minerals [i.e., like the model of Regan and Marsh (1982)], or iron meteoritic material. Models similar to the latter have been considered by Ravat (1989) and Girdler et al. (1992). A two-dimensional model of Hastings (1982) is discussed in Section 8.2.2.

Ravat (1989), guided by a revised susceptibility map for Africa and what was known about the regional geology and tectonics, proposed the model shown in Figure 9.18, assuming a 40-km-thick magnetic layer, whose predicted scalar anomalies are compared to *Magsat* anomalies in Figure 9.19. The model of Figure 9.18 could correspond to an infinite number of depth distributions of susceptibility. One that is in general accord with other data is shown in Figure 9.20. In arriving at this cross section, Ravat (1989) considered evidence from gravity and seismic data indicating a thinned crust and a region of mantle upwarp under the Benue trough and its extensions. Relatively high susceptibilities throughout the crust were assumed in central Africa, and lower susceptibilities in the lower crust under the Congo basin. Lower crustal rocks in the Kasai and Angola shields and the southern Chad basin were assumed to have relatively high susceptibilities, though lower than in central Africa. Note the presence of a high-susceptibility body corresponding to Bangui, after the second suggestion by Green (1976), at 4–7 km (c in Figure 9.20).

Girdler et al. (1992) proposed that the Bangui anomaly is due to the effects of a very large, early Precambrian impact event. Evidence for an impact structure

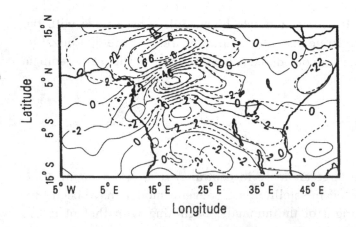

FIGURE 9.19
Observed (continuous) and computed (dashed) *Magsat* anomalies of central Africa generated with the model in Figure 9.18; cylindrical equidistant projection; contour interval 2 nT. (From Ravat, 1989, with permission.)

Central African Provinces Magnetic Model

Along 16°E Longitude

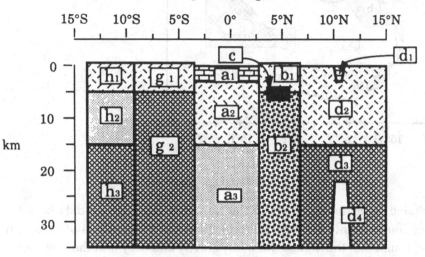

Effective Susceptibilities for Various Crustal Units (in SI Units)

Angola Shield
h_1 = 0.0063 (near-surface low susceptibility granites)
h_2 = 0.0465 (upper crustal high susceptibility granulites characteristic of shields)
h_3 = 0.0628 (high susceptibility lower crustal granulites)

Kasai Shield
g_1 = 0.0063 (near-surface low susceptibility granites)
g_2 = 0.0640 (mafic granulites in both upper and lower crust)

Congo Basin
a_1 = 0.0000 (sedimentary rocks)
a_2 = 0.0063 (near-surface low susceptibility granites)
a_3 = 0.0390 (mafic lower crust)

Central African Shield
b_1 = 0.0126 (near-surface granulites, charnokites, etc.)
b_2 = 0.0754 (mafic granulites in both upper and lower crust)

Bangui
c = 1.0000 (Fe-Ni-rich meteorite or Fe-rich iron formations)

Southern Chad Basin
d_1 = 0.0000 (sedimentary rocks of Benue rift, shown exaggerated)
d_2 = 0.0063 (near-surface low susceptibility granites)
d_3 = 0.0628 (lower crustal mafic granulites)
d_4 = 0.0000 (mantle upwarp or mantle intrusions in the lower crust)

FIGURE 9.20
A possible magnetic depth section along a north–south profile across the magnetic model shown in Figure 9.18. (From Ravat, 1989, with permission.)

FIGURE 9.21
Three-dimensional magnetic model of western and north-central Africa. Numbers inside the map are effective susceptibility contrasts (in SI units) for 40-km-thick polygons; cylindrical equidistant projection: a, Liberian Man shield; b, western Reguibat rise; c, eastern Reguibat rise and eastern Tindouf basin; d, greater Chad basin; X, Taoudeni basin. (From Ravat, 1989, with permission.)

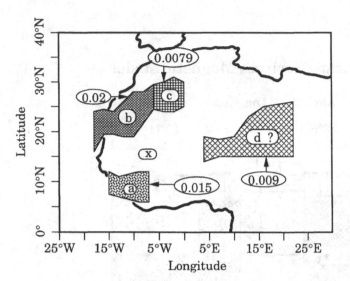

was marshaled from the gradients of the 5-minute topographic data base (in which they found a double-ring structure), from the Bouguer gravity map (in which they found that the highest-intensity gravity anomaly is at the ring center and that the anomalies form part of a circle to the north and west), and from reexamination of the geologic map (in which the anomaly center is close to the Oubangui basin, one of a number of small Precambrian basins). They noted that "the early Precambrian geology gives the impression of a series of small basins within a very large basin with a high in the middle. Such a conglomeration of basins is typical of a large impact structure [S. R. Taylor, personal communication, 1989]."

Girdler et al. (1992) noted that the 800-km diameter of the ring structure indicates the largest impact feature yet found, with an impactor diameter between 80 and 200 km, and that the early Precambrian age is in accord with estimates by Grieve (1987) for the expected size of Precambrian earth impacts. They also called attention to mineralization in the area, consistent with an impact origin: polycrystalline diamonds (carbonados) of crustal origin in an area with no known kimberlite pipes, and numerous iron deposits and heavy minerals, possibly associated with a large iron meteorite.

The basic features of the observed anomaly can be reproduced by a simple disk model. The disk is at 3–7.5 km of depth, is 800 km in diameter, and has magnetization of 10 A/m directed along declination $-18°$ and inclination $25°$. The present-day ambient field has $D = -3°$ and $I = -12°$, so a significant portion of the model magnetization is remanent. The available collateral geologic data, both surface and subsurface, and geophysical data are currently insufficient to determine the validity of the proposed crustal models. Further evidence consistent with the impact hypothesis is given by P. Taylor (personal

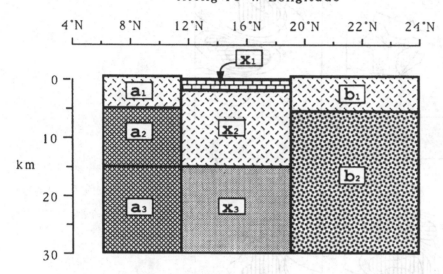

West African Craton Magnetic Model 1
(Constant Thickness Crust)
Along 10°W Longitude

Effective Susceptibilities for Various Crustal Units (in SI Units)

Man Shield
a_1 = 0.0063 (near-surface low susceptibility rocks,
              excluding the iron-formations and other
              short-wavelength features)
a_2 = 0.0440 (uplifted lower crust and/or most of the
              upper crust almost entirely intruded by highly
              magnetic intrusions)
a_3 = 0.0440 (lower crustal mafic granulites)

Taoudeni-Mali Basin
x_1 = 0.0000 (sedimentary rocks)
x_2 = 0.0063 (felsic upper crust)
x_3 = 0.0380 (mafic lower crust)

Western Reguibat Shield
b_1 = 0.0063 (near-surface granitic and metamorphic rocks of
              the Reguibat rise)
b_2 = 0.0580 (uplifted lower crust composed of highly magnetic
              rocks)

Eastern Reguibat Shield and Eastern Tindouf Basin (not shown)
High susceptibility lower crust (0.058) from 15 to 30 km

FIGURE 9.22
A possible magnetic depth section along a north–south profile across the magnetic model shown in Figure 9.21. (From Ravat, 1989, with permission.)

FIGURE 9.23

Model bodies and calculated scalar magnetic anomalies for the West African craton. Bodies in part a represent, from south to north, the Man shield (flecked), the Casamance rift–Taoudeni basalts–Gourma aulacogen corridor (dotted), the Reguibat shield (flecked), and the Canary Island–Essauria rift block (dotted). For parts b–d, see text. Susceptibility contrasts and thicknesses are given in the text. Anomalies are calculated at a 400-km altitude; contour interval 1 nT. Horizontal (vertical) shading emphasizes negative (positive) anomalies. (Reprinted from P. B. Toft, P. T. Taylor, J. Arkani-Hamed, and S. E. Haggerty, Interpretation of satellite magnetic anomalies over the West African Craton. *Tectonophysics, 212,* 21–32, 1992, with kind permission of Elsevier Science NL, Sara Burgerhartstraat 25, 1055 KV Amsterdam, The Netherlands.)

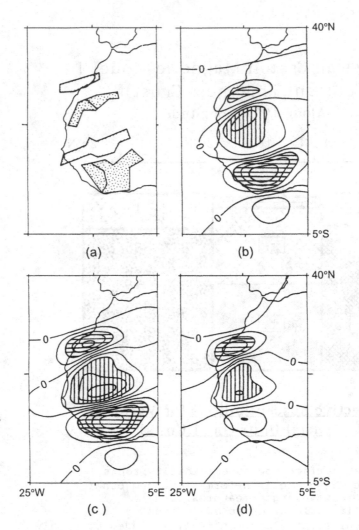

communication, 1997), who calls attention to the work of De and Heaney (1996), who found shock-related deformation features, and Gorshkov et al. (1996), who detected inclusions of native metals, including taenite in carbonados from the Central African Republic. Taenite is a nickle-rich (13–50%) iron-nickel alloy whose only known occurrence is in association with iron meteorites. In any case, the size and amplitude of this magnetic anomaly are unique, suggesting that further study and analysis of the Bangui anomaly region crust are warranted.

9.3.2.2 Northwest Africa

A clear spatial correlation of satellite magnetic anomalies with shields and basins has been noted in northwest Africa (e.g., Hastings, 1982). Ravat (1989) derived the three-dimensional model shown in plan view in Figure 9.21, assuming a 40-km-thick magnetic layer. High susceptibility contrasts are found

(a)

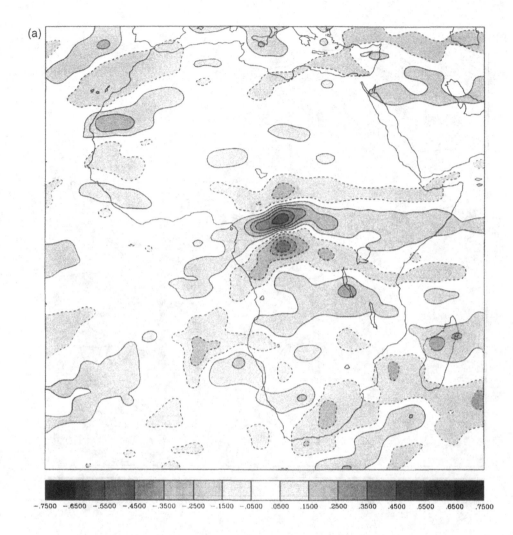

-.7500 -.6500 -.5500 -.4500 -.3500 -.2500 -.1500 -.0500 .0500 .1500 .2500 .3500 .4500 .5500 .6500 .7500

FIGURE 9.24
Magnetization (A/m) in a
40-km-thick layer over the
region 40° S–40° N,
25° W–55° E,
encompassing Africa and its
continental margins.
Negative contours are
dashed. (a) Component in
the direction of the main
field (values range from
−0.44 to 0.53). (*continued*)

in the lower Precambrian–Archean part of the Man and Reguibat shields, in
agreement with the values for ζ in Figure 9.17. One possible cross section is
shown in Figure 9.22, which is guided by the model of Toft and Haggerty (1986).

Toft et al. (1992) investigated a series of models of the shields and basins of
northwest Africa. Based on the work of Toft and Haggerty (1988), they took the
magnetic thickness of the shield areas to be 70 km, claiming (Haggerty, 1986;
Toft et al. 1989) that the continental nuclei has downward thickened roots and
that 70 km is a reasonable depth to the Curie isotherm. Figure 9.23a shows
a plan view of the model bodies, which are assumed to have vertical sides.
The susceptibility contrast in the shields is taken to be 0.01 SI, equivalent to a
magnetization of 0.21–0.24 A/m in the local ambient field. Figure 9.23b shows
the computed scalar field from just the shield areas. The ambient inclination is
35° over the Reguibat shield, resulting in a negative (positive) scalar anomaly
to the north (south).

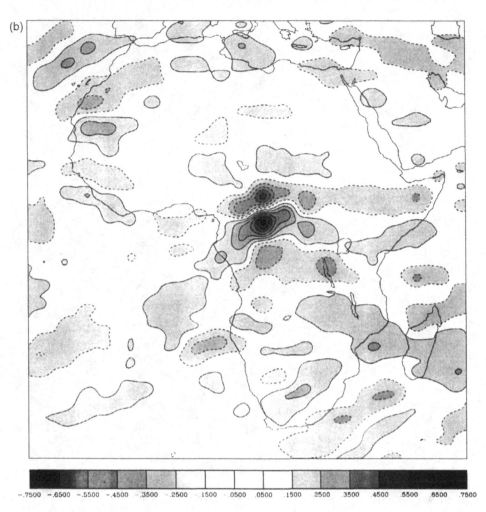

FIGURE 9.24 (*continued*)
(b) Component orthogonal to the main field and in the meridian plane (values range from −0.49 to 0.69). (c) Component orthogonal to the main field and perpendicular to the meridian plane (values range from −0.26 to 0.21). Cylindrical equidistant projection. (Reprinted from K. A. Whaler and R. A. Langel, Minimal crustal magnetizations from satellite data. *Physics of the Earth and Planetary Interiors, 98,* 303–14, 1996, with kind permission of Elsevier Science NL, Sara Burgerhartstraat 25, 1055 KV Amsterdam, The Netherlands.)

The southern positive from the Reguibat shield and northern positive from the Man shield combine to give the positive over the Taoudeni basin. This positive, however, does not adequately reproduce the measured anomaly, which led Toft et al. (1992) to the inclusion of negative susceptibility contrasts in regions thought to be volcanic rifts. The most important such region is associated with the Casamance rift and Gourma aulacogen (Figure 9.23a). Noting the existence of a region of Triassic–Jurassic basaltic dikes and sills and of Cretaceous kimberlite pipes and dikes within the Taoudeni basin between the Casamance rift and Gourma aulacogen, they considered the existence of a structural trough with a magnetization deficiency (Figure 9.23a). A similar trough is located to the north of the Reguibat shield along and inland from the Essauria rift. Both regions of magnetization deficiency are taken to have −0.01 SI susceptibility in a 35-km-thick crust. The resulting scalar anomaly field is shown in Figure 9.23c. The correspondence of the basin anomaly with measurements is now good, but

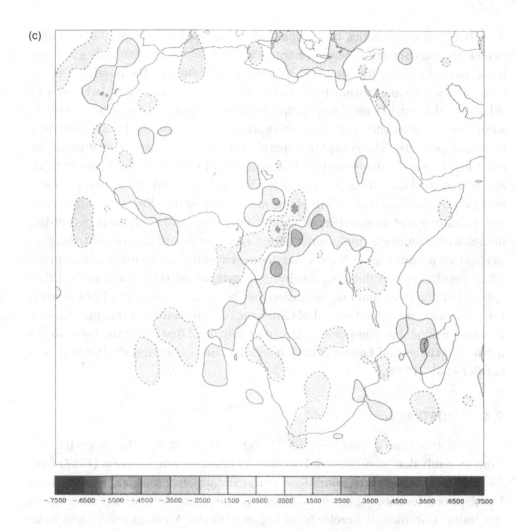

(c)

−.7500 −.6500 −.5500 −.4500 −.3500 −.2500 −.1500 −.0500 .0500 .1500 .2500 .3500 .4500 .5500 .6500 .7500

extending the low-susceptibility region northward could bring the two into better agreement. The amplitude of the model anomaly over the Man shield is thought to be too high.

The contours of Figure 9.23d show the anomaly field from a third model retaining the susceptibility contrasts in the rift and in the Archean segment of the Man shield, reducing the susceptibility contrast of the Reguibat shield to 0.005 SI, and taking the susceptibility contrast of the eastern Man shield to be zero. The Man shield anomaly is then regarded as more satisfactory than that from the model of Figure 9.23c, but the other anomalies are too low in magnitude.

9.3.2.3 Evidence of Large-Scale Remanent Magnetization

Although most inverse solutions assume that the direction of magnetization lies along the ambient main field, several of the forward models described earlier have invoked magnetization in other directions. This is also true for models of some anomalies in Europe, to be discussed in Section 9.5, and also in comparisons of vector measurements with equivalent source models derived from scalar data only (Section 5.8). A further indication of the possible importance of such magnetization is furnished by the minimum-norm inversion of Whaler and Langel (1996), map {80}, reproduced here as Figure 9.24. No assumption is made concerning magnetization direction, and the figure comprises three components. The component parallel to B corresponds to that found by solutions in which magnetization is constrained to be along the ambient field, and it is in good agreement with such solutions. But the other two components, particularly that in the meridian plane, are not negligible. Such a solution does not guarantee that magnetization cannot be purely induced, merely that the model with minimum rms magnetization has a remanent component. Magnetization amplitudes in all three components exceed their formal uncertainties. These results support the view that induced magnetization prevails in the lithosphere but that remanent magnetization may be locally important. For example, the previously discussed model of Girdler et al. (1992) for the Bangui anomaly includes a remanent component estimated at 82% of that along the field. In the model of Whaler and Langel (1996), it is 136%, in a direction similar to that of Girdler et al. (1992).

9.4   AUSTRALIA

Rather than ζ from the SEMM, Figure 9.25a shows the magnetization from an equivalent source model derived by Mayhew and Johnson (1987) from *Magsat* data. The model formalism automatically accounts for the Southern Hemisphere location. Figure 9.25b shows a simplified structural map of Australia. Continental development began with the Archean nucleus in western Australia, exposed in the metamorphic Yilgarn and Pilbara (PI) provinces. Other Archean rocks are found in smaller, isolated locations within, or basement to, younger orogenic belts [e.g., the Gawler craton (Gw) at the southern coast]. Both the Yilgarn and Pilbara provinces are among the major granite-greenstone provinces recognized by Condie (1981). Within the Yilgarn province is a division between a western high-grade gneiss terrane and an eastern low-grade granitoid-greenstone terrane. The Pilbara province generally exhibits low-grade metamorphism tending to amphibolite facies around granitoid batholiths. A major E–W mobile zone is recognized between the Yilgarn and Pilbara provinces.

In Figure 9.25, a solid line, called the Tasman line, is drawn along the edge of proven Precambrian basement, thought to be the location of "earliest Phanerozoic continental breakup" (Powell, 1984). To its west is an area, e on Figure 9.25b, where sedimentary rocks thicken eastward on the craton toward the Tasman line. An indentation of this region into the craton is identified as 1 in the figure. Mayhew and Johnson (1987) noted that the magnetization contours seem to follow this boundary, including the indentation, with more subdued magnetization with a mainly N–S trend to the east, and higher magnetization with an E–W trend to the west.

A region of high magnetization extends to this boundary from the Gawler block (Gw) with an apparent saddle, or relative low, over the Adelaide zone (Ad). Mayhew and Johnson (1987) noted that the Adelaide zone is a region of Precambrian rifting and of present-day seismic activity, anomalous electrical conductivity, and heat flow. They postulated a local rise in the Curie isotherm.

The high previously noted by Mayhew et al. (1980) between the Gawler block and Hammersley basin (H) is defined quite clearly in Figure 9.25a, and they pointed out that this suggests a connection beneath the sediments not apparent in other data. Other associations noted are positive magnetization over the Musgrave block (Mu) and Amadeus basin (Am), and magnetization lows over the Officer basin, the Arunta block (Ar), and the southern part of the Victoria River basin (V).

Figure 9.26 shows a crustal section along the line indicated in southeastern Australia in Figure 9.25b. It extends from the Gawler block eastward to the coast. In the top portion of the figure, the long-dashed line is the vertical integral of magnetization from Figure 9.25a. Heat flow data near the line are indicated by open circles, with boxes giving the range of groups of heat flow measurements projected to the section. To the west of the Tasman line, variations in the magnetization integral are clearly anti-correlated with heat flow: highest over the Gawler block, where heat flow is lowest, lower at the Adelaide zone, where heat flow is highest, and increasing slightly toward the Tasman line, where heat flow seems to have decreased. A relatively deep Curie isotherm is suggested for the Gawler block.

East of the Tasman line, the magnetization integral remains low, with some undulation. A geotherm derived from xenoliths and the range of heat flow measurements in box "b" at the top of the figure indicate high heat flow and a shallow Curie isotherm of about 12 km. Mayhew and Johnson (1987) noted, though, that if the very low magnetization integral in the Murray basin were entirely due to a shallow isotherm, they would expect "thermal manifestations, which are not observed." They concluded that the upper crustal magnetization in segment 2 of the section is less than in the adjacent segment 3.

A detailed model for a crustal section in the Yilgarn block, based on the magnetization distribution of Figure 9.25a, seismic refraction results, and geothermal data, has been proposed by Mayhew et al. (1991). Figure 9.27 shows the location of the section relative to the Yilgarn block and to the magnetization distribution. The model section, shown in Figure 9.28, is based on the preferred seismic section of Mathur (1974), who also fit long-wavelength gravity data by assigning model densities to the seismic layers. Layer velocities and densities

(a)

(b)

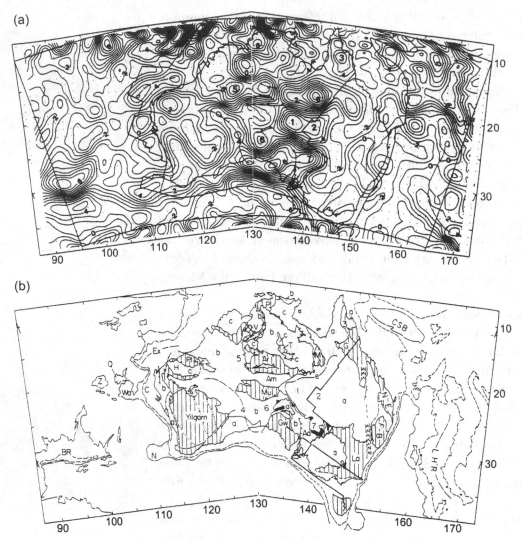

FIGURE 9.25

(a) Equivalent layer magnetization distribution obtained by inversion of *Magsat* total-field anomaly data for Australia; units are tenths of an ampere per meter; contour interval 0.1 A/m. Dots indicate dipole source locations. The distribution represents apparent magnetization contrast in a layer of (arbitrary) thickness 40 km, the top of which is at the earth's surface. The heavy line onshore is the Tasman line, as drawn by Powell (1984); the heavy line offshore is the ocean–continent boundary of Veevers (1984). (b) Sketch map of major structural units in Australia for comparison with Figure 9.25a. The section line of Figure 9.26 is indicated. Onshore symbols: NEFB, New England fold belt; La, Lachlan foldbelt outcrop area; G, Georgetown inlier; M, Mt. Isa block; e, locus of rapid thickening of sediments to the east; Ad, Adelaide zone; Gw, Gawler block; Mu, Musgrave block; Am, Amadeus basin; Ar, Arunta block; T, Tenant Creek block; V, Victoria River basin; P, Pine Creek block; Pl, Pilbara block; H, Hammersley basin. Short-dashed lines indicate sub-province boundaries of Yilgarn indicted by Gee (1979). Cross-hatched lines indicate outcrops of Cambrian flood basalts outlining Victoria River basin. Small blackened areas are outcrops of Precambrian rocks inferred to be related to Gawler rocks. Lines of barbs in eastern Australia indicate a mid-Paleozoic continent-edge volcanic arc (Powell, 1984). Offshore symbols: CSB, Coral Sea basin; LHR, Lord Howe rise; Ex, Exmouth plateau; Wa, Wallaby plateau; Z, Zenith plateau; Q, Quokka rise; C, Carnarvon terrace; N, Naturaliste plateau; BR, Broken ridge; a, b, and c indicate ages of sediment cover (Mesozoic/Cenozoic, late Proterozoic/Paleozoic, early to middle Proterozoic, respectively); 2,000-m and 4,000-m isobaths offshore. Albers equal-area projection. (Reprinted from M. A. Mayhew and B. D. Johnson, An equivalent layer magnetization model for Australia based on *Magsat* data. *Earth and Planetary Science Letters*, *83*, 167–74, 1987, with kind permission of Elsevier Science NL, Sara Burgerhartstraat 25, 1055 KV Amsterdam, The Netherlands.)

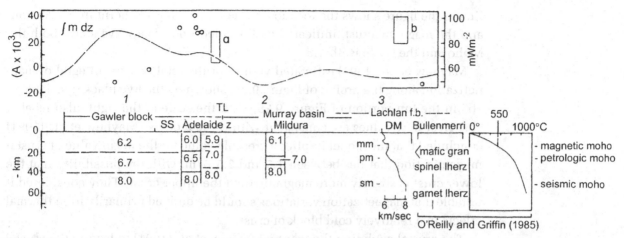

FIGURE 9.26

Crustal section across Gawler block, Adelaide zone, Murray basin, and Lachlan foldbelt. Crustal seismic layering from Finlayson et al. (1984), Schackleford (1978), and Branson et al. (1976). P-wave velocities indicated. At lower right are geotherm and lower-crust/upper-mantle lithologies inferred from xenolith studies (O'Reilly and Griffin, 1985). DM is a projection of the Dartmouth–Marulan seismic line of Finlayson et al. (1979) to the section line. TL is the Tasman line. SS is the Stewart shelf. The dashed line above is the profile of the vertical integral of magnetization inferred from Figure 9.25a (scale at left). Open circles are heat flow determinations (Cull, 1982) near the section line (scale at right). The box labeled "a" encloses a group of heat flow determinations along craton margin; the box labeled "b" is another grouping of values near the section line within the Lachlan foldbelt. Numbers 1–3 refer to three probably distinct crustal provinces. Vertical exaggeration is fivefold. (Reprinted from M. A. Mayhew and B. D. Johnson, An equivalent layer magnetization model for Australia based on *Magsat* data. *Earth and Planetary Sciences Letters*, **83**, 167–74, 1987, with kind permission of Elsevier Science NL, Sara Burgerhartstraat 25, 1055 KV Amsterdam, The Netherlands.)

(in parentheses) are shown in the figure. The particular model adopted from Mathur (1974) assumes an eclogite upper mantle; the line labeled "a/g" is the location of the inferred amphibolite–granulite transition of Glickson and Lambert (1976). The top portion of the figure shows the vertical integral of magnetization contrast, labeled "a" (scale on left), and values of measured heat flow (open circles, scale to far right). They noted that if the variation is the result of differences in the thickness of a magnetic layer, as opposed to lateral variations in magnetization, then the layer must be thinnest in the east (to the right of the section), thicken westward to the magnetization contrast maximum, and then thin again toward the west. In the figure, the Moho is assumed to be the interface between layers with seismic velocities of 7.49 and 8.39 km/s, with rocks below the boundary assumed nonmagnetic. A 550° Curie isotherm is assumed. To the east, the Curie isotherm depth is obtained from the western Australia geotherm of Sass and Lachenbruch (1979), shown to the right in the lower part of the figure; while to the west, the 550° isotherm is assumed to trend to a level determined by an oceanic isotherm consistent with the Indian Ocean age at the Australian west coast, shown to the left in the lower part of the figure. A dashed line in the figure shows the location of the model 550° isotherm in the section, and the magnetic crust, indicated by the ruled lines, is always above both the Moho and the Curie isotherm.

Mayhew et al. (1991) estimated values for the total vertical integral of magnetization along the profile of Figure 9.28, shown by the five black dots labeled "b" in the top portion of Figure 9.28, with the scale at the right, also labeled "b." Heat flow values are consistent with these models. Mayhew et al. (1991) concluded that the model implies a credible value for the mean value of crustal magnetization that lies between 1.5 and 2.5 A/m, with the possibility that the lower crust is slightly more magnetic than the upper crust. They considered it notable that magnetization variations should be derived primarily from thermal effects in a relatively cold block of crust.

The crustal evolution thought by Mayhew et al. (1991) to lead to the crustal section depicted in Figure 9.28 is that the 7.5-km/s layer under the Yilgarn block corresponds to the granulite-phase mafic rocks in the crust–mantle transition zone in the model of O'Reilly and Griffin (1985), described in Section 7.2.1. This is thought to be a relic of a thermal pulse that occurred at the time of Jurassic rifting (Johnson, Powell, and Veevers, 1980) between Australia and India, causing uplift in the western Yilgarn and exposing successively deeper crustal levels toward the continental margin. It is thought that the excess heat is now nearly dissipated and that as the temperature dropped the kinetics of mineral phase transitions slowed (Griffin and O'Reilly, 1987a), so that the mafic granulite layer persists. They hypothesized that the 7.5-km/s-to-8.4-km/s boundary is the granulite–eclogite transition, rather than the petrologic crust–mantle boundary, defined by O'Reilly and Griffin (1985) as the depth where spinel lherzolite becomes dominant. The bottom of the magnetic layer in the eastern Yilgarn would then be due to the disappearance of magnetite at this boundary, whereas in the

FIGURE 9.27
(a) Outcrop area (ruled) of Yilgarn province, southwestern Australia, with boundary faults indicated; gn, western Yilgarn gneiss domain (Gee, 1979). Bar with dots indicates position of section profile. Dashed lines offshore are 1,000- and 3,000-m isobaths. (b) Apparent magnetization contrast from Figure 9.25a; units are tenths of an ampere per meter. (Reprinted from M. A. Mayhew, P. J. Wasilewski, and B. D. Johnson, Crustal magnetization and temperature at depth beneath the Yilgarn block, Western Australia inferred from *Magsat* data. *Earth and Planetary Sciences Letters*, 107, 515–522, 1991, with kind permission of Elsevier Science NL, Sara Burgerhartstraat 25, 1055 KV Amsterdam, The Netherlands.)

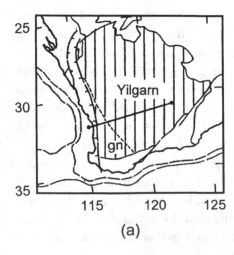

(a)

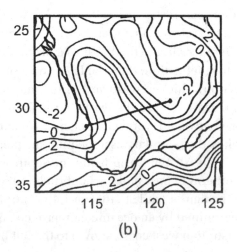

(b)

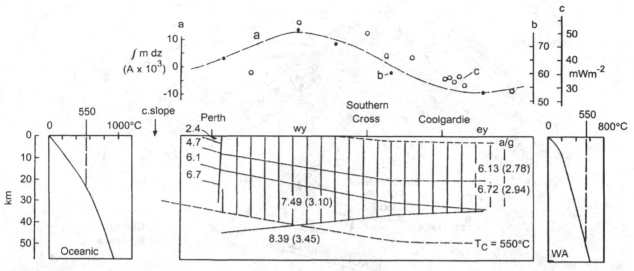

FIGURE 9.28

Crustal section across Yilgarn Province of western Australia. Seismic refraction layering, velocities, and densities (values in parentheses) from Mathur (1974). The symbol a/g indicates the amphibolite–granulite transition of Glickson and Lambert (1976). [Geotherm of western Australia (WA) from Sass and Lachenbruch, 1979; reprinted with kind permission of Academic Press.] The Curie isotherm (550° assumed) must shallow west to the level given by oceanic geotherm. Magnetic crust (*continued*)

FIGURE 9.28 (*continued*)

indicated by vertical ruling. The dashed line labeled "a" is the vertical integral of magnetization contrast found from the magnetization model of Figure 9.25a. Dots labeled "b" are computed values for the vertical integral of magnetization (scale at right) for the two-layer case discussed in the text. Circles labeled "c" are heat flow values from Cull (1982). Vertical exaggeration fivefold; c.slope indicates position of continental slope. (Reprinted from M. A. Mayhew, P. J. Wasilewski, and B. D. Johnson, Crustal magnetization and temperature at depth beneath the Yilgarn block, Western Australia inferred from *Magsat* data. *Earth and Planetary Science Letters, 107,* 515–22, 1991, with kind permission of Elsevier Science NL, Sara Burgerhartstraat 25, 1055 KV Amsterdam, The Netherlands.)

western Yilgarn the Curie isotherm, located above the transition, would be the lower limit of the magnetic crust.

9.5 EUROPE

9.5.1 OVERVIEW

The tectonics of the East European platform are summarized in Figure 9.29. Two shield areas are recognized: Baltic and Ukrainian. Archean granite-greenstone terrane is found in the Kola and Karelian provinces of the Baltic shield and in much of the Ukrainian shield. The Voronezh uplift, or massif, under shallow cover, comprises partly reworked, fault-bounded Archean blocks with surrounding iron-rich early Proterozoic synclinorial bands. In other regions of the platform, the cover (commencing deposition in the late Proterozoic) is 2–4 km thick, with increased thickness in the main depressions, attaining a maximum depth of 20 km in the North Caspian synclinorium.

Figure 9.30 shows ζ for the East European platform. High values (>1.1) occur along a northwest–southeast trend from the Voronezh high, (1), past the Baltic Sea, (2), and extending to the edge of the craton, along a southwest–northeast trend encompassing the Volga–Urals massif, (3), and Glasov synclinorium, (4), and over the northern Karelian province, (5). A high, (1), seems

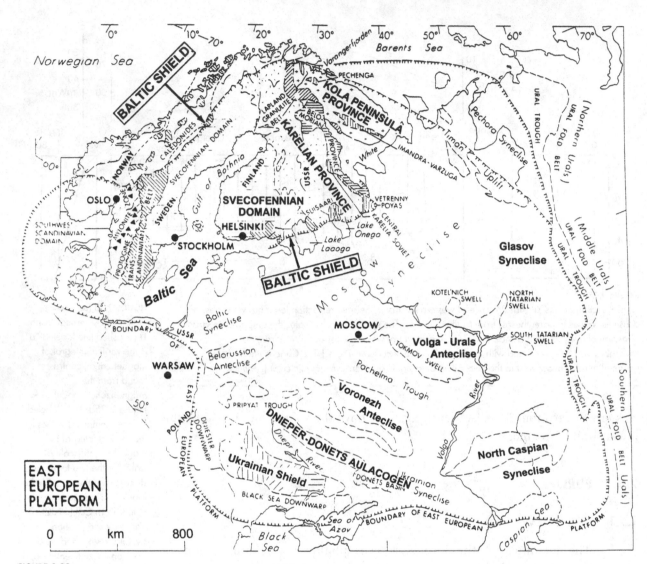

FIGURE 9.29

Main geologic outline and divisions of the East European platform. Locations referred to in text are in bold print. (Adapted from Goodwin, 1991, with permission of Harcourt Brace and Company, Ltd.)

clearly associated with the Voronezh uplift, though not located at its center. This feature is discussed in some detail in Section 9.5.2. No geologic feature is clearly associated with the positive, (2). The basement under features (3) and (4) is thought to comprise Archean and Proterozoic blocks, but is mostly hidden by Phanerozoic cover.

The regions of high ζ are separated by trends of low ζ. One such trend, including (6), (7), and (8), transects the platform in a northwest–southeast direction. A low, (6), and its extension to the northwest overlie the Svecofennian domain, centered particularly in its southeastern segment, as discussed by Ravat et al. (1993) and summarized in Section 9.5.2. Values of ζ are also low along the southwest–northeast trend defined by features (7), (9), and (10), all under Phanerozoic cover. The trends of features (7) and (9) are very similar to

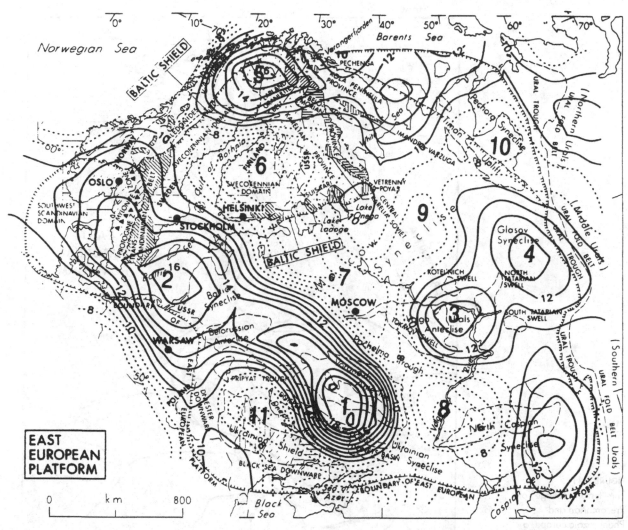

FIGURE 9.30
Contours of ζ in $(10 \cdot \mathrm{SI} \cdot \mathrm{km})$ overlying Figure 9.29.

the nearly underlying trends of graben-like troughs, or aulacogens, noted by Zonenshain, Kuzmin, and Natapov (1990).

No expression in ζ is apparent from the Dnieper–Donets aulacogen, nor is there an anomaly showing close correspondence with the Ukrainian shield. A low, (11), overlies part of that shield, perhaps being associated with the early Proterozoic Odessa–Kanev (or Odessa–Belotserkovsky) block. (The correspondence is more apparent when ζ contours are overlain on Figure 2 of Zonenshain et al., 1990.)

A broad tectonic map of western Europe is shown in Figure 9.31, with contours of ζ shown in Figure 9.32 for comparison. (Note that magnetization models of this region have also been derived by Nolte and Hahn, 1992.) The evolution of Europe is discussed in detail by Blundell, Freeman, and Mueller (1992) and Ziegler (1988). The Tornquist–Teisseyre zone (TTZ), from 50 to 90 km wide,

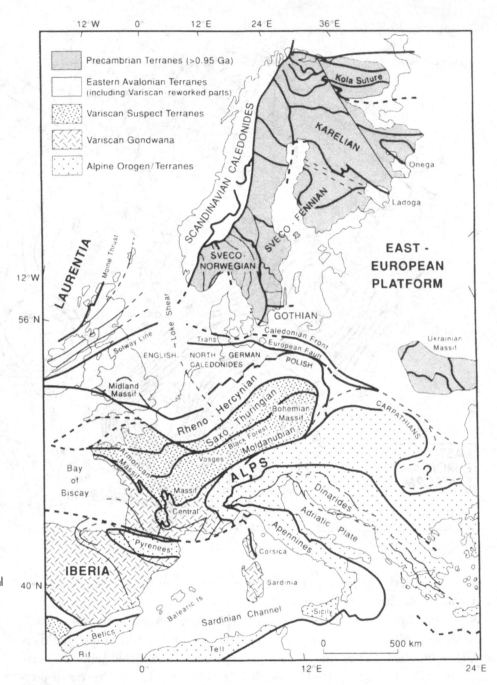

FIGURE 9.31
The "terrane collage" of Precambrian and Phanerozoic Europe, a simplified sketch. Sutures and orogenic fronts are shown as bold lines, internal borders as thin lines or thin broken lines. (From Berthelsen, 1992; reprinted with kind permission from the European Science Foundation.)

is a fundamental boundary between the younger (Paleozoic) central European platform and the Precambrian East European platform. Important differences between the crust on either side of the line can be discerned in the maps of crustal thickness and mantle heat flow shown in Figures 9.33 and 9.34. The crust is generally thinner (30–40 km in thickness) and hotter (50–80 mW/m^2)

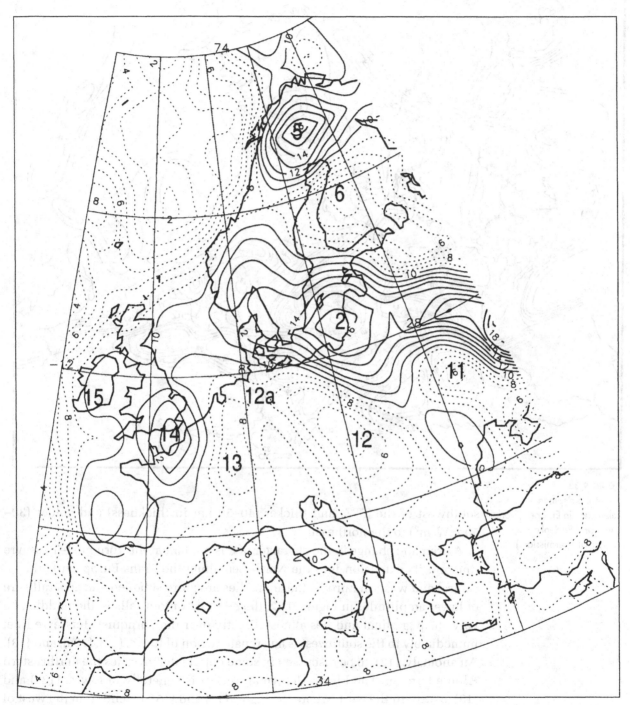

FIGURE 9.32
Contours of ζ in (10· SI · km) for comparison with Figure 9.31.

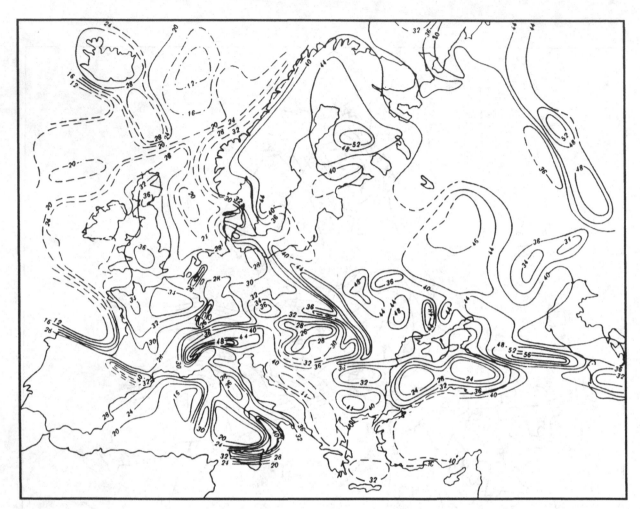

FIGURE 9.33
The Moho depth in kelometers in Europe. (From Giese and Pavlenkova, 1988, with permission.)

southwest of the TTZ, and thicker (40–50 km in thickness) and colder (30–50 mW/m²) to the northeast.

A dramatic change in the character of ζ is evident at the Tornquist–Teisseyre line and its extension into the North Sea along the Trans-European fault. In central and western Europe, the anomalies are more subdued and generally are of lower amplitude. In Figure 9.32, the $\zeta = 1.1$ contour falls in the middle of a very steep gradient and lies almost directly over the Tornquist–Teisseyre line. Immediately to the southwest is an arcuate region of low ζ, (12), (12a), and (13). An anomaly, (12), to be discussed in some detail later, overlies the northeastern Alpine terrane. The Rheno-Hercynian region is generally of low ζ, (12a) and (13), except in the west, where the highs, (14) and (15), extend from the west of France to Ireland.

In part, variations of ζ in Europe seem to reflect variations in heat flow and/or crustal thickness. As seen in Figures 9.33 and 9.34, variations in both crustal

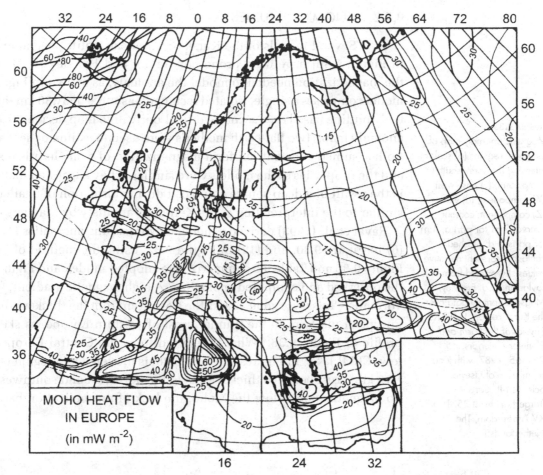

MOHO HEAT FLOW
IN EUROPE
(in mW m^{-2})

FIGURE 9.34
Estimated heat flow at the Moho. (Reprinted from V. Cermak, Crustal temperature and mantle heat flow in Europe. *Tectonophysics, 83,* 123–42, 1982, with kind permission of Elsevier Science NL, Sara Burgerhartstraat 25, 1055 KV Amsterdam, The Netherlands.)

thickness and heat flow are subdued to the east of the Tornquist–Teisseyre line. However, variations in both are noted in the region of two anomalies, (1) and (11). In particular, the crustal thickness shows local variability beneath the low, (11), with a trough between anomalies (1) and (11) and slightly thinner crust under (11) than under (1). Also, the heat flow is marginally greater at the location of (1) compared with (11). Heat flow in the region of the high, (2), also is somewhat greater than under nearby anomalies.

In central and western Europe, anomaly (12), a low, appears associated with a distinct region of thinned crust and high heat flow, especially in its southern part, the Alpine terrane. Again, a low, (13), is located over thinned crust and a region of high heat flow. Between features (12) and (13), under (12a), ζ is higher, the crust is slightly thicker, and, especially, heat flow is lower. Anomaly (14), a high, and the nearby anomaly, (15), overlie a region of lower heat flow and thicker crust. Off the west coast of Italy, ζ is moderately low in a region of very high heat flow and thin crust.

9.5.2 ANOMALY STUDIES

Several models of the Kursk or Voronezh uplift [magnetic anomaly (1) in Figure 9.30] have been proposed. Taylor and Frawley (1987) investigated the magnetization needed to model the anomaly, as shown in Figure 9.35. The model comprises a single, isolated rectangular source body 30 km thick, with the body outline based on a second-vertical-derivative map derived from the scalar anomaly map. Two flat-earth methods were used to estimate the magnetization. Both assumed constant, uniform magnetization of a single prism source body, and the properties of the models were similar. They suggested that the anomaly is the result of a deep source emplaced during formation of a failed aulacogen, similar to the interpretation of the Kentucky anomaly by Mayhew et al. (1982).

Ravat et al. (1993) derived a model encompassing anomalies (1), (2), (5), and (6) in Figure 9.30, and (12) in Figure 9.32, using the method of Section 8.2.3. Figure 9.36 and Table 9.5 summarize the adopted model, and Figure 9.37 shows the observed and computed anomalies. A high magnetic gradient, following the Tornquist–Teisseyre zone, marks the transition from the thicker crust of the East European platform to the thinner crust of western Europe, as sketched in the profile of Figure 9.38. What is referred to as the Central European magnetic low (CEML), "a" in Figures 9.36 and 9.38, is modeled as a region of thinned (by ~10 km) crust between thicker crust to the northeast and southwest. Assuming no magnetization below the Moho, the lack of magnetization between 32 and

FIGURE 9.35
Magnetic anomaly map of the Kursk region: (a) scalar intensity; (b) north–south (X) component; (c) east–west (Y) component; (d) vertical (Z) component; contour interval 2 nT for part a, 4 nT for parts b–d. Dashed contours represent negative values. (Reprinted from P. T. Taylor and J. J. Frawley, *Magsat* anomaly data over the Kursk region, U.S.S.R., *Physics of the Earth and Planetary Interiors, 45,* 255–65, 1987, with kind permission of Elsevier Science NL, Sara Burgerhartstraat 25, 1055 KV Amsterdam, The Netherlands.)

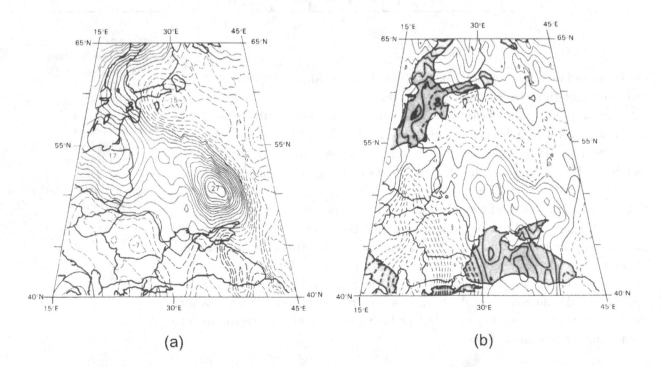

(a)   (b)

40 km is modeled as a negative contrast with the surrounding regions. Crustal thickness estimates are based on the crustal thickness map of Meissner, Wever, and Fluh (1987), which is almost identical with Figure 9.33.

Source "b" in Figure 9.36 is at the location of the Voronezh uplift, where the crust is 40–45 km thick. Ravat et al. (1993) pointed out that seismic studies indicate a high degree of inhomogeneity in the form of discrete high- and low-velocity zones between depths of 6 and 30 km. The Voronezh uplift is a region of immense Proterozoic iron ore deposits, resulting in one of the largest surface magnetic and aeromagnetic anomalies in the world, called the Kursk magnetic anomaly (KMA). They noted that the KMA and satellite anomalies have the same geographic orientation and possibly similar directions of magnetization, but they assumed that the KMA does not produce a significant magnetic contribution at satellite altitude. Ravat et al. (1993), however, noting that the iron formations have large lateral and depth extents, and that Dmitriyevskiy, Afanas'yev, and Frolov (1968) reported mafic, serpentinized ultramafic, gneissic, and amphibolitic upper crustal basement rocks, investigated the possible contributions of those basement rocks to the *Magsat* magnetic anomaly. Aeromagnetic data were used to outline the extent of this source, and its susceptibility (1.25 SI units) was estimated from that of the Kursk iron formations and neighboring rocks (Dmitriyevskiy et al., 1968). The result, source "c" in Figure 9.36, accounts for 25–50% of the observed *Magsat* anomaly.

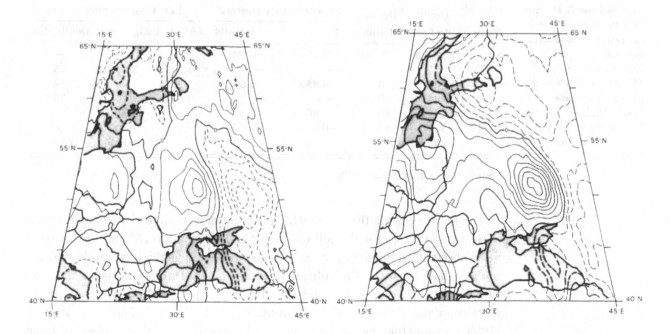

FIGURE 9.36
Areal view of the regional magnetic models of Ravat et al. (1993). a, Central European *Magsat* low (CEML); b, Voronezh region source; c, Kursk iron formations and neighboring rocks; d, origin uncertain; e, central Svecofennian subprovince; f, origin uncertain; g, deep crustal parent body of Kiruna magmatic iron ore. The vertical extents of the sources and their susceptibilities and remanent magnetizations are given in Table 9.5. TT, Tornquist–Teisseyre tectonic zone; LBB, Ladoga–Gulf of Bothnia zone. The thin continuous line in the Baltic shield is the boundary between Archean and lower Proterozoic domains; the thin dashed line is the outline of the central Svecofennian subprovince. (Reprinted from D. N. Ravat, W. J. Hinze, and P. T. Taylor, European tectonic features observed by *Magsat*, *Tectonophysics*, 220, 157–73, 1993, with kind permission of Elsevier Science NL, Sara Burgerhartstraat 25, 1055 KV Amsterdam, The Netherlands.)

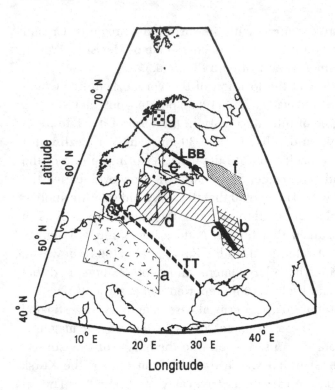

Table 9.5. *Magnetic contrasts and vertical extents of model sources*

Source[a]	Depth to top	Depth to bottom	Susceptibility contrast		Remanent magnetization		
			SI units	cgs units	(A/m)	Inclination	Declination
a	32 km	40 km	−0.086	−0.007	—	—	—
b	5	25	—	—	3.2	20°	25°
c	0	3	0.850	0.067	—	—	—
d	10	30	—	—	0.6	20°	25°
e	30	40	−0.86	−0.007	—	—	—
f	32	40	−0.088	−0.007	—	—	—
g	20	40	0.113	0.009	—	—	—

[a]Sources a–g as in Figure 9.36.

The source, "b," in the Voronezh Uplift region is modeled by remanent magnetization of 3.2 a/m with inclination 20° and declination 25° and Q estimated to be about 10 in the depth range 5–25 km based on geologic grounds. During the mid-late Devonian, the volcanic Donetz graben (Dnieper-Donets aulacogen in Figure 9.29) was formed and the shield divided by the updoming of the Ukrainian and Voronezh highs (Beloussov, 1981; Ziegler, 1988). Ravat et al. (1993) consider that the ambient magnetic field of that period might be imprinted on the rocks experiencing magmatic activity. Paleomagnetic poles from Irving

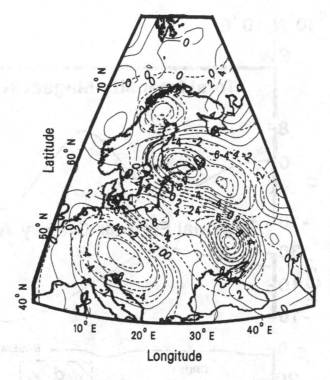

FIGURE 9.37
Comparison of observed (continuous) and computed (dashed) *Magsat* anomalies. The computed anomalies were generated from models in Figure 9.36 and Table 9.5; contour interval 2 nT. (Reprinted from D. N. Ravat, W. J. Hinze, and P. T. Taylor, European tectonic features observed by *Magsat. Tectonophysics, 220,* 157–73, 1993, with kind permission of Elsevier Science NL, Sara Burgerhartstraat 25, 1055 KV Amsterdam, The Netherlands.)

(1977) predict an inclination and declination of about 20° and 48°, respectively, compatible with the model magnetization best matching the observations.

Source "d" is also modeled by remanent magnetization. With direction taken to be the same as for source "b"; a magnitude of 0.6 A/m reproduces the measured anomaly. No geologic evidence is found to justify (or refute) this choice of magnetization. Attempts to reproduce this positive, and associated gradients, by varying sources "a" and "e" were not satisfactory because they required unrealistically extended regions of reversely magnetized sources. Control of the depth extent of source "d" is considered lacking.

The magnetic low over the Gulf of Finland is assumed to be due to lower-than-average magnetization in the rocks of the central portion of the Sveco-fennian block. According to Ravat et al. (1993), the boundaries of the source region, "e," were chosen to coincide with the transition to volcanic belts to the north and south. The rocks of the source region were extensively reworked during its accretion to the Baltic shield at about 1.9 Ga. According to Goodwin (1991), low-pressure amphibolite-phase metamorphism is dominant. Source region "f" also requires low magnetization, although the geologic basis for this magnetization is unclear. This is a region of numerous Phanerozoic cross-cutting rifts and aulacogens.

Ravat et al. (1993) modeled the magnetic high, (5), in the north of Scandi-navia, near Kiruna, with a large, deep source body, "g." However, they pointed

FIGURE 9.38
Profile of geophysical quantities and crustal components across the principal European magnetic trend. TT, Tornquist–Teisseyre tectonic zone; LBB, Ladoga–Gulf of Bothnia zone. Magnetic sources a, d, and e are the same as in Figure 9.36 and Table 9.5. The long-wavelength free-air gravity anomalies are from 50° × 50° spherical harmonic gravity models (GEM T-3S) (Lerch et al., 1992) and are computed at the earth's surface. Basement depths are from Giese and Pavlenkova (1988). Moho depths are from Cermak (1979). (Reprinted from D. N. Ravat, W. J. Hinze, and P. T. Taylor, European tectonic features observed by *Magsat. Tectonophysics, 220,* 157–173, 1993, with kind permission of Elsevier Science NL, Sara Burgerhartstraat 25, 1055 KV Amsterdam, The Netherlands.)

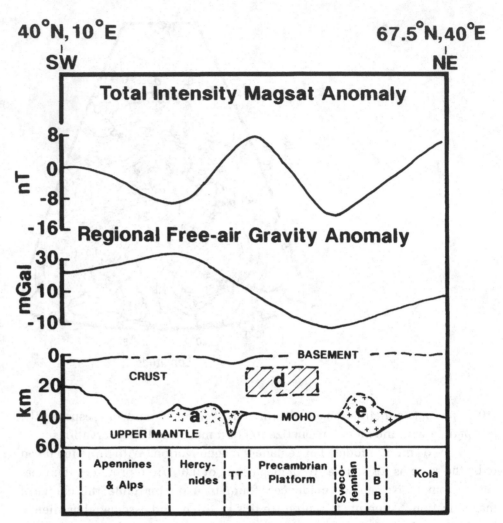

out that this region of the shield is a large metallogenic province. Locally, the Kiruna region is the location of large stratabound/stratiform iron ore deposits in a volcanic/magmatic environment. The iron ore is predominantly magnetite, with 57–71% iron, and is believed to have been derived from magmatic differentiation of deep crustal sources. They suggested that the satellite anomaly may be a combination of near-surface and deep-seated rocks.

An alternative model for anomalies (12), (12a), and (2) in Figure 9.32 has been offered by Taylor and Ravat (1995). In this model they investigated the possibility of reverse magnetization as the source of the negative anomaly (12). Figure 9.39 shows the adopted model in plan view, and Table 9.6 summarizes the model parameters together with those of Ravat et al. (1993). Comparison of parameters should be done with some caution, because the body shapes and areas are not identical in the two models. The depth and thickness of body

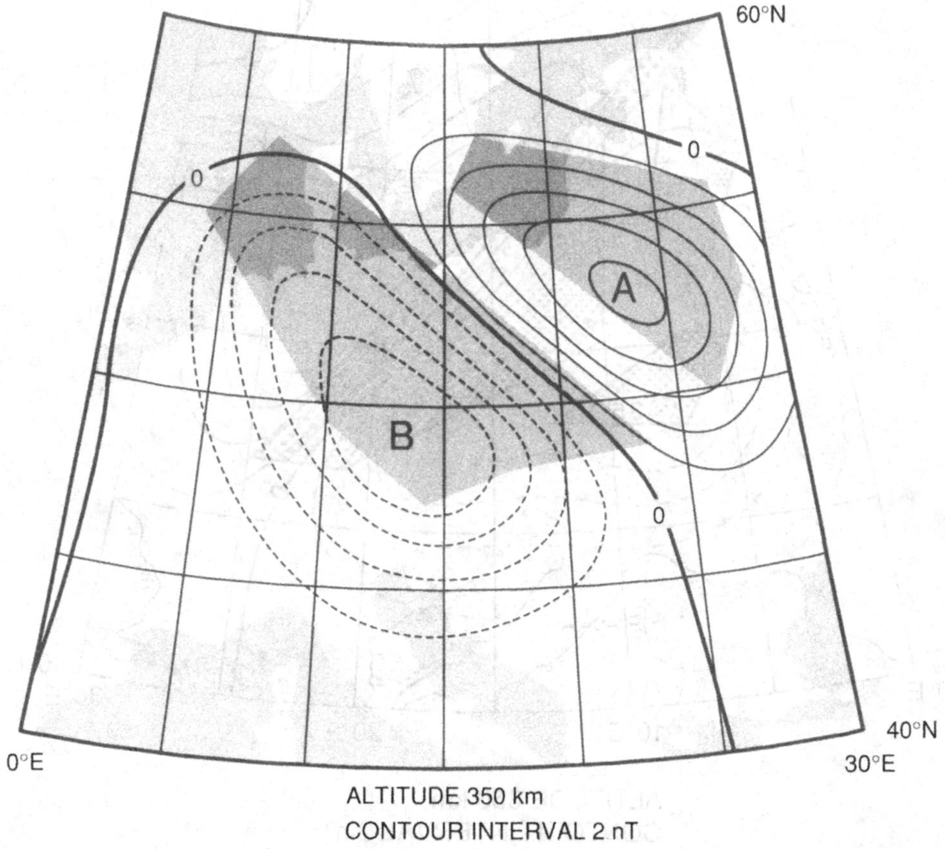

FIGURE 9.39
Satellite altitude synthetic anomaly field produced by modeling observed field at a 350-km altitude. Shaded polygons represent the source bodies. The zero contour is bold; positive contours are solid; negative contours are dashed; contour interval 2 nT; altitude 350 km. Magnetization for both the positive block of the East European platform (labeled A) and the central European block (labeled B) is 3 A/m. The central European block was assigned a reversed remanent magnetization, as summarized in Table 9.6. (Reprinted from P. T. Taylor and D. Ravat, An interpretation of the *Magsat* anomalies of central Europe. *Journal of Applied Geophysics*, 34, 83–91, 1995, with kind permission of Elsevier Science NL, Sara Burgerhartstraat 25, 1055 KV Amsterdam, The Netherlands.)

Table 9.6. *Comparison of Central Europe magnetization models*

Taylor and Ravat (1995)		Ravat et al. (1993)	
Anomaly designation	Model	Anomaly designation	Model
A (positive)	Induced magnetization 3 A/m between 0 and 10 km Dip (I) = 65° Declination (D) = 0°	d	Remanent magnetization 0.6 A/m between 10 and 30 km Dip (I) = 20° Declination (D) = 25°
B (negative)	Remanent magnetization 3 A/m between 0 and 10 km Dip (I) = −50° Declination (D) = 180°	?	Induced magnetization −0.86 SI between 32 and 40 km

FIGURE 9.40
Model and computed field from the model of Pucher and Wonik (1997). The important parts of the *Magsat* anomaly are caused only by an ensemble of source bodies within the Palaeozoic central European block, assuming a constant direction of magnetization of $D = 230°$ and $I = -15°$. The intensities of magnetization vary between DIM = 75 kA and 1 kA [DIM: intensity of magnetization (A/m) × thickness (m)], and they are indicated within the polygonal areas. The northeastern model body with $D = 0°$, and $I = 65°$ is not important in this context. (Reprinted from R. Pucher and T. Wonik, Comment on the paper of Taylor and Ravat, "An interpretation of the *Magsat* anomalies of Central Europe." *Journal of Applied Geophysics, 36,* 213–16, 1997, with kind permission of Elsevier Science NL, Sara Burgerhartstraat 25, 1055 KV Amsterdam, The Netherlands.)

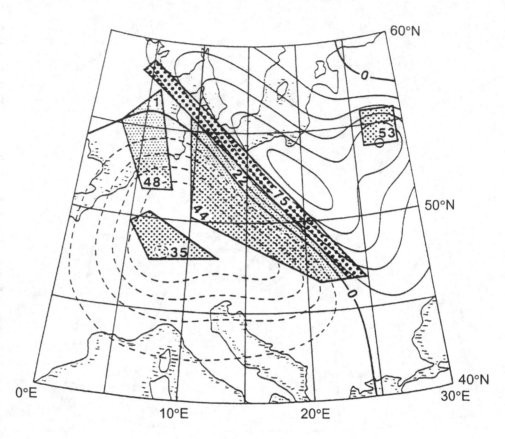

ALTITUDE 350 km
CONTOUR INTERVAL 2 nT

B is chosen to correspond with the midcrustal low-velocity layer of Guterch, Grad, and Perchuc (1986). Consideration of remanent magnetization is based on several grounds:

1. the presence of large amounts of reversely magnetized Permian quartz porphyries under the northern German basins,
2. Henkel's (1994) finding that pyrrhotitic metasedimentary rocks can have susceptibilities up to 2×10^{-3} cgs or 0.025 SI, with values of Q up to 100, implying that remanent magnetization will dominate,
3. Pucher's (1994) mapping of western German anomalies and collection of rock samples, the anomalies being produced by Paleozoic pyrrhotite-bearing metamorphic rocks, several of which were indicated by model studies to be reversely magnetized,
4. the possibility that regional metasedimentary rocks may have been over-printed with (reverse) magnetization during the Permo-Carboniferous reversed superchron (Piper, 1987; Thominski, Wohlenberg, and Bleil, 1993).

On the basis of those considerations, Taylor and Ravat (1995) proposed that a possible origin of the negative anomaly could be "a large number of relatively small reversely magnetized bodies whose magnetic fields coalesce at satellite altitudes."

Pucher and Wonik (1997) agreed with Taylor and Ravat (1995) that the source of anomalies (12), (12a), and (2) is remanent magnetism, but considered that the region east of the Tornquist–Teisseyre zone makes no contribution. They noted a confirmatory result of Worm (1995) in which no decrease in the Koenigsberger ratio, Q, occurred for magnetite and pyrrhotite in a deep drill hole in Germany. With regard to central Europe, they reiterated the importance of the dominance of rocks from the Permian reversed-polarity period, with magnetization acquired at a paleomagnetic latitude, according to Soffel (1985), of $I = -30°$. On that basis they estimated a direction of $D = -30°$, $I = -15°$, from rock magnetism and proposed the model of Figure 9.40, with pyrrhotite the main source (see Section 7.1.2). As pointed out by Taylor and Ravat (1997), Pucher and Wonik (1997) gave no information regarding the magnitude of Q in their model nor regarding the reason for the shallowing of I from the paleomagnetic value. Taylor and Ravat (1997) also regarded the model of Pucher and Wonik (1997) as difficult to reconcile with the magnetic anomaly map of Simonenko and Pashkevich (1990), with its larger-amplitude–shorter-wavelength anomalies to the northeast of the Tornquist–Teisseyre zone.

9.6 OCEANS

9.6.1 GENERAL

Oceanic crust is younger and better understood than continental crust. The dominant magnetic signatures are the striped seafloor spreading anomalies. Because fields from adjacent, oppositely polarized stripes cancel, these mostly result in anomaly fields of smaller amplitude and shorter wavelength than can be detected at satellite altitude. An exception occurs for the longer interval of normal-polarity that occurred in the Cretaceous (i.e., the Cretaceous Quiet Zone, KQZ). A shorter, but still considerably longer than average, normal polarity period also occurred in the Jurassic, but it does not seem to have a signature in the satellite data (Hayling and Harrison, 1986).

In their equivalent source magnetization solution, Hayling and Harrison (1986) noted anomalies correlating with several features that they considered to be dominated by remanent magnetization, namely, the New England seamounts, Corner seamounts, Bahamas platform, Madeira rise, Sierra Leone rise, Gulf of Guinea islands, and Walvis ridge. They found no anomalies at the Azores rise,

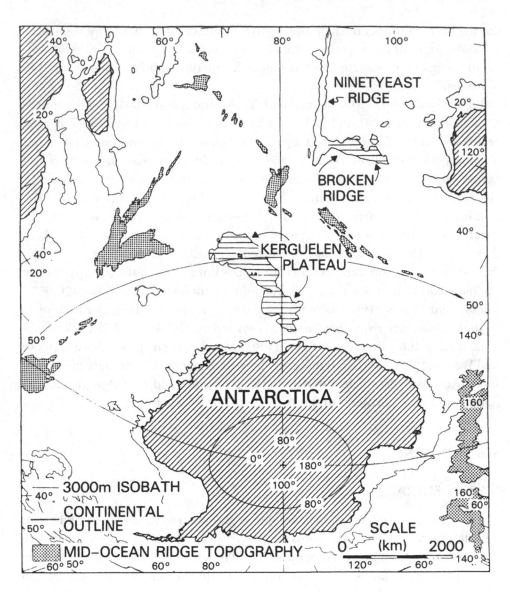

FIGURE 9.41
Location map for the Kerguelen plateau and Broken ridge plateau in the southern Indian Ocean. (Reprinted from L. M. Bradley and H. Frey, Constraints on the crustal nature and tectonic history of the Kerguelen plateau from comparative magnetic modeling using *Magsat* data, *Tectonophysics, 145,* 243–51, 1988, with kind permission of Elsevier Science NL, Sara Burgerhartstraat 25, 1055 KV Amsterdam, The Netherlands.)

Lesser Antilles, Cape Verde islands, or Bermuda platform, locations at which they expected to find induced magnetization in thickened crust. Hayling and Harrison (1986) pointed out that a thick sediment blanket can result in increased temperatures in the underlying crust, elevating the Curie isotherm depth and leading to decreased depth-integrated magnetization, and they calculated that basaltic layers should begin to be affected when the sediment thickness reaches 5 km. Correlations of thick sediments with lower magnetizations were identified for the Laurentian, Amazon, and Congo cones. The continent–ocean contrast and KZQ have been discussed in Sections 8.5 and 7.2.2, respectively.

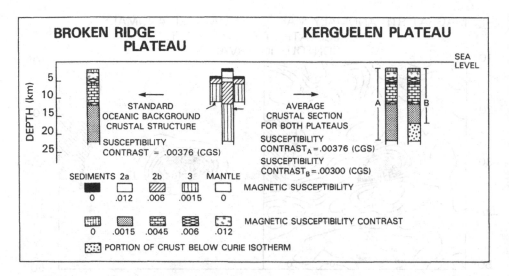

FIGURE 9.42
The magnetic susceptibility contrast for each plateau is determined by first calculating the difference between the magnetic susceptibility for each layer of the plateau and that for the oceanic background. These layer differences (multiplied by the layer thickness) are then summed and divided by the total thickness to calculate a bulk susceptibility contrast for the plateau. For the Kerguelen plateau, model A shows total crustal thickness. Model B shows *magnetic* crustal thickness, reduced because of the raised Curie isotherm, with a corresponding reduced bulk susceptibility contrast (averaged over the total thickness). (Reprinted from L. M. Bradley and H. Frey, Constraints on the crustal nature and tectonic history of the Kerguelen Plateau from comparative magnetic modeling using *Magsat* data, *Tectonophysics, 145,* 243–51, 1988, with kind permission of Elsevier Science NL, Sara Burgerhartstraat 25, 1055 KV Amsterdam, The Netherlands.)

9.6.2   PLATEAUS AND RISES

Magnetic anomalies at satellite altitude from oceanic plateaus and rises are common, but not universal. They have been the subjects of numerous specialized studies that have considered either the submarine crust reflecting a continental heritage or a thickened crust associated with volcanic plateaus. One of the more comprehensive studies was conducted by Johnson (1985), who reported on the dipole anomaly located over the Broken Ridge, a prominent feature in all of the global satellite anomaly maps. Using a model in which the ridge boundaries were determined from its bathymetric contours and its volume from assuming Airy isostatic compensation, he was able to reproduce measured anomaly profiles with a magnetization of 6 A/m directed along the present-day ambient field. This magnetization strength is thought to require a combination of remanent magnetization and viscous remanent magnetization.

Bradley and Frey (1988) applied the methodology of Section 8.2.4 to the Kerguelen plateau and Broken Ridge, located as shown in Figure 9.41 in the southern Indian Ocean. They pointed out that existing seismic data are inadequate to resolve deep crustal structure, and controversy exists over the nature of the crust in these regions. The two plateaus are located on either side of the Southeast Indian ridge (spreading center), leading to the suggestion that a once-larger plateau has been split by seafloor spreading. Active volcanoes on Heard and Kerguelen islands indicate growth of the Kerguelen plateau to the north since the split. These regions have positive RTP magnetic anomalies, each with contrasts of 14–16 nT with the surrounding region. The initially adopted susceptibilities of 0.006 cgs (0.075 SI) for layers 2A and 2B, 0.00151 cgs (0.019 SI) for layer 3A, and zero for layer 3B, for sediments and for the mantle, resulted in a model anomaly contrast for Broken Ridge lower than that

FIGURE 9.43
Broken Ridge plateau. Left:
Magsat (RTP) dawn data
with a contour interval of
2 nT. The anomaly contrast
between the Broken Ridge
anomaly and the large
negative anomaly to the
south of it is between 14
and 16 nT. Right: Model
anomaly for the Broken
Ridge plateau. The
maximum anomaly contrast
calculated is 15.2 nT, with a
peak value in the center of
the plateau and a small
negative anomaly (<−2 nT)
to the south. (Reprinted from
L. M. Bradley and H. Frey,
Constraints on the crustal
nature and tectonic history
of the Kerguelen plateau
from comparative magnetic
modeling using *Magsat*
data. Tectonophysics, 145,
243–51, 1988, with kind
permission of Elsevier
Science NL, Sara
Burgerhartstraat 25, 1055
KV Amsterdam, The
Netherlands.)

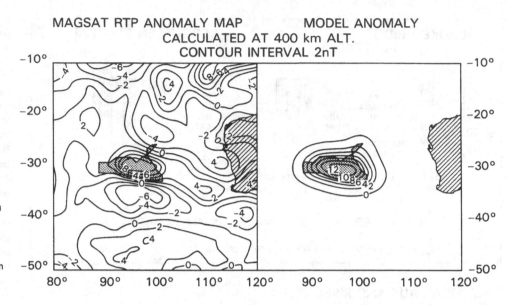

FIGURE 9.43
Broken Ridge plateau. Left: *Magsat* (RTP) dawn data with a contour interval of 2 nT. The anomaly contrast between the Broken Ridge anomaly and the large negative anomaly to the south of it is between 14 and 16 nT. Right: Model anomaly for the Broken Ridge plateau. The maximum anomaly contrast calculated is 15.2 nT, with a peak value in the center of the plateau and a small negative anomaly (<−2 nT) to the south. (Reprinted from L. M. Bradley and H. Frey, Constraints on the crustal nature and tectonic history of the Kerguelen plateau from comparative magnetic modeling using *Magsat* data. Tectonophysics, 145, 243–51, 1988, with kind permission of Elsevier Science NL, Sara Burgerhartstraat 25, 1055 KV Amsterdam, The Netherlands.)

measured. Under the assumption that Kerguelen plateau and Broken Ridge are lithologically similar, because of a common origin, the susceptibility for layer 2A was changed to 0.012 cgs (0.151 SI), based on geologic studies of the Heard and Kerguelen islands that indicated the presence of alkali rather than tholeiitic basalt. Figure 9.42 shows the adopted susceptibility estimates by layer, for the total column, and the contrast with surrounding oceanic lithosphere. Figure 9.43 shows the measured and model anomalies for Broken Ridge. Adoption of the same susceptibility model, A in Figure 9.42, for Kerguelen resulted in a model anomaly contrast of near 25 nT, much larger than the measured 15 nT. Overestimation of susceptibilities of this magnitude is regarded as unlikely. Rather, taking note of the active volcanism at Kerguelen and of thermal models of hotspot regions, they assumed a shallow Curie isotherm depth of 15 km, model B in Figure 9.42. That produced the model anomaly shown in Figure 9.44, together with the measured anomaly. Though certainly not unique, and relatively unconstrained by other data, this seems to be a plausible model. The model susceptibility contrasts are consistent with those expected from oceanic crust, and if the shallow Curie isotherm is correct, the model is consistent with the hypothesis that the two plateaus resulted from the splitting of a single feature by seafloor spreading.

Figures 9.45a and 9.45b show the measured RTP magnetic anomalies and bathymetry for the southwestern Indian Ocean region studied by Fullerton et al. (1994). They call particular attention to two SW–NE-trending regions of positive anomalies, one of which is off the southeastern coast of Africa, with peaks of 6 nT, 5.4 nT, and 6.2 nT over the Agulhas plateau, the southern extension of the Mozambique plateau, and the central Mozambique basin, respectively. To the northwest, the positive extends over the Madagascar ridge. A second

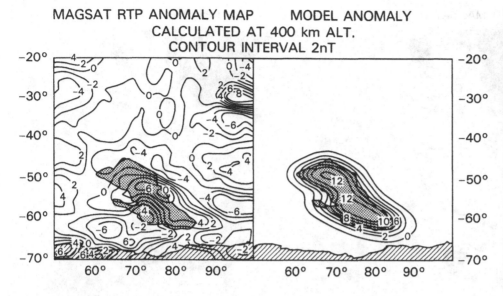

MAGSAT RTP ANOMALY MAP   MODEL ANOMALY
CALCULATED AT 400 km ALT.
CONTOUR INTERVAL 2nT

FIGURE 9.44
Kerguelen plateau. Left:
Magsat (RTP) dawn data
with a contour interval of 2
nT. The anomaly contrast
between the Kerguelen
plateau anomaly and the
surrounding negative
anomalies is 14–16 nT.
Right: Model anomaly for
the Kerguelen plateau. The
maximum anomaly contrast
calculated is 14.2 nT, with a
peak value in the center of
the plateau, and small
negative anomalies (<−2
nT) to the east and west.
(Reprinted from L. M.
Bradley and H. Frey,
Constraints on the crustal
nature and tectonic history
of the Kerguelen plateau
from comparative magnetic
modeling using *Magsat*
data. *Tectonophysics, 145*,
243–51, 1988, with kind
permission of Elsevier
Science NL, Sara
Burgerhartstraat 25, 1055
KV Amsterdam, The
Netherlands.)

SW–NE-trending line of positive anomalies, parallel to the first, is located to the south off the Antarctic coast. There, positive peaks of 4.0 nT, 3.5 nT, and 3.7 nT are located over the Lazarev Sea, the north-central Enderby basin, and the northeastern Enderby basin/Conrad rise, respectively. The north and south anomaly peaks are noted to correspond generally to conjugate north- and south-spreading pairs, that is, the Agulhas plateau and Maud rise, the Mozambique plateau and Astrid ridge, and the Madagascar ridge and Conrad rise. A 5.1 nT positive anomaly over the Del Cano rise/Crozet bank is also included in the model. Figure 9.45c shows the plateaus' and basins' assigned induced and viscous susceptibilities, and Figure 9.46 shows the adopted generalized cross sections and assigned values for susceptibility. Layer 2, defined by seismic velocities in the 3.9–5.8-km/s range, is assigned a susceptibility of 0.075 SI or 0.157 SI, depending on whether tholeiitic or alkali basalts are expected, based on the available Deep Sea Drilling Project (DSDP/ODP) drill-core data. Layer 3 is defined by seismic velocities in the 6.6–7.6-km/s range and is assigned a susceptibility of 0.013 SI, thought appropriate for gabbros. Continental-like granitic layers in the southern Agulhas plateau, with seismic velocities of 5.8–6.4 km/s, are assigned a susceptibility of 0.006 SI. Table 9.7 summarizes the model body thicknesses and contrasts in susceptibility. These values are found to reproduce the measured anomalies outside the KQZ, but not within the KQZ. For KQZ regions, shown shaded in Figure 9.45d, additional natural remanent magnetization (NRM) is assumed. This NRM is assumed to vary spatially in two steps (Figure 9.45d). High NRM and low NRM values are 3.0 and 10.0 A/m, respectively, based on DSDP/ODP drill cores from the Maud rise (3.1 ± 1.0 A/m), Madagascar basin (3.7 ± 3.2 A/m), and Mozambique basin (10.4 ± 2.5 A/m). To reproduce the measured anomalies, the NRM is distributed uniformly

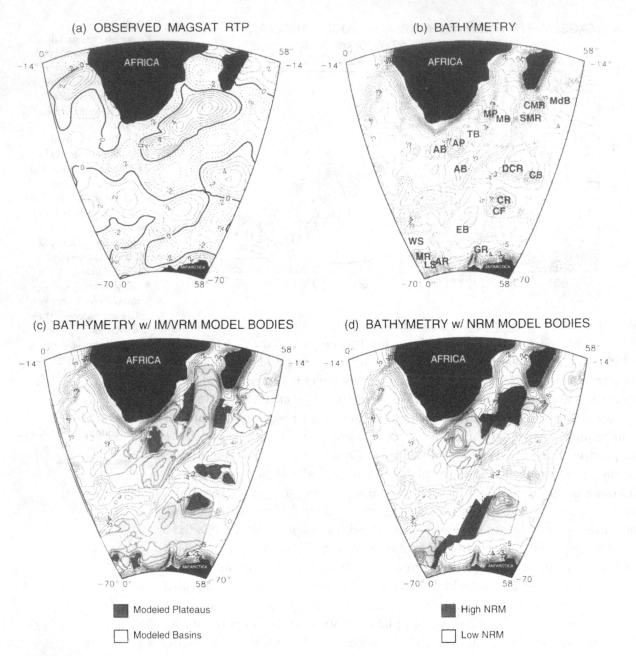

(a) OBSERVED MAGSAT RTP

(b) BATHYMETRY

(c) BATHYMETRY w/ IM/VRM MODEL BODIES

(d) BATHYMETRY w/ NRM MODEL BODIES

■ Modeled Plateaus

□ Modeled Basins

■ High NRM

□ Low NRM

FIGURE 9.45
(see caption at foot of p. 375.)

throughout the crust. Modeled and measured RTP anomalies are compared in Figure 9.47. The correspondence is generally good.

Toft and Arkani-Hamed (1992) examined volcanic plateaus and the effects of the KQZ in the Pacific Ocean. The plateaus considered were the Shatsky rise, Hess rise, Manihiki plateau, and Mid-Pacific Mountains. All were considered to be Airy-isostatic-compensated, and the crustal thickness in excess

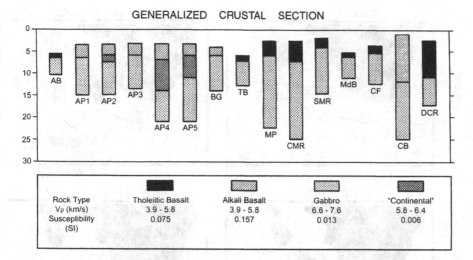

GENERALIZED CRUSTAL SECTION

Rock Type	Tholeiitic Basalt	Alkali Basalt	Gabbro	"Continental"
V_P (km/s)	3.9 - 5.8	3.9 - 5.8	6.6 - 7.6	5.8 - 6.4
Susceptibility (SI)	0.075	0.157	0.013	0.006

FIGURE 9.46
Generalized crustal sections of modeled induced plus viscous remanent magnetization bodies. See Table 9.7 for model body parameters. Model body designations as in Figure 9.45. (From Fullerton et al., 1994. with permission.)

Table 9.7. *Model body parameters*

Model body	Name	Thickness (km)	Susceptibility contrast, SI	Conjugate
AB	Agulhas/Mozambique basin	4.8	−0.06333	Weddell Sea/Enderby basin
AP1	Agulhas plateau	11.5	0.01420	Maud rise
AP2	Agulhas plateau	11.5	0.00716	—
AP3	Agulhas plateau	10.3	0.01156	—
AP4	Agulhas plateau	17.8	0.00140	—
AP5	Agulhas plateau	17.8	0.00081	—
BG	Background	10.0	0.00000	—
TB	Transkei basin	6.8	−0.03644	Lazarev Sea
MP	Mozambique plateau	19.8	0.00239	Astrid ridge
CMR	Central Madagascar ridge	22.3	0.00754	—
SMR	South Madagascar ridge	12.6	−0.00993	Conrad rise
MdB	Madagascar basin	5.9	−0.4599	—
CF	Conrad rise (flank)	8.7	−0.02463	—
CB	Crozet bank	24.0	0.06145	—
DCR	Del Cano rise	14.8	0.01973	—

Source: From Fullerton et al. (1994), with permission.

FIGURE 9.45 (opposite page)
Magsat data, bathymetric data, and modeled bodies in the southwestern Indian Ocean. (a) *Magsat* RTP anomalies from Baldwin and Frey (1991); contour interval 1.0 nT. (b) Smoothed (0.5°) bathymetry derived from Digital Bathymetric Data Base 5 (DBDB5); contour interval 500 m. Designated features: AB, Agulhas basin; AP, Agulhas plateau; TB, Transkei basin; MP, Mozambique plateau; MB, Mozambique basin; CMR, central Madagascar ridge; SMR, southern Madagascar ridge; MdB, Madagascar basin; WS, Weddell Sea; EB, Enderby basin; CR, Conrad rise; CF, Conrad rise (flank); DCR, Del Cano rise; CB, Crozet bank; MR, Maud rise; LS, Lazarev Sea; AR, Astrid ridge; GR, Gunnerus ridge. (c) Induced plus viscous remanent magnetization bodies overlying bathymetry. (d) KQZ crustal regions and natural remanent magnetization bodies overlying bathymetry. Note that unshaded areas were not modeled. Orthographic projections. (From Fullerton et al., 1994, with permission.)

FIGURE 9.47

Observed *Magsat* RTP anomalies and models of the African marginal bodies; contour interval 1.0 nT. (a) *Magsat* RTP anomalies. (b) Induced plus viscous remanent magnetization model with model bodies shaded. Light shading corresponds to ocean basins, and dark shading to submarine plateaus. (c) Natural remanent magnetization model with KQZ crust and model bodies shaded. Light shading corresponds with "low" magnetization areas, and dark shading to "high" magnetization areas. (d) Combined induced plus viscous remanent and natural remanent magnetization model. (From Fullerton et al., 1994, with permission.)

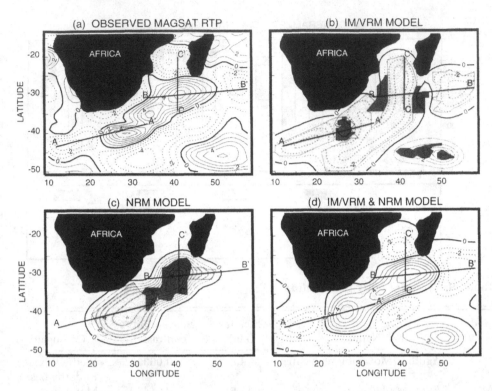

of the surrounding crust was computed accordingly, with magnetization contrasts modeled from the topographic contrast plus roots. Figure 9.48 summarizes the model for the Shatsky rise. For the NRM model, normal- and reversed-polarity anomaly positions were extrapolated from anomaly lineaments in the surrounding area, with directions estimated from an appropriate published pole position, and with the NRM assumed to be 4 A/m, based on DSDP/ODP results. This is regarded as an overestimate, because the magnetization of the roots likely is less than that assumed, but it is desired to find an upper bound to the resulting anomaly at satellite altitude. As seen from Figure 9.48d, the resulting anomaly peak-to-peak amplitude of about 3 nT, smaller and different in shape than the measured anomaly shown in Figure 9.48f. NRM probably is an insignificant contributor to the satellite altitude anomaly. An induced magnetization contrast of 2 A/m resulted in the model anomaly shown in Figure 9.48e, which compares favorably with the measured anomaly. In the local main-field strength, this amounts to a susceptibility contrast of about 0.061 SI, which is a combination of strictly induced magnetization and viscous magnetization acquired since the last field reversal.

9.7   SUMMARY

The foregoing discussions of the relationships among satellite magnetic anomalies and the surface geology of selected continental and oceanic regions,

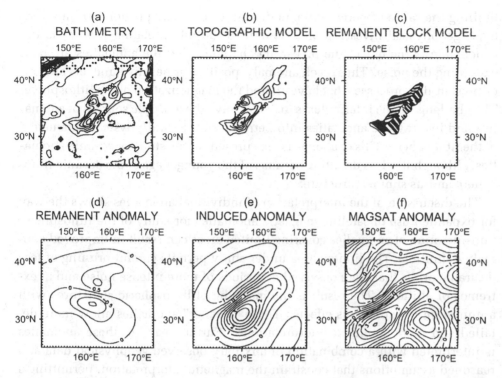

(a) BATHYMETRY

(b) TOPOGRAPHIC MODEL

(c) REMANENT BLOCK MODEL

(d) REMANENT ANOMALY

(e) INDUCED ANOMALY

(f) MAGSAT ANOMALY

FIGURE 9.48
Shatsky rise.
(a) Bathymetry; Contour interval 500 m. Dots show west edge of the Pacific KQZ. (b) Topographic model. (c) Normal-magnetization (solid) and reversed-magnetization (open) zones of Shatsky rise, derived from Nakanishi et al. (1989). (d) Calculated filtered anomaly due to NRM of 4 A/m in swell and root, with reversal polarity structure shown in Figure 9.54c; contour interval 0.5 nT. (e) Calculated filtered anomaly due to effective induced magnetization contrast of 2 A/m; contour interval is 0.5 nT. (f) *Magsat* anomaly field over Shatsky rise and surrounding area; contour interval 0.5 nT. (Internal ticks are equal-area grid units at spacing of 1 grid unit = 55.6 km.) (From Toft and Arkani-Hamed, 1992, with permission.)

as well as interpretations of specific satellite magnetic anomalies, illustrate the utility of these anomalies in understanding the nature and evolution of the lithosphere. However, this discussion also highlights the problems in applying them to lithospheric studies. Many interesting correlations are noted between anomalies and surface geology, but also it is clear that the numerous exceptions make generalizations very questionable. This is particularly evident for the continents. The long geologic history of most continental segments, with their overlapping tectonism, greatly complicates interpretation. This situation is exacerbated by (1) the horizontal tectonism caused by plate interactions resulting in major vertical changes in the lithosphere, unpredictable from the surface geology, (2) the possible presence of remanent magnetization of largely unknown directional attributes, because of the unknown ages of acquisition and subsequent modifications by plate movements and rotation, (3) the complex effects of geologic processes upon the magnetizations of rocks, and (4) the lack of vertical resolution for satellite magnetics over the thickness of the magnetic lithosphere, and the limited horizontal resolution of discrete anomalies, measured in hundreds of kilometers. Considering the potential complications from these factors, it is not surprising that similar surface geologic features often have somewhat different magnetic signatures. Rather, it is encouraging that many continental tectonic features and structural trends have corresponding anomalies. The precise sources of these anomalies are not interpretable from the anomalies alone. However, the results of the interpretations are useful in limiting the range of permissible models and in isolating regions for more intense geophysical investigations. Central to the interpretation process is the conclusion borne out in the preceding chapters: that the anomalies are real, exceeding the noise. Thus, each anomaly, positive or negative, reflects ancient or present-day processes that have altered the magnetization of the lithosphere. The challenge to the interpreter is to deconvolve the multiple sources of anomalies and invert those anomalies into petrologic/structural/physical parameters of the lithosphere. This challenge is also present in the study of oceanic anomalies, although it is less daunting because of the younger age of the oceanic lithosphere and its simpler structure.

The discussion of the interpretation of individual anomalies shows the way for expanded use of satellite magnetic anomalies for study of the lithosphere. Consideration of anomalies covering limited areas can result in improved anomaly definition. The limited area under study permits careful pruning of less desirable orbits that were observed at periods of more intense noise and at extremes of altitude. The result is the mapping of lithospheric signatures with a minimum of extraneous effects. These "cleaner" anomalies allow more detailed and more confident interpretations. Further, each of these anomalies is interpreted with a combination of auxiliary observed geophysical data and reasoned assumptions that constrain the magnetic interpretation, permitting a more detailed and realistic interpretation.

APPENDIX 9.1

REFERENCES TO REGIONAL STUDIES

The following citations are categorized by region. They include publications dealing with map generation, model derivation, and, where deemed appropriate, related studies. Some papers are difficult to categorize, and others are cited for more than one region.

1. *Canada*

   Arkani-Hamed et al. (1984, 1985a,b)

   Hall et al. (1985)

   Langel et al. (1980a)

   Noble (1983)

   Pilkington and Roest (1996)

2. *United States*

   Allenby and Schnetzler (1983)

   Black (1981)

   Braile et al. (1989)

   Butler (1984)

   Carmichael and Black (1986)

   Counil et al. (1989)

   Langel (1990b)

   Lugovenko and Pronin (1982, 1984)

   Mayhew (1979, 1982a,b, 1984)

   Mayhew and Galliher (1982)

   Mayhew et al. (1982, 1984, 1985a)

   Purucker et al. (1997)

   Purucker et al. (1998)

   Ridgway (1984)

   Ridgway and Hinze (1986)

   Ruder (1986)

   Ruder and Alexander (1986)

   Schnetzler (1985)

   Schnetzler and Allenby (1983)

   Sexton et al. (1982)

   Starich (1984)

   Thomas (1984)

   von Frese et al. (1981b, 1982a,b)

   Whaler and Langel (1996)

   Won and Son (1982)

   Zietz et al. (1970)

3. *Africa*

   Ajakaiye et al. (1985, 1986)

   Arkani-Hamed and Strangway (1985c)

   Baldwin and Frey (1991)

   Boukeke (1994)

   Dorbath et al. (1985)

   Frey et al. (1983)

   Galdéano (1981, 1983)

   Girdler et al. (1992)

   Green (1973, 1976)

   Hastings (1982)

   Kuhn and Zaaiman (1986)

   Langel and Whaler (1996)

   Lugovenko and Pronin (1982)

   Phillips and Brown (1985)

   Ravat (1989)

   Ravat et al. (1992)

   Regan and Marsh (1982)

   Regan et al. (1973, 1975)

   Toft and Haggerty (1986, 1988)

   Toft et al. (1989, 1992)

   von Frese et al. (1986, 1987)

   Whaler (1994)

   Whaler and Langel (1996)

4. *Australia*

Dooley and McGregor (1982)
Kelso et al. (1993)
Mayhew and Johnson (1987)
Mayhew et al. (1980, 1991)

O'Reilly and Griffin (1985)
Purucker et al. (1997)
Tarlowski et al. (1996)
Wellman et al. (1984)

5. *Europe/Russia*

Arkani-Hamed and Strangway (1986b)
Berti and Pinna (1987)
Cain et al. (1989)
DeSantis et al. (1989, 1990)
Lugovenko and Pronin (1982)
Lugovenko and Matushkin (1984)

Nolte and Hahn (1992)
Pucher and Wonik (1997)
Ravat et al. (1993)
Shapiro et al. (1986)
Taylor and Frawley (1987)
Taylor and Ravat (1995, 1997)
Taylor et al. (1992)

6. *South America*

Arkani-Hamed and Strangway (1985c)
Frey et al. (1983)
Galdéano (1981, 1983)
Hinze et al. (1982)
Longacre (1981)
Longacre et al. (1982)
Parrott (1985)

Ravat et al. (1991, 1992)
Renbarger (1984)
Ridgway (1984)
Ridgway and Hinze (1986)
von Frese et al. (1986, 1987, 1989)
Yanagisawa and Kono (1984)
Yuan (1983)

7. *Caribbean*

Counil (1987)
Counil et al. (1989)

Ridgway (1984)
Ridgway and Hinze (1986)

8. *India*

Agarwal et al. (1986)
Agrawal et al. (1986)
Arur et al. (1985)
Bapat et al. (1987)
Basavaiah et al. (1989)
Chowdhury et al. (1989)
Dewey et al. (1988)
Mishra (1977, 1984)
Mishra and Venkatraydu (1985)
Negi et al. (1985, 1986a,b 1987a)
Oraevsky et al. (1994)

Pandey and Negi (1987)
Qureshy and Midha (1986)
Rajaram (1993a,b)
Rajaram and Singh (1986)
Rajaram and Langel (1992)
Rotanova et al. (1995)
Singh (1989)
Singh and Rajaram (1990)
Singh et al. (1986, 1989, 1991, 1992)
Singh, Rastogi, and Adam (1993)
Thakur (1988)
von Frese et al. (1989)

9. *Middle East*
   Arkani-Hamed and Strangway
       (1986b)

10. *China/Tibet/Southeast Asia*
    Achache et al. (1987)   Hamoudi et al. (1995)
    An et al. (1992)   McGue (1988)
    Arkani-Hamed et al. (1988)   Rajaram and Langel (1992)

11. *Japan*
    Nakagawa and Yukutake (1984,   Nakatsuka and Ono (1984)
        1985)   Yanagisawa and Kono (1984)
    Nakagawa et al. (1985)   Yanagisawa et al. (1982)

12. *Continental reconstructions*
    Arkani-Hamed and Strangway   Pandey and Negi (1987)
        (1985c)   Ravat (1989)
    Frey et al. (1983)   Ravat et al. (1992)
    Galdéano (1981, 1983)   von Frese et al. (1986, 1987)

13. *Antarctica*
    Alsdorf (1991)   Pandey and Negi (1987)
    Alsdorf et al. (1994)   Ritzwoller and Bentley (1982,
    Bentley (1991)   1983)
    Bormann et al. (1986)   Takenaka et al. (1991)
    Ghidella et al. (1991)   von Frese et al. (1992, 1997)
    Kovatch (1990)

14. *Arctic*
    Alsdorf et al. (1997)   Haines (1985b)
    Coles (1985)   Langel and Thorning (1982a,b)
    Coles et al. (1982)   Sweeney and Weber (1996)
    Coles and Taylor (1990)   Taylor (1983a,b)
    Forsyth et al. (1986)

15. *Oceans – General*
    Harrison et al. (1986)   Harrison (1987)

15a. *Atlantic*
     Alsdorf (1991)   Raymond (1989)
     Hayling and Harrison (1986)   Raymond and LaBrecque (1987)
     LaBrecque and Raymond (1985) Yañez and LaBrecque (1997)

15b. *Pacific*
     LaBrecque and Cande (1984)   LaBrecque et al. (1985)

15c. *Indian*

Negi et al. (1987b)   Taylor (1991)

16. *Magnetization of oceanic lithosphere*

Arkani-Hamed (1988, 1989, 1991)   Pozzi and Dubuisson (1992)
Arkani-Hamed and Strangway   Raymond and LaBrecque (1987)
   (1986c)   Toft and Arkani-Hamed (1992,
Cohen (1989)   1993)
Cohen and Achache (1994)   Yañez and LaBrecque (1997)

17. *Cretaceous quiet zones*

Arkani-Hamed (1988, 1989, 1991)   Raymond (1989)
Fullerton et al. (1989, 1994)   Raymond and LaBrecque (1987)
Hayling and Harrison (1986)   Toft and Arkani-Hamed (1992)
LaBrecque and Raymond (1985)   Yañez and LaBrecque (1997)
Pal (1990)

18. *Subduction zones/trenches*

Arkani-Hamed and Strangway   Counil (1987)
   (1986c, 1987)   Counil and Achache (1987)
Clark et al. (1985)   Vasicek et al. (1988)

19. *Oceanic plateaus and rises*

Antoine and Moyes (1992)   LaBrecque and Cande (1984)
Bradley and Frey (1988)   Schmitz et al. (1982)
Frey (1985)   Toft and Arkani-Hamed (1992,
Fullerton et al. (1989, 1994)   1993)
Johnson (1985)

20. *Continent–ocean boundaries/contrast*

Arkani-Hamed (1990, 1993)   Hinze et al. (1991)
Bradley and Frey (1991)   Meyer et al. (1983, 1985)
Cohen (1989)   Mörner (1986)
Cohen and Achache (1994)   Purucker et al. (1996b)
Counil et al. (1991)   Toft and Arkani-Hamed (1993)
Hayling (1991)

GLOSSARY OF SYMBOLS

The following is a list of symbols used in the equations. In some cases, symbols used only within a single section are not listed here. In other cases, within particular sections of the book a symbol will take on a local meaning different from that given here (that should be clear from the context). Because of the diversity of the subject matter, some symbols will take on more than one meaning. Such multiple meanings are given in the list, and the meaning should be clear from the context. Numbers in parentheses after the definitions indicate the defining equations or sections.

$\mathbf{A}$	three-vector magnetic field from the earth's crust/lithosphere (i.e., anomaly field) (1.5)
A^i, A_i	ith component of magnetic anomaly field
A^s	scalar anomaly field (1.10), (1.11)
$A_1(\mathbf{r})$, $A_2(\mathbf{r})$	two values for some anomaly component at the same location (6.4)
$\mathcal{A}_n$	coefficients in expansion of $P_n(z)$ (4.11), (4.58)
A_f^k, B_f^k	amplitude coefficients in Fourier analysis (4.17)
$\mathcal{A}_{nm}$	coefficients in spherical harmonic expansion (4.21), (5.208)
A_λ, AA_λ	amplitude of λ wavelength (4.1)
a	mean radius of the earth (6,371.2 km) (2.3); slope of nonlinear regression (6.4)
$\mathbf{a}$	vector of polynomials in crossover analysis (4.5); N-vector of anomaly component values (5.42)
$\mathbf{B}(\mathbf{r}, t)$, $\mathbf{B}$	magnetic field vector or magnetic induction vector (1.1), (1.2)

383

$\mathbf{B}_m(\mathbf{r}, t)$, $\mathbf{B}_m$	magnetic field from the earth's core, the main magnetic field (1.5)
$B(\mathbf{r}, t)$, B	scalar magnitude of $\mathbf{B}$
$\mathbf{B}_0$	field from centered dipole (5.198)
$\mathbf{B}_1$	nondipole part of main field (5.198)
B_r, B_θ, B_ϕ	components of magnetic field
$\mathbf{B}_n$	magnetic field of degree n (2.4)
B_y	eastward component of the interplanetary magnetic field
B_z	southward component of the interplanetary magnetic field
$\mathbf{B}_s$	magnetic field of spacecraft (Section 3.5.2)
b	intercept of nonlinear regression (6.4)
$\mathbf{b}$	N-vector of computed magnetic field values (5.83), (5.157)
$\hat{\boldsymbol{b}}$	unit vector in direction of ambient field or main field (5.37)
$\mathcal{C}$	arbitrary real constant
C_n	condition number of matrix
C_{nm}, C_k^p, CG, CH	spherical harmonic coefficients for cosine terms (4.21), (4.63), (5.102)
$\mathbf{c}$	vector of measurements or data (5.48), (5.49)
$\mathbf{D}$	disturbance magnetic field or magnetic field from magnetospheric, ionospheric, and coupling currents (1.5)
D	magnetic declination (2.1); sometimes indicates magnetic disturbance
D^s	scalar disturbance field (1.10), (1.11)
$\mathbf{D}st$	symmetric part of disturbance magnetic field $\mathbf{D}$ (2.12)
$\mathbf{DS}$	nonsymmetric part of disturbance magnetic field $\mathbf{D}$ (2.12)
Dst	horizontal component of $\mathbf{D}st$ at equator; used as a magnetic index
$\mathbf{d}$	vector of powers of distance in crossover analysis (4.5)
d_{ij}	exponential factor in rectangular-coordinate potential function (5.77)

$E(\ )$	expectation value
$\mathbf{e}$	measurement error in magnetic field (1.5); N-vector of measurement error (5.49), (5.52)
$\mathbf{F(x)}$	arbitrary vector-valued function (5.63)
$\mathcal{F}(\mathbf{r})$	arbitrary function of $\mathbf{r}$ (5.120)
F	matrix for inner-product weighting and for inclusion of damping in minimization (5.119), (5.124)
$\left.\begin{array}{l} f(x),\ F(k) \\ f(x,y),\ F(k,j) \\ g(x),\ G(k) \end{array}\right\}$	discrete Fourier transform pairs (4.12), (4.13); or functions of x, y.
$f(p),\ f_p$	$f(x_p)$ (4.12), (4.14)
f_n	factor for introducing norm-dependent functions into an inner product (5.127), (5.130)
f	arbitrary continuous function
G	matrix operator (5.53); universal gravitational constant (8.1)
$\mathbf{G}_k(\mathbf{r},\mathbf{r}')$	kernel function (5.136)
$G_{nm},\ H_{nm}$	coefficients in a spherical harmonic analysis (not the main field, see g_n^m) (4.25), (4.41), (4.43)
$\mathcal{G}$	inner product function (5.119)
$g_n^m,\ h_n^m$	Gauss coefficients in spherical harmonic expansion of geomagnetic potential functions representing the main and crustal fields (2.3)
g	arbitrary continuous function
$\mathbf{H}$	magnetic intensity vector (1.15)
H	horizontal component of magnetic field
H^x	matrix relating data to model parameters (i.e., matrix of partial derivatives of model equation with respect to the model parameters) (5.52); the superscript x indicates which model, where $x = m, p, q, z, h$, or b, as follows:

H^m	equivalent source model (5.17), (5.30), (5.41)
H^p	global spherical harmonic potential analysis (5.101)
H^q	spherical cap harmonic analysis (Section 5.3.1)
H^z	rectangular harmonic analysis (Section 5.3.2)

H^h global spherical harmonic component analysis (5.104)

H^b collocation (5.83), (5.89)

$\mathcal{H}$	Hilbert space (Section 5.6.3, Appendix 2 of Chapter 5)
h	height or altitude (4.1), (5.80)
$\mathbf{h}$	N-vector of spherical harmonic coefficients of component representation (5.104)
$\hbar$	arbitrary continuous function
I	inclination of the earth's magnetic field (2.1); identity matrix
$\mathbf{J}$	current density (1.2)
j	integer index
K	matrix of inner products of (depleted) basis functions and data functions (5.170); number of subdivisions of line interval in Fourier analysis (4.12)
$K_n^m, K1, K2, \ldots$	constants, specified by context
$\mathcal{K}$	magnetic susceptibility matrix (5.23)
k	wavenumber; integer index
L	linear operator (a superscript may indicate which operator) (5.139); half-line interval in rectangular harmonic analysis (5.77)
$\mathcal{L}$	matrix mapping generalized inverse solution to predicted anomaly field values (5.157)
l^α	elements of a Hilbert space mapping the potential function into field components via the inner product (5.140)
l_n^m	spherical harmonic coefficients for l^α (5.141)
ℓ	vector of inner products of l^α (5.156)
$\mathbf{M}$	annihilator magnetization (Section 5.1.5)
$\mathbf{M}$	magnetization = dipole moment per unit volume (1.12), (5.18)
$\mathbf{M}_r$	natural remanent magnetization (1.18)
$\mathbf{M}_i$	induced magnetization (1.18)
m	integer index; order of Legendre function; magnitude of $\mathbf{m}$

m	N-vector of magnetic moment magnitudes; sometimes dipole moment (1.12)
$\mathbf{m}_j$	dipole moment at location j (1.12), (Section 5.1.1.1), (5.6)
m_r, m_θ, m_ϕ	components of magnetic moment (5.10)
n	integer index; degree of Legendre polynomial or Legendre function
n^*	maximum degree/order of spherical harmonic expansion representing the main field
n_c	degree below which R_n is dominated by the main field and above which R_n is dominated by the lithospheric field (Section 4.3)
$\hat{\mathbf{n}}$	unit vector normal to a surface
P	percentage of trace retained in principal-component analysis (Section 5.5.2)
$P_n(\cos\theta)$, $P_n(\theta)$, P_n	Legendre polynomials (usual usage) (4.10); power in nth degree of harmonic analysis (4.22); $P_n(\theta)$ and P_n used as shorthand forms
$P_n^m(\cos\theta)$, $P_n^m(\theta)$, P_n^m	Schmidt quasi-normalized associated Legendre functions of degree n and order m (2.3); $P_n^m(\theta)$ and P_n^m used as shorthand forms
p	integer index; polynomial in crossover analysis (4.4), (4.5)
p	N-vector of spherical harmonic coefficients from a potential function (5.100), (5.101)
Q	Koenigsberger ratio (1.19); matrix operator (5.53), (5.59); matrix of inner products of (depleted) basis functions in generalized inversion (5.170); kernel function (8.7)
q	integer index
q	transformed model parameter vector in principal-component analysis (5.113)
q_n^m, s_n^m	Gauss coefficients representing fields of magnetospheric origin (2.3)
R	a ratio of some sort (subscripts may be appended); resolution matrix (5.55); rotation matrix (sometimes with subscripts, superscripts, or primes to indicate coordinate systems)

R_n	root-mean-square (rms) of the magnetic field strength over the earth's surface from the nth-degree spherical harmonics (2.6)
R_{nm}	signal-to-noise ratio in harmonic correlation analysis
r	radius in polar coordinates; magnitude of $\mathbf{r}$
$\mathbf{r}$	radius vector
$\hat{\mathbf{r}}$	unit vector in direction of $\mathbf{r}$
$\mathbf{r}_i$	radius vector for ith position (5.1)
$\mathbf{r}_{ij}$	$\mathbf{r}_i - \mathbf{r}_j$, difference between radius vectors (5.1)
r_{ij}	magnitude of $\mathbf{r}_{ij}$ (5.1)
r	distance function in collocation (5.94)
r_c	correlation length in collocation (5.95), (5.96), (5.97)
$\mathbf{S}$	magnetic field from SEMM magnetization (8.15)
$\mathbf{S}, Sq$	regular diurnal magnetic field during periods of low magnetic activity, thought to be mainly ionospheric in origin
S_{nm}, S_n^m, SG, SH	spherical harmonic coefficients for sine terms (4.21), (5.102)
$\mathcal{S}_n^m$	surface harmonic (4.2)
$\mathcal{S}^2$	bounding constant (5.163)
T	rotation matrix (3.2), (3.3); absolute temperature (7.1)
$\mathbf{t}$	transformed noise vector in principal-component analysis (5.113); coefficients to be determined in generalized inverse problem (5.152)
U	matrix of eigenvectors in principal-component analysis (5.112)
V	matrix of eigenvectors in principal-component analysis (5.112)
$V, \delta V$	volume (1.12), (5.18)
$\mathbf{V}_N$	N-dimensional vector space (Section 5.2.1)
V_d	covariance matrix of data errors (5.58)
V_x	covariance matrix, with subscript indicating which parameter, e.g.:

x	Parameter
m	magnetic moments (5.45)
p	spherical harmonic coefficients of potential function (Section 5.4.1)
h	spherical harmonic coefficients of component representation (5.104)
q	spherical cap harmonic coefficients (Section 5.3.1)
z	rectangular harmonic coefficients (Section 5.3.2)
ε	estimation error (5.58), (5.61), (5.86) etc.
V_p	seismic P-wave velocity (Section 7.2.1)
V_1, V_2, V_3	covariance matrix in collocation (5.91), (5.92), (5.93)
$\mathbf{v}$	arbitrary vector, sometimes with subscript or prime to indicate coordinate systems
$\mathbf{w}$	arbitrary vector, sometimes with subscript or prime to indicate coordinate systems
W	weight matrix (5.44)
X	northward component of the magnetic field (2.2)
x_f^k, y_f^k	amplitude coefficients in Fourier analysis (4.17), (5.77)
$\mathbf{X}$	vector space (Section 5.2.1)
x	cartesian variable
x	$x = a^2/rr'$ in generalized inversion analysis (5.158)
$\mathbf{x}$	vector of model parameters (5.52)
$\mathbf{x}^0$	a priori estimate of $\mathbf{x}$ (5.53)
X_0	a priori covariance of $\mathbf{x}^0$ (5.58)
Y	eastward component of the magnetic field (2.2)
y	cartesian variable
$\mathbf{y}$	vector of field differences at crossover location (4.6), (4.7)
Z	vertical (downward) component of the magnetic field (2.2)
z	cartesian variable
$\bar{\jmath}$	$\bar{\jmath} = a/r$ in generalized inversion analysis (5.158)
$\mathbf{z}$	transformed data vector in principal-component analysis (5.113)

Greek symbols

α, β, γ	angle designations
α	used for a component designation (i.e., r, θ, or ϕ)
α_n, β_n	Fourier coefficients (5.88)
γ_{ij}	angle between radial vectors $\mathbf{r}_i$ and $\mathbf{r}_j$ (5.3)
γ_{nm}, δ_{nm}	constants in susceptibility calculation (5.219)
Γ	form factor for dipole field (5.201)
$\Delta\mathbf{B}$	residual field when estimate of main field is removed from a measurement (1.6)
ΔB	measured scalar field magnitude minus estimate of main field magnitude; not the absolute value of $\Delta\mathbf{B}$ (1.7), (1.9)
Δh	increment in height; increment in crustal thickness (4.1), (5.25)
ΔW	increment of area (5.25)
$\Delta X, \Delta Y, \Delta Z$	residuals of magnetic components (for a spherical earth, equal to $-\Delta B_\theta$, ΔB_ϕ, $-\Delta B_r$); sometimes used to mean A^i, the anomaly components
δ_{nk}	Kronecker delta (equals 1 if $n = k$, equals 0 otherwise)
$\delta_{r\theta}^{ij}$	direction cosine between $\hat{\mathbf{r}}_i$ and $\hat{\boldsymbol{\theta}}_j$ (Section 5.1.1.2, Table 5.1)
$\varepsilon, \varepsilon_x$	estimation error (5.57), (5.85)
ζ_h^k, ζ_{gh}^k	phase angles in spectral decomposition (4.17), (4.21)
$\Delta\zeta^k$	difference between phase angles
η	combination model and measurement error (1.6)
θ	colatitude, usually geocentric; may be subscripted to designate a particular coordinate system
θ_d	dipole or geomagnetic colatitude (2.7)
$\hat{\boldsymbol{\theta}}$	unit vector in θ direction
κ	scalar susceptibility (5.24)
κ_m	volume magnetic susceptibility (1.16)
$\kappa_{n,m}$	coefficients in approximate susceptibility spherical harmonic expansion (5.217)
λ	wavelength (4.1)
λ^d	Lagrange multiplier; damping factor in ridge regression or in minimum-norm estimation (5.121), (5.125)

λ_d	local time in dipole coordinate system (4.41)
λ_i	eigenvalue
Λ	eigenvalue matrix (transformed model matrix) in principal-component analysis (5.114); Gram matrix in generalized inversion (5.152)
Λ_{ij}	elements of Λ
μ	absolute magnetic permeability (1.17)
μ_0	permeability of free space ($4\pi \times 10^{-7}$ H/m) (1.2)
μ_{ij}	cosine of γ_{ij}, the angle between $\mathbf{r}_i$ and $\mathbf{r}_j$ (5.2), (5.3)
ρ	correlation coefficient, with subscripts used to indicate what is being correlated
σ	standard deviation; source (rock) density (8.1)
σ_μ	standard error of the mean (6.1)
τ	scalar anomaly value transformed by dipole form factor (5.205)
τ_{nm}	coefficients in spherical harmonic expansion of τ (5.206)
τ_d	local time variable in dipole coordinate system (4.46)
ϕ	east longitude, usually geocentric; may be subscripted to designate a particular coordinate system
$\hat{\phi}$	unit vector in ϕ direction
ϕ_d	dipole longitude (2.10)
χ^2	quantity to be minimized in nonlinear regression (6.5)
Ψ, Ψ_x, ψ	potential function, sometimes with subscript attached to indicate a particular field (i.e., a for anomaly, s for Sq, etc.)
ξ_{nm}, ζ_{nm}	constants in iterative potential calculation (5.211)
$\omega_{n,m}$	expansion coefficients (5.226)

Modifying symbols

^	a circumflex over a variable indicates a unit vector
$\langle , \rangle$	inner product (5.50)
$\langle \rangle$	taking some sort of average
—	a bar under a variable indicates an *estimate* of that quantity
0	a zero subscript or superscript may indicate an a priori quantity, or a starting quantity in an iterative process

COMMON
ABBREVIATIONS

AE, *AL*, *AU*	auroral electrojet magnetic activity indices
Aj	geographic sector magnetic activity index
Am	magnetic activity index, midlatitude
An	Northern Hemisphere magnetic activity index
ap	magnetic activity index, midlatitude
As	Southern Hemisphere magnetic activity index
ATS	attitude transfer system
CRM	chemical remanent magnetization
DGRF	Definitive Geomagnetic Reference Field
DIM	depth-integrated magnetization
EE	equatorial electrojet; may be the current itself or the fields therefrom
ES	equivalent source
FAC	field-aligned current
GPS	Global Positioning System
GSFC	Goddard Space Flight Center
GSFC ( )	designation of spherical harmonic model derived at GSFC using multiple data sources, with date of derivation in parentheses
IGRF	International Geomagnetic Reference Field
IMF	interplanetary magnetic field
Kp	magnetic activity index, midlatitude (Section 2.2.4)

KQZ	Cretaceous Quiet Zone
MEA	mean equatorial anomaly
MGST ( )	designation of spherical harmonic model derived using data from *Magsat*, with date of derivation in parentheses
MIF	mean ionospheric field correction
MLT	magnetic local time
MPDF	mean polar disturbance field
NASA	National Aeronautics and Space Administration
NRM	natural remanent magnetization
POGOs	Polar Orbiting Geophysical Observatories
PTRM	partial thermoremanent magnetization
RHA	rectangular harmonic analysis
RTP	reduced to pole
SCHA	spherical cap harmonic analysis
SEMM	standard earth magnetization model
SHCA	spherical harmonic correlation analysis
Sq	generic terminology for quiet day magnetic field variations
TRM	thermoremanent magnetization
UT	Universal Time
VRM	viscous remanent magnetization

REFERENCES

Abragam, A., *Principles of Nuclear Magnetism*, Oxford University Press, 1961.

Abramowitz, M., and I. A. Stegun, *Handbook of Mathematical Functions*, Applied Mathematics Series, vol. 55, National Bureau of Standards, Washington, DC, 1964.

Achache, J., A. Abtout, and J. L. LeMouël, The downward continuation of *Magsat* crustal anomaly field over Southeast Asia, *J. Geophys. Res., 92*, 11584–96, 1987.

Acuna, M., The *Magsat* precision vector magnetometer, *Johns Hopkins APL Technical Digest, 1*, 210–13, 1980.

Acuna, M., C. S. Scearce, J. B. Seek, and J. Scheifele, *The Magsat Vector Magnetometer – A Precision Fluxgate Magnetometer for the Measurement of the Geomagnetic Field*, NASA TM 79656, 1978.

Agarwal, A. K., B. P. Singh, R. G. Rastogi, and S. Srinivasan, On utility of space-borne vector magnetic measurements in crustal studies, *Phys. Earth Planet. Int., 41*, 260–8, 1986.

Agrawal, P. K., N. K. Thakur, and J. G. Negi, A deep structural ridge beneath central India, *Geophys. Res. Lett., 13*, 491–4, 1986.

Ajakaiye, D. E., D. H. Hall, and T. W. Millar, Interpretation of aeromagnetic data across the central crystalline shield area of Nigeria, *Geophys. J. Royal Astron. Soc., 83*, 503–17, 1985.

Ajakaiye, D. E., D. H. Hall, T. W. Millar, P. J. T. Verheijen, M. B. Awad, and S. B. Ojo, Aeromagnetic anomalies and tectonic trends in and around the Benue Trough, Nigeria, *Nature, 319*, 582–5, 1986.

Albouy, Y., and R. Godivier, *Carte Gravimétrique de la République Centrafricaine, 1/2,000,000*, Office de la Recherche Scientifique et Technique Outre-Mer, Bondy, France, 1981.

Alldredge, L. R., Rectangular harmonic analysis applied to the geomagnetic field, *J. Geophys. Res., 86*, 3021–6, 1981.

Allenby, R. J., and C. C. Schnetzler, U.S. crustal structure, *Tectonophysics, 93*, 13–31, 1983.

Alsdorf, D. E., Statistical processing of *Magsat* data for magnetic anomalies of the lithosphere, M.S. thesis, Ohio State University, 1991.

Alsdorf, D. E., R. R. B. von Frese, J. Arkani-Hamed, and H. C. Noltmier, Separation of lithospheric, external, and core components of the south polar geomagnetic field at satellite altitudes, *J. Geophys. Res., 99*, 4655–68, 1994.

Alsdorf, D., P. Taylor, R. R. B. von Frese, R. Langel, and J. Frawley, Arctic and Asia lithospheric satellite magnetic anomalies, *Phys. Earth Planet. Int.*, 1997.

An, Z., S. Ma, D. Tan, D. R. Barraclough, and D. J. Kerridge, A spherical cap harmonic model of the satellite magnetic anomaly field over China and adjacent areas, *J. Geomag. Geoelectr., 44*, 243–52, 1992.

Andreasen, G. E., and I. Zietz, *Magnetic Fields for a 4 × 6 Prismatic Model*, U.S. Geological Survey professional paper 666, 1969.

Antoine, L. A. G., and A. B. Moyes, The Agulhas *Magsat* anomaly: implications for continental break-up of Gondwana, *Tectonophysics, 212*, 33–44, 1992.

Arkani-Hamed, J., Remanent magnetization of the oceanic upper mantle, *Geophys. Res. Lett.*, *15*, 48–51, 1988.

Arkani-Hamed, J., Thermoviscous remanent magnetization of oceanic lithosphere inferred from its thermal evolution, *J. Geophys. Res.*, *94*, 17421–36, 1989.

Arkani-Hamed, J., Magnetization of the oceanic crust beneath the Labrador Sea, *J. Geophys. Res.*, *95*, 7101–10, 1990.

Arkani-Hamed, J., Thermoremanent magnetization of the oceanic lithosphere inferred from a thermal evolution model: implications for the source of marine magnetic anomalies, *Tectonophysics, 192*, 81–96, 1991.

Arkani-Hamed, J., The bulk magnetization contrast across the ocean–continent boundary in the east coast of North America, *Geophys. J. Int.*, *115*, 152–8, 1993.

Arkani-Hamed, J., and J. Dyment, Magnetic potential and magnetization contrasts of Earth's lithosphere, *J. Geophys. Res.*, *101*, 11401–25, 1996.

Arkani-Hamed, J., and W. J. Hinze, Limitations of the long-wavelength components of the North American magnetic anomaly map, *Geophysics, 55*, 1577–88, 1990.

Arkani-Hamed, J., R. A. Langel, and M. Purucker, Magnetic anomaly maps of Earth derived from POGO and *Magsat* data, *J. Geophys. Res.*, *99*, 24075–90, 1994.

Arkani-Hamed, J., and D. W. Strangway, Intermediate-scale magnetic anomalies of the Earth, *Geophysics, 50*, 2817–30, 1985a.

Arkani-Hamed, J., and D. W. Strangway, Lateral variations of apparent susceptibility of lithosphere deduced from *Magsat* data, *J. Geophys. Res.*, *90*, 2655–64, 1985b.

Arkani-Hamed, J., and D. W. Strangway, An interpretation of magnetic signatures of aulacogens and cratons in Africa and South America, *Tectonophysics, 113*, 257–69, 1985c.

Arkani-Hamed, J., and D. W. Strangway, Band-limited global scalar magnetic anomaly map of the Earth derived from *Magsat* data, *J. Geophys. Res.*, *91*, 8193–203, 1986a.

Arkani-Hamed, J., and D. W. Strangway, Magnetic susceptibility anomalies of lithosphere beneath eastern Europe and the Middle East, *Geophysics, 51*, 1711–24, 1986b.

Arkani-Hamed, J., and D. W. Strangway, Effective magnetic susceptibility of the oceanic upper mantle derived from *Magsat* data, *Geophys. Res. Lett., 13*, 999–1002, 1986c.

Arkani-Hamed, J., and D. W. Strangway, An interpretation of magnetic signatures of subduction zones detected by *Magsat*, *Tectonophysics, 133*, 45–55, 1987.

Arkani-Hamed, J., D. W. Strangway, D. J. Teskey, and P. J. Hood, Comparison of *Magsat* and low-level aeromagnetic data over the Canadian shield: implications for GRM, *Can. J. Earth Sci.*, *22*, 1241–7, 1985a.

Arkani-Hamed, J., W. E. S. Urquhart, and D. W. Strangway, Delineation of Canadian sedimentary basins from *Magsat* data, *Earth Planet. Sci. Lett.*, *70*, 148–56, 1984.

Arkani-Hamed, J., W. E. S. Urquhart, and D. W. Strangway, Scalar magnetic anomalies of Canada and northern United States derived from *Magsat* data, *J. Geophys. Res.*, *90*, 2599–608, 1985b.

Arkani-Hamed, J., S. K. Zhao, and D. W. Strangway, Geophysical interpretation of the magnetic anomalies of China derived from *Magsat* data, *Geophys. J.*, *95*, 347–59, 1988.

Arndt, N. T., and S. L. Goldstein, An open boundary between lower continental crust and mantle: its role in crust formation and crustal recycling, *Tectonophysics, 161*, 201–12, 1989.

Arnow, J. A., K. D. Nelson, J. H. McBride, J. E. Oliver, L. D. Brown, and S. Kaufman, Location and character of the Late Paleozoic suture beneath the southeastern U.S. coastal plain: evidence from new COCORP profiling, *EOS, Trans. AGU, 66*, 359, 1985.

Arur, M. G., P. S. Bains, and J. Lal, Anomaly map of *Z* component of the India sub-continent from magnetic satellite data, *Proc. Indian Acad. Sci. (Earth Planet. Sci.), 94*, 111–15, 1985.

Austin, C. B., and G. R. Keller, A crustal structure study of the Mississippi embayment, In: *An Integrated Geophysical and Geological Study of the Tectonic Framework of the 38th Parallel Lineament in the Vicinity of Its Intersection with the Extension of the New Madrid Fault Zone*, ed. L. W. Braile, W. J. Hinze, J. L. Sexton, G. R. Keller, and E. G. Lidiak, pp. 101–33, U.S. Nuclear Regulatory Commission Report NUREG/CR-1014, NRC, Washington, DC, 1979.

Backus, G. E., Non-uniqueness of the external geomagnetic field determined by surface intensity measurements, *J. Geophys. Res., 75*, 6337–41, 1970.

Backus, G. E., Determination of the external geomagnetic field from intensity measurements, *Geophys. Res. Lett., 1*, 21, 1974.

Baldwin, R., and H. Frey, *Magsat* crustal anomalies for Africa: dawn and dusk data differences and a combined data set, *Phys. Earth Planet. Int., 67*, 237–50, 1991.

Banerjee, S. K., Physics of rock magnetism, In: *Geomagnetism*, vol. 3, ed. J. A. Jacobs, pp. 1–30, Academic Press, London, 1989.

Bapat, V. J., B. P. Singh, and M. Rajaram, Application of ridge-regression in inversion of low latitude magnetic anomalies derived from space measurements, *Earth Planet. Sci. Lett., 84*, 277–84, 1987.

Barton, C. E., International Geomagnetic Reference Field: the seventh generation, *J. Geomag. Geoelectr., 49*, 157–206, 1997.

Basavaiah, N., M. Rajaram, and B. P. Singh, Comments on latitudinal dependence of *Magsat* anomalies in the B field and associated inversion instabilities, *Phys. Earth Planet. Int., 55*, 26–30, 1989.

Beloussov, V. V., *Continental Endogenous Regimes*, Mir, Moscow, 1981.

Benkova, N. P., and Sh. Sh. Dolginov, The survey with *Cosmos 49*, In: *World Magnetic Survey, 1957–1969*, ed. A. J. Zmuda, pp. 75–8, IAGA bulletin no. 28, IUGG Publication Office, Paris, 1971.

Bentley, C. R., Configuration and structure of the subglacial crust, In: *The Geology of Antarctica*, vol. 9, ed. R. J. Tingey, pp. 335–64, Clarendon Press, Oxford, 1991.

Berckhemer, H., A. Rauen, H. Winter, H. Kern, A. Kontny, M. Lienert, G. Nover, J. Pohl, T. Popp, A. Schult, J. Zinke, and H. C. Soffel, Petrophysical properties of the 9-km deep crustal section at KTB, *J. Geophys. Res., 102*, 18337–61, 1997.

Bernard, J., J.-C. Kosik, G. Laval, R. Pellat, and J.-P. Philippon, Représentation optimale du potentiel géomagnétique dans le repère d'un dipole décentre, incliné, *Ann. Geophys., 25*, 659–65, 1969.

Berthelsen, A., Mobile Europe, In: *A Continent Revealed – The European Geotraverse*, ed. D. J. Blundell, R. Freeman, and St. Mueller, pp. 11–32, Cambridge University Press, 1992.

Berti, G., and E. Pinna, Tentativo di definizione delle province magnetiche d'Italia, In: *Lithospheric Structure of the Ionian Basin from Gravity and Magnetic Data*, vol. 2, ed. G. Berti, pp. 895–916, Consiglio Nazionale delle Ricerche, Atti Del 6° Convegno, Roma, 1987.

Bevington, P. R., *Data Reduction and Error Analysis for the Physical Sciences*, McGraw-Hill, New York, 1969.

Bhattacharyya, B. K., Continuous spectrum of the total-magnetic-field anomaly due to a rectangular prismatic body, *Geophysics, 31*, 97–121, 1966.

Bhattacharyya, B. K., Some general properties of potential fields in space and frequency domain: a review, *Geoexploration, 5*, 127–43, 1967.

Bhattacharyya, B. K., Computer modeling in gravity and magnetic interpretation, *Geophysics, 43*, 912–29, 1978.

Björck, A., and T. Elfving, Accelerated projection methods for computing pseudoinverse solutions of systems of linear equations, *BIT, 19*, 145–63, 1979.

Black, R. A., Geophysical processing and interpretation of *Magsat* satellite magnetic anomaly data over the U.S. midcontinent, M.S. thesis, University of Iowa, 1981.

Blackwell, D. D., Heat flow and energy loss in the western United States, In: *Cenozoic Tectonics and Regional Geophysics of the Western Cordillera*, ed. R. B. Smith and G. P. Eaton, pp. 175–208, Geological Society of America Memoir 152, GSA, Boulder, 1979.

Blakely, R. G., *Potential Theory in Gravity and Magnetic Applications*, Cambridge University Press, 1995.

Blundell, D. J., R. Freeman, and St. Mueller (eds.), *A Continent Revealed – The European Geotraverse*, Cambridge University Press, 1992.

Bormann, P., P. Bankwitz, E. Bankwitz, V. Damm, E. Hurtig, H. Kämpf, M. Menning, H.-J. Paech, U. Schäfer, and W. Stackegrandt, Structure and development of the passive continental margin across the Princess Astrid coast, East Antarctica, *J. Geodyn., 6*, 347–73, 1986.

Bosum, W., U. Casten, F. C. Fieberg, I. Heyde, and

H. C. Soffel, Three-dimensional interpretation of the KTB gravity and magnetic anomalies, *J. Geophys. Res.*, *102*, 18307–21, 1997.

Boukeke, D.-B., Structures Crustales D'Afrique centrale déduites des anomalies gravimétriques et magnétiques: le domaine Précambrien de la République Centrafricaine et du Sud Cameroun, Ph.D. thesis, Universite de Paris Sud, Centre D'Orsay, 1994.

Bradley, L. M., and H. Frey, Constraints on the crustal nature and tectonic history of the Kerguelen Plateau from comparative magnetic modeling using *Magsat* data, *Tectonophysics*, *145*, 243–51, 1988.

Bradley, L. M., and H. Frey, *Magsat* magnetic anomaly contrast across Labrador Sea passive margins, *J. Geophys. Res.*, *96*, 16161–8, 1991.

Braile, L. W., W. J. Hinze, R. R. B. von Frese, and G. R. Keller, Seismic properties of the crust and uppermost mantle of the conterminous United States and adjacent Canada, In: *Geophysical Framework of the Continental United States*, ed. L. C. Pakiser and W. D. Mooney, pp. 655–80, Geological Society of America Memoir 172, GSA, Boulder, 1989.

Branson, J. C., F. J. Moss, and F. J. Taylor, *Deep Crustal Reflection Seismic Test Survey, Mildura, Victoria, and Broken Hill, N.S.W.*, Bureau of Mineral Resources Report 193, BMR, Canberra, 1976.

Burke, K., and J. F. Dewey, Plume-generated triple junctions: key indicators in applying plate tectonics to old rocks, *J. Geol.*, *81*, 406–33, 1973.

Butler, R., Azimuth, energy, *Q*, and temperature variations on P wave amplitudes in the United States, *Rev. Geophys.*, *22*, 1–36, 1984.

Cain, J. C., S. J. Hendricks, R. A. Langel, and W. V. Hudson, A proposed model for the International Geomagnetic Reference Field – 1965, *J. Geomag. Geoelectr.*, *19*, 335–55, 1967.

Cain, J. C., B. Holter, and D. Sandee, Numerical experiments in geomagnetic modeling, *J. Geomag. Geoelectr.*, *42*, 973–87, 1990.

Cain, J. C., and R. A. Langel, The geomagnetic survey by the polar orbiting geophysical observatories *OGO-2* and *OGO-4*, 1965–1967, In: *World Magnetic Survey*, ed. A. J. Zmuda, pp. 65–75, IAGA bulletin no. 28, IUGG Publication Office, Paris, 1971.

Cain, J. C., D. R. Schmitz, and L. Muth, Small-scale features in the Earth's magnetic field observed by *Magsat, J. Geophys. Res.*, *89*, 1070–6, 1984.

Cain, J. C., Z. Wang, C. Kluth, and D. R. Schmitz, Derivation of a geomagnetic model to *n* = 63, *Geophys. J.*, *97*, 431–41, 1989.

Campbell, W. H., *Quiet Daily Geomagnetic Fields*, Birkhausen Verlag, Basel, 1989a.

Campbell, W. H., The regular geomagnetic field conditions during quiet solar conditions, In: *Geomagnetism*, vol. 3, ed. J. A. Jacobs, pp. 385–460, Academic Press, London, 1989b.

Campbell, W. H., and E. R. Schiffmacher, Quiet ionospheric currents of the Northern Hemisphere derived from geomagnetic field records, *J. Geophys. Res.*, *90*, 6475–86, 1985.

Cande, S. C., J. L. LaBrecque, R. L. Larson, W. C. Pitman III, X. Golochenko, and W. Haxby, *Magnetic Lineations of the World's Ocean Basins*, Special map, American Association Petroleum Geologists, AAPG, Houston, 1989.

Carlson, R. L., N. I. Christensen, and R. P. Moore, Anomalous crustal structures in ocean basins: continental fragments and oceanic plateaus, *Earth Planet. Sci. Lett.*, *51*, 171–80, 1980.

Carmichael, R. S., and R. A. Black, Analysis and use of *Magsat* satellite magnetic data for interpretation of crustal structure and character in the U.S. midcontinent, *Phys. Earth Planet. Int.*, *44*, 333–47, 1986.

Carnahan, B., H. A. Luther, and J. O. Wilkes, *Applied Numerical Methods*, Wiley, New York, 1969.

Cermak, V., Heat flow map of Europe, In: *Terrestrial Heat Flow in Europe*, ed. V. Cermak and L. Rybach, pp. 3–41, Springer-Verlag, Berlin, 1979.

Cermak, V., Crustal temperature and mantle heat flow in Europe, *Tectonophysics*, *83*, 123–42, 1982.

Chandler, F. W., *The Structure of the Richmond Gulf Graben and the Geological Environments of Lead-Zinc Mineralization and of Iron-Manganese Formations in the Nastapoka Group, Richmond Gulf Area, New Quebec, Northwest Territories, Current Research, Part A*, Geological Survey of Canada paper 82-1A, GSC, Ottawa, 1982.

Chandler, F. W., and E. J. Schwarz, *Tectonics of the Richmond Gulf Area, Northern Quebec – A Hypothesis, Current Research, Part C*, Geological Survey of Canada paper 80-1C, GSC, Ottawa, 1980.

Chandler, V. W., and K. Carlson Malek, Moving-window Poisson analysis of gravity and magnetic data from the Penokean Orogen, east-central Minnesota, *Geophysics, 56*, 123–32, 1991.

Chandler, V. W., J. S. Koski, L. W. Braile, and W. J. Hinze, Utility of correlation studies in gravity and magnetic interpretation, Contract report to Goddard Space Flight Center from the Department of Geosciences, Purdue University, Contract NAS 5-22816, February 1977.

Chandler, V. W., J. S. Koski, W. J. Hinze, and L. W. Braile, Analysis of multisource gravity and magnetic anomaly data sets by moving-window application of Poisson's theorem, *Geophysics, 46*, 30–9, 1981.

Chapman, S., and J. Bartels, *Geomagnetism*, Clarendon Press, Oxford, 1940.

Chapman, S., and A. T. Price, The electric and magnetic state of the interior of the earth as inferred from terrestrial magnetic variations, *Phil. Trans. R. Soc. London, A229*, 427–60, 1930.

Chappell, B. W., and A. J. R. White, Two contrasting granite types, *Pacific Geology, 8*, 173–4, 1974.

Chowdhury, K., L. K. Das, and R. N. Bose, Geophysical lineaments over some geological provinces of India and their tectonic implications, In: *Regional Geophysical Lineaments: Their Tectonic and Economic Significance*, ed. M. N. Qureshy and W. J. Hinze, pp. 251–62, Geological Society of India, Memoir 12, GSI, Bangalore, 1989.

Clark, D. A., and D. W. Emerson, Notes on rock magnetization characteristics in applied geophysical studies, *Exploration Geophysics, 22*, 547–55, 1991.

Clark, S. C., H. Frey, and H. H. Thomas, Satellite magnetic anomalies over subduction zones: the Aleutian arc anomaly, *Geophys. Res. Lett., 12*, 41–4, 1985.

Cohen, Y., Traitements et interprétations de données spatiales en géomagnétisme: etude des variations laterales d'aimantation de la lithosphére terrestre, Ph.D. thesis, University of Paris VII and Institute de Physique du globe de Paris, June 1989.

Cohen, Y., and J. Achache, New global vector magnetic anomaly maps derived from *Magsat* data, *J. Geophys. Res., 95*, 10783–800, 1990.

Cohen, Y., and J. Achache, Contribution of induced and remanent magnetization to long-wavelength oceanic magnetic anomalies, *J. Geophys. Res., 99*, 2943–54, 1994.

Coles, R. L., *Magsat* scalar magnetic anomalies at northern high latitudes, *J. Geophys. Res., 90*, 2576–82, 1985.

Coles, R. L., G. V. Haines, G. Jansen van Beek, A. Nandi, and J. K. Walker, Magnetic anomaly maps from 40° N to 83° N derived from *Magsat* satellite data, *Geophys. Res. Lett., 9*, 281–4, 1982.

Coles, R. L., and P. T. Taylor, Magnetic anomalies, In: *The Arctic Ocean Region*, vol. L, ed. A. Grantz, L. Johnson, and J. F. Sweeney, pp. 119–32, *The Geology of North America*, vol. 8, Geological Society of America, Boulder, 1990.

Condie, K. C., *Archean Greenstone Belts*, Elsevier, Amsterdam, 1981.

Condie, K. C., *Plate Tectonics and Crustal Evolution*, 3rd ed., Pergamon Press, Elmsford, NY, 1989.

Constable, C., R. Parker, and P. B. Stark, Geomagnetic field models incorporating frozen flux constraints, *Geophys. J. Int., 113*, 419–33, 1993.

Cook, F. A., D. S. Albaugh, L. D. Brown, S. Kaufman, J. E. Oliver, and R. D. Hatcher, Thin-skinned tectonics in the crystalline southern Appalachians: COCORP seismic-reflection profiling of the Blue Ridge and Piedmont, *Geology, 7*, 563–7, 1979.

Cook, F. A., and J. E. Oliver, The Late Precambrian–Early Paleozoic continental edge in the Appalachian orogen, *Am. J. Sci., 281*, 993–1008, 1981.

Copson, E. T., *An Introduction to the Theory of Functions of a Complex Variable*, Oxford University Press, 1935.

Cordell, L., Regional positive gravity anomaly over the Mississippi embayment, *Geophys. Res. Lett., 4*, 285–7, 1977.

Counil, J.-L., Contribution du géomagnétisme à l'étude des hétérogénéités latérales de la croûte et du manteau supérieur, Ph.D. thesis, University of Paris VII and Institute de Physique du globe de Paris, January 1987.

Counil, J.-L., and J. Achache, Magnetization gaps associated with tearing in the Central America subduction zone, *Geophys. Res. Lett., 14*, 1115–18, 1987.

Counil, J.-L., J. Achache, and A. Galdéano, Long-wavelength magnetic anomalies in the Caribbean: plate boundaries and allochthonous continental blocks, *J. Geophys. Res., 94*, 7419–31, 1989.

Counil, J., Y. Cohen, and J. Achache, A global

continent–ocean magnetization contrast: spherical harmonic analysis, *Earth Planet. Sci. Lett., 103*, 354–64, 1991.

Courtillot, V., J. Ducruix, and J. L. LeMouël, Sur une accélération récente de la variation séculaire du champ magnétique terrestre, *C.R. Acad. Sci. Paris, D287*, 1095–8, 1978.

Creer, K. M., I. G. Hedley, and W. O'Reilly, Magnetic oxides in geomagnetism, In: *Magnetic Oxides*, ed. R. Cruik, pp. 649–89, Wiley, New York, 1976.

Cull, J. P., An appraisal of Australian heat-flow data, *J. Aust. Geol. Geophys., 7*, 11–21, 1982.

Dampney, C. N. G., The equivalent source technique, *Geophysics, 45*, 39–53, 1969.

Davis, T. N., and M. Sugiura, Auroral electrojet activity index AE and its universal time variations, *J. Geophys. Res., 71*, 785–801, 1966.

Davis, W. M., and J. C. Cain, Removal of DS from Pogo satellite data (abstract), *EOS, Trans. AGU, 54*, 242, 1973.

De, S., and P. Heaney, A microstructural study of carbonados from the Central African Republic, *EOS, Trans. AGU (Suppl.), 77*, S143, 1996.

Dean, W. C., Frequency analysis for gravity and magnetic interpretation, *Geophysics, 23*, 97–127, 1958.

DeSantis, A., Translated origin spherical cap harmonic analysis, *Geophys. J. Int., 106*, 253–63, 1991.

DeSantis, A., O. Battelli, and D. J. Kerridge, Spherical cap harmonic analysis applied to regional field for Italy, *J. Geomag. Geoelectr., 42*, 1019–36, 1990.

DeSantis, A., D. J. Kerridge, and D. R. Barraclough, A spherical cap harmonic model of the crustal magnetic anomaly field in Europe observed by *Magsat*, In: *Geomagnetism and Paleomagnetism*, ed. F. J. Lowes, D. W. Collinson, J. H. Parry, S. K. Runcorn, D. C. Tozer, and A. Soward, pp. 1–17, Kluwer, Dordrecht, 1989.

Dewey, J. F., R. M. Shackleton, C. Chengfa, and S. Yiyin, The tectonic evolution of the Tibetian Plateau, *Phil. Trans. R. Soc. London, A327*, 379–413, 1988.

Dietz, R. S., J. C. Holden, and W. P. Sproll, Geotectonic evolution and subsidence of Bahama Platform, *Bull. Geol. Soc. Am., 81*, 1915–28, 1970.

Dmitriyevskiy, V. S., N. S. Afanas'yev, and S. M. Frolov, Study of the physical properties of crystalline rocks in the southeast Voronezh anteclise, In: *Geological Council*, ed. N. A. Plaksenko, pp. 247–53,

Voronezh University Publishing House, Voronezh, 1968.

Dooley, J. C., and P. M. McGregor, Correlative geophysical data in the Australian region for use in the *Magsat* project, *Bull. Aust. Soc. Explor. Geophys., 13*, 63–7, 1982.

Dorbath, C., L. Dorbath, R. Gaulon, and D. Hatzfeld, Seismological investigation of the Bangui magnetic anomaly region and its relation to the margin of the Congo craton, *Earth Planet. Sci. Lett., 75*, 231–44, 1985.

Drummond, B. J., Seismic constraints on the chemical composition of the crust of the Pilbara Craton, northwest Australia, *Revista Brasileira de Geociencias, 12*, 227–36, 1982.

Ducruix, J., V. Courtillot, and J.-L. LeMouël, The late 1960's secular variation impulse, the eleven year magnetic variation and the electrical conductivity of the deep mantle, *Geophys. J. Royal Astron. Soc., 61*, 73–94, 1980.

Dunlop, D. J., Magnetism in rocks, *J. Geophys. Res., 100*, 2161–74, 1995.

Durrheim, R. J., and W. D. Mooney, Archean and Proterozoic crustal evolution: Evidence from crustal seismology, *Geology, 19*, 606–9, 1991.

Ervin, C. P., and L. D. McGinnis, Reelfoot rift: reactivated precursor to the Mississippi embayment, *Geol. Soc. Am. Bull., 86*, 1287–95, 1975.

Farthing, W. H., The *Magsat* scalar magnetometer, *Johns Hopkins APL Technical Digest, 1*, 205–9, 1980.

Farthing, W. H., and W. C. Folz, Rubidium vapor magnetometer for near earth orbiting spacecraft, *Rev. Sci. Inst., 38*, 1023–30, 1967.

Feldstein, Y. I., Some problems concerning the morphology of auroras and magnetic disturbances at high latitudes, *Geomagn. Aeron. 3*, 183–92, 1963.

Ferguson, J., R. J. Arculus, and J. Joyce, Kimberlite and kimberlitic instusives of southeastern Australia: a review, *J. Aust. Geol. Geophys., 4*, 227–41, 1979.

Finlayson, D. M., J. P. Cull, and B. J. Drummond, Upper mantle structure from the Trans-Australian Seismic Refraction data, *J. Geol. Soc. Aust., 21*, 447–58, 1984.

Finlayson, D. M., C. Prodehl, and C. D. N. Collins, Explosion seismic profiles and implications for crustal evolution in southeastern Australia, *J. Aust. Geol. Geophys. 4*, 242–52, 1979.

Fischbach, E., H. Kloor, R. A. Langel, A. T. Y. Lui, and M. Peredo, New geomagnetic limits on the photon mass and on long-rage forces coexisting with electromagnetism, *Phys. Rev. Lett.*, *73*, 514–17, 1994.

Fletcher, R., Conjugate gradient methods for indefinite systems, In: *Numerical Analysis Dundee 1975*, ed. A. Dold and B. Eckmann, pp. 73–89, *Lecture Notes in Mathematics*, vol. 506, Springer-Verlag, Berlin, 1976.

Forbes, A. J., General instrumentation, In: *Geomagnetism*, vol. 1, ed. J. A. Jacobs, pp. 51–142, Academic Press, London, 1987.

Forbes, J. M., The equatorial electrojet, *Rev. Geophys. Space Phys.*, *19*, 469–504, 1981.

Forsyth, D. A., P. Morel-a-L'Huissier, I. Asudeh, and A. G. Green, Alpha Ridge and Iceland – products of the same plume? *J. Geodyn.*, *6*, 197–214, 1986.

Frey, H., *Magsat* scalar anomaly distribution: the global perspective, *Geophys. Res. Lett.*, *9*, 277–80, 1982a.

Frey, H., *Magsat* scalar anomalies and major tectonic boundaries in Asia, *Geophys. Res. Lett.*, *9*, 299–302, 1982b.

Frey, H., Satellite-elevation magnetic anomalies over oceanic plateaus (abstract), *EOS, Trans. AGU, 64*, 214, 1983.

Frey, H., *Magsat* and POGO anomalies over the Lord Howe Rise: evidence against a simple continental crustal structure, *J. Geophys. Res.*, *90*, 2631–9, 1985.

Frey, H., R. Langel, G. Mead, and K. Brown, POGO and Pangaea, *Tectonophysics, 95*, 181–9, 1983.

Fullerton, L. G., H. V. Frey, J. H. Roark, and H. H. Thomas, Evidence for a remanent contribution in *Magsat* data from the Cretaceous Quiet Zone in the South Atlantic, *Geophys. Res. Lett.*, *16*, 1085–8, 1989.

Fullerton, L. G., H. V. Frey, J. H. Roark, and H. H. Thomas, Contributions of Cretaceous Quiet Zone natural remanent magnetization to *Magsat* anomalies in the Southwest Indian Ocean, *J. Geophys. Res.*, *99*, 11923–36, 1994.

Furlong, K. P., and D. M. Fountain, Continental crustal underplating: thermal considerations and seismic-petrologic consequences, *J. Geophys. Res.*, *91*, 8285–94, 1986.

Galdéano, A., Les mesures magnétiques du satellite *Magsat* et la dérive des continents, *C.R. Acad. Sci. Paris, 293*, 161–4, 1981.

Galdéano, A., Acquisition of long wavelength magnetic anomalies pre-dates continental drift, *Phys. Earth Planet. Int., 32*, 289–92, 1983.

Galliher, S. C., and M. A. Mayhew, On the possibility of detecting large-scale crustal remanent magnetization with *Magsat* vector magnetic anomaly data, *Geophys. Res. Lett.*, *9*, 325–8, 1982.

Garcia, A., J. M. Torta, J. J. Curto, and E. Sanclement, Geomagnetic secular variation over Spain 1970–1988 by means of spherical cap harmonic analysis, *Phys. Earth Planet. Int., 68*, 65–75, 1991.

Gee, R. D., Structure and tectonic style of the Western Australia shield, *Tectonophysics, 58*, 327–69, 1979.

Geological Survey of Canada, *Metamorphic Map of the Canadian Shield*, map 1475A, scale 1 : 3,500,000, 1978.

Ghidella, M. E., C. A. Raymond, and J. L. LaBrecque, Verification of crustal sources for satellite elevation magnetic anomalies in West Antarctica and the Weddell Sea and their regional tectonic implications, In: *Geological Evolution of Antarctica*, eds. M. R. A. Thomson, J. A. Crame, and J. W. Thomson, pp. 243–50, Cambridge University Press, 1991.

Giess, P., and N. I. Pavlenkova, Structural maps of the Earth's crust for Europe, *Izv. Earth Phys., 24*, 767–75, 1988.

Girdler, R. W., P. T. Taylor, and J. J. Frawley, A possible impact origin for the Bangui magnetic anomaly (Central Africa), *Tectonophysics, 212*, 45–58, 1992.

Glickson, A. Y., and I. B. Lambert, Vertical zonation and petrogenesis of the early Precambrian crust in Western Australia, *Tectonophysics, 58*, 3278–369, 1976.

Godivier, R., and L. Le Donche, *Réseau magnétique ramené au 1er Janvier 1956: République Centrafricaine, Tchad Méridional*, Office de la Recherche Scientifique et Technique Outre-Mer, Paris, 1956.

Golub, G. H., and C. F. Van Loan, *Matrix Computations*, 2nd ed., Johns Hopkins University Press, Baltimore, 1989.

Goodwin, A. M., *Precambrian Geology*, Academic Press, London, 1991.

Gorshkov, A. I., S. V. Titkov, A. M. Pleshakov, A. V.

Sivtsov, and L. V. Bershov, Inclusions of native metals and other mineral phases into Carbonado from the Ubangi Region (Central Africa), *Geology of Ore Deposits, 38*, 114–19, 1996.

Goswold, W. D., Preliminary heat flow data from Nebraska (abstract), EOS, *Trans. AGU, 61*, 1193, 1980.

Goyal, H. K., R. R. B. von Frese, W. J. Hinze, and D. N. Ravat, Statistical prediction of satellite magnetic anomalies, *Geophys. J. Int., 102*, 101–11, 1990.

Grant, F. S., Aeromagnetics, geology, and ore environments. I. Magnetite in igneous, sedimentary and metamorphic rocks: an overview, *Geoexploration, 23*, 303–33, 1984/1985a.

Grant, F. S., Aeromagnetics, geology, and ore environments. II. Magnetite and ore environments, *Geoexploration, 23*, 335–62, 1984/1985b.

Grauch, V. J. S., Limitations on digital filtering of the DNAG magnetic data set for the conterminous US, *Geophysics, 58*, 1281–96, 1993.

Green, A. G., Part I: Interpretation of project MAGNET data (1959 to 1966) for Africa and the Mozambique Channel, Ph.D. dissertation, University of Newcastle-upon-Tyne, 1973.

Green, A. G., Interpretation of project MAGNET aeromagnetic profiles across Africa, *Geophys. J. Royal Astron. Soc., 44*, 203–8, 1976.

Green, D. H., and A. E. Ringwood, A comparison of recent experimental data on the gabbro–garnet granulite–eclogite transition, *J. Geol., 80*, 277–88, 1972.

Grieve, R. A. F., Terrestrial impact structures, *Annu. Rev. Earth Planet. Sci., 15*, 245–70, 1987.

Griffin, W. L., and S. Y. O'Reilly, Is the Moho the crust–mantle boundary? *Geology, 15*, 241–4, 1987a.

Griffin, W. L., and S. Y. O'Reilly, The composition of the lower crust and the nature of the continental Moho – xenolith evidence, In: *Mantle Zenoliths*, ed. P. H. Nixon, pp. 413–30, Wiley, New York, 1987b.

Griffin, W. L., S. Y. Wass, and J. D. Hollis, Ultramafic xenoliths from Bullenmerri and Gnotuk maars, Victoria, Australia: petrology of a subcontinental crust–mantle transition, *J. Petrol., 25*, 53–87, 1984.

Gubbins, D., and J. Bloxham, Geomagnetic field analysis. III. Magnetic fields on the core–mantle boundary, *Geophys. J. Royal Astron. Soc., 80*, 696–713, 1985.

Gudmundsson, G., Spectral analysis of magnetic surveys, *Geophys. J. Royal Astron. Soc., 13*, 325–7, 1967.

Guterch, A., M. Grad, and E. Perchuc, Deep structure of the Earth's crust in the contact zone of the Paleozoic and Precambrian platforms in Poland (Tornquist-Teisseyre zone), In: *The European Geotraverse, Part 2*, ed. D. A. Galson, and St. Mueller, *Tectonophysics, 128*, 251–79, 1986.

Haggerty, S. E., Mineralogical constraints on Curie isotherms in deep crustal magnetic anomalies, *Geophys. Res. Lett., 5*, 105–8, 1978.

Haggerty, S. E., The aeromagnetic mineralogy of igneous rocks, *Can. J. Earth Sci., 16*, 1281–93, 1979.

Haggerty, S. E., Diamond genesis in a multiply constrained model, *Nature, 320*, 34–8, 1986.

Haggerty, S. E., and P. B. Toft, Native iron in the continental lower crust: petrological and geophysical implications, *Science, 229*, 647–9, 1985.

Hahn, A., H. Ahrendt, J. Jeyer, and J.-H. Hufen, A model of magnetic sources within the Earth's crust compatible with the field measured by the satellite *Magsat, Geol. Jb., A75*, 125–56, 1984.

Hahn, A., and W. Bosum, *Geomagnetics: Selected Examples and Case Histories*, Gebrüder Borntraeger, Berlin, 1986.

Haines, G. V., Spherical cap harmonic analysis, *J. Geophys. Res., 90*, 2583–91, 1985a.

Haines, G. V., *Magsat* vertical field anomalies above 40° N from spherical cap harmonic analysis, *J. Geophys. Res., 90*, 2593–8, 1985b.

Haines, G. V., Spherical cap harmonic analysis of geomagnetic secular variation over Canada 1960–1983, *J. Geophys. Res., 90*, 12563–74, 1985c.

Haines, G. V., Regional magnetic field modeling: a review, *J. Geomag. Geoelectr. 42*, 1001–18, 1990.

Haines, G. V., and L. R. Newitt, Canadian Geomagnetic Reference Field, 1985, *J. Geomag. Geoelectr., 38*, 895–921, 1986.

Hall, D. H., Long-wavelength aeromagnetic anomalies and deep crustal magnetization in Manitoba and northwestern Ontario, Canada, *J. Geophys., 40*, 403–30, 1974.

Hall, D. H., I. A. Noble, and T. W. Millar, Crustal structure of the Churchill–Superior boundary zone between 80° and 98° W longitude from *Magsat* anomaly maps and stacked passes, *J. Geophys. Res., 90*, 2621–30, 1985.

Hamoudi, M., J. Achache, and Y. Cohen, Global *Magsat* anomaly maps at ground level, *Earth Planet. Sci. Lett., 133*, 533–47, 1995.

Harrison, C. G. A., The crustal field, In: *Geomagnetism*, vol. 1, ed. J. A. Jacobs, pp. 513–610, Academic Press, London, 1987.

Harrison, C. G. A., H. M. Carle, and K. L. Hayling, Interpretation of satellite elevation magnetic anomalies, *J. Geophys. Res., 91*, 3633–50, 1986.

Hastings, D. A., Preliminary correlations of *Magsat* anomalies with tectonic features of Africa, *Geophys. Res. Lett., 9*, 303–6, 1982.

Hatcher, R. D., Tectonics of the western Piedmont and Blue Ridge, southern Appalachians: review and speculation, *Am. J. Sci., 278*, 276–304, 1978.

Hayden, K. J., Airborne gravity and magnetic surveys over rugged topography: a case study for the Appalachians, M.S. thesis, Dept. of Geological Sciences, Ohio State University, 1996.

Hayling, K. L., Magnetic anomalies at satellite altitude over continent–ocean boundaries, *Tectonophysics, 192*, 129–43, 1991.

Hayling, K. L., and C. G. A. Harrison, Magnetization modeling in the north and equatorial Atlantic Ocean using Magsat data, *J. Geophys. Res., 91*, 12423–43, 1986.

Heirtzler, J. R., G. O. Dickson, E. M. Herron, W. C. Pitman III, and X. Le Pichon, Marine magnetic anomalies, geomagnetic field reversals and motions of the ocean floor and continents, *J. Geophys. Res., 73*, 2119–36, 1968.

Henderson, R. G., and L. Cordell, Reduction of unevenly spaced potential field data to a horizontal plane by means of finite harmonic series, *Geophysics, 36*, 856–66, 1971.

Henkel, H., Petrophysical properties (density and magnetization) of rocks from the northern part of the Baltic Shield, *Tectonophysics, 192*, 1–19, 1991.

Henkel, H., Standard diagrams of magnetic properties and density – a tool for understanding magnetic petrology, *J. Appl. Geophys., 32*, 43–53, 1994.

Heppner, J. P., and N. C. Maynard, Empirical electric field models, *J. Geophys. Res., 92*, 4467–89, 1987.

Hildebrand, F. B., *Advanced Calculus for Applications*, Prentice-Hall, Englewood Cliffs, NJ, 1976.

Hildenbrand, T. G., R. J. Blakely, W. J. Hinze, G. R. Keller, R. A. Langel, M. Nabighian, and W. Roest, Aeromagnetic survey over U.S. to advance geomagnetic research, *EOS, Trans. AGU, 77*, 265, 269, 1996.

Hildenbrand, T. G., M. F. Kane, and W. Stauder, *Magnetic and Gravity Anomalies in the Northern Mississippi Embayment and Their Spatial Relationship to Seismology*, U.S. Geological Survey, map mf-914, 1977.

Hinze, W. J., R. R. B. von Frese, M. B. Longacre, L. W. Braile, E. G. Lidiak, and G. R. Keller, Regional magnetic and gravity anomalies of South America, *Geophys. Res. Lett., 9*, 314–17, 1982.

Hinze, W. J., R. R. B. von Frese, and D. N. Ravat, Mean magnetic contrasts between oceans and continents, *Tectonophysics, 192*, 117–27, 1991.

Hinze, W. J., and I. Zietz, The composite magnetic-anomaly map of the conterminous United States, In: *The Utility of Regional Gravity and Magnetic Anomaly Maps*, ed. W. J. Hinze, pp. 1–24, Society of Exploration Geophysicists, Tulsa, 1985.

Hoerl, A. E., and W. K. Kennard, Ridge regression: biased estimation for nonorthogonal problems, *Technometrics, 12*, 55–67, 1970a.

Hoerl, A. E., and W. K. Kennard, Ridge regression: applications to nonorthogonal problems, *Technometrics, 12*, 69–82, 1970b.

Hoffman, P. F., United plates of America, the birth of a craton: early Proterozoic assembly and growth of Laurentia, *Annu. Rev. Earth Planet. Sci., 16*, 543–603, 1988.

Hoffman, P. F., Precambrian geology and tectonic history, In: *The Geology of North America – An Overview*, ed. A. W. Bally, and A. R. Palmer, pp. 447–512, Geological Society of America, Boulder, 1989.

Hrvoic, I., Dispersion type nuclear magnetic resonance magnetometer for weak field measurements, Canadian patent 932801, 1973.

Hughes, T. J., and G. Rostoker, Current flow in the magnetosphere and ionosphere during periods of moderate activity, *J. Geophys. Res., 82*, 2271, 1977.

Irving, E., *Paleomagnetism*, Wiley, New York, 1964.

Irving, E., Drift of the major continental blocks since the Devonian, *Nature, 270*, 304–9, 1977.

Isaaks, E. H., and R. M. Srivastava, *An Introduction to Applied Geostatistics*, Oxford University Press, 1989.

Jackson, A., Accounting for crustal magnetization in

models of the core magnetic field, *Geophys. J. Int., 103*, 657–73, 1990.

Jackson, D. D., The use of *a priori* data to resolve non-uniqueness in linear inversion, *Geophys. J. Royal Astron. Soc., 57*, 137–57, 1979.

Jackson, J. D., *Classical Electrodynamics*, 2nd ed., Wiley, New York, 1975.

Jacobs, J. A. (ed.), *Geomagnetism*, 4 vols. Academic Press, London, 1987–91.

Jenkins, G. M., and D. G. Watts, *Spectral Analysis and Its Applications*, Holden-Day, San Francisco, 1968.

Johnson, B. D., Viscous remanent magnetization model for the Broken Ridge satellite magnetic anomaly, *J. Geophys. Res., 90*, 2640–6, 1985.

Johnson, B. D., C. McA. Powell, and J. J. Veevers, Early spreading history of the Indian Ocean between India and Australia, *Earth Planet. Sci. Lett., 47*, 131–43, 1980.

Jones, M. B., Correlative analysis of the gravity and magnetic anomalies of Ohio and their geologic significance, M.S. thesis, Ohio State University, 1988.

Kaiser, J. F., Nonrecursive digital filter design using the I_0–sinh window function, In: *Proceedings of the 1974 IEEE International Symposium on Circuits and Systems*, pp. 20–3, IEEE, New York, 1974.

Kawasaki, K., and J. C. Cain, Effects of field-aligned and associated currents on spherical harmonic geomagnetic models, *J. Geomag. Geoelectr., 44*, 167–80, 1992.

Kearey, P., and F. J. Vine, *Global Tectonics*, Blackwell, London, 1990.

Keller, G. R., E. G. Lidiak, W. J. Hinze, and L. W. Braile, The role of rifting in the tectonic development of the midcontinent, U.S.A., *Tectonophysics, 94*, 391–412, 1983.

Kelso, P. R., and S. K. Banerjee, An experimental study of the temperature dependence of viscous remanent magnetization of coarse grained natural and synthetic magnetite, *EOS, Trans. AGU, 72 (Suppl. 44)*, 137, 1991.

Kelso, P. R., S. K. Banerjee, and C. Teyssier, Rock magnetic properties of the Arunta Block, Central Australia, and their implication for the interpretation of long-wavelength magnetic anomalies, *J. Geophys. Res., 98*, 15987–99, 1993.

Kernevez, N., D. Duret, M. Moussavi, and J.-M. Leger, Weak field NMR and ESR spectrometers and magnetometers, *IEEE Trans. Mag., 28*, 3054–9, 1992.

Kernevez, N., and H. Glenat, Description of a high-sensitivity CW scalar DNP-NMR magnetometer, *IEEE Trans. Mag., 27*, 5402–4, 1991.

King, E. R., and I. Zietz, The New York–Alabama lineament: geophysical evidence for a major crustal break in the basement beneath the Appalachian basin, *Geology, 6*, 312–18, 1978.

Klitgord, K. D., P. Popenoe, and H. Schouten, Florida: a Jurassic transform plate boundary, *J. Geophys. Res., 89*, 7753–72, 1984.

Kovatch, G. T., Geologic analysis of *Magsat* observations over Antarctica and surrounding marine areas, M.S. thesis, Ohio State University, 1990.

Krutikhovskaya, Z. A., and I. K. Pashkevich, Magnetic model for the Earth's crust under the Ukrainian shield, *Can. J. Earth Sci., 14*, 2718–28, 1977.

Krutikhovskaya, Z. A., and I. K. Pashkevich, Long-wavelength magnetic anomalies as a source of information about deep crustal structure, *J. Geophys., 46*, 301–17, 1979.

Ku, C. C., A direct computation of gravity and magnetic anomalies caused by 2- and 3-dimensional bodies of arbitrary shape and arbitrary magnetic polarization by equivalent point method and a simplified cubic spline, *Geophysics, 42*, 610–22, 1977.

Kuhn, G. J., and H. Zaaiman, Long wavelength magnetic anomaly map for southern Africa from *Magsat, Trans. Geol. Soc. S. Afr., 89*, 9–16, 1986.

LaBrecque, J. L., and S. C. Cande, Intermediate-wavelength magnetic anomalies over the Central Pacific, *J. Geophys. Res., 89*, 11124–34, 1984.

LaBrecque, J. L., S. C. Cande, and R. D. Jarrard, Intermediate-wavelength magnetic anomaly field of the North Pacific and possible source distributions, *J. Geophys. Res., 90*, 2549–64, 1985.

LaBrecque, J. L., and C. A. Raymond, Seafloor spreading anomalies in the *Magsat* field of the North Atlantic, *J. Geophys. Res., 90*, 2565–75, 1985.

Lachenbruch, A. H., and J. H. Sass, Heat flow in the United States and the thermal regime of the crust, In: *The Nature and Physical Properties of the Earth's Crust*, ed. J. G. Heacock, pp. 325–675, Geophysical Monograph 9, American Geophysical Union, Washington, DC, 1977.

Ladd, J. W., and R. E. Sheridan, Seismic stratigraphy

of the Bahamas, *Am. Assoc. Pet. Geol. Bull.*, *71*, 719–36, 1987.

Lancaster, E. R., T. Jennings, M. Morrissey, and R. A. Langel, *Magsat Vector Magnetometer Calibration Using Magsat Geomagnetic Field Measurements*, NASA/GSFC TM 82046, November 1980.

Langel, R. A., *Processing of the Total Field Magnetometer Data from the OGO-2 Satellite*, GSFC report X-612-67-272, 1967.

Langel, R. A., A study of high latitude magnetic disturbance, Ph.D. thesis, Technical note BN-767, Institute for Fluid Dynamics and Applied Mathematics, University of Maryland, 1973.

Langel, R. A., Near-Earth magnetic disturbance in total field at high latitudes. 2. Interpretation of data from *OGO 2, 4*, and *6*, *J. Geophys. Res.*, *79*, 2373–921, 1974.

Langel, R. A., The main geomagnetic field, In: *Geomagnetism*, vol. 1, ed. J. A. Jacobs, pp. 249–512, Academic Press, London, 1987.

Langel, R. A., Real and artificial linear features in satellite magnetic anomaly maps, In: *Regional Geophysical Lineaments, Their Tectonic and Economic Significance*, ed. M. N. Qureshy and W. J. Hinze, pp. 165–70, Geological Society of India, Memoir 12. GSI, Bangalore, 1989.

Langel, R. A., Global magnetic anomaly maps derived from POGO spacecraft data, *Phys. Earth Planet. Int.*, *62*, 208–30, 1990a.

Langel, R. A., Study of the crust and mantle using magnetic surveys by *Magsat* and other satellites, *Proc. Indian Acad. Sci. (Earth Planet. Sci.)*, *99*, 581–618, 1990b.

Langel, R. A., International Geomagnetic Reference Field: the sixth generation, *J. Geomag. Geoelectr.*, *44*, 679–707, 1992.

Langel, R. A., The use of low altitude satellite data bases for modeling of core and crustal fields and the separation of external and internal fields, *Surveys in Geophysics*, *14*, 31–87, 121–7, 1993.

Langel, R. A., An investigation of a correlation/covariance method of signal extraction, *J. Geophys. Res.*, *100*, 20137–57, 1995.

Langel, R. A., B. J. Benson, and R. M. Orem, *The Magsat Bibliography (Revision 1)*, NASA Technical Memorandum 100776, February 1991.

Langel, R., J. Berbert, T. Jennings, and R. Horner, *Magsat* data processing: a report for investigators,

NASA TM 82160, Goddard Space Flight Center, November 1981.

Langel, R. A., R. L. Coles, and M. A. Mayhew, Comparison of magnetic anomalies of lithospheric origin measured by satellite and airborne magnetometers over western Canada, *Can. J. Earth Sci.*, *17*, 876–7, 1980a.

Langel, R. A., J. A. Conrad, T. J. Sabaka, and R. T. Baldwin, Adjustments of UARS, POGS, and DE-1 satellite magnetic field data for modeling of Earth's main field, *J. Geomag. Geoelectr.*, *49*, 393–413, 1997.

Langel, R. A., and R. H. Estes, A geomagnetic field spectrum, *Geophys. Res. Lett.*, *9*, 250–3, 1982.

Langel, R. A., and R. H. Estes, Large-Scale, near-Earth magnetic fields from external sources and the corresponding induced internal field, *J. Geophys. Res.*, *90*, 2487–94, 1985a.

Langel, R. A., and R. H. Estes, The near-Earth magnetic field at 1980 determined from *Magsat* data, *J. Geophys. Res.*, *90*, 2495–510, 1985b.

Langel, R. A., R. H. Estes, and G. D. Mead, Some new methods in geomagnetic field modeling applied to the 1960–1980 epoch, *J. Geomag. Geoelectr.*, *34*, 327–49, 1982a.

Langel, R. A., R. H. Estes, G. D. Mead, E. B. Fabiano, and E. R. Lancaster, Initial geomagnetic field model from *Magsat* vector data, *Geophys. Res. Lett.*, *7*, 793–6, 1980b.

Langel, R. A., R. H. Estes, and T. J. Sabaka, Uncertainty estimates in geomagnetic field modeling, *J. Geophys. Res.*, *94*, 12281–99, 1989.

Langel, R. A., G. Ousley, J. Berbert, J. Murphy, and M. Settle, The *Magsat* mission, *Geophys. Res. Lett.*, *9*, 243–5, 1982b.

Langel, R. A., J. D. Phillips, and R. J. Horner, Initial scalar magnetic anomaly map from *Magsat*, *Geophys. Res. Lett.*, *9*, 269–72, 1982c.

Langel, R. A., M. Rajaram, and M. Purucker, The equatorial electrojet and associated currents as seen in *Magsat* data, *J. Atm. Terr. Phys.*, *55*, 1233–69, 1993.

Langel, R. A., T. J. Sabaka, R. T. Baldwin, and J. Conrad, The near-Earth magnetic field from magnetospheric and quiet day ionospheric sources and how it is modeled, *Phys. Earth Planet. Int.*, *78*, 235–67, 1996.

Langel, R. A., C. C. Schnetzler, J. D. Phillips, and R. J.

Horner, Initial vector magnetic anomaly map from *Magsat, Geophys. Res. Lett., 9,* 273–6, 1982d.

Langel, R. A., E. V. Slud, and P. J. Smith, Reduction of satellite magnetic anomaly data, *J. Geophys., 54,* 207–12, 1984.

Langel, R. A., and R. A. Sweeney, Asymmetric ring current at twilight local time, *J. Geophys. Res., 76,* 4420–7, 1971.

Langel, R. A., and L. Thorning, Satellite magnetic field over the Nares Strait region, In: *Nares Strait and the Drift of Greenland: A Conflict in Plate Tectonics,* ed. P. R. Dawes and J. W. Kerr, pp. 291–3, *Meddr. Grönland, Geosci., 8,* 1982a.

Langel, R. A. and L. Thorning, A magnetic anomaly map of Greenland, *Geophys. J. Royal Astron. Soc., 71,* 599–612, 1982b.

Langel, R. A., and K. A. Whaler, Maps of the lithospheric magnetic field at Earth's surface from scalar satellite data, *Geophys. Res. Lett., 23,* 41–4, 1996.

Lanzerotti, L. J., R. A. Langel, and A. D. Chave, Geoelectromagnetism, In: *Encyclopedia of Applied Physics,* ed. G. L. Trigg, VCH Publishers, New York, 1993.

Lawson, C. L., and R. J. Hanson, *Solving Least Squares Problems,* Prentice-Hall, Englewood Cliffs, NJ, 1974.

Leite, L. W. B., Application of optimization methods to the inversion of aeromagnetic data, Ph.D. dissertation, St. Louis University, 1983.

LeMouël, J. L., J. Ducruix, and C. H. Duyen, The worldwide character of the 1969–1970 impulse of the secular variation rate, *Phys. Earth Planet. Int., 28,* 337–50, 1982.

Lerch, F., et al., *Geopotential Models of the Earth from Satellite Tracking, Altimetry and Surface Observations: GEM T-3 and GEM T-3S,* NASA TM 10455, 1992.

Lincoln, J. V., Geomagnetic and solar data, *J. Geophys. Res., 85,* 4306, 1980.

Longacre, M. B., Satellite magnetic investigation of South America, M.S. thesis, Purdue University, 1981.

Longacre, M. B., W. J. Hinze, and R. R. B. von Frese, A satellite magnetic model of northeastern South American aulacogens, *Geophys. Res. Lett., 9,* 318–21, 1982.

Lotter, C. J., Stable inversions of *Magsat* data over the geomagnetic equator by means of ridge regression, *J. Geophys., 61,* 77–81, 1987.

Lowes, F. J., Vector errors in spherical harmonic analysis of scalar data, *Geophys. J. Royal Astron. Soc., 42,* 637–51, 1975.

Lowes, F. J., and G. V. Haines, Some problems with spherical cap harmonic analysis (abstract), In: *Seventh Scientific Assembly,* IAGA bulletin no. 55, Part C, p. 413, IAGA, Buenos Aires, 1993.

Luenberger, D. G., *Optimization by Vector Space Methods,* Wiley, New York, 1969.

Lugovenko, V. N., and B. A. Matushkin, On the nature of the Earth's anomalous magnetic field, *Izv., Earth Phys., 20,* 705–7, 1984.

Lugovenko, V. N., and V. P. Pronin, Correlation of magnetic, gravitational, and thermal fields over continents, *Gerlands Beitr. Geophysik (Leipzig), 91,* 346–54, 1982.

Lugovenko, V. N., and V. P. Pronin, Combined correlation analysis of geophysical fields to study the North American continent, *Gerlands Beitr. Geophysik (Leipzig), 93,* 89–94, 1984.

Lugovenko, V. N., V. P. Pronin, L. V. Kosheleva, and S. S. Shultz, Correlation connection between the anomalous magnetic and gravitational fields for regions with different types of the Earth's crust, Preprint N13a(627), Institute of Terrestrial Magnetism, Ionosphere, and Radio Wave Propagation, Academy of Sciences, USSR, 1986.

McCamy, K., and R. P. Meyer, Crustal results of fixed multiple shots in the Mississippi embayment, In: *The Earth Beneath the Continents,* ed. J. S. Steinhart and T. J. Smith, pp. 370–81, Geophysical Monograph 10, American Geophysical Union, Washington, DC, 1966.

McGue, C. A., Tectonic analysis of the geopotential field anomalies of south Asia and adjacent marine areas, M.S. thesis, Ohio State University, 1988.

McPherron, R. L., Physical processes producing magnetospheric substorms and magnetic storms, In: *Geomagnetism,* vol. 4, ed. J. A. Jacobs, pp. 593–740, Academic Press, London, 1991.

Maeda, H., T. Iyemori, T. Araki, and T. Kamei, New evidence of a meridional current system in the equatorial ionosphere, *Geophys. Res. Lett., 9,* 337–40, 1982.

Maeda, H., T. Kamei, T. Iyemori, and T. Araki,

Geomagnetic perturbations at low latitudes observed by *Magsat, J. Geophys. Res., 90*, 2481–6, 1985.

Makarova, Z. A. (ed.), *Map of the Anomalous Magnetic Field of the Territory of the USSR and Adjacent Water Areas, 1/2,500,000* (18 sheets), USSR Ministry of Geology, Order of Lenin All-Union Scientific Research Geology Institute, Leningrad, 1974

Malin, S. R. C., Worldwide distribution of geomagnetic tides, *Phil. Trans. R. Soc. London, A274*, 551–94, 1973.

Malin, S. R. C., Z. Düzgît, and N. Baydemîr, Rectangular harmonic analysis revisited, *J. Geophys. Res., 101*, 28205–9, 1996.

Malin, S. R. C., B. M. Hodder, and D. R. Barraclough, Geomagnetic secular variation: a jerk in 1970, In: *Publicado en volumen commemorativo 75 aniversario del observatorio del Ebro* (75th anniversary volume of Ebro Observatory), ed. J. O. Cardus, pp. 239–56, Tarragona, 1983.

Marks, K. M., and D. T. Sandwell, Analysis of geoid height versus topography for oceanic plateaus and swells using nonbiased linear regression, *J. Geophys. Res., 96*, 8045–55, 1991.

Marquardt, D. W., Generalized inverses, ridge regression, biased linear estimation, and nonlinear estimation, *Technometrics, 12*, 591–612, 1970.

Mateskon, S. R., Gravity and magnetic terrain effects computed by Gaussian quadrature integration, M.S. thesis, Department of Geology and Mineralogy, Ohio State University, 1985.

Mathur, S. P., Crustal structure in southwestern Australia from seismic and gravity data, *Tectonophysics, 24*, 151–82, 1974.

Mayaud, P. N., *Derivation, Meaning and Use of Geomagnetic Indices*, Geophysical Monograph 22, American Geophysical Union, Washington, DC, 1980.

Mayhew, M., *A Method of Inversion of Satellite Magnetic Anomaly data*, Report X-922-77-260, NASA, Goddard Space Flight Center, Greenbelt, MD, 1977.

Mayhew, M. A., Inversion of satellite magnetic anomaly data, *J. Geophys., 45*, 119–28, 1979.

Mayhew, M. A., An equivalent layer magnetization model for the United States derived from satellite altitude magnetic anomalies, *J. Geophys. Res., 87*, 4837–45, 1982a.

Mayhew, M. A., Application of satellite magnetic anomaly data to Curie isotherm mapping, *J. Geophys. Res., 87*, 4846–54, 1982b.

Mayhew, M. A., *Magsat* anomaly field inversion for the U.S., *Earth Planet. Sci. Lett., 71*, 290–6, 1984.

Mayhew, M. A., Curie isotherm surfaces inferred from high-altitude magnetic anomaly data, *J. Geophys. Res., 90*, 2647–54, 1985.

Mayhew, M. A., R. H. Estes, and D. M. Myers, Remanent magnetization and three-dimensional density model of the Kentucky anomaly region, Contract report, contract NAS 5-27488, for Goddard Space Flight Center, by Business and Technological Systems, Inc., February 1984.

Mayhew, M. A., R. H. Estes, and D. M. Myers, Magnetization models for the source of the "Kentucky anomaly" observed by *Magsat, Earth Planet. Sci. Lett., 74*, 117–29, 1985a.

Mayhew, M. A., and S. C. Galliher, An equivalent layer magnetization model for the United States derived from *Magsat* data, *Geophys. Res. Lett., 9*, 311–13, 1982.

Mayhew, M. A., and B. D. Johnson, An equivalent layer magnetization model for Australia based on *Magsat* data, *Earth Planet. Sci. Lett., 83*, 167–74, 1987.

Mayhew, M. A., B. D. Johnson, and R. A. Langel, An equivalent source model of the satellite-altitude magnetic anomaly field over Australia, *Earth Planet. Sci. Lett., 51*, 189–98, 1980.

Mayhew, M. A., B. D. Johnson, and P. J. Wasilewski, A review of problems and progress in studies of satellite magnetic anomalies, *J. Geophys. Res., 90*, 2511–22, 1985b.

Mayhew, M. A., H. H. Thomas, and P. J. Wasilewski, Satellite and surface geophysical expression of anomalous crustal structure in Kentucky and Tennessee, *Earth Planet. Sci. Lett., 58*, 395–405, 1982.

Mayhew, M. A., P. J. Wasilewski, and B. D. Johnson, Crustal magnetization and temperature at depth beneath the Yilgarn block, Western Australia inferred from *Magsat* data, *Earth Planet. Sci. Lett., 107*, 515–22, 1991.

Meissner, R., Th. Wever, and E. R. Fluh, The Moho in Europe – implications for crustal development, *Ann. Geophys., 5B(4)*, 357–64, 1987.

Merrill, R. T., and M. A. McElhinny, *The Earth's Magnetic Field*, Academic Press, London, 1983.

Mestraud, J.-L., *Carte Géologique de la République*

Centrafricaine, 1/500,000, Bureau de Recherches géologiques et miniéres, Orleans, 1964.

Meyer, J., J.-H. Hufen, M. Siebert, and A. Hahn, Investigations of the internal geomagnetic field by means of a global model of the Earth's crust, *J. Geophys.*, *52*, 71–84, 1983.

Meyer, J., J.-H. Hufen, M. Siebert, and A. Hahn, On the identification of Magsat anomaly charts as crustal part of the internal field, *J. Geophys. Res.*, *90*, 2537–42, 1985.

Meyerhoff, A. A., and C. W. Hatten, Bahamas salient of North America: tectonic framework, stratigraphy, and petroleum potential, *Am. Assoc. Pet. Geol. Bull.*, *58*, 1201–39, 1974.

Mishra, D. C., Possible extensions of the Narmada–Son lineament towards Murray Ridge (Arabian Sea) and the eastern syntaxial bend of the Himalays, *Earth Planet. Sci. Lett.*, *36*, 301–8, 1977.

Mishra, D. C., Magnetic anomalies – India and Antarctica, *Earth Planet. Sci. Lett.*, *71*, 173–80, 1984.

Mishra, D. C., and M. Venkatraydu, *Magsat* scalar anomaly map of India and a part of Indian Ocean – magnetic crust and tectonic correlation, *Geophys. Res. Lett.*, *11*, 781–4, 1985.

Mooney, W. D., and R. Meissner, Multi-genetic origin of crustal reflectivity: a review of seismic reflection profiling of the continental lower crust and Moho. In: *Continental Lower Crust*, ed. D. M. Fountain, R. Arculus, and R. W. Kay, pp. 45–80, Elsevier, Amsterdam, 1992.

Mooney, W. D., and C. S. Weaver, Regional crustal structure and tectonics of the Pacific coastal states: California, Oregon, and Washington, In: *Geophysical Framework of the Continental United States*, ed. L. C. Pakiser and W. D. Mooney, pp. 129–61, Geological Society of America Memoir 172, GSA, Boulder, 1989.

Moritz, H., *Advanced Physical Geodesy*, Herbert Wichmann Verlag, Karlsruhe, 1980.

Morley, L. W., A. S. MacLaren, and B. W. Charbonneau, *Magnetic Anomaly Map of Canada*, Map 1255A, Geological Survey of Canada, Ottawa, 1968.

Mörner, N.-A., The lithospheric geomagnetic field: origin and dynamics of long wavelength anomalies, *Phys. Earth Planet. Int.*, *44*, 366–72, 1986.

Mueller, R. D., W. R. Roest, J.-Y. Royer, L. M. Gahagan, and J. G. Sclater, *A Digital Age Map of the Ocean Floor*, SIO reference series 93-30, Scripps Institute of Oceanography, La Jolla, 1993.

Mullins, H. T., and G. W. Lynts, Origin of the northwestern Bahama Platform: review and reinterpretation, *Bull. Geol. Soc. Am.*, *88*, 1447–61, 1977.

Nagata, T., *Rock Magnetism*, 2nd ed., Maruzan, Tokyo, 1961.

Nakagawa, I., and T. Yukutake, Spatial properties of the geomagnetic field over the area of the Japanese Islands deduced from *Magsat* data, *J. Geomag. Geoelectr.*, *36*, 443–53, 1984.

Nakagawa, I., and T. Yukutake, Rectangular harmonic analysis of geomagnetic anomalies derived from *Magsat* data over the area of the Japanese islands, *J. Geomag. Geoelectr.*, *37*, 957–77, 1985.

Nakagawa, I., T. Yukutake, and N. Fukushima, Extraction of magnetic anomalies of crustal origin from *Magsat* data over the area of the Japanese islands, *J. Geophys. Res.*, *90*, 2609–16, 1985.

Nakanishi, M., K. Tamaki, and K. Kobayashi, Mesozoic anomaly lineations and seafloor spreading history of the northwestern Pacific, *J. Geophys. Res.*, *94*, 15437–62, 1989.

Nakatsuka, N., and Y. Ono, Geomagnetic anomalies over the Japanese island region derived from *Magsat* data, *J. Geomag. Geoelectr.*, *36*, 455–62, 1984.

Nazarova, K. A., Serpentinized peridotites as a possible source for oceanic magnetic anomalies, *Marine Geophysical Researches, 16*, 455–62, 1994.

Neel, L., Théorie du Traînage magnétique des ferromagnétiques en grains fins avec applications aux Terres Cuites, *Ann. Geophys.*, *5*, 99–136, 1949.

Neel, L., Some theoretical aspects of rock magnetism, *Adv. Phys.*, *4*, 191–243, 1955.

Negi, J. G., P. K. Agrawal, and O. P. Pandey, Large variation of Curie depth and lithospheric thickness beneath the Indian subcontinent and a case for magnetothermometry, *Geophys. J. Royal Astron. Soc. 88*, 763–75, 1987a.

Negi, J. G., P. K. Agrawal, and N. K. Thakur, Vertical component *Magsat* anomalies and Indian tectonic boundaries, *Proc. Indian Acad. Sci. (Earth Planet. Sci.), 94*, 35–41, 1985.

Negi, J. G., N. K. Thakur, and P. K. Agrawal, Crustal magnetization-model of the Indian subcontinent through inversion of satellite data, *Tectonophysics, 122*, 123–33, 1986a.

Negi, J. G., N. K. Thakur, and P. K. Agrawal, Prominent *Magsat* anomalies over India, *Tectonophysics*, *122*, 345–56, 1986b.

Negi, J. G., N. K. Thakur, and P. K. Agrawal, Can depression of the core–mantle interface cause coincident *Magsat* and geoidal "lows" of the Central Indian Ocean? *Phys. Earth Planet. Int.*, *45*, 68–74, 1987b.

Ness, N. F., Magnetometers for space research, *Space Sci. Rev.*, *11*, 111–222, 1970.

Nielsen, O. V., J. R. Petersen, F. Primdahl, P. Brauer, B. Hernando, A. Fernandez, J. M. G. Merayo, and P. Ripka, Development, construction and analysis of the "Ørsted" fluxgate magnetometer, *Meas. Sci. Technol.*, *6*, 1099–115, 1995.

Noble, I. A., Magsat anomalies and crustal structure of the Churchill-Superior boundary zone, M.Sc. thesis, University of Manitoba, 1983.

Nolte, H. J., and A. Hahn, A model of the distribution of crustal magnetization in central Europe compatible with the field of magnetic anomalies deduced from *Magsat* results, *Geophys. J. Int.*, *111*, 483–96, 1992.

Nur, A., and Z. Ben-Avraham, Oceanic plateaus, the fragmentation of continents, and mountain building, *J. Geophys. Res.*, *87*, 3644–61, 1982.

Olsen, N., Ionospheric F-region currents at middle and low latitudes estimated from *Magsat* data, *J. Geophys. Res.*, *102*, 4563–76, 1997.

Opdyke, N. D., D. S. Jones, B. J. MacFadden, D. L. Smith, P. A. Mueller, and R. D. Shuster, Florida as an exotic terrane: paleomagnetic and geochronologic investigation of Lower Paleozoic rocks from the subsurface of Florida, *Geology*, *15*, 900–3, 1987.

Oraevsky, V. N., N. M. Rotanova, A. L. Kharitonov, and O. D. Pugacheva, Anomalous magnetic field over the Russian–Indian region derived from *Magsat* satellite data, *Geomagn. Aeron.*, *33*, 816–21, 1994.

O'Reilly, W., *Rock and Mineral Magnetism*, Blackie & Son, Glasgow, 1984.

O'Reilly, S. Y., and W. L. Griffin, A xenolith-derived geotherm for southeastern Australia and its geophysical implications, *Tectonophysics*, *111*, 41–63, 1985.

Pake, G. E., *Paramagnetic Resonance*, W. A. Benjamin, New York, 1962.

Pakiser, L. C., and I. Zietz, Transcontinental crustal and

upper-mantle structure, *Rev. Geophys.*, *3*, 505–20, 1965.

Pal, P. C., The Indian Ocean *Magsat* anomalies and strong geomagnetic field during Cretaceous 'quiet' zone, *Phys. Earth Planet. Int.*, *64*, 279–89, 1990.

Pandey, O. P., and J. G. Negi, Signals of degeneration of the sub-crustal part of the Indian lithosphere since the break-up of Gondwanaland, *Phys. Earth Planet. Int.*, *48*, 1–4, 1987.

Panofsky, W. K. H., and M. Phillips, *Classical Electricity and Magnetism*, Addison-Wesley, Reading, MA, 1962.

Parker, R. L., Understanding inverse theory, *Annu. Rev. Earth Planet. Sci.*, *5*, 35–64, 1977.

Parker, R. L., and S. P. Huestis, The inversion of magnetic anomalies in the presence of topography, *J. Geophys. Res.*, *79*, 1587–93, 1974.

Parker, R. L., and L. Shure, Efficient modeling of the Earth's magnetic field with harmonic splines, *Geophys. Res. Lett.*, *9*, 311–13, 1982.

Parker, R. L., L. Shure, and J. A. Hildebrand, The application of inverse theory to seamount magnetism, *Rev. Geophys.*, *25*, 17–40, 1987.

Parrott, M. H., Interpretation of *Magsat* anomalies over South America, M.S. thesis, Purdue University, 1985.

Percival, J. A., J. K. Mortensen, R. A. Stern, and K. D. Card, Giant granulite terranes of northeastern Superior Province: the Ashuanipi complex and Minto block, *Can. J. Earth Sci.*, *29*, 2287–308, 1992.

Petters, S. W., *Regional Geology of Africa, Lecture Notes in Earth Sciences*, vol. 40, Springer-Verlag, Berlin, 1991.

Phillips, R. J., and C. R. Brown, The satellite magnetic anomaly of Ahaggar: evidence for African plate motion, *Geophys. Res. Lett.*, *12*, 697–700, 1985.

Pilkington, M., and W. R. Roest, An assessment of long-wavelength anomalies over Canada, *Can. J. Earth Sci.*, *33*, 12–23, 1996.

Pindell, J. L., Alleghenian reconstruction and subsequent evolution of the Gulf of Mexico, Bahamas, and proto-Caribbean, *Tectonics*, *4*, 1–39, 1985.

Piper, J. D. A., *Palaeomagnetism and the Continental Crust*, Open University Press, Milton Keynes, 1987.

Piper, J. D. A., Palaeomagnetism, In: *Geomagnetism*,

vol. 3, ed. J. A. Jacobs, pp. 31–162, Academic Press, London, 1989.

Plouff, D., Gravity and magnetic fields of polygonal prisms and application to magnetic terrain corrections, *Geophysics, 41*, 727–41, 1976.

Poisson, S. D., Mémoire sur la théorie du magnétisme, *Mémoires de l'Académie Royale des Sciences de l'Institut de France*, 247–348, 1826.

Potemra, T. A., Birkeland currents: present understanding and some remaining questions, In: *High-Latitude Space Plasma Physics*, ed. B. Hultqvist and T. Hagfors, p. 335, Nobel Symposium, vol. 54, 1982.

Powell, C. Mc. A., Terminal fold-belt deformation: relationship of mid-Carboniferous megakinks in the Tasman fold belt to coeval thrusts in cratonic Australia, *Geology, 31*, 545–9, 1984.

Pozzi, J. P., and G. Dubuisson, High temperature viscous magnetization of oceanic deep crustal- and mantle-rocks as partial source for *Magsat* magnetic anomalies, *Geophys. Res. Lett., 19*, 21–4, 1992.

Press, W. H., and S. A. Teukolsky, Straight-line data with errors in both coordinates, *Computers in Physics, 6*, 274–6, 1992.

Press, W. H., S. A. Teukolsky, W. T. Vetterling, and B. P. Flannery, *Numerical Recipes*, Cambridge University Press, 1992.

Pucher, R., Pyrrhotite-induced aeromagnetic anomalies in western Germany, *J. Appl. Geophys., 32*, 32–42, 1994.

Pucher, R., and T. Wonik, Comment on the paper of Taylor and Ravat "An interpretation of the *Magsat* anomalies of Central Europe," *J. Appl. Geophys., 36*, 213–16, 1997.

Purucker, M. E., The computation of vector magnetic anomalies: a comparison of techniques and errors, *Phys. Earth Planet. Int., 62*, 231–45, 1990.

Purucker, M. E., The correction of attitude jumps in the *Magsat* data set, Science Applications Research contract report, Science Applications Research, Greenbelt, MD, 1991.

Purucker, M. E., R. A. Langel, M. Rajaram, and C. Raymond, Global magnetization models with *a priori* information, *J. Geophys. Res.*, 103, 2563–2584, 1998.

Purucker, M. E., T. J. Sabaka, and R. A. Langel, Conjugate gradient analysis: a new tool for studying satellite magnetic data sets, *Geophys. Res. Lett., 23*, 507–10, 1996.

Purucker, M. E., T. J. Sabaka, R. A. Langel, and N. Olsen, The missing dimension in *Magsat* and POGO anomaly studies, *Geophys. Res. Lett., 24*, 2909–2912, 1997.

Qureshy, M. N., and R. K. Midha, Deep crustal signatures in India and contiguous regions from satellite and ground geophysical data, In: *Reflection Seismology: The Continental Crust*, ed. M. Barazangi, and L. Brown, pp. 77–94, Geodynamics Series no. 14. American Geophysical Union, Washington, DC, 1986.

Rajaram, M., *Magsat*'s contribution to geophysical surveys, *Adv. Space Res., 13(11)*, 33–42, 1993a.

Rajaram, M. *Magsat*'s view of the Earth, In: *Space Applications in Earth System Science*, pp. 23–30, Indian Geophysical Union, Hyderabad, 1993b.

Rajaram, M., and R. A. Langel, Magnetic anomaly modeling at the Indo-Eurasian collision zone, *Tectonophysics, 212*, 117–27, 1992.

Rajaram, M., and B. P. Singh, Spherical Earth modelling of the scalar magnetic anomaly over the Indian region, *Geophys. Res. Lett., 13*, 961–4, 1986.

Rangarajan, G. K., Indices of geomagnetic activity, In: *Geomagnetism*, vol. 3, ed. J. A. Jacobs, pp. 323–84, Academic Press, London, 1989.

Rastogi, R. G., The equatorial electrojet: magnetic and ionospheric effects, In: *Geomagnetism*, vol. 3, ed. J. A. Jacobs, pp. 461–526, Academic Press, London, 1989.

Ravat, D. N., *Magsat* investigations over the greater African region, Ph.D. thesis, Purdue University, 1989.

Ravat, D. N., and W. J. Hinze, Considerations of variations in ionospheric field effects in mapping equatorial lithospheric *Magsat* magnetic anomalies, *Geophys. J. Int., 113*, 387–98, 1993.

Ravat, D. N., W. J. Hinze, and P. T. Taylor, European tectonic features observed by *Magsat, Tectonophysics, 220*, 157–73, 1993.

Ravat, D. N., W. J. Hinze, and R. R. B. von Frese, Lithospheric magnetic property contrasts within the South American plate derived from damped least-squares inversion of satellite magnetic data, *Tectonophysics, 192*, 159–68, 1991.

Ravat, D., W. J. Hinze, and R. R. B. von Frese, Analysis of *Magsat* magnetic contrasts across Africa and South America, In: *Lithospheric Analysis of*

Magnetic and Related Geophysical Anomalies, ed. R. R. B. von Frese and P. T. Taylor, *Tectonophysics*, *212*, 59–76, 1992.

Ravat, D. N., R. A. Langel, M. Purucker, J. Arkani-Hamed, and D. E. Alsdorf, Global vector and scalar *Magsat* magnetic anomaly maps, *J. Geophys. Res.*, *100*, 20111–36, 1995.

Raymond, C. A., Satellite elevation magnetic anomalies and their use in tectonic studies, Ph.D. thesis, Columbia University, 1989.

Raymond, C. A., and J. L. LaBrecque, Magnetization of the oceanic crust: thermoremanent magnetization or chemical remanent magnetization, *J. Geophys. Res.*, *92*, 8077–88, 1987.

Regan, R. D., and J. C. Cain, The use of geomagnetic field models in magnetic surveys, *Geophysics*, *40*, 621–9, 1975.

Regan, R. D., J. C. Cain, and W. M. Davis, A global magnetic anomaly map, *J. Geophys. Res.*, *80*, 794–802, 1975.

Regan, R. D., W. M. Davis, and J. C. Cain, The detection of "intermediate" size magnetic anomalies in *Cosmos 49* and *OGO 2, 4, 6* data, *Space Research*, *13*, 619–23, 1973.

Regan, R. D., D. W. Handschumacher, and M. Sugiura, A closer examination of the reduction of satellite magnetometer data for geological studies, *J. Geophys. Res.*, *86*, 9567–73, 1981.

Regan, R. D., and B. D. Marsh, The Bangui magnetic anomaly: its geological origin, *J. Geophys. Res.*, *87*, 1107–20, 1982.

Renbarger, K. S., A crustal structure study of South America, Ph.D. thesis, University of Texas at El Paso, 1984.

Reynolds, R. L., J. G. Rosenbaum, M. R. Hudson, and N. S. Fishman, Rock magnetism, the distribution of magnetic minerals in the Earth's crust, and aeromagnetic anomalies, In: *Geologic Application of Modern Aeromagnetic Surveys*, ed. W. F. Hanna, pp. 24–45, USGS Bulletin 1924, 1990.

Ridgway, J. R., Preparation and intepretation of a revised *Magsat* satellite magnetic anomaly map over South America, M.S. thesis, Purdue University, August 1984.

Ridgway, J. R., and W. J. Hinze, *Magsat* scalar anomaly map of South America, *Geophysics*, *51*, 1472–9, 1986.

Rikitake, T., and S. Sato, The geomagnetic *Dst* field of the magnetic storm on June 18–19, 1936, *Bull. Earthquake Res. Inst., Tokyo Univ.*, *35*, 7–21, 1957.

Ritzwoller, M. H., and C. R. Bentley, *Magsat* magnetic anomalies over Antarctica and the surrounding oceans, *Geophys. Res. Lett.*, *9*, 285–8, 1982.

Ritzwoller, M. H., and C. R. Bentley, Magnetic anomalies over Antarctica measured from *Magsat*, In: *Antarctic Earth Science – 4th International Symposium*, ed. R. L. Olivier, pp. 604–7, Cambridge University Press, 1983.

Rotanova, N. M., V. N. Oraevsky, and A. L. Kharitonov, *Magsat* vector and scalar anomalous magnetic fields over Russian–Indian area, *J. Geomag. Geoelectr.*, *47*, 283–93, 1995.

Ruder, M. E., Interpretation and modeling of regional crustal structure of the southeastern United States using raw and filtered conventional and satellite gravity and magnetic data, Ph.D. thesis, Department of Geosciences, Pennsylvania State University, 1986.

Ruder, M. E., and S. S. Alexander, *Magsat* equivalent source anomalies over the southeastern United States: implications for crustal magnetization, *Earth Planet. Sci. Lett.*, *78*, 33–43, 1986.

Runcorn, S. K., On the interpretation of lunar magnetism, *Phys. Earth Planet. Int.*, *10*, 327–35, 1975.

Sailor, R. V., A. R. Lazarewicz, and R. F. Brammer, Spatial resolution and repeatability of *Magsat* crustal anomaly data over the Indian Ocean, *Geophys. Res. Lett.*, *9*, 289–92, 1982.

Sandwell, D. T., and K. R. MacKenzie, Geoid height versus topography for oceanic plateaus and swells, *J. Geophys. Res.*, *94*, 7403–18, 1989.

Sass, J. H., D. D. Blackwell, D. S. Chapman, J. H. Costain, E. R. Decker, L. A. Lawver, and C. A. Swanberg, Heat flow from the crust of the United States, In: *Physical Properties of Rocks and Minerals*, ed. Y. S. Touloukian, W. R. Judd, and R. F. Roy, pp. 503–48, McGraw-Hill, New York, 1981.

Sass, J. H., W. H. Diment, A. H. Lachenbruch, B. V. Marshall, R. J. Monroe, T. H. Moses, Jr., and T. C. Urban, *A New Heat-Flow Contour Map of the Conterminous United States*, U.S. Geological Survey, open-file report 76–756, 1976.

Sass, J. H., and A. H. Lachenbruch, Thermal regime of the Australian continental crust, In: *The*

Earth, Its Origin, Structure and Evolution, ed. M. W. McElhinney, pp. 301–52, Academic Press, London, 1979.

Schackleford, P. R. J., The determination of crustal structure in the Adelaide Geosyncline using quarry blasts as seismic sources, M.Sc. thesis, Department of Physics, University of Adelaide, 1978.

Schmitz, D., J. B. Frayser, and J. C. Cain, Application of dipole modeling to magnetic anomalies, *Geophys. Res. Lett., 9*, 307–10, 1982.

Schmitz, D. R., J. Meyer, and J. C. Cain, Modeling the Earth's geomagnetic field to very high degree and order, *Geophys. J., 97*, 421–30, 1989.

Schnetzler, C. C., An estimation of continental crust magnetization and susceptibility from *Magsat* data for the conterminous U.S., *J. Geophys. Res., 90*, 2617–20, 1985.

Schnetzler, C. C., Satellite measurements of the Earth's crustal magnetic field, *Adv. Space Res., 9*, 5–12, 1989.

Schnetzler, C. C., and R. J. Allenby, Estimation of lower crust magnetization from satellite derived anomaly field, *Tectonophysics, 93*, 33–45, 1983.

Schnetzler, C. C., P. T. Taylor, and R. A. Langel, *Mapping Magnetized Geologic Structures from Space: The Effects of Orbital and Body Parameters*, NASA TM 86134, Goddard Space Flight Center, Greenbelt, MD, August 1984.

Schnetzler, C. C., P. T. Taylor, R. A. Langel, W. J. Hinze, and J. D. Phillips, Comparison between the recent U.S. composite magnetic anomaly map and *Magsat* anomaly data, *J. Geophys. Res., 90*, 2543–8, 1985.

Schumacher, R., *Introduction to Magnetic Resonance*, W. A. Benjamin, New York, 1970.

Sexton, J. L., W. J. Hinze, R. R. B. von Frese, and L. W. Braile, Long-wavelength aeromagnetic anomaly map of the conterminous United States, *Geology, 10*, 364–9, 1982.

Shapiro, V. A., A. V. Tsirulsky, N. V. Fedorova, F. I. Nikonova, A. G. Dyakonova, A. V. Chursin, and L. O. Turmina, The anomalous magnetic field and its dynamics used to study the deep structure and modern geodynamic processes of the Urals, *J. Geodyn., 5*, 221–35, 1986.

Sheridan, R. E., J. T. Crosby, G. M. Bryan, and P. L. Stoffa, Sratigraphy and structure of southern Blake Plateau, northern Florida straits, and northern Bahama Platform from multichannel seismic reflection data, *Am. Assoc. Pet. Geol. Bull., 65*, 2571–93, 1981.

Sheridan, R. E., H. T. Mullins, J. A. Austin, Jr., M. M. Ball, and J. W. Ladd, Geology and geophysics of the Bahamas, In: *The Atlantic Continental Margin, U.S.*, ed. R. E. Sheridan and J. A. Grow, pp. 329–64, The Geology of North America, vol. I-2, Geological Society of America, Boulder, 1988.

Shive, P. N., Suggestions for the use of SI units in magnetism, *EOS, Trans. AGU, 67*, 25–6, 1986.

Shive, P. N., R. J. Blakely, B. R. Frost, and D. M. Fountain, Magnetic properties of the lower continental crust, In: *Continental Lower Crust*, ed. D. M. Fountain, R. Arculus, and R. W. Kay, pp. 145–77, Elsevier, Amsterdam, 1992.

Shive, P. N., B. R. Frost, and A. Peretti, The magnetic properties of metaperidotite rocks as a function of metamorphic grade: implications for crustal magnetic anomalies, *J. Geophys. Res., 93*, 12187–95, 1988.

Shor, G. G., Jr., H. H. Kirk, and H. W. Menard, Crustal structure of the Melanesian area, *J. Geophys. Res., 76*, 2562–86, 1971.

Shuey, R. T., and A. S. Pasquale, End corrections in magnetic profile interpretation, *Geophysics, 38*, 507–12, 1973.

Shuey, R. T., D. K. Schellinger, E. H. Johnson, and L. B. Alley, Aeromagnetics and the transition between the Colorado Plateau and the Basin and Range Provinces, *Geology, 1*, 107–10, 1973.

Shure, L., Modern mathematical methods in geomagnetism, Ph.D. dissertation, Institute of Geophysics and Planetary Physics, University of California, San Diego, 1982.

Shure, L., R. L. Parker, and G. E. Backus, Harmonic splines for geomagnetic modelling, *Phys. Earth Planet. Int., 28*, 215–29, 1982.

Siebert, M., and J. Meyer, Geomagnetic activity indices, In: *The Upper Atmosphere: Data Analysis and Interpretation*, ed. G. K. Hartmann and R. Leitinger, pp. 887–911, Springer-Verlag, Berlin, 1996.

Simonenko, T. N., and I. K. Pashkevich, *Magnetic Anomaly Map of Europe, 1:5,000,000*, S.I. Subbotin Institute of Geophysics, G. V. Plekhanov

Mining Institute, Academy of Sciences of the Ukrainian SSR, Leningrad, 1990.

Singh, B. P., *Magsat* in lineament studies: results from Indian region, *Mem. Geol. Soc. India, 12*, 181–8, 1989.

Singh, B. P., A. K. Agarwal, and R. G. Rastogi, On the nature of residual trend in *Magsat* passes after removal of core and external components, *Annales Geophysicae, 4*, 653–8, 1986.

Singh, B. P., N. Basavaiah, and M. N. Qureshy, A preliminary study of satellite-derived gravity and magnetic anomalies of India and contiguous regions, *Tectonophysics, 212*, 129–39, 1992.

Singh, B. P., N. Basavaiah, M. Rajaram, and G. Geetharamanan, A method of obtaining solutions with only positive dipole moments on inversion of satellite magnetic anomalies, *Phys. Earth Planet. Int., 58*, 95–102, 1989.

Singh, B. P., and M. Rajaram, *Magsat* studies over the Indian region, *Proc. Indian Acad. Sci. (Earth Planet. Sci.), 99*, 619–37, 1990.

Singh, B. P., M. Rajaram, and V. J. Bapat, Definition of the continent–ocean boundary of India and the surrounding oceanic regions from *Magsat* data, *Tectonophysics, 192*, 145–51, 1991.

Singh, R. P., A. Rastogi, and A. Adam, Combined interpretation of *Magsat* and surface geophysical data, *Adv. Space Res., 13*, 43–50, 1993.

Slichter, C. P., *Principles of Magnetic Resonance*, Harper & Row, New York, 1963.

Smithson, S. B., R. A. Johnson, and Y. K. Wong, Mean crustal velocity; a critical parameter for interpreting crustal structure and crustal growth, *Earth Planet. Sci. Lett., 53*, 323–32, 1981.

Soffel, H., Palaeomagnetism and archaeomagnetism, In: *Landolt-Börnstein – Numerical Data and Functional Relationships in Science and Technology*, vol. 2b, ed. K. Fuchs and H. Soffel, p. 196, Springer, Berlin, 1985.

Spector, A., and F. S. Grant, Statistical models for interpreting aeromagnetic data, *Geophysics, 35*, 293–302, 1970.

Stacey, F. D., A generalized theory of thermoremanence, covering the transitions from single-domain to multi-domain magnetic grains, *Phil. Mag., 7*, 1887–900, 1962.

Stacey, F. D., The physical theory of rock magnetism, *Adv. Phys., 12*, 46–133, 1963.

Stacey, F. D., and S. K. Banerjee, *The Physical Principles of Rock Magnetism*, Elsevier, Amsterdam, 1974.

Starich, P. J., The south-central United States magnetic anomaly, M.S. thesis, Purdue University, 1984.

Stern, D. P., and J. H. Bredekamp, Error enhancement in geomagnetic models derived from scalar data, *J. Geophys. Res., 80*, 1776–82, 1975.

Stern, D. P., R. A. Langel, and G. D. Mead, Backus effect observed by *Magsat*, *Geophys. Res. Lett., 7*, 941–4, 1980.

Stewart, G. W., *Introduction to Matrix Computations*, Academic Press, New York, 1973.

Stroud, A. H., and D. Secrest, *Gaussian Quadrature Formulas*, Prentice-Hall, Englewood Cliffs, NJ, 1966.

Sugiura, M., Hourly values of equatorial *Dst* for the IGY, *Ann. Int. Geophys. Year, 35*, 9–45, 1964.

Sweeney, J. F., and J. R. Weber, Progress in understanding the age and origin of the Alpha Ridge, Arctic Ocean, *J. Geodyn., 6*, 237–44, 1986.

Szeto, A. M. K., and W. H. Cannon, On the separation of core and crustal contributions to the geomagnetic field, *Geophys. J. Royal Astron. Soc., 82*, 319–29, 1985.

Takenaka, J., M. Yanagisawa, R. Fuji, and K. Shibuya, Crustal magnetic anomalies in the Antarctic region detected by *Magsat*, *J. Geomag. Geoelectr., 43*, 525–38, 1991.

Talwani, M., Computation with the help of a digital computer of magnetic anomalies caused by bodies of arbitrary shape, *Geophysics, 30*, 797–817, 1965.

Talwani, M., and J. R. Heirtzler, *Computation of Magnetic Anomalies Caused by Two-Dimensional Structures of Arbitrary Shape*, Stanford University Publications of the Geological Sciences, Computers in the Mineral Industries, 1964.

Tarling, D. H., *Paleomagnetism*, Chapman & Hall, London, 1983.

Tarlowski, C., A. J. McEwin, C. V. Reeves, and C. E. Barton, Dewarping the composite aeromagnetic anomaly map of Australia using control traverses and base stations, *Geophysics, 61*, 696–705, 1996.

Taylor, P. T., Magnetic data over the Arctic from aircraft and satellites, *Cold Regions Sci. Techn., 7*, 35–40, 1983a.

Taylor, P. T., Nature of the Canada Basin – implications from satellite-derived magnetic anomaly data, *J. Alaska Geol. Soc., 2*, 1–8, 1983b.

Taylor, P. T., Investigation of plate boundaries in the eastern Indian Ocean using *Magsat* data, *Tectonophysics, 192*, 153–8, 1991.

Taylor, P. T., and J. J. Frawley, *Magsat* anomaly data over the Kursk region, U.S.S.R., *Phys. Earth Planet. Int., 45*, 255–65, 1987.

Taylor, P. T., W. J. Hinze, and D. N. Ravat, The search for crustal resources: *Magsat* and beyond, *Adv. Space Res., 12*, 5–15, 1992.

Taylor, P. T., and D. Ravat, An interpretation of the *Magsat* anomalies of central Europe, *J. Appl. Geophys., 34*, 83–91, 1995.

Taylor, P. T., and D. Ravat, Reply to comments by R. Pucher and T. Wonik, *J. Appl. Geophys., 36*, 217–19, 1997.

Thakur, N. K., Evaluation of Indian coastal response: an integrated approach, *Phys. Earth Planet. Int., 56*, 285–93, 1988.

Thomas, H. H., Petrologic model of the northern Mississippi Embayment based on satellite magnetic and ground-based geophysical data, *Earth Planet. Sci. Lett., 70*, 115–20, 1984.

Thomas, H. H., A model of ocean basin crustal magnetization appropriate for satellite elevation anomalies, *J. Geophys. Res., 92*, 11609–13, 1987.

Thominski, H. P., J. Wohlenberg, and U. Bleil, The remagnetization of Devono-Carboniferous sediments from the Ardenno-Rhenish Massif, *Tectonophysics, 225*, 411–31, 1993.

Thompson, G. A., and M. L. Zoback, Regional geophysics of the Colorado Plateau, *Tectonophysics, 61*, 149–81, 1979.

Tivey, M. A., Vertical magnetic structure of ocean crust determined from near-bottom magnetic field measurements, *J. Geophys. Res., 101*, 20275–96, 1996.

Toft, P. B., and J. Arkani-Hamed, Magnetization of the Pacific Ocean lithosphere deduced from *Magsat* data, *J. Geophys. Res., 97*, 4387–406, 1992.

Toft, P. B., and J. Arkani-Hamed, Induced magnetization of the oceanic lithosphere and ocean–continent magnetization contrast inferred from *Magsat* anomalies, *J. Geophys. Res., 98*, 6267–82, 1993.

Toft, P. B., and S. E. Haggerty, A remanent and induced magnetization model of *Magsat* vector anomalies over the West African Craton, *Geophys. Res. Lett., 13*, 341–4, 1986.

Toft, P. B., and S. E. Haggerty, Limiting depth of magnetization in cratonic lithosphere, *Geophys. Res. Lett., 15*, 530–3, 1988.

Toft, P. B., D. V. Hills, and S. E. Haggerty, Crustal evolution and the granulite to eclogite phase transition in xenoliths from kimberlites in the West African Craton, *Tectonophysics, 161*, 213–31, 1989.

Toft, P. B., P. T. Taylor, J. Arkani-Hamed, and S. E. Haggerty, Interpretation of satellite magnetic anomalies over the West African Craton, *Tectonophysics, 212*, 21–32, 1992.

Torta, J. M., A. Garcia, and A. DeSantis, New representation of geomagnetic secular variation over restricted regions by means of SCHA: application to the case of Spain, *Phys. Earth Planet. Int., 74*, 209–17, 1992.

Torta, J. M., A. Garcia, and A. DeSantis, A geomagnetic reference for Spain at 1990, *J. Geomag. Geoelectr., 45*, 573–88, 1993.

Uchupi, E., J. D. Milliman, B. P. Luyendyk, C. O. Bowin, and K. O. Emery, Structure and origin of southeastern Bahamas, *Bull. Am. Assoc. Pet. Geol., 55*, 687–704, 1971.

van der Sluis, A. and H. A. van der Vorst, Numerical solution of large, sparse linear algebraic systems arising from tomographic problems, In: *Seismic Tomography*, ed. G. Nolet, pp. 49–83, Reidel, Dordrecht, 1987.

Vasicek, J. M., H. V. Frey, and H. H. Thomas, Satellite magnetic anomalies and the Middle America trench, *Tectonophysics, 154*, 19–24, 1988.

Veevers, J. J., *Phanerozoic Earth History of Australia*, Oxford University Press, 1984.

Venkatakrishnan, R., and S. J. Culver, Plate boundaries in west Africa and their implications for Pangean continental fit, *Geology, 16*, 322–5, 1988.

Verhoogen, J., The origin of thermoremanent magnetization, *J. Geophys. Res., 64*, 2441–9, 1959.

von Frese, R. R. B., D. E. Alsdorf, J.-H. Kim, T. M. Stepp, D. R. H. O'Connell, K. J. Hayden, and W.-S. Li, Regional geophysical imaging of the Antarctic lithosphere, In: *Recent Progress in Antarctic Earth Science*, ed. Y. Yoshida et al., pp. 465–74, TERRAPUB, Tokyo, 1992.

von Frese, R. R. B., W. J. Hinze, and L. W. Braile, Spherical Earth gravity and magnetic anomaly analysis by equivalent point source inversion, *Earth Planet. Sci. Lett., 53*, 69–83, 1981a.

von Frese, R. R. B., W. J. Hinze, and L. W. Braile, Regional North American gravity and magnetic anomaly correlations, *Geophys. J. Royal Astron. Soc., 69*, 745–61, 1982a.

von Frese, R. R. B., W. J. Hinze, L. W. Braile, and A. J. Luca, Spherical earth gravity and magnetic anomaly modeling by Gauss-Legendre quadrature integration, Contract report to NASA/Goddard Space Flight Center, January 1980.

von Frese, R. R. B., W. J. Hinze, L. W. Braile, and A. J. Luca, Spherical-Earth gravity and magnetic anomaly modeling by Gauss-Legendre quadrature integration, *J. Geophys., 49*, 234–42, 1981b.

von Frese, R. R. B., W. J. Hinze, C. A. McGue, and D. N. Ravat, Use of satellite magnetic anomalies for tectonic lineament studies, *Mem. Geol. Soc. India, 12*, 171–80, 1989.

von Frese, R. R. B., W. J. Hinze, R. Olivier, and C. R. Bentley, Regional magnetic anomaly constraints on continental breakup, *Geology, 14*, 68–71, 1986.

von Frese, R. R. B., W. J. Hinze, R. Olivier, and C. R. Bentley, Satellite magnetic anomalies and continental reconstructions, In: *Gondwana Six: Structure, Tectonics, Geophysics*, ed. G. D. McKenzie, pp. 9–15, Geophysical Monograph 40, American Geophysical Union, Washington, DC, 1987.

von Frese, R. R. B., W. J. Hinze, J. L. Sexton, and L. W. Braile, Verification of the crustal component in satellite magnetic data, *Geophys. Res. Lett., 9*, 293–5, 1982b.

von Frese, R. R. B., J. W. Kim, L. Tan, D. E. Alsdorf, C. A. Raymond, and P. T. Taylor, Satellite-measured magnetic anomaly fields of the Antarctic lithosphere, In: *Terra Antarctica*, 1997.

von Frese, R. R. B., D. N. Ravat, W. J. Hinze, and C. A. McGue, Improved version of geopotential field anomalies for lithospheric investigations, *Geophysics, 53*, 375–85, 1988.

Walker, J. S., *Fourier Analysis*, Oxford University Press, 1988.

Wang, Z., Understanding models of the geomagnetic field by Fourier analysis, *J. Geomag. Geoelectr., 39*, 333–47, 1987.

Wasilewski, P. J., Magnetic properties of mantle xenoliths and the magnetic character of the crust-mantle boundary, In: *Mantle Xenoliths*, ed. P. J. Nixon, pp. 577–88, Wiley, New York, 1987.

Wasilewski, P., and D. M. Fountain, The Ivrea Zone as a model for the distribution of magnetization in the continental crust, *Geophys. Res. Lett., 9*, 333–6, 1982.

Wasilewski, P. J., and M. A. Mayhew, Crustal xenolith magnetic properties and long wavelength anomaly source requirements, *Geophys. Res. Lett., 9*, 329–32, 1982.

Wasilewski, P. J., and M. A. Mayhew, The Moho as a magnetic boundary revisited, *Geophys. Res. Lett., 19*, 2259–62, 1992.

Wasilewski, P., H. H. Thomas, and M. A. Mayhew, The Moho as a magnetic boundary, *Geophys. Res. Lett., 6*, 541–4, 1979.

Wasilewski, P. J., and R. Warner, The xenolith record: insights into the magnetic lithosphere, In: *Magnetism: Rocks to Superconductors*, ed. K. V. Subbarao, pp. 65–80, Memoir 29, Geological Society of India, Bangalore, 1994.

Wellman, P., A. S. Murray, and M. W. McMullan, Australian long-wavelength magnetic anomalies, *J. Aust. Geol. Geophys., 9*, 297–302, 1984.

Whaler, K. A., Downward continuation of *Magsat* lithospheric anomalies to the Earth's surface, *Geophys. J. Int., 116*, 267–78, 1994.

Whaler, K. A., and R. A. Langel, Minimal crustal magnetizations from satellite data, *Phys. Earth Planet. Int., 48*, 303–19, 1996.

Williams, H., and R. D. Hatcher, Suspect terranes and accretionary history of the Appalachian orogen, *Geology, 10*, 530–6, 1982.

Won, I. J., and K. H. Son, A preliminary comparison of the *Magsat* data and aeromagnetic data in the continental U.S., *Geophys. Res. Lett., 9*, 296–8, 1982.

Worm, H.-U., Comment on "Can remanent magnetization in the deep crust contribute to long-wavelength magnetic anomalies?" by P. N. Shive, *Geophys. Res. Lett., 16*, 595–7, 1989.

Worm, H., Rock magnetic results of the KTB drilling, Personal communication to Pucher and Wonik, 1995.

Yanagisawa, M., and M. Kono, Magnetic anomaly maps obtained by means of the mean ionospheric field correction, *J. Geomag. Geoelectr., 36*, 417–42, 1984.

Yanagisawa, M., and M. Kono, Mean ionospheric field

correction for *Magsat* data, *J. Geophys. Res., 90*, 2527–36, 1985.

Yanagisawa, M., M. Kono, T. Yukutake, and N. Fukushima, Preliminary interpretation of magnetic anomalies over Japan and its surrounding area, *Geophys. Res. Lett., 9*, 322–4, 1982.

Yañez, G. A., and J. L. LaBrecque, Age-dependent three-dimensional magnetic modeling of the North Pacific and North Atlantic oceanic crust at intermediate wavelengths, *J. Geophys. Res., 102*, 7947–61, 1997.

Yarger, H. L., Kansas basement study using spectrally filtered aeromagnetic data, In: *The Utility of Regional Gravity and Magnetic Anomaly Maps*, ed. W. Hinze, pp. 213–32, Society of Exploration Geophysicists, Tulsa, 1985.

Yuan, D.-W., Relation of *Magsat* and gravity anomalies to the main tectonic provinces of South America, Ph.D. thesis, University of Pittsburgh, 1983.

Zaaiman, H., and G. J. Kuhn, The application of the ring current correction model to *Magsat* passes, *J. Geophys. Res., 91*, 8034–8, 1986.

Ziegler, P. A., *Evolution of the Arctic–North Atlantic and the Western Tethys*, Memoir 43, American Association of Petroleum Geologists, Tulsa, 1988.

Zietz, I., G. E. Andreasen, and J. C. Cain, Magnetic anomalies from satellite magnetometer, *J. Geophys. Res., 75*, 4007–15, 1970.

Zietz, I., E. R. King, W. Geddes, and E. G. Lidiak, Crustal study of a continental strip from the Atlantic Ocean to the Rocky Mountains, *Geol. Soc. Am. Bull., 77*, 1427–48, 1966.

Zonenshain, L. P., M. I. Kuzmin, and L. M. Natapov, *Geology of the USSR: A Plate-Tectonic Synthesis*. Geodynamics Series vol. 21, American Geophysical Union, Washington, DC, 1990.

AUTHOR INDEX

SUBJECT INDEX

accuracy
  anomaly field estimation, 61, 95, 185, 227, *see* anomaly map, error estimates for
  attitude or vector, 40, 44
  data or measurement, 5, 42, 44, 48, 50, 112, 164, 174
  inversion model, 146, 172, 174
  main field model, 13, 23, 61, 236–7
  satellite position, 44–5
addition theorem, 157, 160, 166
AE, *see* magnetic indices
aeromagnetic, *see* surface survey data
  in crustal study, 11, 15, 322–3, 324, 330, 332, 363
  interpretation of, 58, 316
  magnetization estimate from, 263, 315
  reduction/analysis of, 22, 63–4, 83
  survey, 236, 237, 238–9
  compared to satellite survey, 4, 184, 239–40, 241, 274
  maps from, 185, 187, 188, 198; reproduced, 240, 241
Africa, 297, 372, 374, 376, 379
  anomaly (study) in, 92, 148, 173, 246, 248, 275–7, 311, 338–50
  Bangui (Central African Republic), 11–13, 39, 138, 281, 282, 341–4, 346
  field from removed from

ionospheric field estimate, 100, 104
  anomaly map of, 163–4, 187–8, 191, 195, 196
  reproduced, 276, 342, 346; magnetization, 347–9; susceptibility times thickness (ζ), 339
airborne (magnetic instrument), 3, 9
aircraft, 9, 21, 40, 233, 281, 342
airmag (aeromagnetic anomaly map, Table 6.1), 187, 188
ambient field, *see* magnetic field sources, main field
  change over model area, 281, 282
  deviation from magnetization direction, 269, 344, 349–50
  governs ionospheric current, 94
  as inducing field, 118, 123, 146, 364
  magnetic domains along, 254, 255
  magnetization/magnetic moment in direction of, 120, 123, 173, 248, 256, 262, 265, 306, 350, 371
  in/at magnetometer, 41, 42, 47
  susceptibility-magnetization equivalence in, 308, 347
amphibolite, 259, 264, 318, 319, 337, 350, 353, 355, 365
annihilator, 124, 198
anomaly (fields), definition, 6–7, 234

anomaly map, marine magnetic, 242
anomaly map (satellite magnetic)
  comparisons, *see* map comparisons
  complications of satellite, 9–14
  data discontinuity, effect of, 54–5
  descriptions, 109–10, 184 ff., 275–7, 287–8, 322, 329, 371
  low/mid latitudes, 194–215; polar, 215–23; regional, 244–8
  error estimates for, 200, 201, 204–5, 208, 211, 213, 221–3, 226–7
  gridding, 137
  introduced, 3, 7, 8, 9–14
  main field contamination of, 61–6, 102, 201–2, 203
  maps
    low-mid latitude: scalar, 10, 200, 202, 207, Plate 1; vector, 98, 99, 105, 214, 215, Plate 2
    magnetization, 244, 347–9, 352, 354
    polar (scalar), 218, 219, 221, 224, Plate 3
    regional: scalar, 199, 235, 240, 241, 247, 276, 289, 324, 331, 342, 362, 365, 377; RTP, 289, 372, 373, 374, 376; vector, 362–3
  susceptibility, susceptibility contrast, 245